OLYMPIC
TRIUMPH AND DISASTER
& TITANIC

OLYMPIC & TITANIC

TRIUMPH AND DISASTER

MARK CHIRNSIDE

The History Press

Above: An early view of *Olympic* at Plymouth shows off her impressive profile. (Daniel Klistorner/ J. Kent Layton/ Mark Chirnside collection)

Front cover: *Olympic* leaving Southampton on her maiden voyage in 1911. (© TopFoto)

Rear cover: *Titanic*'s propellers and rudder are well clear of the water as the stern continues to rise out of the water in the final stages of the sinking. She is depicted here a couple of minutes before her final plunge into the abyss. (HFX Studios / Tom Lynskey / Levi Rourke)

First published 2026

The History Press
97 St George's Place, Cheltenham,
Gloucestershire, GL50 3QB
www.thehistorypress.co.uk

British Library Cataloguing in Publication Data.
A catalogue record for this book is available from the British Library.

ISBN 978 1 80399 673 8

Typesetting and origination by The History Press
Printed in India by Thomson Press

MIX
Paper | Supporting responsible forestry
FSC® C010615

EU Authorised Representative: Easy Access System Europe
Mustamäe tee 50, 10621 Tallinn, Estonia
gpst.request@easproject.com

The commanders must distinctly understand that the issue of the following instructions does not, in any way, relieve them from entire responsibility for the safe and efficient navigation of their respective vessels; and they are also enjoined to remember that, whilst they are expected to use every diligence to secure a speedy voyage, *they must run no risk which might by any possibility result in accident to their ships. It is to be hoped that they will ever bear in mind that the safety of the lives and property entrusted to their care is the ruling principle that should govern them in the navigation of their ships, and no supposed gain in expedition, or saving of time on the voyage, is to be purchased at the risk of accident* [original emphasis]. The company desires to maintain for its vessels a reputation for safety, and only looks for such speed on the various voyages as is consistent with safe and prudent navigation.

'Regulations for the Safe and Efficient Navigation of the Company's Steamships', White Star Line – Oceanic Steam Navigation Company Limited, 1873.

CONTENTS

INTRODUCTION

Olympic and *Titanic* started out identically. On 30 April 1907, Harland & Wolff recorded an order from the White Star Line for two new ships: 'Yard Number 400' and 'Yard Number 401'. The shipbuilder worked from the same sets of construction plans and, when a design decision was taken, it was taken to apply to both ships. Over time, they diverged. This was particularly true once *Olympic* entered service. Lessons learned from *Olympic* were applied to *Titanic*'s design, including making her first-class accommodation even more luxurious, but a newspaper article headline calling her '*Olympic*'s Sister' helps illustrate that the focus at the time was on *Olympic*.

Immediately after *Olympic*'s launch, the *Belfast News-Letter* wrote, 'She was not built to fail; a higher destiny awaits her'. She went on to have a long and successful career in peace and war. Her sister *Titanic* had precisely the same promise and potential.

Both ships were commanded by Captain Edward J. Smith. On *Olympic*'s maiden voyage, he took the decision to slow down for a short period due to fog. She arrived safely in New York. On *Titanic*'s maiden voyage, his decision in different circumstances to maintain course and speed, trusting that any ice would be sighted in time to avoid it, proved fatal.

It is impossible to understand either ship without understanding the broader context. Part I explores the development of the White Star Line and the International Mercantile Marine (IMM), as well as competing steamship lines and the North Atlantic passenger trade leading up to 1912. We see some of the day-to-day interactions, with White Star chairman J. Bruce Ismay's wife, Florence, travelling by Cunard. She reported a pleasant voyage but Cunard were less impressed when an individual in the White Star Line's Liverpool offices (probably J. Bruce Ismay himself) cracked jokes about their ship and the extent of the vibration. We see all the challenges that Harland & Wolff faced as the shipbuilder worked through its design concepts and began construction on an unprecedented scale. Each ship was substantially larger than any other constructed previously and yet both were being built at the same time, by the same builder, in Belfast. White Star faced their own logistical and practical challenges in raising the finance and battling to get the necessary port facilities in place.

Part II takes us through *Olympic*'s early months in service. We follow her triumphant maiden voyage and the successes of the summer of 1911. She set record after record for the number of first-class passengers she carried. Her success led to White Star ordering a third sister ship.

The *Hawke* collision on 20 September 1911 was a serious setback which took *Olympic* out of service at the height of the season, setting the scene for a mammoth legal battle. It involved many of the same individuals who sailed on *Titanic* the following year. We see her commander and officers testifying in court. Adding to the drama, while *Olympic* was being repaired at Belfast in November 1911, an argument between workers involving a gun and a physical altercation took place.

White Star's rivals, including observers from Cunard, the French Line, the Hamburg America Line (HAPAG) and the Holland America Line, all took a keen interest in *Olympic*. In particular, we see Leonard Peskett, on behalf of Cunard, and Willem Frederik Piek, on behalf of Holland America, sail on her. We follow their experiences in detail. This rare material gives insights into the reality of sailing on *Olympic* in 1911. Their comments are free from the 'spin' of period advertisements, but perhaps going to the other extreme in some cases.

Part III sees *Olympic* return to service and battle a harsh winter, even by the standards of the North Atlantic. During the late summer of 1911 and early months of 1912, we see J. Bruce Ismay and rival shipping lines looking to cooperate with each other, discussing options for working together and coming to an arrangement about insuring their largest liners. There were even suggestions about whether they might place their largest liners in a form of mutual ownership. The pace of competition was forcing the major shipping lines to raise and invest more and more money in building bigger and better ships. In January 1912, Ismay himself signalled his intention to retire as IMM's president in 1913.

Titanic is nearing completion and, despite the coal strike and significant disruption, she sails as planned on her maiden voyage. News of the disaster filters out to a stunned world. When *Olympic* attempts to sail on her next crossing from Southampton, many of her stokehold crew are concerned about the adequacy of the additional lifeboats on board and refuse to sail. Their replacements include many non-union men, which leads to a further strike. Her voyage ends up being cancelled.

The summer of 1912 sees *Olympic* gradually regaining her popularity with the travelling public, eventually returning to Belfast in October 1912 for a comprehensive refit aimed at restoring public confidence.

This is the story of two sister ships with a shared potential. Through circumstances, one of their stories was a triumph; the other's was a disaster.

CONCEPTION TO REALITY

The success of the White Star Line ... is a remarkable example of the potency in business of one necessary man. The late Mr. T.H. Ismay was a benevolent autocrat, and a man whose pre-eminent ability in the management of shipping was fully recognised by his competitors ...

'Thirty-Three Years of the White Star Line', *The Practical Engineer – Engineers' Gazette Supplement* (*c*. 1904) pp.177–78

We have out of our original capital built 64,000 tons of shipping, which have cost over two millions sterling; and we have now ... in all 53,033 tons of shipping such as there is not to be found in any part of the world. With that position, we have great cause to congratulate ourselves and the managers. But how has this marked success been attained? First of all, I may fairly say, in the presence of Sir Edward Harland and his partners, that it is in a considerable degree attributable to the excellence of the build of our steamers. We all know what they are; how they were built regardless of cost, so that they are absolutely as perfect in strength and finish as the great talent of their builders could make them.

Thomas H. Jackson, Chair (in the absence of Gustav Schwabe), Oceanic Steam Navigation Company: Presentation to the Managers, 16 September 1885

These ships [*Lusitania* and *Mauretania*] ought to draw all the passengers ... If also we could run on the three ship service, so that we should have the fast ships on two weeks out of every three, we should sweep the market of passengers.

William Watson, Cunard Chairman, 19 October 1907

They [Harland & Wolff] have *carte blanche* to build the ship and put everything of the very best into that ship, and after they have spent all the money they can on her they add on their commission to the gross cost of the ship, which we pay them.

J. Bruce Ismay, White Star Chairman, 30 April 1912

The length of steamships has now reached about 800 feet, and there is no reason, from a technical point of view, that we should not have the thousand-foot steamship ... The present piers are 800 feet long. They would have to be longer with longer vessels, and then, again, with greater length comes greater beam, and ... it would mean another change. We have reached the limit, I believe, with our present facilities.

Emile Boas, Hamburg America Line (HAPAG) General Manager, 3 March 1907

You will no doubt have seen published in the newspapers an account of a new ship 1,000 feet long that the White Star Company are going to build. It seems pretty authentic this time, as it has not been contradicted by the White Star, but on the other hand confirmed.

... Of course it has appealed to my own imagination as to where all the passengers are to come from to fill these enormous carriers.

William Watson, Cunard Chairman, 19 April 1908

1

THE NORTH ATLANTIC RUN

THE OCEANIC STEAM NAVIGATION COMPANY

> Marking a new era in steam navigation, in marine architecture, and in the history of the ocean ferry, the colossal new steamship *Oceanic* of the White Star Line came into port yesterday from her maiden voyage. She was accorded a welcome in keeping with her distinction as the giantess of the seas.[1]

The 'splendidly proportioned' steamer that New Yorkers greeted on 13 September 1899 marked a triumph for her owners. 'Her passengers were found in a state of great enthusiasm,' according to one reporter. They had enjoyed 'perfect weather' and *Oceanic* was very comfortable. Her engines 'caused absolutely no vibration. One could hardly tell whether or not they were going.' William James Pirrie, Chairman of the Belfast shipbuilders Harland & Wolff, was pleased. 'She has done all that we ask,' he said. She was designed as a 'fast mail boat' and a crack express liner, but not a speed record holder. Pirrie insisted, 'The only record she is built to beat is that for regular running. She is to swing back and forth across the Atlantic with the regularity of a pendulum.' One of the reporters pressed him, 'If there was a cyclone, would she arrive on time?' Pirrie explained that the engines were powerful enough for the liner to maintain her schedule in spite of severe weather conditions.

One of White Star's veterans, surgeon Dr J.C.H. Beaumont, noted that the new *Oceanic*'s 'hull and machinery were evidently of the very best, and the fittings of her public and staterooms were of the finest design and quality'. In New York, whenever she came into sight, the harbour pilots 'vied with each other to have the honour of taking up to port the beautiful "White Star Yacht" as they called her. Not only did she look a fine model but she was an excellent sea boat, even in bad weather, and a reliable timekeeper'.[2] Her name harked back to the company's pioneering steamer *Oceanic*, introduced in 1871.

The Oceanic Steam Navigation Company had made rapid progress since being founded in 1869. Thomas Henry Ismay purchased the name, flag and goodwill of the bankrupt White Star Line for the sum of £1,000 in 1867. He opted to enter the highly competitive North Atlantic trade. Roy Anderson explained:

> It is said that this important decision was taken during a game of billiards. The famous story goes that the game was at the home of Liverpool financier Gustav Christian Schwabe, uncle of Gustav Wolff, the junior partner of the Belfast shipbuilders Harland & Wolff, and holder of a big stake in the Bibby Line of steamers ... Schwabe invited Ismay to start a new line, promising to give him the financial backing. There was one condition: the new company's ships would be ordered from Harland & Wolff.[3]

Oceanic docking in New York. She was White Star's 'Ship of the Century' and their last express liner until *Olympic* in 1911. (Library of Congress Prints & Photographs Division)

By the end of the summer of 1869, new vessels were already under agreement with Harland & Wolff. They would sail under the trading name of the White Star Line. Ismay formed a business partnership with William Imrie. The Oceanic Steam Navigation Company's capital of £400,000 was divided into 400 shares of £1,000 each. Thomas Henry Ismay and George Hamilton Fletcher were the two largest individual shareholders and had fifty shares each. Gustav Schwabe (occupation: 'gentleman') had fifteen. The capital was increased to £500,000 in April 1872, with the issue of 100 new shares, and £750,000 in December 1872, with the issue of another 250 shares. By 1880, the capital for all 750 shares had been subscribed in full.

Their first new ship, *Oceanic*, created a stir when she made her maiden voyage in March 1871. *Oceanic*'s hull was slender, with her unusual length-to-breadth ratio of more than ten to one. She was a handsome steamer. Her design was innovative and had plenty of features for passenger comfort, including large first-class staterooms with fresh and saltwater supplies, electric call bells for stewards and individual chairs in the dining saloon rather than long benches.[4] 'The vessel is a masterpiece of naval architecture, and in every particular of size and workmanship,' wrote the *New York Herald*.[5] Third class was comfortable: 'there are separate and distinct lavatories below for married men and women, single women and single men'.[6] Although mechanical problems delayed her maiden voyage, she soon earned a popular reputation.[7]

Her sister ships followed: *Atlantic* made her maiden voyage in June 1871, *Baltic* in September 1871 and *Republic* in February 1872. The longer *Celtic* made her debut in October 1871 and *Adriatic* arrived in April 1872.

White Star's new fleet performed well. In 1872, their westbound crossings averaged nine days, eighteen hours and fifty-five minutes, which was over thirteen hours less than Cunard's. The Inman Line's average was more than a day longer, while the National and Guion Lines both took more than two and a half days more. White Star's average eastbound crossings took eight days, twenty hours and fifty-five minutes.[8]

Thomas Henry Ismay. *The Marine Engineer*, 1899. (Author's collection)

J. Bruce Ismay. *Harper's Magazine*, 1891. (Author's collection)

Rival shipping lines 'scrambled to modernise their now outdated fleets'.[9] Ismay, as the firm's principal manager, explained, 'I personally look after the fleet. There are general managers, marine superintendent, superintendent engineer and superintendent of the victualling department under me.' He set high standards and gave 'instructions there shall be no stint'.[10]

ATLANTIC DISASTER

On 20 March 1873, *Atlantic* left Liverpool for her eighteenth round trip to New York under Captain J.A. Williams' command. After calling at Queenstown to take on additional passengers, she had 957 passengers and crew on board. She never made it. On 31 March 1873, her commander ordered the ship's course changed to take her to Halifax, Nova Scotia, and 'in the early morning of 1 April she was wrecked on the coast of Nova Scotia'. There was a heavy loss of life.[11] It was an enormous blow.

The enquiry into the disaster, held at Halifax, found that the ship had been short of coal, which had forced Captain Williams to divert course. Williams was in the chartroom shortly before the ship struck rocks. His officers had not called him when they had been ordered to. There was a 'total neglect to obtain soundings [the depth of water], or use the most ordinary precautions that ought to be used in approaching the coast'. He had not made any allowance for a potential error in his calculated position, instead ordering the ship to be run at high speed for several hours as she approached the coast:

> The ship struck upon a rock ... She seems to have struck nearly square on, with her head to the northward, her bow remaining fast, but her stern immediately swinging round to the eastward. In a few minutes after striking, several hundred of the passengers and crew rushed upon deck, but at the end of that time she listed to seaward, her deck becoming nearly perpendicular, when all access to the stairways leading from the saloon and steerage was cut off, and those remaining between decks were drowned by the ship filling with water. The conduct of Captain Williams and his officers during this time of trial after the ship struck, seems to have been all that could be demanded of men in their situation; their efforts to save life appear to have been characterised by judgement, coolness, and bravery, but unfortunately all human efforts at such a time were of comparatively little avail.[12]

Ismay and the company's managers were so surprised at the inquiry's findings that they sought a further investigation at Liverpool. They argued that she had not been short of coal, pointing out that she had 960 tons of coal available, compared to a maximum of 810 tons and an average of 670 tons used on the voyage.[13] However, they did not get the outcome they wished. They launched a strong protest to the Board of Trade on 18 November 1873.

Chief surveyor Thomas W. Traill and London surveyor J. McFarlane Gray examined the evidence, issuing their report on 6 December 1873. They noted that Halifax (on the westbound crossing) or Queenstown (on the eastbound one) were sometimes used as 'auxiliary ports' and the major shipping lines had provided coal depots at both of them. During the winter of 1872–73, 'no fewer than *ten* [original emphasis] of the transatlantic steamers, originally bound for New York, had to put into Halifax short of fuel'. They consisted of three Cunard steamers, two Guion steamers and one steamer from the other major lines. *Atlantic* was 'the first steamer of the White Star Line that had ever put into a port for auxiliary fuel'.

After an exhaustive analysis, where Traill and Gray looked at 'the quality, the quantity, and the sufficiency of the supply of coal', they found that 'the supply of coal was sufficient for the voyage to New York'. They thought:

> The statements of Foxley, the chief engineer, with respect to quantities [of coal] are utterly

WRECK OF THE ATLANTIC, SKETCHED THE SECOND DAY AFTER THE WRECK.

The sorry sight of White Star's *Atlantic*, wrecked on the rocks. *Illustrated London News*, 1873. (Author's collection)

> worthless, and cannot be depended on for any purpose … It is only upon the statement of that witness that the assessors at Halifax, and also those at Liverpool, have formed an opinion that the *Atlantic* was short of fuel.

The White Star Line had made sure the coal supply was adequate: 'at the time the vessel's course was altered for Halifax, there still remained sufficient coal onboard to have taken her to New York and to leave seventy tons in the bunkers, even if the weather did not improve'. A mistaken conclusion about the coal supply led to the decision to divert to Halifax; and poor seamanship led to tragedy.

Nothing in her design or construction had doomed *Atlantic*: human error was at fault. Her structural design and high state of repair did not escape notice during regular inspections. William Crees Taylor, principal surveyor to the Board of Trade at Liverpool, commented on *Atlantic*'s condition. She had been in dry dock on 12 December 1872, so that her three-bladed propeller could be replaced by a four-bladed one. He examined her exterior:

> Her general condition outside then was very good … No material repair was done to her then. She showed no symptom of weakness. I never saw a fairer bottom in a dry dock … She was a very long ship; but a fine shaped one … I examined the rivets of the *Atlantic*, 12 December; they were firm and staunch, and the butts not weeping. I examined her generally, and found her in excellent order.

Another surveyor, W.H. Bisset, spoke of her strength:

> I have been [for] five years surveyor for the Board of Trade, and 28 years looking after ships, fitting out and going to sea in different capacities. I think the *Atlantic* was one of the most substantial ships that I ever was onboard.
>
> The *Atlantic* was very strongly built.
>
> I would have no scruple in passing [certifying as seaworthy] any ship ten times her own breadth.[14]

Edward James Harland, senior partner in Harland & Wolff, was proud of the engineering achievement demonstrated by *Oceanic* and *Atlantic*. He explained in May 1873:

> We did not make the engines, but we designed them; they were made by Messrs. Forester & Co., of Liverpool …
>
> The engines were compound, the most economical at present known. And so satisfactorily did they work in the *Oceanic* and *Atlantic* that we are at present building larger vessels for the same owners on identically the same plan. They were built for the New York trade expressly.
>
> If the whole of the expansion power was on, the vessel could still be worked against a head wind. Even at half boiler power against a gale of wind they would make from five to six knots.[15]

AN EMBARRASSMENT OF RICHES

Britannic and *Germanic* appeared not long afterwards. *Britannic*'s maiden voyage took place in June 1874 and *Germanic*'s in May 1875. They were larger, faster and more powerful. *Britannic* 'caught on' – passengers 'would alter their dates in order to sail in her'. She developed a reputation for safety and comfort.[16]

White Star's activities were not confined to the North Atlantic. In 1872–73, their attempt to establish a South American steamship service was unsuccessful against 'intense competition' from the Pacific Steam Navigation Company. Their sailing ships focused on South America, Australia and New Zealand. *Oceanic*, *Gaelic* and *Belgic* were chartered to the Occidental & Oriental Steamship Company: *Oceanic* sailed from Liverpool for Hong Kong in April 1875. Early in the 1880s, White Star partnered with Shaw, Savill & Albion on a steamship service to New Zealand. They provided the ships and crew for their partner to manage. (In later years, White Star took an increasing shareholding in Shaw, Savill & Albion, with new ships built for the trade and jointly owned.)[17]

The failure of their South American venture demonstrated that success in the steamship business was hard won. When White Star entered the North Atlantic steamship trade, they were up against formidable, long-established competitors. They succeeded so well that, in 1878, Cunard's John Burns suggested both lines merge. Ismay did not find the suggestion attractive. He was critical of his rival. Cunard nearly doubled their tonnage of ships on the New York trade between 1880 and 1885, which simply reduced their profit per ton. Burns' argument was that the expansion of the fleet would pay off in the future.[18]

Speaking in 1885, Ismay looked back and argued that White Star had 'made quite a new departure in ocean travelling. We aimed at comfort; that was our speciality … Our ships have all the latest improvements, and are admirably designed and built regardless of expense.' He discussed the perils of the sea, 'the chief danger is not from icebergs, as some people think, but from collisions with other ships'. Despite the rules of the road at sea, fog was perilous, as well as navigational errors 'estimating the approach or position of the lights at sea. As long as men are human and fallible, accidents must happen; look at the accidents which happen on shore.'[19]

In September the same year, about seventy of the company's shareholders and managers gathered on board *Adriatic* for a presentation to Thomas Henry Ismay. The chairman relayed a record of success:

> The company has had to fight its way from the beginning against the keenest competition, and we have attained a position which I feel we may be all very justly proud of today. We have out of our original capital built 64,000 tons of shipping, which have cost over two millions sterling.

Harland & Wolff had built a fleet 'regardless of cost, so that they are absolutely as perfect in strength and finish as the great talent of their builders could make them … our managers showed good judgement in going to the right builders'. He praised Ismay's:

> … coolness and good judgement in not being led into following in the wake of those who have rushed into large and costly steamers … when

> he thinks the time has come to build larger and faster boats we, as shareholders, shall be safe in his hands, for he will consider our interests before either the profit or the fame of the managers.

In a toast to the builders, Ismay said they had given:

> ... us boats in which speed was obtained upon lines as profitable to the owners as creditable to the builders. They neither trebled consumption or coal nor reduced cargo capacity, and, bearing these points in mind, what they have done is yet to be beaten.

Harland praised the line's focus:

> Quality was more important than cost ... In good times you can afford to run almost any 'trap'; but when bad times come, and the 'trap' begins to give out, and when you want to obtain the most work at the least cost, you find your vessel is unable to do it.

White Star's fleet had competed 'successfully against newer, and faster, and larger vessels, with their splendid load of nothing in them, and producing less than that as a dividend to their shareholders'. The *Daily Post* said:

> In these times of commercial stagnation it is both astounding and instructive to hear that one of the great steamship lines has been obliged to modify its articles of association in order that its managers may be able to get rid of their embarrassment of riches by paying bigger dividends than were legal under the original constitution of the company.

They argued 'the growth and prosperity of large steamship companies, and all other important commercial enterprises, are matters in which all the people of Liverpool are interested'. In fact:

> ... the only complaint which the Liverpool public have to make against Mr. Ismay is that he has not deemed it possible to place his business talents at the service of the public. More than once local Liberals have desired that so eminently suitable a man should represent in Parliament the commercial interests of this city.[20]

When the time did come to order more ships, Ismay saw an opportunity. He proposed to build two express liners under Admiralty supervision that could be made available quickly for government service in the event of war. In return, he sought government support. White Star ordered 'two ships of a speed and strength superior to any merchant ship afloat',[21] but passenger comfort was not neglected. *Teutonic* and *Majestic* entered service in 1889 and 1890.[22] One reporter noted:

> The *Majestic* has quarters for three hundred first class passengers, all of whom can be seated at once in the main saloon. The great size of the vessel would admit of staterooms for a larger number of first class passengers, but her owners had resolved to give the passengers plenty of room. The staterooms are large, well lighted, and well ventilated, and contain every facility for comfort.[23]

In August 1891, *Teutonic* made a record crossing westbound and gained the Blue Riband that her sister had won only days earlier. She was the last White Star liner to achieve the prestigious accolade. She covered 2,778 miles over five days, sixteen hours and thirty-one minutes, at an average speed of 20.35 knots.[24] Although Cunard responded and *Teutonic* lost the record, the two ships continued to be successful.

White Star saw some generational changes in its third decade of existence. Ismay's sons, James Ismay and J. Bruce Ismay, were admitted as partners in Ismay, Imrie & Co. – the management company of the White Star Line – on 1 January 1891.[25]

J. Bruce Ismay was born on 12 December 1862. He attended the prestigious schools Elstree and Harrow, where he was fond of football and tennis, then served an apprenticeship with his father's company before becoming the White Star Line's New York representative.[26] Although he was shy and sensitive, he 'put on a façade of brusqueness to avoid appearing over-sensitive', which could be off-putting for those who did not know him well.[27] Wilton J. Oldham noted one unfortunate incident that he remembered for years afterwards:

> During the summer holidays he had occasionally accompanied his father to his office; on these occasions he always hung his overcoat and hat with his father's in the office. When he reported for his first day's work he did this as usual, but after he had received his instructions and left the room, T.H. Ismay rang the bell for one of the clerks and said to him: 'Please inform the new office boy that he is not to leave his hat and coat lying about in my office'. This snub in front of his father's staff so humiliated the boy that he rarely wore an overcoat again.[28]

He married an American lady, Julia Florence Schieffelin, in 1888. Roland Shelley commented on his 'commanding presence':

> Mr. Ismay disliked publicity immensely, and was of a rather shy and reserved nature, but in his private life he had more than a few friends who enjoyed his company, for though he had rather a caustic whit, they knew – as the outside world did not – the kindly act that he performed without one touch of ostentation.[29]

One morning, Ismay had a cheque made out for £500 to a children's hospital in Liverpool (one of many charitable donations) and forbade any news of the donation to appear in the press. He was 'quietly spoken and reserved' as well as 'intolerant of inefficiency', but 'if one put on a bold front and gave one's own opinion without reserve (as very few did) he respected one for it'.[30]

Ismay's friend Harold Sanderson also had family connections in the steamship business. In 1888, he had succeeded his father as the head of the firm Sanderson & Son, situated on New York's old State Street. Following a two-year spell as the general manager of Wilson & Sons, based in Hull, Sanderson became the White Star Line's general manager in 1895. Four years later, he also became a partner in Ismay, Imrie & Co.[31]

Times were challenging for all the shipping lines. In 1892, Cunard described a 'disastrous' year: freights were 'exceptionally low' nearly all year, then westbound immigrant traffic was 'practically extinguished' from September to December as the United States government enforced quarantine restrictions to keep out cholera. The next year, they found trade 'unprofitable and disappointing', with first-class business 'smaller than it has been for a number of years'. A coal strike resulted in increased expenses. During 1894, westbound immigration fell to 'less than half' the level of 1893, despite a reduction in fares. First-class traffic maintained its low level but second class 'fell off considerably'. And then there were freights, which were 'lower than ever'.

By 1896, Cunard's business was 'on the whole, more satisfactory'. Even so, the Atlantic lines suffered the next year:

> ... from the loss of passenger traffic between Europe and the United States which marks 1897 as the worst of recent years. In first cabin business the total falling off for British and continental lines compared with 1896 was about 14 per cent, in second cabin 5 per cent and in steerage 18 per cent.

Westbound third-class numbers fell by 76,000 passengers.[32]

J. Bruce Ismay referred to difficult trading conditions in correspondence with the Holland America Line early in February 1895. His Dutch colleague described the recent loss of Norddeutscher Lloyd's *Elba*, which sank after the steamship *Crathie* collided with her in the North Sea. There was a heavy loss of life and few survivors. Ismay agreed that 'the *Elbe* is a fearful accident and emphasizes in a singular degree the risks of a shipowner's business', adding, 'the pity of it is in the Atlantic trade that so great a work should be carried out with such poor financial results'.[33]

White Star's development over its first three decades was therefore impressive. In 1876, net profits exceeded £100,000 for the first time and, by 1880, they were over £200,000. The company paid a healthy dividend to its shareholders of either 10 or 15 per cent of its paid-up share capital every year from 1879 onwards. This meant that a shareholder who had paid £1,000 for a share in the company received either £100 or £150 each year as a dividend payment. Cunard's shareholders were not so fortunate. Their company did not pay a dividend from 1883 to 1886 or from 1894 to 1895, and the dividend never exceeded 6 per cent of their paid-up share capital.

Ismay believed in 'rigid secrecy' when it came to White Star's financial results and he benefited from the fact that his was a private company whose shares were not traded publicly. They did not need to release the sort of public financial information that Cunard did. He was able to track his competitors but they could not track him in the same way. If his shareholders had known the full extent of White Star's prosperity, they might well have 'clamoured for higher dividends'. Instead, Ismay kept the profits and put them to use in expanding the fleet, generating even more profit.[34]

White Star suffered another blow when *Naronic* vanished. She left Liverpool for New York on 11 February 1893, never to be seen again. Harland & Wolff had only recently completed the single-screw 6,594-gross-ton steamer at a cost of £112,625, but the ship and her cargo were not insured. After battling 'a succession of gales from the north west and south west, with high mountainous seas' in late February and into March, on 4 March 1893 the steamship *Coventry* reported sighting two lifeboats, the second of which was definitely one of the lost steamer's. She was lost with everyone and everything on board: twelve passengers, sixty crew 'and a general cargo of 3,572 tons'.[35]

Investigators were baffled. They eliminated one possibility after another. There was no obvious explanation. She had been seaworthy on leaving port; had sufficient stability; had a properly stowed cargo; had 'nothing in the cargo which was likely to explode or ignite spontaneously'; was equipped with 'proper appliances onboard to extinguish fire'; and had been 'fully provided with life saving apparatus'. Her watertight subdivision was up to the highest standards. 'Her scantlings [thickness of structural components of the hull] were up to, and in some cases indeed above, those required by Lloyds' rules.' Captain Roberts was an experienced man and other officers and crew had been used to the ship. Captain Thompson, *Naronic*'s first captain, 'spoke of her in the highest possible manner, and had, when in command of her, on several occasions encountered bad weather, and had never found her at all tender or to require stiffening'. In fact, 'on one occasion with seven hundred head of cattle on deck she encountered a whole gale from the eastward, but she behaved so well and was so steady that he did not lose one of them'.[36]

'It is difficult to attribute her loss to want of stability and sudden capsizing,' they reported. There was no evidence of other vessels disappearing at the same time or in the same area, making a collision seem unlikely. Her engines and propelling machinery were in excellent condition, making it 'difficult to suppose that a defect in them was the cause of her loss'. There was no evidence or indication that anything in her construction or operation had been lacking. She had completed six previous round-trip voyages to America and back without any problems. On 25 July 1893, the Board of Trade's report concluded that unless any more information came to light, 'the probable cause of the loss of this vessel remains a mere matter of speculation, and adds one more to the mysteries of the sea'.[37]

White Star was not alone in its mishaps. Cunard went through an unfortunate run of events in the late 1880s. *Sidon* struck rocks off the Spanish coast and was wrecked in October 1885; *Oregon* foundered after a collision with a schooner in March 1886; *Demerara* sailed for Gibraltar in December 1887 and was never heard from again; *Nantes* was lost after a collision in November 1888 and *Malta* was stranded off the coast of Cornwall and lost in October 1889. After *Demerara*'s loss, Cunard's directors said she was 'well found in every respect, and was commanded by one of the most trustworthy and experienced of the company's masters, whose loss, as well as that of the others on board, the directors deeply deplore'. She, too, became another of the sea's mysteries.[38]

THE GERMAN FOUR STACKERS

Britain's maritime reputation was increasingly under strain. The German shipping lines were expanding rapidly. In 1897, the new Norddeutscher Lloyd (NDL) liner *Kaiser Wilhelm der Grosse* steamed across the Atlantic and captured the coveted Blue Riband speed record.[39] She was designed for luxury as well. Her first-class passenger accommodation included a 'large music room' for the ladies and a smoking room and Vienna Café for the gentlemen.[40] John Malcolm Brinnin described her interiors as 'outsized magnificence'. The 'designers overawed and overwhelmed' the passenger with 'noble apartments of cathedral proportions'.[41] *Kronprinz Wilhelm* (1901), *Kaiser Wilhelm II* (1903) and *Kronprinzessin Cecilie* in 1907 each brought further improvements.[42]

The Hamburg America Line (HAPAG) responded by putting the faster *Deutschland* into service in 1900, but she demonstrated the practical and economic difficulties with pursuing the highest possible speed. She suffered from severe vibration and 'passengers did not take kindly to being shaken continuously throughout a five-day crossing'.[43] HAPAG's focus switched to luxury and comfort, much as White Star's had done several years earlier.

Oceanic represented a happy balance, combining 'the luxury of the express steamer and the capacity of the slower leviathans, with a speed approaching that attained by the everyday greyhound of the period, but without attempting to outdo the best that has been accomplished in speed'.[44] It was a sad fact that Thomas Henry Ismay passed away shortly after she entered service.[45]

Nonetheless, the White Star Line actively expanded their fleet in the late 1890s and early 1900s with a succession of larger ships which were intermediate steamers. These included the 'Big Four': *Celtic* (1901), *Cedric* (1903), *Baltic* (1904) and *Adriatic* (1907). They were several knots slower than *Oceanic* and 7 to 8 knots slower than the fastest ships afloat, but they were extremely economical to run and had a huge cargo-carrying capacity. Each was also significantly cheaper to build than *Oceanic*.

'The long, lean, slippery "greyhound" carries many passengers, much coal, and little freight,' reflected one observer:

> Here, on the other hand, is a ship [*Cedric*] which carried many passengers and much freight, while its coal space and fuel bills allow an astonishing reduction … the earning power of such an all-round merchantman must be far superior to that of the headlong smasher of Atlantic records, and that especially in winter, when the west winds howl, and passengers are few, the big, steady going ship with a good cargo in her hold is much the better investment.[46]

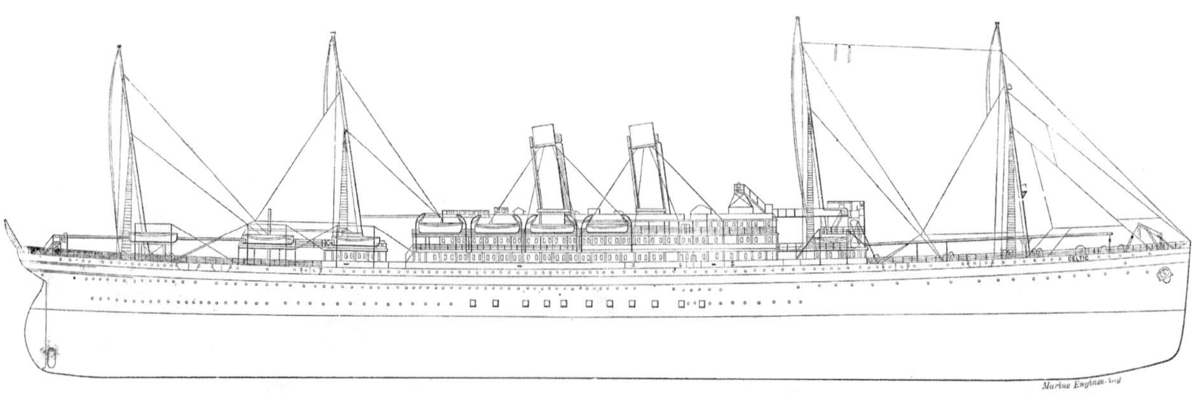

Celtic had an enormous carrying capacity for passengers and cargo, as well as being very economical to operate. Her relatively deep draught is clearly visible. Although she did not share the elegance of *Oceanic*, her two funnels and four masts represented a balanced profile. *Marine Engineering*, 1901. (Author's collection)

J. Bruce Ismay spoke with a reporter in New York early in 1901, saying that *Oceanic* was so successful that the White Star Line would not build any faster ships.[47]

When *Celtic* arrived in New York for the first time in August 1901, she was praised for being 'as sturdy as the Rock of Gibraltar'.[48] She had taken over eight days to cross the ocean and her average speed was less than 15 knots. *Celtic* was 'not so elegant or luxurious as the *Oceanic*, the *Deutschland* or the *St Paul*'. Eighteen months later, *Cedric* followed in her older sister's wake. A smoke room steward explained:

> When we left Queenstown, I placed a small wine glass filled with champagne on the edge of a sideboard on the port side of the smoke room. I never touched that glass all the way across, and when we got to Sandy Hook today the glass had not moved half an inch and not a drop of wine had been spilled. Now, that proves that the *Cedric* is a wonder, doesn't it?[49]

Captain Herbert James Haddock said that the bad weather had not bothered *Cedric*. 'On this ship,' he claimed, 'you would hardly know you were at sea unless you happened to take a walk on deck or looked out of your stateroom window.' (She had seen less than a year's service when rumours spread that she had been rammed and sunk by a Lamport & Holt Line ship. Friends and relatives of *Cedric*'s passengers 'flocked to the office of the White Star Line'. They were reassured by company officials, who were quoted as saying that *Cedric* 'is so built that she is practically unsinkable'.)[50]

Surgeon Dr Beaumont praised 'the steady *Celtic* class' in contrast to the older *Majestic*:

> As well as pounding heavily into head seas, or rolling badly, the *Majestic* had a pet trick of kicking in the opposite direction to what one expected – which is known in the service as being 'vicious'. However, she did ease up sometimes, and went along merrily at nearly 20 knots an hour. The passengers were few but got to know each other pretty well, and were happy enough in the smoke room or deck if they found the dining saloon too much for them.[51]

They even had their admirers serving in the Cunard fleet. Purser J. McCubbin, serving on board Cunard's ageing *Etruria*, observed:

> I think I may venture to say that the liners best patronised are those like the *Saxonia* and *Cedric*, which are the largest and roomiest, though somewhat slower than less bulky boats. The comfort of these ships is in proportion to the high esteem in which they are held for all-round steadiness and roominess of accommodation.[52]

Perhaps he was thinking of the Norddeutscher Lloyd record-breakers when he wrote:

> … more attention has been given to the speed of the boat, the splendour more than the comfortable accommodation of the Saloon and such details, instead of the absolute comfort of the passenger which is the point which influences that individual more than any other, in the choice of the ship by which he shall sail. The majority of passengers will be found to give their opinion, to the effect, that 'one day more or less at sea doesn't matter, provided one is comfortable'.[53]

Baltic was enlarged because her sisters were so successful. She arrived in New York for the first time on 8 July 1904. Captain Edward J. Smith – known fondly as 'E.J'. – appeared 'delighted' with his command. 'I tried to see how she would work coming around the tail of the Southwest Spit, and, as the channel was clear, I sent her around at full speed. She behaved admirably.'

There was no doubt that White Star were on to a winning formula. *Cedric* had 'carried more first cabin passengers trip for trip than any other liner sailing from New York' in 1903.[54] On 11 September 1904, *Cedric* arrived in New York with 2,723 passengers; *Celtic* followed thirteen days later with 2,957 passengers, an all-time company record; and at the end of the month, *Baltic* brought in 2,890 passengers. At the height of the season, White Star's three ships had delivered 8,570 passengers in the space of a mere three weeks, aided by their enormous third-class passenger capacities. First-class figures were strong: *Baltic* carried 410 first-class passengers. By 1905, *Baltic* was carrying around 30,000 passengers annually. Two years later, she carried her 100,000th passenger. Each sister ship was profitable and successful, but the White Star Line itself was undergoing one of the most significant changes of its long history.

A 'FINANCIAL COLOSSUS'

John Pierpont ('J.P.') Morgan was described as a 'financial colossus', who became 'the most powerful banker in the world'. When the New York Stock Exchange fell sharply from mid-October 1907 and there were bank runs, Morgan played an important role in helping restore confidence. His financial interests, including railways in the United States and the enormous US Steel Corporation, formed in 1901, 'helped transform a largely agrarian society into a modern industrial state'.[55]

He may have been sceptical of the shipping business but, if so, he changed his mind in the second half of 1900. His firm, J.P. Morgan & Co., acted as banker for Clement Griscom's International Navigation Company (established in New Jersey), assisting its merger with the Atlantic Transport Company, advancing cash for new ships and then bringing in new shipping lines.[56]

On 28 November 1900, Griscom, Bernard Baker, A.B. Widener and J.P. Morgan held a meeting 'to discuss and outline an agreement' for the merger. The agreement was made on 10 December 1900. By the end of April 1901, agreement was also reached to purchase the Leyland Line. During the summer of 1901, newspapers were 'full of rumours':

> … that the Leyland fleet would ultimately be transferred to the American flag; that the Leyland fleet was acquired to work in connection with the Steel Trust; that there was to be a consolidation of the Leyland, the Atlantic Transport, and the International Navigation Company; that Mr. Morgan was negotiating for the purchase of the Dominion Line; that Mr. Morgan intended to build a number of fast passenger ships to beat anything afloat; that the White Star and Cunard Lines had decided to combine for mutual protection.[57]

This did nothing to quell a growing sense of unease in Britain. Morgan set his sights on the White Star Line. The *Liverpool Daily Post* reported on 30 April 1901:

> Mr. Morgan and the group of capitalists for whom he operates are resolved to make the Atlantic and Pacific oceans American lakes so far as commerce is concerned, declared a friend of Mr. Morgan's last night … Aided by the American railways, the trust can make low rates for carriage of passengers and cargo, and from the very first will be in a position to take business from rival lines. America is anxious to re-establish her merchant marine, and the Government will give liberal subsidies to aid in restoring the flag to its old position in shipping.[58]

Rumours began to circulate that an entire new fleet of vessels was going to be constructed:

> Four of the boats are to be high speed greyhounds of the *Deutschland* class, and six are to be great seven day boats like the *Oceanic*. A number of fast cargo boats are also to be constructed. All are to be built in America … Within the next five years it is expected to expend £10,000,000 in new vessels, and a large sum in buying craft already in commission. Cargo and passenger rates will be reduced, and through tickets to points in the United States will be sold at very low rates. One fast boat and one seven day boat will be liners deluxe, carrying only first class passengers, and having in stock theatre and music hall, for which high class performers will be engaged. The vessels will be floating palaces, and nothing will be spared in the furnishing that conduces to luxury. By reducing the cost

> of ocean travel the Morgan syndicate expects to increase the number of passengers on the Atlantic ferry vastly.[59]

Following Thomas Henry Ismay's death, his estate – including 150 shares, or 20 per cent of the line – was in the process of being settled by the trustees. His shareholding represented the single largest block of shares. Pirrie, holder of the second-largest block, took an 'active interest in the sale of White Star' and came to favour an arrangement with the Americans. White Star was an attractive asset. The Atlantic Transport Company was the 'more prosperous of the American-owned lines' but White Star's profit was 'about fifty per cent more per ton'. It was 'very efficiently operated' and 'by far the most profitable British Line operating in the North Atlantic'.[60]

In September 1901, J. Bruce Ismay 'ridiculed the report that the White Star Line had been purchased by J. Pierpont Morgan'. Prior to leaving New York, he said, 'A statement that the company contemplated the establishment of a line to the moon … would be about as true as the reported sale.'[61]

On 25 January 1902, the *New York Times* quoted him as saying, 'Money can't buy the White Star Line. It's not for sale. An offer was made for it some time ago, but was not entertained. We are satisfied with the condition of this property and its dividends. It's a good thing to hold.' Outside the public press, Ismay 'refused to give figures' for White Star's expenditure on new ships to the Americans and 'kept coming up with new conditions and demands'. Late in April 1902, Morgan advised his associates, 'It does not seem to me safe to continue to make concessions placing ourselves at [the] mercy of Ismay.'[62] (Morgan had already established a 'community of interest' between the large German shipping companies HAPAG and Norddeutscher Lloyd. Albert Ballin met him in New York in February 1902 and the two men signed a ten-year agreement, which defined areas of mutually beneficial cooperation and profit sharing.[63] The combine also acquired a controlling interest in the Holland America Line.)

J. Bruce Ismay and other family shareholders eventually acquiesced. On 1 May 1902, it was reported that the White Star Line had issued an official statement 'last evening':

> The White Star Line entered into a provisional agreement with the Morgan company in February, with the object of securing community of interest between the American, Atlantic Transport, Dominion, Leyland, Red Star, and White Star Lines. The agreement was binding on the White Star Line, but was optional as regards Mr. Morgan until April 30, upon which date the agreement expired unless extended by mutual consent. Mr. Morgan today has given notice to the White Star Line of his intention to exercise the option and of proceeding with the arrangement.
>
> It is now in order for the agreement to be submitted to the shareholders … The shareholders of the White Star Line will meet in May.[64]

Pirrie told reporters, 'We are not advocating anything to our own disadvantage as Britishers or individuals. Apart from nationality, if the suggested partnership arrangement comes to ahead, I have no doubt that it will benefit both Englishmen and Americans.' Later, speaking before the Steamship Subsidies Committee in the House of Commons, he recalled that 'the community of interests, was established purely for commercial purposes and on business principles … I consider the deal was in every way satisfactory to the English companies … and our own national interests were never overlooked.' They 'had seized the opportunity of making a friendly and profitable arrangement for all our interests, both commercial, national, and even political, as nothing can do so much to promote good relations between the two great Anglo-Saxon communities as the community of business interests'.[65]

On 14 May 1902, the *Northern Whig* covered statements and interviews Pirrie had given to *The Times* and *Daily Telegraph*. Faced with public concern at British shipping coming under American control, Pirrie welcomed 'the sudden interest', given the 'apathy and indifference displayed by the British public in their mercantile marine'. He argued that he had three interests:

> The first is the interests of the country – and I wish, by the way, you would invent some comprehensive word which would, unlike Briton, include Irishmen – my second is in shipbuilding, and my third in shipowning. In which of these capacities could I possibly be an assenting party to a scheme which threatened injury to Imperial interests and ruin to British shipbuilders and shipowners?

Confronted with concerns about monopolies, Pirrie said 'a large part of the savings' achieved by the combine would be in 'the pockets of the public'. Tickets would be interchangeable so that:

> ... passengers who have booked by one line can travel by one of the others. In winter there will, I fully expect, be two fast steamers sailing every week to and from America, whereas at present during some of the winter months the fast ships are lying up, and the mails are carried by slow steamers.

The public 'has a very natural dread of monopolies' because they worried about price increases, but he highlighted the results of the US Steel Corporation, presented at its first annual meeting in February 1902. It had achieved 'great economies in manufacture' and consumers benefited because they did not increase prices. Meanwhile, investors got a 'fair return' on their capital and the 'stability in prices, both of new material and finished products, has had a reassuring effect on the trade, and has contributed greatly towards restoring confidence in the general business situation'. Those benefits were 'what we hope to effect by organising this community of interests in the shipping world'.

Faced with Morgan's offer, the White Star shareholders approved the sale at their meeting on 18 May 1902. For each share of £1,000, the holder was to receive:

> ... in cash from the American combine the sum of £4,143 18s 3d, together with preference shares bearing interest at 6 percent, to the value of £6,713 1s 4d, and ordinary shares to the value of £3,356 10s 8d ... There is to be deducted from each share the sum of £52 8s, which is to go to a fund for compensation to the officials of the White Star Line. The net sum to be received by each shareholder is therefore £14,265 18s 3d.[66]

There is a strong argument that IMM paid far more than the company was worth. The valuation was based on a generous multiple of White Star's annual earnings and it was calculated from its financial results in 1900, which was an exceptionally good year.

Planning for the future and questions over exactly what the new company would look like within the combine occupied J. Bruce Ismay and his colleagues. On 4 June 1902, he wrote to Pirrie at Harland & Wolff about the proposed sister ship for *Oceanic*, intended to be named *Olympic*, 'Of course had we been left to ourselves there is little doubt ... we should have gone [ahead] with an improved *Oceanic*, but under the altered circumstances we don't know where we are.'

Pirrie agreed, 'Under the circumstances I think it would hardly be justifiable to decide at present to go on with the proposed *Olympic*.'

It appears Harland & Wolff had intended to assign 'Yard Number 356' to the ship, but it was used for the Union Castle Line's *Kenilworth Castle* (laid down early in October 1902).

Ismay's list of 'Matters for Discussion' with his American colleagues, dated 24 September 1902, reflected his sharp eye for detail and business expertise. 'Is it the intention to make Southampton the port for the fast mail and passenger service of the IMM Co. or not?' he asked. 'There are advantages and disadvantages in doing so, and the question needs careful consideration. It must be remembered that doing so would practically leave the Cunard Line control of the fast passenger business from the North of England.' The mail contract was another consideration: 'It is worth about £70,000 a year. The decision on this point will have a material bearing on the question of fast steamers making Southampton or

An IMM share certificate for the combine's 'common stock', dating from the late 1920s. The White Star Line's existing shareholders received a combination of cash, preference shares in IMM and ordinary (common) shares in IMM after the US combine purchased a controlling interest in the British shipping line in 1902. (Author's collection)

Liverpool their port, as the Post Office authorities will wish the mail boats to call at Queenstown.' The port has 'very large Irish emigration and if the IMM Co. do not serve it properly, the business will get into the hands of the Cunard company'. He wanted to know 'exactly what are the commitments of the various companies of the proposed combine' to building new tonnage. (In June 1902, White Star itself was making payments of about £70,000 a month to Harland & Wolff to cover the cost of new tonnage.)

Ismay's focus was not only on the North Atlantic operations. White Star's working agreement with Shaw, Savill & Albion had eight years to run but their partners were arguing that 'under the altered circumstances the contract is rendered null and avoid'. Ismay disputed this. He also argued, 'The New Zealand trade is a valuable one, and it is anticipated will become more so, and should be closely followed.' He drew attention to the Pacific, 'An enormous trade will be done across the Pacific.'

White Star had three steamers – *Gaelic*, *Coptic* and *Doric* – chartered to Occidental & Oriental Steamship Co. 'at £2,000 a month, leaving a profit of about £45,000 per annum'. They were satisfactory ships but 'could be much improved, and it might be well to at once build tonnage to replace the present steamers'.

There were particular advantages to the scale of combine. Ismay foresaw that all coal purchase contracts could be under a single individual, enabling them 'to make good contracts' with better bargaining power. There were challenges, too, including remuneration to managers. 'This matter should be considered carefully,' Ismay wrote.[67] (He also worried about the amount of debt the new combine would hold, suggesting each company produce a forecast of their financial position as at 31 December 1902.)

He received lots of correspondence. On 18 September 1902, 'a Loyal Englishman' wrote to him, asking, 'I take the liberty to ask a question ere you sign giving the "Yankees" control of your line. Will you, and those you represent, feel proud of being dictated to, for ever hereafter, by Morgan and his "swift" set? Can not Englishmen get together and manage their affairs?'

Despite the public controversy, by year's end the transaction had been completed. White Star now fitted into a somewhat complicated structure. Ismay, Imrie & Company[68] were the largest shareholders in the business prior to 1902 and acted as 'managers of the Oceanic Steam Navigation Company'.[69] White Star was the trading name of the Oceanic Steam Navigation Company, whose directors were J. Bruce Ismay (as chairman and managing director), Harold Sanderson and William Pirrie. After 1902, practically all of the Oceanic Steam Navigation Company's shares were held by the International Navigation Company, registered in Liverpool with four directors: J. Bruce Ismay (chairman), Harold Sanderson, Charles Torrey and Henry Concanon.

The International Navigation Company also held almost all the shares of the British & North Atlantic Steam Navigation Company; the Mississippi & Dominion Steamship Company (Dominion Line); the Atlantic Transport Company (Atlantic Transport Line); and 'practically the whole of the issued ordinary share capital and about one-half of the preference share capital of Frederick Leyland Company (Leyland Line)'. The shares of the International Navigation Company were, in turn, held by the International Mercantile Marine Company (IMM), which was an American entity registered in the state of New Jersey. (IMM had itself changed its name from the International Navigation Company when the shipping combine was formed.) IMM's first annual report covered the calendar (and financial) year 1903. The company's chairman, Clement Griscom, was one of twelve directors, including J. Bruce Ismay and William Pirrie. Ismay chaired the Finance Committee. Although IMM recorded a profit, the management cautioned shareholders that the year's operations reflected 'to a very limited extent' the anticipated benefits of reorganisation.

J.P. Morgan was not happy with IMM's performance. He believed J. Bruce Ismay would be able to improve matters. As Ismay explained to Harold Sanderson, Morgan:

> ... stated that he did not mind losing money, but he did object to doing so owing to poor organisation. He felt things were on a wrong basis, and that, in his opinion, I was the only man he knew that could straighten matters out. He was prepared to make me president and give me unlimited control, and anything I decided was to be final.[70]

In December 1904, an American investment banker writing in the *Washington Post* gave a vote of confidence in Ismay's abilities:

> At the time the [IMM] combine was formed by J.P. Morgan & Co. it was estimated that the earning capacity could be increased $10,000,000 over the previous four-year average, when the properties were operated under separate management. A good deal of criticism has been passed upon the extravagant management since the combine was formed; and it is said that Mr. Ismay, the present executive head, has instituted the most rigorous economies. It was Mr. Ismay's management which made the White Star Line one of the foremost Atlantic steamship lines, and one of the most prosperous from a financial standpoint.[71]

He thought IMM would receive support in the United States, because the British government provided an 'immense subsidy' to Cunard 'in opposition to the lines owned by American interests'. White Star was certainly doing well. During 1904, they 'carried a larger number of first class passengers both to and from New York than any other Atlantic line'.[72]

***

Competition was intensifying. HAPAG's *Amerika* arrived in New York for the first time on 20 October 1905. The *New York Times* reported how closely she resembled *Celtic*, *Cedric* and *Baltic*, '*Amerika* was [also] built at Belfast by the Harland & Wolff company.'[73] She was very similar in size and speed, but Ballin insisted on an even higher standard of first-class passenger accommodation, including a large gymnasium, photographer's dark room, children's room, stateroom telephone service, first-class lift ('elevator') and an à la carte restaurant, an exclusive extra-tariff facility where passengers could eat what they wanted, when they wanted. *Amerika*'s deck names included the Kaiser Deck, highlighting her German ownership. The Washington Deck set an American theme and contained some of the finest suites on the ship: the two *Kaiser-Zimmer* Imperial suites; a number of *Staats-Zimmer* suites; and then the Roosevelt Deck below housed several large, comfortable *Luxuskammern Cabines de Luxe*.[74]

In 1906, the slightly larger *Kaiserin Auguste Victoria* appeared.[75] Ballin recognised the economic benefit of intermediate steamers and combined it with passenger accommodation intended to outdo the 'Big Four'.

Amerika was successful. She averaged 1,850 passengers on each crossing under a 'regular service' on the Hamburg to New York route, including very strong third-class numbers, whereas *Deutschland* averaged 746 passengers, operating their 'express service' on the same route.

White Star's *Adriatic* was laid down only five months after *Baltic*, but her construction was paused for an extended period.[76] Her passenger accommodation was improved. She had a gymnasium, Turkish and electric baths and a plunge bath in first class. They made significant improvements in second class, adding an additional deckhouse on the upper promenade deck to house a spacious smoke room; beneath it, a large ladies' room was double the size of the library on her sister ships; and the dining saloon was enlarged to span the width of the ship. Her second-class passengers enjoyed spacious public rooms that were equal to those in first class a generation earlier.

'The Beginning of the New Era at Southampton: the *Adriatic* Leaving for the first voyage of the new mail service' on 5 June 1907. It is tempting to speculate whether one of the officers stationed at the stern is First Officer William Murdoch. *Illustrated London News*, 1907. (Author's collection)

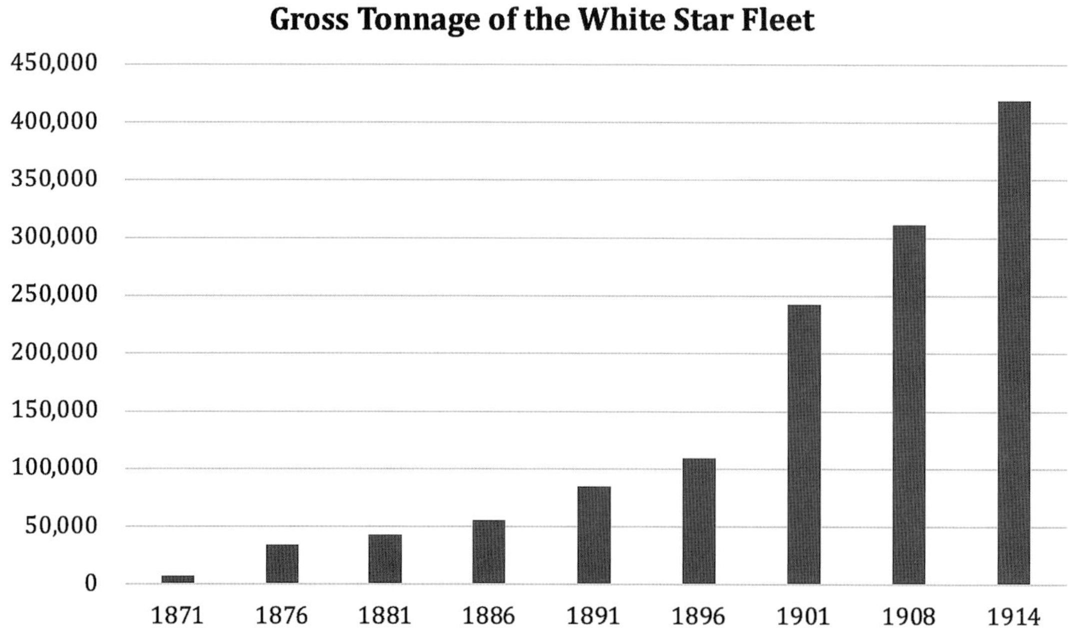

The gross tonnage of the White Star Line's fleet grew strongly from 1871 to 1914. The increase is all the more impressive given that *Titanic* was not included in the 1914 figure. Cunard's fleet stood at 244,466 gross tons in June 1908, whereas White Star's was 311,403 tons in September 1908. (Author's collection)

White Star's offices at Oceanic House, 1 Cockspur Street, London SW1, were decked out in 1911 to celebrate King George V's coronation. *Olympic* featured prominently. Construction was completed several years earlier. In the late 1900s, Cunard, HAPAG, Norddeutscher Lloyd and Harland & Wolff had premises on the same street. (© Heritage-Images/ TopFoto)

Her propelling machinery was also enlarged so that she had more power and slightly more speed. She came into New York for the first time in May 1907, as reported by the *New York Times*:

> She is a stately and dignified craft, and among all the saloon passengers who crossed in her there was not one who did not agree with the officers that there is not a better vessel afloat when it comes to steadiness in rough seas, easy going in smooth ones, and general all around comfort in all kinds of weather. The liner is one of the most luxuriously appointed afloat, and is in a class by herself in that she is equipped with a modern Turkish Bath apartment, in addition to all the other innovations.

Captain Smith described his many years at sea as 'uneventful', telling a reporter:

> When any one asks me how I can best describe my experiences of nearly forty years at sea I merely say uneventful. Of course, there have been winter gales and storms and fog and the like, but in all my experience I have never been in an accident of any sort worth speaking about. I have seen but one vessel in distress in all my years at sea, a brig, the crew of which was taken off in a small boat in charge of my third officer. I never saw a wreck and have never been wrecked, nor was I ever in any predicament that threatened to end in disaster of any sort.
>
> The love of the ocean that took me to sea as a boy has never left me. In a way, a certain amount of wonder never leaves me, especially as I observe from the bridge a vessel plunging up and down in the trough of the seas, fighting her way through and over great waves. A man never outgrows that.[77]

He said he 'found a crowded London street much more awe-inspiring' than the ocean, which was 'a great domain of well-defined paths, with degrees of latitude and longitude slipping by like telegraph poles'.[78] Smith opined that the travelling public preferred a large, comfortable vessel of average speed, 'High speed eats up money mile by mile, and extreme high speed is suicidal.' He expected advances in shipbuilding, but thought 'speed would not increase with size'. He had confidence in his ship's safety:

> I will not assert that she is unsinkable, but I can say confidently that, whatever the accident, this vessel would not go down before time had been given to save the life of every person on board. I will go a bit further. I will say that I cannot imagine any condition that would cause the *Adriatic* to founder. I cannot conceive of any fatal disaster happening to this ship. Modern shipbuilding has reduced that danger to a minimum.[79]

'Don't forget,' one of his officers suggested, 'when you write of the captain's uneventful life to put in that it is the great captain who doesn't let things happen.'[80]

HARLAND & WOLFF

Harland & Wolff and the White Star Line were closely tied together. White Star 'obtained through that business intimacy the greatest ships in the world … [and Harland & Wolff] gained prestige and fame from a policy that set at nought all previous conceptions regarding the size limit of vessels, and defied the pessimism of less farsighted rivals'.[81] By July 1898, the shipyard had expanded to the point that Margaret Pirrie argued that 'at least one eighth of the entire population of Belfast is dependent on our works'.[82]

Edward James Harland bought the Belfast shipyard from Robert Hickson & Company in 1858, paying £5,000.[83] Three years later, he took on Gustav Wilhelm Wolff as a business partner.[84] Harland reputedly kept a lump of chalk in one of his pockets at all times:

> In the old days he used to spend his Sundays inspecting the great ships on the stocks on the Queen's Island. When his eagle eye detected a blemish in a plate out came the chalk and the flaw was marked with a cross, round which was a circle. On Monday morning Harland's cross

> and circle revealed the defect to the managers and men in charge of the work. It may be imagined that the terror of the cross and circle was a keen incentive to thoroughness on the part of the whole staff. Everybody knew that a piece of careless or slipshod work might be pounced upon by the vigilant eye of the Head. Indeed, Harland's all-seeing eye was a proverb on the Island. Fabulous powers were ascribed to it. Nothing could evade its searching scrutiny …
>
> I never knew any man so fiercely and ferociously thorough in every fragment and fraction of life. He always carried a small ivory foot-rule in his pocket, and it was always flying out to test the dimensions of anything, from a collar to a sock. He designed his own clothes, collars, shirts, socks and boots. His tailor, his shirtmaker and his bootmaker had an anxious time when he gave an order. If they fell into the slightest deviation from the pattern, he had no mercy. Back went the whole consignment, and all the work had to be done over again … he had a principle, and he stuck to it.[85]

These two ambitious men, who were both talented engineers, introduced significant innovations. One of these was demonstrated in the design of two ships completed for John Bibby & Sons in 1861. Their length was almost ten times their breadth. Harland foresaw 'the greater carrying power and accommodation, both for cargo and passengers, that would be gained by constructing the new vessels of increased length, without any increase in the beam'. His design compensated fully for the greater length. The upper deck was made entirely of iron, 'in this way, the hull of the ship was converted into a box girder of immensely increased strength'.[86]

White Star were equally happy. Thomas Henry Ismay said, 'Never was more beauty or greater strength and stability wrought in iron' than when the first *Oceanic* was completed. He continued, 'To Harland & Wolff belongs the honour of having placed upon the seas a type of steamer which was unknown until their genius created it, and which has since been accepted as the standard of the highest class of ocean going steamers.'[87]

Remarkably, Harland & Wolff were working on at least one White Star liner every day between 1869 and 1919.[88] 'The ships were all built on a "cost plus" basis. Harland & Wolff were given absolute *carte blanche* to produce the finest possible ships, which they did. To the total sum spent was added a percentage, which was their profit.'[89] Put simply, the shipbuilder had 'no motive to resort to cheap or inferior materials or work'.[90] They calculated 'the actual cost' of each ship, 'including wages, materials, direct expenses [etc.]' and took 'a commission on the whole as their profit', which was charged at 5 per cent for building new ships and their machinery; 10 per cent for 'new boilers and/or engines for other than new ships'; and 15 per cent for 'repairs, renewals, alterations and ship work' for existing ships which needed new propelling machinery. Each ship had its own 'cost account' to keep track of the wages paid to the workers.[91]

Gustav Schwabe, Gustav Wolff's uncle, invested in the White Star Line from the start, as did Edward Harland, and William Pirrie later became a director and shareholder. There were personal and financial ties between the shipyard and its customer throughout their relationship. Cunard's Lord Inverclyde pointed out that such close ties were essential because of the 'cost plus' arrangement.

> The difficulty in building without a contract is where the shipbuilder is not also interested as a shipowner … They are very apt to mount up the cost, regardless of what it will come to, not only because the higher the cost the better their percentage, but also because they can spend what they choose in getting the results aimed at.[92]

Harland & Wolff used their judgement to deliver a succession of high-quality ships, without spending extravagantly or producing ships that would be so costly as to be unable to earn their cost and produce a profit.

William Pirrie, at school at Belfast's Royal Academical Institution, persuaded his mother to arrange for him to become a gentleman apprentice at the Queen's Island yard. He started on

23 June 1862 and played a growing role in the business. As an apprentice, Pirrie was described as 'diligent, painstaking, thorough and earnest, and he seemed to make up his mind from the start that he would learn all that could be known concerning the building of ships ... No detail was too small or apparently too insignificant to escape his attention.' It was perhaps this attention to detail that served him so well:

> That he worked hard and prospered exceedingly is, moreover, amply proved by the fact that when only twenty-two he had gained sufficient knowledge and experience to be made chief draughtsman, in which position he was responsible for the designs of some important steamships. In the intervening seven years he had passed through all departments of the works and become thoroughly conversant with all the shipbuilding and engineering operations carried out in them. Some of his love of shipping and his ability to grasp details in connection with it and with shipbuilding had doubtless been inherited by him from his grandfather, William Pirrie, who was not only a shipowner of high standing, but who was largely responsible for straightening and deepening the channel through Belfast Lough from the sea to the city, which was opened in 1849.[93]

Walter Henry Wilson and Alexander Basil Wilson joined Pirrie as the three 'managing directors of the company' in August 1874. Walter was the senior of the three, with his salary fixed at £1,000 a year compared to the £800 of each of his two colleagues.[94] When either Edward Harland or Gustav Wolff ceased to be a director, Pirrie and the Wilsons would each 'receive one-third of the percentage' that would have been payable to them. (Rosa Harland once commented, 'My husband builds the ships, but Mr. Pirrie gets the orders for them.')[95]

On 22 December 1885, the Queen's Island Shipbuilding & Engineering Company was incorporated 'for the purpose of acquiring the business, property, and goodwill of the firm of Harland & Wolff', with Harland & Wolff as managers. The company's capital consisted of 350 shares of £1,000 each. Three years later, the company's official name was changed to Harland & Wolff, Limited.[96]

Following Edward Harland's death in 1895, Walter Wilson passed away in 1904 and Gustav Wolff retired two years later. He had not been actively involved for some time. Pirrie, who became a lord in 1906, remained and acted as the chairman and principal of the business.[97] Although in 'supreme command', there were six managing directors below him who were each responsible for a particular department in 1909: Alexander Carlisle, general manager; William John Pratten, consulting engineer; John W. Kempster, electrical engineer; Thomas Andrews, naval architect; George Cuming, managing engineer; and Robert Crighton, manager of the new Harland & Wolff works at Southampton. They were 'an uncommonly able body of men, who are accustomed to do and, if need be, to dare'.[98]

During a programme of improvement in 1906–08, a number of the slipways were reconstructed so that even larger ships could be built. It included two new large slipways in an area formerly occupied by three. They were served by the huge overhead gantry constructed by Sir William Arrol & Co. Ltd and it was equipped with electrical cranes from Stothert & Pitt. 'The space covered is more than 850 feet long by 270 feet broad, while the height is 180 feet.'[99]

These larger berths were accompanied by improved dry docking facilities. The main contract for a huge dry dock was signed in 1903 and work began the following year. Over seven years, a small army of workers removed over 300,000 cubic yards of sand and clay as part of an enormous excavation. About 76,000 cubic yards of concrete, 24,000 cubic yards of brickwork and 26,000 cubic yards of cut granite were required to form the masonry basin of the dry dock itself. Although the original 'contract called for the completion of the work in three years and four months' from the start of construction, two 'serious subsidences in the large Alexandra graving dock alongside delayed the work'.[100]

Sir Edward J. Harland paid £5,000 for 'the shipyard at the Queen's Island, Belfast, together with steam engine, boiler plant, tools, machinery and other appliances' in use. He chaired the Belfast Harbour Commissioners from 1875 to 1887, was Mayor of Belfast in 1885 and 1886, and a Member of Parliament for North Belfast from 1887 to 1895. He died suddenly on 24 December 1895. *The Shipping World*, 1896. (Author's collection)

Gustav Wilhelm Wolff became Harland's business partner when he invested £500 in 1861. He 'was an excellent engineer' devoted to his work, but later entered politics, serving as the Conservative Member of Parliament for East Belfast from 1892. He retired from Parliament in 1910 and died on 17 April 1913. *The Marine Engineer & Naval Architect*, 1913. (Author's collection)

ALDERMAN·PIRRIE, LORD MAYOR OF BELFAST.

Lord Pirrie's talent was apparent when he became a junior partner in Harland & Wolff at a relatively young age. His 'life was so full', with a wide range of interests. He was Lord Mayor of Belfast in 1896 and 1897, High Sheriff of County Antrim in 1898 and of County Down in 1899. In the years leading up to 1914, he was essentially in control of Harland & Wolff. *The Shipping World*, 1896. (Author's collection)

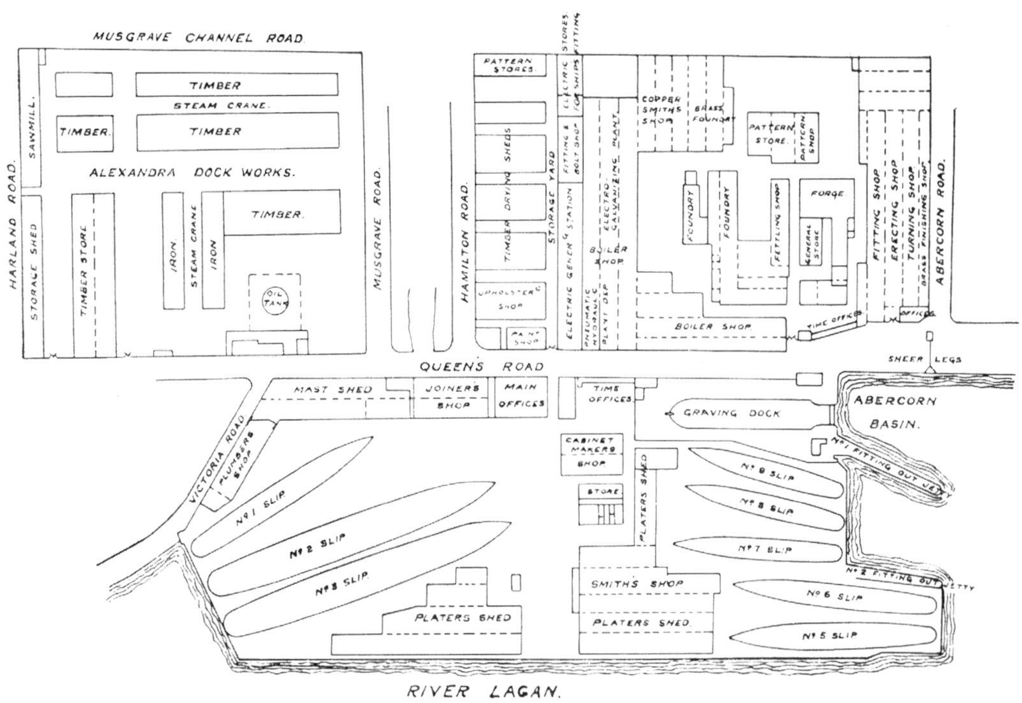

Fig. 3.—Plan of the Queen's Island Works.

Harland & Wolff's extensive works at Belfast enabled them to build the two largest ships yet constructed and to build them side by side, simultaneously. The scale of Slipways 2 and 3 is readily apparent. *The Shipbuilder*, 1911. (Ioannis Georgiou collection)

The enormous 'Arrol gantry', constructed by Sir William Arrol & Co. Ltd for Harland & Wolff. In June 1909, *The Engineer* called the structure 'a magnificent piece of work, not only in design, but in construction and equipment'. It was 'primarily designed' by Harland & Wolff, 'We called it a gantry simply because there was no other word for it ...; but a gantry in the true sense of the word it is, for it is a fixed structure made for other things to move about on.' It remained a feature of the Belfast skyline and was in use for decades afterwards. (Roy Derouin Collection)

The Alexandra dry dock itself was 'one of the largest in the world' – 80ft wide at the entrance, 50ft wide on the dock floor and 800ft long on the dock floor. The new one was 96ft wide at the entrance, 100ft wide on the dock floor and 850ft long on the floor 'when the caisson gate is placed in its normal position at the entrance'. The dock's length could be extended by 'floating the gate out and placing it against the outer face of the dock entrance', increasing the length of the dock floor to 887ft 6in. Even then, there were concerns it might not be long enough in the future. 'The construction of the dock has been carried out in such a way as to permit of the permanent lengthening of the structure to the extent of 200 feet,' which would 'comfortably accommodate the much-talked-of 1,000-foot ship of the future'.[101]

Three 1,000hp pumping engines, working together, took one hour and forty minutes to 'pump the dock dry'. Auxiliary pumps were also available. The dock gate replaced hinged, swinging entrance gates seen on older, smaller dry docks, 'a massive, rectangular structure, which, when the dock is open, is contained in a recess on the west side of the entrance'. Two lines of heavy rollers on the dock floor moved it 'across the entrance by special hydraulic appliances, the opening or closing being done in about five minutes'. For the longest ships using the dry dock, the caisson was floated out and placed against the outer entrance sill.[102]

★★★

In March 1907, White Star's *Suevic* ran onto the rocks off the Lizard, Cornwall, England. Salvagers were able to sever the stern from the stranded bow section, then tow it to Southampton. Harland & Wolff built a new bow in Belfast, which was towed to Southampton and joined to the stern in the Trafalgar Dry Dock. It was a stunning accomplishment. *Engineering*, 1907. (Author's collection)

Harland & Wolff's expansion over the decades was evident from its wage bills. In 1859, the shipyard's annual wage bill was over £20,000, rising to over £48,000 ten years later. By 1911, wages for the shipyard had risen to almost £734,000 and an additional £346,000 for the newer engine works – a total wage bill of almost £1,080,000.[103] While Pirrie planned for the future, the shipyard's workforce endured a long and arduous routine:

> Working Hours: Start at ten minutes to eight, finish at 5.30 p.m. Half hour lunch break and ten minutes break at 10 a.m. hence, the starting time of ten minutes to eight was actually buying a tea break. Saturday was a part of the working week, the hours being ten minutes to eight until 12.30 p.m.
>
> It was a nine hour day, forty-nine hour working week. Working these hours, a tradesman would earn £2 per week and if he worked all night Friday and all day Saturday, he could boost his earnings up to £5. There were no special amenities or modern facilities such as canteens or showers.
>
> Holidays: Two days at Christmas, two days at Easter, and one week in the summer (July). These were not paid holidays. Working men's wives used to save one shilling a week from their husband's wages to help cover these 'holiday periods'. The same applied at a launch in the shipyard. Those who were not required that day were told the previous day not to report for work. Therefore, when we read that 'the shipyard held holiday for the launch', what it really meant, was that some married men's families had to endure a certain amount of hardship (the loss of a day's pay) people being so poor and money and work being so scarce.[104]

The thousands of men they employed at Southampton, Glasgow and Liverpool took their total workforce to 'about 25,000 men'. The White Star Line's ambitious shipbuilding schemes relied upon Harland & Wolff workers' labour.

THE COMPETITION

'The policy of the Cunard Company is to procure the necessary capital to make it the strongest shipping company in the country,' wrote Sir Christopher Furness, one of Cunard's prominent shareholders, in a letter to the company's chairman, Lord Inverclyde, on 18 May 1902.[105] On the same day, the White Star Line's shareholders approved the IMM scheme. The Blue Riband had been in German hands for half a decade. Cunard was under threat.

The Cunard Steamship Company had been incorporated in May 1878, but its origins preceded the White Star Line. The British & North American Royal Mail Steam Packet Company and the British & Foreign Steam Navigation Company were under the same management but operated as different companies, keeping separate books and accounts until 'they were united under an arrangement by which Messrs. Cunard, Burns and MacIver became the sole partners, as well as joint managers'. After forty years operating as a private company, Cunard opted to go public and raise new share capital in 1880, 'The growing wants of the company's transatlantic trade demand the acquisition of additional steam ships of great size and power, involving a cost for construction which may best be met by a large public company.' They had a reasonable story to tell investors, 'The average net profit on the capital employed has exceeded eight per cent per annum over the whole period [1868–79]. But for the great depression of trade, especially with America, during the last few years, this average rate would, of course, have been much higher.' They had an existing fleet of 'the highest class steamers', with more on the way, 'capable of taking advantage of the improved general trade ... and of the special revival in American commerce'.[106]

Following the debut of White Star's *Teutonic* and *Majestic*, Cunard regained the Blue Riband with *Campania* and *Lucania*. In November 1894, one of their officials was pleased that their 'splendid' performance had 'been the subject of comment in the press of the whole country

… not only the good passages they have made, but their wonderful regularity'. There was some discussion about starting a Southampton service, 'We cannot give up Queenstown, not only on account of the mails, but also for the large number of passengers that embark and land there adding very considerably to our revenue.' White Star's 'Mr. Ismay had always spoken about Southampton' and the American Line had left Liverpool in favour of the southern port:

> I believe he has expressed the opinion that it would be a good thing for the Cunard company to send the *Campania* and *Lucania* there. I don't know what the result might be for the Cunard company, but I am sure of this that it would be a very good thing for the White Star company, as it would give them the monopoly of the very large percentage of saloon [first-class] passengers from the Midland and Northern part of England.

In May 1895, he worried about exaggerated reports of uncomfortable vibration on *Campania* and *Lucania* and a false rumour that some of *Campania*'s hull plating had been 'loosened':

> There is a persistent attempt made to damage the character of these ships, especially of the *Campania*, but who the authors are we do not know. Surely our competitors would not condescend to anything so malicious. However they will make their mark and prove themselves the most popular ships, as they are the best, in the trade.

Late in April 1896, the Cunard official said it was 'far from pleasant' to compare *Majestic*'s first-class passenger list for her last crossing to *Lucania*'s, 'There must be some special influence at work to keep the *Majestic* and *Teutonic* in such favour with the public.' White Star had apparently 'worked the ships up during the winter, and secured the passengers some time ahead'. He thought Cunard's own ships were 'superior' in their 'speed … steadiness … and the catering', but Cunard's New York agent made a remark that *Teutonic* and *Majestic* were 'the fashion'.[107]

At the turn of the century, Cunard's profits lagged behind White Star's by a considerable margin. An exceptional year generated £538,080 of net profit in 1900, followed by £195,849 in 1901 and £247,150 in 1902. Over the three years as a whole, White Star's profits were 70 per cent greater, and in 1901, the last full year before it came under the control of American interests, they were more than double Cunard's.[108]

Late in October 1901, Cunard were pondering their position. They had not taken a firm decision to build a new ship to compete for the express traffic from Liverpool to New York. Building two new ships would cost at least £1.5 million, 'which is practically the whole of our capital … this would require some courage'. The cost of new tonnage had risen considerably in just a few years. A new steamer already being built for the Boston route would 'absorb all our surplus funds. The matter is receiving great consideration'. Times seemed to be getting tougher: 'It is very depressing to see such low freights and the passengers falling off and saloon [first class] rates declining.'

For the six months ending 30 June 1902, J. Bruce Ismay noted the comparative first-class passenger carryings of White Star and its rivals. Westbound, White Star had carried 4,285 passengers compared to 3,138 for Norddeutscher Lloyd and 2,642 for Cunard; eastbound, they carried 5,918 passengers compared to 5,860 for Norddeutscher Lloyd, 5,069 for HAPAG and 4,613 for Cunard. 'We expect, all well, to further improve our position before the end of the year.'[109]

Cunard needed to build new express ships and they needed the capital to finance them. Lord Inverclyde saw an opportunity to appeal to the government for assistance. New liners would restore national pride by setting speed records and they would be available to the government for use in the event of war. A great deal of discussion went on throughout the spring and summer of 1902, but Inverclyde felt that there was too much discussion and not enough action. He warned Joseph Chamberlain, Secretary of State for the Colonies, that Cunard's very existence was in doubt, in a letter dated 1 August 1902:

> ... the delay had prevented the Company carrying out its policy with reference to shipbuilding, thereby paralysing it in the face of the severe competition with which it was threatened by the United American [i.e. IMM] and German Lines. The result was that if the Government did nothing, the Company must face either absorption or annihilation.[110]

Inverclyde argued that his company was 'practically the only important purely British Steamship Company on the Atlantic, having regard to the assurances given by Lord Selborne [First Lord of the Admiralty] that the Government strenuously objected to the Cunard Company joining Mr. Morgan's combination'. The government wanted to do all it could so that Inverclyde could justify himself to the Cunard shareholders. Inverclyde had warned that 'the cost of building and running vessels of the speed desired by the Admiralty would render them unprofitable' unless the government was prepared to subsidise them.[111]

A committee paper prepared for the British Cabinet and printed on 6 August 1902 set out the government's aims.[112] They agreed that it was necessary 'to save the remaining North Atlantic Lines, and especially the Cunard Company, from the danger of being forcibly absorbed into the Morgan Combination'. At 'the least possible expense', they wanted to 'provide for the construction of at least two fast steamers of 24 to 25 knots'. These would serve as armed merchant cruisers in wartime. They also wanted to 'avoid unnecessary friction with the Morgan Combination, and in particular to abstain from action likely to stir up national animosities or provoke a war of subsidies with the United States'.

Several schemes were mentioned. One ambitious proposal included amalgamating Cunard with other shipping lines and ensuring that 'the best liners of the White Star Company' were acquired 'by the Admiralty under their powers of compulsory purchase and transferred to the new company'. Three large passenger steamers could then be built, costing around £1,050,000 each, to form a 'weekly express service to and from New York' at 25 knots. The government worried about committing to an ongoing subsidy to Cunard and wanted to see 'whether it is possible to come to an amicable arrangement with Mr. Morgan' before any compulsory purchase of White Star's ships:

> The navy in war will have two functions to fulfil: to destroy the fleets of the enemy and to protect British trade ...
>
> [...]
>
> If we did find ourselves at war with Germany, we have no ships existing or projected, either of the Royal Navy or of the Mercantile Marine, which could catch these four German steamers. The fastest cruisers in the navy will be 23 knots. These ships are of 23.5 knots and over. The biggest cruiser in the navy does not carry 3,000 tons of coal. These ships could carry much more in time of war, so that, although no doubt re-coaling would be difficult for them to organise, once coaled they could go for a long time and cover an immense distance without re-coaling. No experience of the past can give a guide to the impression on the public mind and the effect on British trade if it were known that we possessed absolutely no ship which could by any possibility except that of a sheer fluke catch these German ships. The result might be disastrous.

Two options were available: either to build cruisers for that particular purpose or to subsidise 'merchant steamers to be built of equal or greater speed'. The cost of building 'special cruisers' would 'be as great or greater than that of a battleship, and yet the ships would be weak in fighting power compared with a battleship'. They would also be too large for:

> ... general naval use and would make a great drain on the manning resources of the navy. The most effective and economical method of making provision against this danger would be by subsidizing merchant cruisers to be specially built to match these German boats, and slightly improve upon their speed.[113]

The argument was compelling and so the government came to an arrangement with Cunard. By the end of September 1902, Inverclyde could report, 'I have concluded negotiations with His Majesty's Government on behalf of the Cunard company.' The terms included:

1. The Cunard company are to build two large steamers for the Atlantic trade of high speed.
2. The agreement is to remain in force for twenty years from the completion of the second of these vessels.
3. The Cunard company pledges itself, until the expiry of the agreement, to remain a purely British undertaking, and that under no circumstances shall the management of the company be in the hands of, or the shares or the vessels of the company held by, other than British subjects.[114]

Cunard would be able to borrow the necessary money from the government at the favourable rate of 2.75 per cent, significantly less expensive than the more typical 4.5 per cent if the company had been able to borrow commercially. The loan would be repaid over twenty years and the government would pay an annual subsidy of £150,000. Their two new large steamers would become *Lusitania* and *Mauretania*. (When Treasury officials learned about the low interest rate they were 'rather shocked', but 'it was too late to withdraw'. The government borrowed the money at 3 per cent to relend to Cunard at 2.75 per cent, then the 'difference was charged to Admiralty votes'.)[115]

J.P. Morgan met government representatives on 8 August 1902. He suggested that IMM 'should be allowed to tender for two similar fast vessels'. It was explained to him that 'the government at

An early artist's illustration of *Lusitania*, published in *The Engineer* in June 1904. It depicts her with three funnels rather than the four she was built with. Another interesting point is that she seems to have at least ten lifeboats on the starboard side (and presumably the same number on the port side). It's hard to tell which details were simply speculation! (Author's collection)

present only required two ships with such a high rate of speed as 25 knots'. Morgan 'appeared to think that the action of the government was unfriendly to the combination [IMM]'. They assured him that the ships were 'not to be built to create trade competition, but merely for the purposes of naval defence'. Morgan 'intimated in strong terms his determination to maintain the British nationality of the English ships'. IMM might be an American company, but the ships would continue to be registered and operated from Britain. In return, Morgan hoped that IMM's companies would receive 'equally good treatment as regards tenders for ships, naval and mail subsidies as that accorded to other lines'.[116]

They came to an agreement. The White Star Line and other companies were to be 'treated on equal terms' with Cunard and other lines when it came to assigning government contracts (such as postal agreements). White Star would remain British, with a majority of the directors being British subjects. None of the company's ships would be transferred to a foreign registry unless the Board of Trade consented. The government retained the right to acquire or hire ships, while a proportion of any new tonnage was to be built for the British flag. The agreement was intended to last 'until September 1922, and thereafter, subject to five years' notice on either side'.[117]

The results of the government's financial backing were *Lusitania* and *Mauretania*, both completed in 1907. They were the largest liners in the world by a considerable margin, the first to exceed 30,000 gross tons. Their propelling machinery was by far the most powerful afloat. They were among the first passenger liners to use steam turbines, driving quadruple screws and enabling them to set new records for speed. Both were designed expressly as 'fast mail boats'. Their speed was only viable due to state support. *Lusitania* was built by John Brown's at Clydebank; *Mauretania* by Swan, Hunter & Wigham, Richardson at Wallsend. (It set up something of a friendly rivalry between the Scottish-built *Lusitania* and her English sister.)

Cunard's agreement with John Brown's called 'Attention … to the Provisions of the Official Secrets Act 1889', and included the usual provisions: 'builders to guarantee all workmanship and material for twelve months after delivery of the vessel' and Cunard were 'to have power to make any alteration or amendment in the specification and plans, which will not prejudicially affect the fulfilment of the obligations assumed by the builders'.[118]

The 'principal dimensions' included a length between perpendiculars of 760ft and a moulded breadth of 87ft. Their focus on speed necessitated keeping weight to a minimum:

> As the weight of all materials to be worked into the hull, fittings, and machinery is of vital importance, it is essential in getting out the designs of the various parts that the greatest care should be taken to avoid unnecessary weight, every part to be as light as possible, consistent with the necessary strength.[119]

To keep the weight down and help improve stability, high-tensile steel was used in the upper hull structure rather than the usual mild steel. This innovative feature enabled them to use less steel to obtain the same strength. (There was some comment at the time that the high-tensile steel plates were hydraulically riveted with steel rivets because the rivets were weaker than the hull plates they connected together. This was particularly significant because this plating was in one of the areas of the hull exposed to the greatest stresses. However, the shipbuilders believed that the use of hydraulic riveting and the use of rivets 'at least equal to the requirements of Lloyd's rules for mild steel plates of equivalent strength to the high tensile steel used' mitigated this concern.) Their watertight subdivision was up to high standards. In May 1907, previewing the new *Lusitania*, the *New York Times* explained, 'It may be claimed for her that she is as unsinkable as a ship can be'.[120]

Once in service, both ships proved their speed. *Mauretania* was slightly faster but both ships suffered from a degree of vibration. In the case of *Lusitania* in particular, she needed extensive stiffening and structural reinforcement in the second-class areas nearer to the stern. As they settled into service and their engineers gained experience operating their machinery, both ships increased their speed steadily. Further improvements – trialling different propeller configurations – helped to reduce vibration and increase efficiency.

Late in April 1908, one observer compared the two ships:

> On the voyage home so far there has been no strong wind. At the start wind and sea were following us, and the smoke went nearly straight up from the funnels. The sea was rough but did not trouble the *Mauretania*. There has been little pitching and but slight lazy rolling. We have done some splendid steaming; from Sandy Hook we have averaged over 24½ knots. The *Lusitania* averaged 24.08 [knots] on the passage out.
>
> I have been continuously about the ships noting, not merely their behaviour, but the attitudes of mind and body of passengers.
>
> My broad conclusion is that on both ships the means taken to prevent troublesome vibration have been successful, and that no reasonable cause of complaint remains. The actual vibration is less than in any other fast Atlantic steamer I have travelled upon, and it is quite *local* [original emphasis], affecting only very small portions of the length.
>
> As between the two ships I consider the *Mauretania* has an advantage, but the difference is not great. The pitching of the *Lusitania* may have tended to produce slightly greater vibration, but after allowing for this I think the *Mauretania* is somewhat freer from local tremors.[121]

Towards the end of January 1908, Vernon Brown, Cunard's agent in New York, was advised that J. Bruce Ismay's wife Florence would be sailing on *Lusitania* and that he should ensure her party were made comfortable. The company offered her the port side Regal Suite, which had not been booked. Florence Ismay wrote to Cunard afterwards and said that she had 'had a comfortable voyage'. Brown was later told by a director that someone in the White Star Line's Liverpool office, 'I rather think Mr. Ismay, said she had never had such a shaking up in her life, but I think it was really said as a joke.' Chairman William Watson fretted:

> The question of vibration we shall always hear about, to some extent it is exaggerated – to some extent it is made more apparent by loose windows or loose fittings, or various things of that sort, which we shall hope to make right. Everybody appears to be on the look out for every fault. Coming over and running before westerly seas she rolled from 3 to 11 [degrees], taking 22 seconds to complete the roll and back again. Everybody was most comfortable, and yet Captain Watt tells me that the moment she began to roll they were surprised that it should be so and made out that it was twice as bad as it really was – this we shall live down no doubt.[122]

The issue of rolling came up a month later:

> [It] seems to be talked about by passengers, or rather talked about by people who are not passengers.
>
> [...]
>
> This week I know of some half dozen men who had heard that these ships rolled so much – five of them went to the White Star, and one of them said he would not go in the fast ships but booked a passage in the *Carmania*. They seem to hear of things like your friend the Vice-President of the railway, and immediately accept them.
>
> I mentioned chaffingly to Bruce that they had taken some of our passengers, and did he know why, because they say the ships rolled? Bruce replied that all ships rolled, and then he added, 'but I will tell you what it is: in these broad ships they appear to roll more than they do, and it makes passengers think they roll. Now this is the sort of thing that comes about – a man says "that they don't roll much, we know by the instruments that they only roll a very small number of degrees, but they are so broad and so high that they describe it greater".' And this is exactly the way in which people speak of our steamers. Like the gossips of old in the *School for Scandal* – 'mind you I don't think that they roll but they do say that the ships being so broad appear to roll

very much. Don't say I said so, but that is what I hear'.[123]

At the end of May 1908, Watson argued, 'They are the ships that will make the money. They are robbing [...] the other ships, but we have got them and we must make the most of them.'

Both ships were popular. *Lusitania* initially did slightly better with passengers, but *Mauretania* equalled her by 1909 and was in the lead thereafter. In 1911, they carried almost 80,000 passengers between them.[124] Financial returns were strong. *Lusitania* generated a gross profit of £244,820 and *Mauretania* achieved £297,430 in 1910. Even after paying insurance, office and repair expenses, as well as setting aside money for depreciation, their net profits were £107,752 and £146,622.

The company's change of fortune was remarkable. Financial historian Francis Hyde wrote that, without government support, the company might not have survived the 'difficult years' from 1903 to 1908. The strength of competition against Cunard 'would probably not only have reduced receipts below the level of costs, but would, because of the high cost of ships, have wiped out the company's reserves bringing Cunard to the verge of bankruptcy'.[125] Instead, they were prospering.

***

White Star made the decision to switch their express service from Liverpool to Southampton. In February 1907, they published an 'Important Notice' regarding their 'Southampton–Cherbourg–New York' service:

> The White Star Line, as recently announced in the Press, has decided to transfer its Wednesday Royal and United States Mail Service from Liverpool to Southampton in June next.
>
> The reason for taking this important step is simply and solely to meet the growing demand of travellers that facilities should be provided to enable them to embark and disembark at either a Continental or British port, thus obviating the necessity of crossing the English Channel.
>
> The steamers allocated to the Southampton route, and which will carry on the Mail Service, are the *Oceanic*, *Majestic*, *Teutonic* and *Adriatic*. They will be despatched from Southampton on Wednesdays, and will, after calling at Cherbourg, proceed to Queenstown, where, in order to carry out the company's obligations to the Post Office, the ships will embark the mails, which will meet them on arrival, and having taken onboard the Queenstown passengers will sail at once for New York, avoiding the long detention at Queenstown to which they have hitherto been subjected. Returning, they will proceed direct from New York to Plymouth, thence to Cherbourg, and end the voyage at Southampton.
>
> The first departure will be that of the palatial twin-screw steamer *Adriatic*, now being completed at the well-known Belfast shipbuilding yard of Messrs. Harland & Wolff Limited, the advent of which in the North Atlantic trade is awaited with so much interest by the shipping world. She will sail from Liverpool for New York, via Queenstown, on her maiden voyage, on May 8, the *Teutonic*, *Oceanic* and *Majestic* following in the sequence named, and the regular weekly departure from Southampton will be established with the sailing of the *Adriatic* from that port on June 5.[126]

They were now competing more directly for continental passengers. They needed to build new ships to maximise the opportunity. Ismay's biographer, Wilton J. Oldham, described an evening when the White Star Line's chairman and his wife went to dine with Lord and Lady Pirrie:

> Mr. and Mrs. Bruce Ismay went one night to dine with Lord and Lady Pirrie at their London home, Downshire House, Belgrave Square. During that evening Ismay and his host, who both believed in the 'big ship' with great comfort and moderate speed, first discussed the *Olympic* and *Titanic* which were to be followed by a third.[127]

The order for the first two ships had already been recorded by Harland & Wolff on 30 April 1907, which demonstrates that the dinner cannot have been in summer 1907 as popularly believed.[128] It seems certain both men had already anticipated the need some years earlier. They recognised that it was simply not practical on a commercial basis to try and build ships faster than the new Cunarders.

All the major shipping lines recognised that reality. On 21 October 1907, *Lloyd's List & Shipping Gazette* quoted comments by Norddeutscher Lloyd's Dr Wiegand. He thought additional speed would be 'zero' value, 'With our present speed of 22 to 23 knots we arrive at New York by day, whereas we should arrive at night if our steamers should make 24 or 25 knots [...] To bring us any practical advantage [...] the speed of a steamer [...] would have to make 25½ to 26 knots.' The speed of their existing express ships was suitable for the scheduled calls at Southampton and Cherbourg, which were timed to join up with the train connections to London and Paris.

Early in November 1907, the *Western Morning News* quoted Albert Ballin saying that *Mauretania* was 'wonderful' but that it would be 'hopeless' for HAPAG to attempt to build 'a faster ship'. He thought 'there are plenty of people for the faster boats and plenty for the slower ones'. The French Line, increasingly interested in the North Atlantic passenger traffic, opted for a new ship that was similar in size to *Adriatic*. She was intended to be a similar speed to the Norddeutscher Lloyd liners.

Ironically, Cunard faced the same problem. Building a running mate to match *Lusitania* or *Mauretania*'s speed was not viable without further government support. In September 1907, Watson was worried 'about the inability to sail these ships on a three-ship service'. 'If also we could run on the three ship service, so that we should have the fast ships on two weeks out of every three, we should sweep the market of passengers,' he wrote the following month. In 1908 and 1909, they were considering whether to proceed with new 'first class mail boats' or 'intermediate steamers'. 'Do the results obtained from existing large, fast steamers, and large intermediate steamers, justify the consideration of building still larger vessels than those already built?' asked their naval architect, Leonard Peskett, in January 1909.

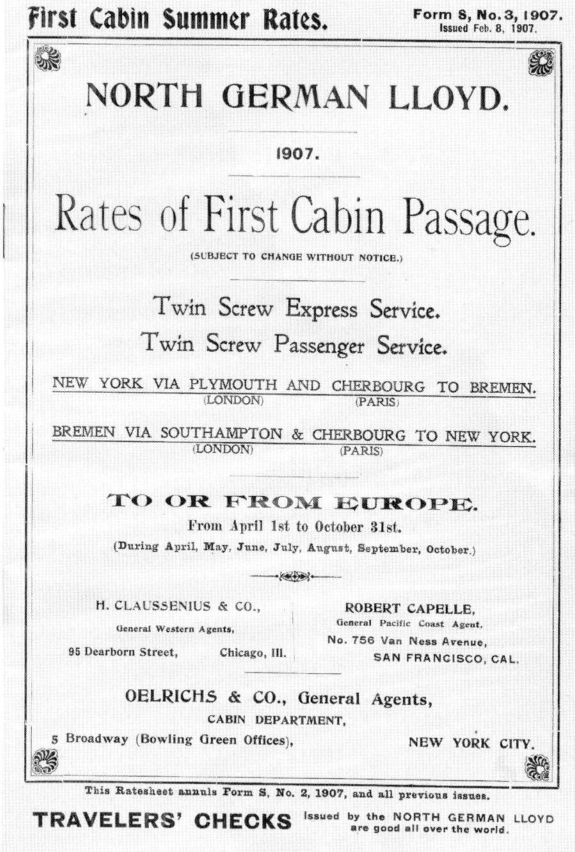

Norddeutscher Lloyd's advertisement for first-class ticket prices, published 7 February 1907. Their express ships were formidable competitors for the continental passenger traffic. In the high season, westbound passengers travelling from Bremen, Paris or London paid $1,500 for one of *Kronprinzessin Cecilie*'s 'Imperial suites', with 'parlour, bedroom, bath and toilet room, private dining room and maid's quarters [accommodating] 2 adults and 1 maid'. (Author's collection)

They were discussing much the same strategic choices as those facing White Star. The success of intermediate ships such as HAPAG's *Amerika* and *Kaiserin Auguste Victoria*, Holland America's *Rotterdam* (completed in 1908) and White Star's own *Adriatic*, setting new standards of comfort, demonstrated they could attract passengers without being the fastest liners afloat. White Star therefore made their choice. They opted for 'first class mail boats' slightly faster than *Oceanic*, in order to maintain a creditable express service. The difference between 21 knots and 25 knots over a 3,000-mile crossing equated to only an extra day at sea. Their new ships would take a day longer than Cunard but a day less than the intermediate ships from their European rivals.

Rotterdam set high standards of first-class accommodation when she entered service in 1908, including a very high number of private bathrooms for first-class passengers. She is seen here docked at in Hoboken, New Jersey, captured by a Detroit Publishing Company photographer. (Library of Congress Prints & Photographs Division)

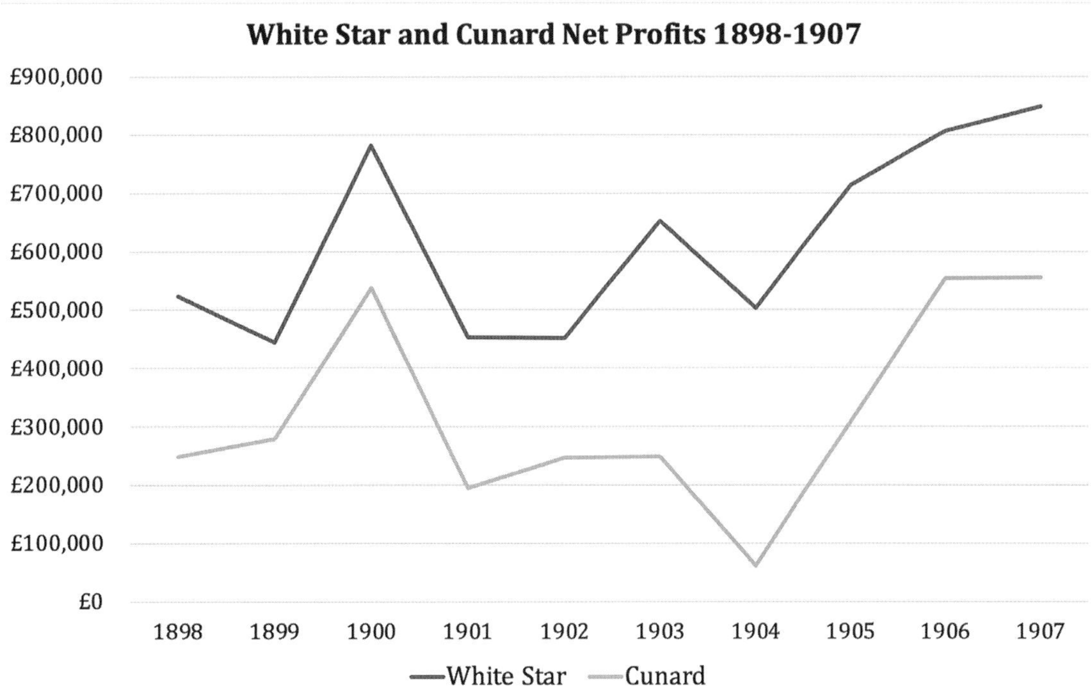

From 1898 to 1907, the profits of White Star and Cunard were both on a rising trend. However, White Star generated significantly more than Cunard. (Author's collection)

They also wanted to do better in passenger comfort, with the most expansive first and second-class accommodation yet to go to sea. First class, in particular, would include an expansion of the leisure facilities seen on *Adriatic*.

These goals dictated the need for larger ships, because they needed more space for passenger accommodation. In turn, more space was required for the propelling machinery necessary to deliver a service speed of 21 knots. Once again, White Star were going to be ordering the largest liners in the world.

Shipping Line	Total Passengers
Norddeutscher Lloyd (Bremen)	208,181
Hamburg America Line (Hamburg)	186,700
Cunard Line (Liverpool)	133,709
White Star Line (Liverpool and Southampton)	128,834
French Line (Havre)	107,892
Red Star Line (Antwerp)	95,540
Holland America Line (Rotterdam)	63,125
Anchor Line (Glasgow)	53,377
American Line (Southampton)	44,090

The top nine passenger lines carrying passengers to and from New York on the European berth, in 1910. They are ranked in order of total passengers carried. The figures for Norddeutscher Lloyd and the Hamburg America Line have been consolidated because they each operated an 'express' and 'regular' service from Bremen and Hamburg, respectively. The White Star Line's figures have been consolidated to reflect both the Liverpool and Southampton passenger services. Although the White Star Line was fourth by the total number of passengers carried, they were third by the total number of first-class passengers. Similarly, the Holland America Line was fifth by the total number of first-class passengers. The two German lines were particularly strong in carrying large numbers of third-class passengers (North Atlantic Passenger Conference, 1910).

2

YARD NUMBERS 400 AND 401

Harland & Wolff were only too pleased to make White Star's plans a reality. The exceptional scale of the new ships marked the project out. Each ship would be larger than any that had previously been built. They would also be constructed side by side at the same time.

Cunard had their own Naval Architects' Department, led by Leonard Peskett, whose team designed *Lusitania* and *Mauretania*. They developed designs and invited different shipyards to tender for the work. Peskett and the chosen shipbuilders also worked together closely with Lloyd's Register to make sure their designs met the classification society's standard. The shipbuilders consulted Cunard and made their own suggestions, based on their experience. By contrast, White Star laid down their own general requirements, but it was Harland & Wolff who produced the overall designs for their new liners. They already had a fairly good idea of the general requirements for the new ships, in terms of their size, speed and capacities. There were many considerations to take into account:

> Experts embark upon a series of calculations relating to weights, displacements, and speeds in order that it may be certain that the requirements put forward by the White Star Line can be satisfactorily embodied within the limits of a ship suitable for the trade. Such practical points as the depth of water [...] in the approaches to Southampton [...] have to be borne in mind. When a ship of the large size required by the White Star Line is in contemplation, the question of the size of the available dry docks may be a determining factor.[1]

The 'order to proceed' with the two new liners was given on 30 April 1907. Their shipyard numbers were assigned: 400 for the first liner and 401 for the second. It was a fitting way for Harland & Wolff to mark their progress with their 400th ship.

By any standard of size, they surpassed all existing competitors. Following his discussions with J. Bruce Ismay, Lord Pirrie laid down the necessary length, breadth, depth and general dimensions required. He also decided the 'modelling' – the basic hull form, proportions and appearance.[2]

Alexander Carlisle took a greater role. Carlisle was Pirrie's first cousin and his sister was Lady Pirrie. After attending Belfast's Royal Academical Institution, where his father served as the headmaster, he joined Harland & Wolff as an apprentice in 1870.[3] He climbed steadily through the ranks. Carlisle's 'sterling qualities' included 'his thorough masterful work and management of men'. He was a 'familiar figure' in 'Belfast and throughout the whole of Ireland, as well as in those business and social circles in Great Britain, on the Continent, and abroad, where he was so well known and deservedly held in great popularity'. The *Weekly Telegraph* credited him with playing 'a foremost part in the progress' of Harland & Wolff over four decades, 'He has seen an increase in accommodation, in size, in speed,

L-R: Alexander Carlisle. (© Smith Archive/Alamy Stock Photo); Thomas Andrews. (George Behe collection); Edward Wilding. (Charles Haas collection)

and, above all things, in comfort and in beauty of the vessels designed and built by his famous firm, such as only the historian of the future will be properly able to appreciate.' The newspaper said he 'had tempting offers' from rival shipbuilders, 'but his allegiance to Harland & Wolff was too staunch and true' for him to leave the firm.[4]

Carlisle's job was to flesh out Pirrie's proposals. He worked out all the requirements for 'the details, the decorations, the equipments, and general arrangements'.[5] The design work of all the fundamental points had been completed when he decided to leave Harland & Wolff, departing at the end of June 1910. (On 11 April 1911, Pirrie reported that Carlisle's 'resignation of his position as general manager and a managing director of the company was final but that he had entered into an arrangement with him under which his services in an advisory capacity would be available at any time in case of need'. The following month, John W. Kempster was appointed as chairman of the managing directors.)[6]

Thomas Andrews, Pirrie's nephew, was also closely involved. Aged 11, Andrews went to the Royal Belfast Academical Institution, joining Harland & Wolff on 1 May 1889 as an apprentice. 'His uncle told him he would require to work his own way up the ladder; no favours.' By 1904, he had been promoted to assistant chief designer; the following year, aged only 32, he was the head of the Design Department.[7] He also became one of the firm's managing directors in March 1907, and married Helen Reilly on 24 June 1908. Their first child, Elizabeth, was born two years later.[8]

Many of the improvements the new ships embodied were the direct result of his hard work and attention to the smallest of details. 'A shipyard is not a happy hunting ground for drones or for tenderfeet – the working of inch-thick steel plates, and the launching of 10,000 tons dead weight, and the handling of an army of 20,000 [*sic*] do not conduce to the softening of a man's nature,' wrote one of his contemporaries. 'There was no excuse for bad work at Harland's, and some may have thought Andrews a hard man. Certainly he expected – and got – a good deal from his people, but he never spared himself.' On one occasion, his contemporary wrote:

> I well remember visiting the drawing office one morning just after six. Naturally, none of the office staff was there, but only the old chap clearing up after the previous day's work. I asked him if Mr. Andrews had come down. 'Yes,' said he, 'Mr. Andrews was down early early,' a descriptive enough expression signifying [...] that he had been on the spot at 4 a.m.![9]

Andrews took over much of the work when Carlisle retired on 30 June 1910. On 2 November 1911, Pirrie wrote to Andrews, 'As from the 1st July last your salary, as chief of the designing department and irrespective of your renumeration of £1,000 per [annum] as a managing director, has been increased from £1,000 to £1,250.'[10] Andrews' duties heading the Design Department were explained by his colleague, Edward Wilding:

> … to interview and consult with owners for the preparation of designs for proposed vessels, and, when they have designated their requirements, to prepare designs embodying these requirements for discussion with them; if necessary to discuss the designs with them and to modify these as required until an approved design is obtained. Then to prepare the preliminary structural plans for the guidance of the draftsmen attached to the shipyard, whose duty is will be to develop them; and during the development, to act as the consultant.[11]

Once construction was under way, it was necessary to visit the slipways to ensure that the ships were being built in accordance with the plans; advise on any practical difficulties that might be encountered; discuss the progress and any modifications with the owners; 'and finally, when the ship is completed, to attend the trials and see that everything has been carried out in accordance with the general requirements'.[12] (Andrews' work was appreciated by White Star. On 1 April 1903, he had been awarded a cheque for £200 'in recognition of services rendered to the company' and with 'our best wishes for your future success'.)

Edward Wilding was responsible for much of the scientific work involved in the design process. Born in 1875, he spent his early years on Merseyside. (His father, Henry Wilding, worked in shipping and later took over the chairmanships of the Leyland and Dominion Lines, after IMM was formed.) Edward was schooled at the Liverpool Institute High School. He started as a cadet, aged 15, doing a year of intensive training.

In 1891, he entered the Royal Navy as an engineer student, joining the Royal Naval Engineering College at Devonport. Two years later, he accepted a studentship in naval construction, then did three years at the Royal Naval Engineering College following the 'standard Admiralty course'. ('My training was that of the naval architects for the British government.') He spent three years as a probationary assistant constructor, attending the course at the Royal Naval College in Greenwich, London, from 1896–99. Then he was stationed at Portsmouth dockyard, where he was attached 'to one of the battleships of the then Channel Squadron in the British fleet' to gain sea experience. His extensive knowledge came in useful.

He was appointed an assistant constructor in the Royal Corps of Naval Constructors, then attached to the Admiralty's department studying scientific questions relating to warship design. He served for four years as an assistant to Dr R.E. Froude, two and a half of them as a senior.

Wilding moved to work on merchant ship construction, joining Harland & Wolff in February 1904. He worked on the scientific side of the Design Department. His progress was 'rapid'. Two years later, he began to serve as a naval architect in charge of the scientific work.[13] He was the head of the calculating section of the design department, supervising 'calculations and technical estimates'.[14] His responsibilities included making sure the ships had adequate strength and were able to meet the design objective of their watertight subdivision.

Pirrie instructed Wilding in writing to prepare 'the first plan' for the new ships, based on the intended dimensions and appearance. Wilding prepared 'Design "A"'. The design concept then went through a number of revisions, working through designs including A, B, C and D: 'the ship was growing; she grew bigger'.[15]

Pirrie's initial specification did not include the number of decks. Certainly, their designs grew significantly. Harland & Wolff's annual report for the year ending 30 June 1907 included their 'new work in hand'. Yard Number 400 had a projected gross tonnage of 38,000 and propelling machinery of 41,500 indicated horsepower ('triple screw'), but no name against her. (Her sister was not listed at all.) Those projections grew to 43,300 gross tons and 50,000 indicated horsepower in the annual report two years later. She ended up even larger.[16] (In January 1910, Cunard believed that the new White Star ships would be 'sixty feet longer than *Mauretania*' with a length of 820ft between perpendiculars – excluding the overhang of the counter stern – and an implied overall length of about 850ft.)[17]

The growth in size reflected the White Star Line's requirements, but even a larger ship had its limitations. The need for powerful propelling machinery restricted the cargo capacity. (Leonard Peskett calculated for Cunard that a ship 800ft long with a breadth of 88ft, designed for 21 knots, would need to carry 'about 5,000 tons of coal and 9,000 tons' of cargo and stores. If she was designed for 19.5 knots, she would need to carry only 4,400 tons of coal, but would be able to carry 12,500 tons of cargo and stores as well as more passengers.) These were the inevitable sort of trade-offs in the design process.

***

Journalists learned about the plans. On 10 August 1907, the *New York Times* wrote about a total of 'nine new Atlantic liners' being built for IMM, including for the Atlantic Transport Line. John Lee, one of IMM's four vice presidents, was asked if White Star were building two new liners to compete with Cunard: 'We have two new steamers building for the Southampton–Cherbourg service of the "*Oceanic*" class, but they will not be any faster.' They would monitor the Cunarders' progress and 'it may make a difference in the machinery and speed of our new steamers'. Whether it was deliberate or not, Lee's use of the term '*Oceanic* class' downplayed their plans for substantially larger liners.

A flurry of reports followed early the following month. On 7 September 1907, the *Manchester Courier* noted a report in the *Glasgow Herald* that 'the new White Star boat to be built at Belfast [...] will be 840 feet long'. Three days later, the *Belfast Evening Telegraph* reported that 'the huge vessel [...] will (an unusually well-informed London correspondent hears) be 840 feet long [...] The beam will be quite 86 feet.' They said she would be 'of the intermediate class, and will be fitted with reciprocating engines', with the exhaust steam being used by a Parsons turbine. On 13 September 1907, the *Daily Telegraph* reported that its Belfast correspondent:

> [had] called at Queen's Island today for the purpose of securing additional particulars of the new White Star liner [...] The officials, however, absolutely declined to give any further information on the subject. Preparations are being pushed forward for laying the keel, but the complete plans have not, I understand, been arranged.

White Star may have tried to direct attention away from Cunard's *Lusitania*. On 12 September 1907, the day before *Lusitania* arrived in New York on her maiden voyage, the *New York Times* published a short article headlined 'Bigger than the *Lusitania*: White Star Line Decides to Build Vessel – Speed to be 22 knots':

> London, Sept. 11. – A Belfast correspondent telegraphs that Harland & Wolff have officially admitted that they are working upon plans for a new White Star liner which is to be bigger than the *Lusitania*. The keel will be laid in a few months. The vessel will be fitted with reciprocating engines and turbines.
>
> According to information from another source, the new boat [*sic*] will be of 40,000 tons register. It is probable that her speed will be only twenty-two knots, the cost of every extra knot over twenty knots being so excessive that steamship companies are averse to high speeds.[18]

The plans sounded impressive, but all eyes were on *Lusitania*'s progress rather than a ship whose keel had not yet been laid.[19] (On 18 September 1907, *Lloyd's List* sounded sceptical, writing, 'The story that a White Star liner of 840 feet was projected is, to put it mildly, premature.' They noted, 'Experience goes to show that without government aid of some kind 840 feet leviathans to excel the *Lusitania* in every detail are not the things to build.' The report stated incorrectly that 'no leviathan White Star liner has been ordered', even though the orders for two ships had been placed almost five months previously.)

In March 1908 an interesting article appeared in the *New York Times*.[20] A drawing of how the new ships would compare to White Star's first *Oceanic* and Cunard's *Lusitania* showed them 1,000ft long and sporting three funnels. The deckhouses looked like an enlarged version of *Adriatic*'s. 'It is a rapid age in which we live,' the reporter began:

> So new are the new Cunarders that one has not had time to lose wonder over these gigantic liners, to cease to marvel at their immensity of length and tremendous power. It looked as though these 790-foot steamships were to be the last word in marine construction and that their 68,000 horsepower marked the limit of their motive strength.

However, J. Bruce Ismay was all too happy to talk to the press. Surprisingly, there were many references to the new liners being 1,000ft long. The reporter exaggerated matters, but Ismay said, 'These two steamships are intended to be far ahead of anything yet designed':

> This is the first time in its history that the White Star Line has been able to enter the field of ship construction without a handicap. Hitherto we have been restricted by the limitations of our former home terminal – that of Liverpool; and in planning for new ships it had always been necessary for us to keep in mind the fact that our vessels must be kept within certain limitations. But now we have moved our terminal to Southampton, that restriction no longer exists, and so, for the first time, we are now able to enter the field without any handicap of this nature, Southampton being a spacious harbour and its waters so wide and deep that so far as that port is concerned we may build ships of any size. It is true that docking facilities on this side must be considered. The longest piers on the North River, where our vessels now dock, are but 850 feet in length, but we are quite sure that longer ones will be constructed.

He was coy about giving away details, 'I will repeat what I just said that they will be far ahead of anything that has yet been projected.' They would be triple-screw steamers propelled by reciprocating engines and a turbine driving the central propeller:

> Extraordinary speed will not be sought for in these larger vessels. About twenty knots an hour, I should say, will be their gait. We have some very good reasons for not filling them up with engines and coal bunkers. There is always a certain percentage of people who are always in a hurry, but we do not believe that that percentage is large, nor is there any reason to believe that it is considerably increasing. To the ordinary voyager a day more or less is not a matter of extreme importance, but two or three additional knots an hour is a matter of extreme importance to the operating company. That the vast majority of ocean travellers are not insistent on high speed seems to be proved by the popularity of such vessels as the *Amerika* of the Hamburg-American Line and our *Adriatic*.

'Where, anyhow, is this era by era increase in the size of ships to end?' wondered the reporter:

> In a comparison that was made when the new Cunarders were first contracted for some interesting calculations were made. These showed that if the rate of increase in steamship dimensions should be maintained for the next hundred years at the same ratio that they increased from 1807 to 1907 the ship launched at the end of the next century would have a speed of 6,527 knots a day, and would be able to cross from New York to England in about fifteen hours. The vessel would be nearly a mile in length and would have accommodations for 33,000 passengers.

An artist's illustration of the 'Projected White Star Liner', 1,000ft long, and how she would compare with *Lusitania* and the White Star Line's first *Oceanic*. Her length was greatly exaggerated. How accurately the rest of her appearance, including the three-funnel profile and separate superstructures, reflected Harland & Wolff's working design concepts at that time is not known. *New York Times*, 1908. (Digital restoration by James Samwell, 2011)

On 17 April 1908, the *Daily Telegraph* carried a report from their Liverpool correspondent. They had made inquiries at the White Star Line's offices and 'learnt the facts of the whole matter', 'the company are now discussing plans and designs for a very large steamer, not necessarily 1,000 feet in length, to be called, probably, the *Olympic*'. He had been told 'no order has been placed with any shipbuilder' and 'there is certainly no intention to build two phenomenally large steamers'. The same day, the name *Olympic* was used in an American press report. She was described as a 1,000ft steamer:

> Liverpool, April 16. – The new 1,000-foot steamship the construction of which is to be commenced later in the year for the White Star Line will be named the *Olympic*.
>
> The present intention is to give it a speed of twenty knots. Harland & Wolff of Belfast will build it, and it will be designed expressly to suit the requirements of the Southampton-New York trade.
>
> It is possible the White Star company will decide to order two boats of this class instead of one.[21]

Shortly afterwards, her sister was named:

> Liverpool, April 22. – Two new White Star Line steamers, which will be larger than any vessels ever before built, will be laid down at Belfast next June.
>
> The exact measurements of the steamers have not been given out, but the boats will be over 840 feet in length and more than 78 feet in breadth, and will have a gross tonnage of 45,000 to 50,000. The ships are to be fitted with combination turbine and reciprocating engines guaranteed to maintain a minimum speed of 21 knots.
>
> The names of the vessels are to be *Olympic* and *Titanic*.[22]

White Star had considered using the name *Olympic* for a sister ship to *Oceanic*, but it was a moot point once the order was cancelled in June 1902. When Cunard had been discussing names for *Lusitania* and *Mauretania* in 1904, one of the names proposed was *Olympia*. The name was a variation, ending with Cunard's familiar 'ia' nomenclature instead of White Star's 'ic'. (It is unlikely in the extreme that White Star would have used the name *Olympic* if their rival had already got a competitor called *Olympia*.)[23]

'Names selected at random and applied indiscriminately to ocean liners would be likely to prove extremely confusing to the casual student of marine nomenclature seeking to determine the ships' identity at sea,' explained one American observer in 1903:

> To avoid the perplexity of such a selection the steamship companies have adopted what must be called an ingenious plan. By adhering strictly to certain prefixes or suffixes originally hit upon in naming their ships, most of the companies have

> developed a series of names whose significance can scarcely be mistaken even by the veriest landlubber. No matter what flag a ship flies or what its physical characteristics may be, by its name one may know in what fleet it sails.[24]

It was a topical choice. London hosted the Olympic Games in 1908. On 29 October 1910, the *Larne Times & Weekly Telegraph* argued the name 'was very aptly chosen', noting the Olympic athletes were subject to 'severe testing and special selection […] in itself an acknowledgement of the highest efficiency and heroic qualities':

> [The] name *Olympic* has therefore been fittingly reserved for a vessel that is the result of the highest intelligence and skill, combined with the greatest experience. Moreover, the selection of the name was happily made in the year of the Franco-British Exhibition, at which the Olympic Games in the Stadium at Shepherds Bush proved a feature of worldwide interest.

In June 1911, the *Cork Examiner* said *Olympic* 'stands for excellence and triumph'. Her sister *Titanic* sounded similarly impressive. Their names tied in to the Olympians and Titans of Greek mythology.

***

By 29 July 1908, design work had progressed far enough for Harland & Wolff to present their 'Design "D"' concept to the White Star Line. The concept was very detailed, representing presumably more than a year's work. Unlike the three-funnelled ship illustrated in the *New York Times* four months earlier, the colour 'Design "D"' concept showed a four-funnelled liner with a continuous superstructure. Her external profile was moving away from the 'Big Four' to something more in common with *Lusitania* or *Mauretania*. The boat, promenade (A) and bridge (B) decks formed the superstructure. Beneath them, the shelter deck (C), saloon deck (D), upper deck (E), middle deck (F) and lower deck (G) contained the passenger accommodation of all three classes.

The concept included a detailed breakdown of the number of berths for first, second and third class, including the number of seats available in each dining saloon. However, plans for the first-class staterooms amidships on B-Deck were vague compared to the detailed layouts for all other areas. The entire section was labelled 'first class staterooms', but the staterooms were not drawn in. It appears they were still discussing the precise arrangements of the plushest first-class suite accommodation. Elsewhere in first, second and third class, the accommodation was fully fleshed out.

Ismay and Sanderson liked what they saw and approved in principle what Harland & Wolff had prepared for them on 31 July 1908. (It may be that there were further design concepts or variations beyond 'D' which were not selected.) White Star's objective was 'to get the best ship we possibly could', according to Ismay, 'we wanted the best ship crossing the North Atlantic when we built her'. They were satisfied the shipbuilder's proposals met that standard.

The formal instruction was 'a contract letter', in Edward Wilding's words, 'a sheet of quarto paper making a formal agreement to do it, and that is all'. Ismay himself gave the direction to Harland & Wolff. On the basis of other 'orders to proceed' recorded by the shipbuilder,[25] it probably looked much like this:

> 31 July 1908
>
> Nos 400 & 401
>
> Two triple screw passenger and cargo vessels to be proceeded with at Belfast under the above numbers for the Oceanic Steam Navigation Company, Limited, the principal dimensions being as follows:
>
> Length between perpendiculars 850'0"
> Breadth moulded 92'0"
> Depth moulded 64'6"

> Each vessel having combination machinery consisting of two direct-acting triple expansion reciprocating engines and a low-pressure turbine engine.
>
> Surveys and Certificates: Board of Trade Survey for Passenger Certificate, Master's and Crew Spaces and Tonnage. To comply with the requirements of American Emigration Laws.
>
> The vessels to be laid down on the following slips:
>
> No. 400 on No. 2 Slip
> No. 401 on No. 3 Slip
>
> [Signed]
> Ismay
> Pirrie

There was plenty of further work to do, refining the concept bit by bit until they had their final design. Ismay recalled:

> Harland & Wolff prepare the plans. They are then submitted to us, to the directors of the White Star Line or to the manager of the White Star Line. They are carefully gone through with the representatives from the shipbuilders. They try to make suggestions to improve those plans. They are taken back and thoroughly thrashed out again, and they are submitted, I should be afraid to say how often. You see, when you build a ship you have to start building her probably five or six years before you want her.[26]

When they were held in Belfast, the conferences included between eight and ten people from Harland & Wolff, as well as a similar number from White Star, 'the principals in each case being Lord Pirrie and Mr. Bruce Ismay'. Fewer people attended when they met in London, whether they met at White Star's London offices at 1 Cockspur Street or Harland & Wolff's office on the same road. Once again, J. Bruce Ismay represented White Star – alongside his fellow director, general manager Harold Sanderson – and Lord Pirrie represented Harland & Wolff, with Alexander Carlisle.[27] Thomas Andrews would always attend, unless he was 'away on other business'. Edward Wilding attended when the conferences were held in Belfast.

During the course of these conferences, the White Star representatives included 'practically all' their managers, superintendent engineers, marine superintendents and victualling superintendents 'at one meeting or another'. Among them were Henry Concanon, Arthur Cauty, Edward Lionel Fletcher, W.J. Willett Bruce and Captains Murray, Thornton and Bartlett.[28] Their input and practical experience was valuable. The White Star Line officials discussed 'such things as sanitary arrangements, ventilation, upholstering, decoration, fittings, the arrangement of stewards' accommodations and officers' accommodations, arrangements for working the ship, and things of that sort'.

It is easy to imagine these discussions between owner and shipbuilder. When the Holland America Line were working with Harland & Wolff on the specification for their *Nieuw Amsterdam*, there was a lot of correspondence back and forth where the owner had a suggestion and wanted the shipbuilder's opinion or vice versa. The Dutch line wanted 'to have the round sidelights in [the] sides of first class [dining] saloon of [the] largest size possible, say 18 inches as promised'. On 14 December 1903, Harland & Wolff responded:

> Sidelights with 15 inch clear glass and 16¼ inch opening are the largest we have made, and the same as fitted in the *Oceanic*, *Celtic* & *Cedric*, and we would not recommend you to exceed this size, as we fear with 18 inch, there would be trouble with the glasses breaking.

Then they wanted 'electric table lamps with shades' to be fitted in the grill room, 'which we intend to name "Ladies Tearoom"', and Harland & Wolff agreed.

During White Star's conferences, there were sometimes differences of opinion about particular points, such as 'whether a Georgian or a Louis Quatorze room was a proper way to decorate a first class room'. Wilding said Ismay was the 'deciding voice, in the sense that the decision had to come from him eventually; but he always listened to what you had to say, and if he found you were right he agreed'. Ismay certainly had the authority to overrule his line's department heads, but preferred to take a more consensual approach as 'one does not run one's business on those lines'.[29] Matters of construction such as the structural design of the ship were left to Harland & Wolff.[30] White Star 'never' overruled the advice of their shipbuilder, who Ismay viewed as 'one of the leading firms in the whole world'.

***

'Everything was taken to be doubled. It always meant the two ships,' Carlisle recalled. 'Anything which was taken up for one applied to the second ship the same.' They were duplicates of each other and the shipbuilder worked from the same sets of plans.[31] *Olympic* and her sister shared precisely the same projected dimensions: a length overall a few inches shorter than 883ft; a length between perpendiculars (at the waterline) of 850ft; and a moulded breadth of 92ft. Although they were nearly 100ft longer than *Mauretania*, they were only a few feet wider.

Their projected gross tonnage grew to over 45,000 tons each, which marked an increase over *Mauretania* of around 42 per cent.[32] Gross tonnage was actually a measure of size and the total enclosed space, rather than the weight of the ship. The weight was measured by the ship's displacement, which was the amount of water that the ship displaced when she was loaded to her designed draught in the water. Despite the substantial proportional increase in gross tonnage, the projected displacement increased by a much smaller proportion. It was only about 17 per cent greater than *Mauretania*'s.

In March 1914, *The Shipbuilder* reviewed the succession of large Atlantic liners that Harland & Wolff had built since *Oceanic*:

> These successive increases in size [...] have been confined within reasonable limits, and the aim underlying the design had been at all times not only to ensure a dividend-earning ship, but to make certain that the vessels would stand up to the greatest of Atlantic gales, and in other respects would conduce to the comfort of passengers.

Olympic's structural design was tried and tested. Harland & Wolff produced an enlarged hull with the graceful profile of *Oceanic*, drawing upon all their experience with large ships they had built previously.

By April 1912, seventeen ships over 650ft long had been launched: Edward Wilding noted that they had constructed eight of them. The proportional increase in depth, compared to previous large ships such as *Rotterdam* and *Adriatic*, was less than the increase in gross tonnage. *Olympic*'s construction was generally traditional, incorporating additional strengthening that was 'more than commensurate' with her increased size. 'One had a pretty good idea of what the stresses were,' Wilding recalled, 'but we were proposing to deal with a very large ship, and we were not proposing to run any chances that spending a few hundred pounds would avoid.'

At the bottom of the ship, the flat keel bar was 3in deep and 19.5in in width. The strong, cellular double bottom was over 5ft deep, increased underneath the reciprocating engine room. For half the ship's length, it extended up the ship's side by 7ft above the keel. This was at the turn of the bilge (where the ship's bottom met the side). It provided additional protection if the ship went aground with a slight list and was a significant improvement on earlier ships such as *Oceanic*. Proportionally, it extended as much as the double bottom on *Lusitania* and *Mauretania*. *Olympic*'s hull had 'a flat bottom'. ('The *Lusitania* and *Mauretania* are comparatively flat bottomed, but nothing like the model of the *Olympic*,' said Commander E.L. Ashley Foakes, nautical advisor to the Post Office since August 1908. 'It is well known that the Harland [&] Wolff model is entirely different to either the […] *Mauretania* or the […] *Lusitania*.')[33] The channel frames extended up the ship's side to carry the steel plating. At intervals, Harland & Wolff used heavy web frames (30in deep) to provide substantial additional strength, which reflected the ship's great size.

Notes on the midsection plan – essentially a transverse 'cutaway' through the ship showing the main structural components – specified improvements based on experience with previous ships such as *Laurentic* and *Adriatic*. The plan, dated 29 June 1908, contained great detail.

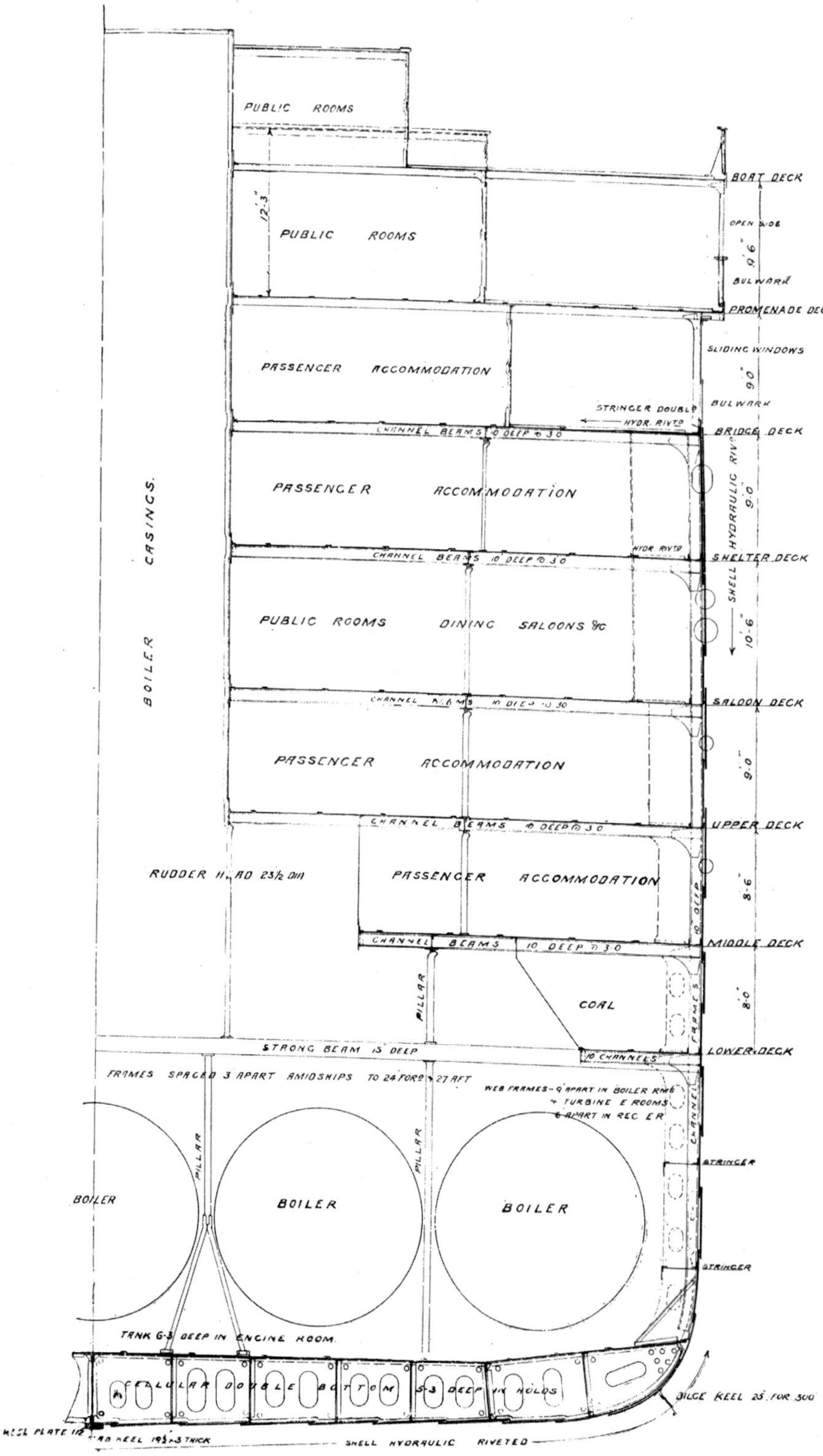

Fig. 14.—Midship Section of "Olympic" and "Titanic."

This simplified midship section gives a good idea of the hull form and the decks. *Olympic* and *Titanic* had no 'tumble home', in that the hull did not narrow significantly towards the upper decks, as it did on some other liners. The deckhouses on the boat, promenade and bridge decks formed the superstructure, which were more lightly constructed than the strong hull beneath. *The Shipbuilder*, 1911. (Ioannis Georgiou collection)

Laurentic, which had a similar three-propeller configuration as *Olympic* but on a much smaller scale, had 'additional keelsons' fitted towards the stern to provide additional strengthening. (At the stern, these structural components helped to strengthen the hull against the pulsation of the propellers; at the bow, against the stresses by plunging into waves.) The same were specified for *Olympic*. At the bow, the hull was strengthened to prevent panting (where a ship's hull might flex in and out like an accordion when the bow plunged into a head sea in severe conditions) and 'damage

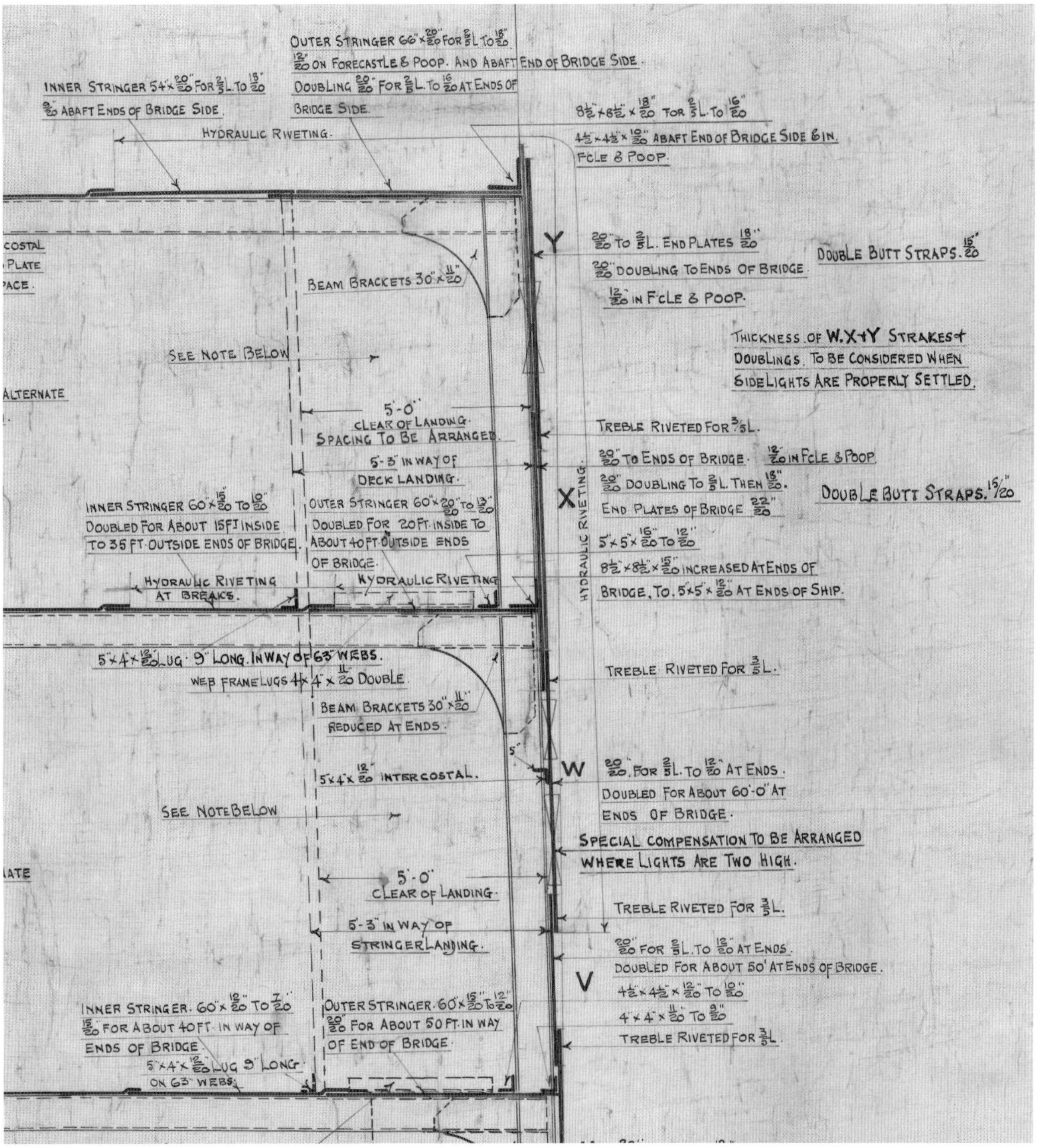

An extract from Harland & Wolff's midship-section plan, which showed all of the key structural elements of the ship's hull, from the keel to the double bottom, deck plating, side plating and numerous columns, stanchions and beams supporting the decks. This section shows the strakes (lines of plating) at C and D-Deck level: Y, X, W and V. The side plating of C-Deck formed the ship's sheer strake at the top of the hull, which was strengthened specifically to endure the high stresses imposed on such a large ship. It was immensely strong, doubled with 1in-thick plates that were riveted hydraulically with steel rivets. The structural design was submitted to the Board of Trade for approval in June 1908. (HOYFM.2010.73.2 © National Museum NI, Harland & Wolff Collection, Ulster Folk & Transport Museum)

when meeting thin harbour ice'. (Lessons could be learned from all aspects of previous ships' design and operation, even if only to confirm that something that worked in theory was satisfactory in practice. For example, Cunard's experience with the double riveting in the bottom plating of *Lusitania* and *Mauretania* had been completely satisfactory with 'no signs of weakness'. Even though *Olympic*'s design incorporated treble riveting, Cunard felt that double riveting was satisfactory on larger vessels, excepting the use of treble-riveted landings on each side where the ship would rest on docking blocks.)

Harland & Wolff had a good understanding of what they needed to work to in order to ensure the ship's structural strength. The ship's hull could be considered as one long 'hollow box' girder, subjected to many stresses and strains along its length, depending on the distribution of weight, buoyancy and sea conditions. One common standard was to ensure that the tension stress on mild-steel vessels was kept to about 10 tons per square inch or below on the upper hull structure.[34] *Olympic*'s was 9.89 tons on the sheer strake, very much in line with other large vessels of the time.[35] Although she was larger and heavier than any of the ship's predecessors, the design calculations took this into account because they were based upon their displacement and size.

Edward Lewis Attwood, author of *Textbook of Theoretical Naval Architecture* and an assistant constructor for the Royal Navy, explained the bending stresses that the hull had to endure. When a ship was on the crest of a wave amidships, the bow and stern had a tendency to drop (known as 'hogging'). Alternatively, with the trough of the wave amidships, the middle had a tendency to drop (known as 'sagging').[36] *Olympic* and her sister were designed to withstand a trochoidal wave: a theoretical wave which was the worst that could be encountered. This wave would be equal to the ship's length between peak and trough, and one-twentieth of its length in height (about 43ft). It would subject the hull to its maximum stress. Similarly, she was wider than any other vessels, requiring ample strength to cope with the sideways stresses on the hull when the ship was rolling.

The keel and double bottom structure provided substantial strength. The hull plating was doubled around the turn of the bilge and along the side of the shelter (C) deck amidships (the sheer strake), to provide longitudinal strength. The deck plating on the bridge (B) deck – which formed the top of the structural hull girder – was also heavier. These features were the result of Harland & Wolff massing material at the lower and upper extremities of the hull girder. Unlike the Cunarders, they did not use high-tensile steel to save weight. Thomas Andrews explained that 'the stability of these vessels would be so much greater than that of the *Lusitania* and *Mauretania* that lightness in the upper works was not a vital necessity with them as it had been in the case of these two ships'.[37]

Although White Star ships were not built under the supervision of Lloyd's Register, the steel used throughout was tested to their standards. (Edward Wilding also believed that their design would be accepted for Lloyd's classification without any changes.) The tests were reported to the Board of Trade by Harland & Wolff:

> The steel material used in the construction of the hull of this vessel has been tested and passed by Lloyds at the maker's works, and the usual certificates furnished stamped by Lloyds' surveyor [...] being a guarantee that the tests prescribed in the society's rules have been carried out to the satisfaction of the surveyor, and that the results are within the limits laid down.

Following Harland & Wolff's experience with previous large ships, they felt that 'riveting was the weak point in big ships' and the company 'adopted, to an unusual extent, hydraulic

Opposite: This dramatic illustration of *Lusitania*, drawn by Charles De Lacy, depicts her against a '60 foot tidal wave of very rare occurrence'. In fact, ships such as *Lusitania* and *Olympic* encountered even larger waves on several occasions. *Illustrated London News*, 1907. (Author's collection)

60 FEET. TIDAL WAVE OF VERY RARE OCCURENCE.
HEIGHT OF PROMENADE DECK 30 FEET.
HEIGHT OF WAVES. HURRICANE. 28. FEET.
HEIGHT OF WAVES. STRONG GALE 18. TO. 28. FEET.
HEIGHT OF WAVES GALE. 10. TO. 18 FEET
WATER LINE

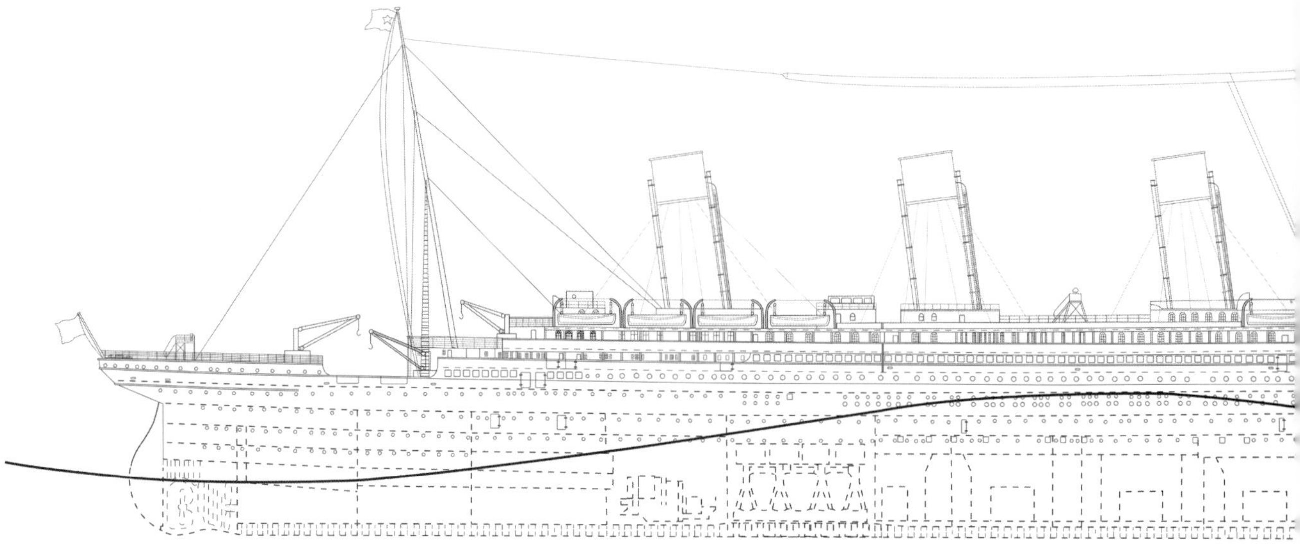

riveting wherever possible, to ensure the rivets being thoroughly well closed'. Although 'a slow and expensive affair [...] it was done'. The specifications included hydraulic riveting for the keel, as far fore and aft as possible, and the bottom shell plating to the upper turn of the bilge for the centre three-fifths of the ship's length.[38] The plating at the side of the hull was also hydraulically riveted, including treble-riveted joints in the plating that formed the sheer strake. It was amidships and in these areas that the hull plating was at its thickest (several inches in some areas). Steel rivets were used for the hydraulic riveting and iron rivets were driven by hand. Each fifth of the ship towards the bow and stern was generally riveted with iron, which could be driven more easily by hand where access was sometimes restricted and the hull curved narrower towards each end of the ship.[39]

Edward Wilding noted that it was proposed initially to dry dock the ships using a single line of blocks beneath the keel, as with smaller liners. That 'required extra strength, because ships of about 90 feet beam are usually docked on bilge blocks [port and starboard] as well as centre line [keel] blocks; consequently [...] we had to make the ship strong enough so that the sides should not drop'.

The fifteen main watertight bulkheads which extended from one side of the ship to the other, dividing the hull into separate compartments, also contributed to the ship's strength. They were made of heavier plating than usual and 'very large stiffeners [...] were used to support the plating, and the stiffeners themselves were elaborately bracketed at top and bottom to the general structure of the ship'. The specifications exceeded Lloyd's standards and, the Board of Trade noted, 'the strength of the bulkheads throughout appears ample'.[40] Compared to Lloyd's requirements issued after the ships were completed, the watertight bulkhead plating was still 'from 10 to 20 per cent in excess', while the stiffening was 75 to 100 per cent greater.

Unlike a land-based structure, there were a number of different circumstances that needed to be accommodated, depending on whether the ship was at sea, in port or dry docked:

> Loads on ship structure[s] are tertiary (point loads such as keel blocks, pier piling impacts, tug bow pushes, equipment foundations), secondary (panel loads like water pushing up on the bottom of the hull) and primary (entire hull bends like a girder). For very much smaller ships, the tertiary loads predominate. For larger ships, the need to resist primary loads dominates.[41]

The deckhouses above B-Deck were built of lighter construction. Unlike the hull beneath, this superstructure was not designed to bear any significant stress. It was built on top of the hull.

Olympic compared to a trochoidal wave, with the peak amidships. This would subject her to significant hogging stresses, with the bow and stern unsupported. She was designed to withstand such sea conditions with ample strength in reserve. This illustration is based on one that Cunard produced showing *Aquitania* under those conditions. (© Lionel Codus, 2024)

This permitted large windows, providing light and air for the first and second-class promenades and public rooms. (Rectangular windows could not be used in the hull because of the stress concentrations that would form at the corners.) However, as the hull flexed beneath in a seaway, long superstructures 'worked' or moved. This was potentially an issue because the new ships' superstructures were long and continuous.

Professor J. Meuwissen wrote for the *Shipbuilder* journal in 1907, pointing out that liners such as HAPAG's *Deutschland* had experienced problems. This was because they had been built with continuous long superstructures on top of the structural hull:

> Building superstructures [...] over the main [hull] structure is rational enough when they are in short lengths; but if continuous they should only be adopted in the case of vessels of smaller dimensions, where the longitudinal stresses are of less importance.

To solve the problem, Harland & Wolff fitted two expansion joints, which were essentially 'cuts' in the decks dividing the superstructure into three sections. They were intended to protect the superstructure from stresses imposed by the movement of the hull beneath.

Cosmetic considerations reportedly lay behind an important change to *Olympic*'s profile in the summer of 1909. The 'Design "D"' concept from July 1908 had shown four funnels and a foremast, but this changed after J. Bruce Ismay arrived in New York on 30 June 1909. Hours after *Oceanic* docked, he watched *Lusitania* depart and took an admiring look at her profile. The story goes that *Lusitania*'s four enormous funnels and two masts impressed him so much 'that he decided then and there to add a mainmast to the White Star leviathans, both of which were designed to carry only a foremast'.[42] (In October 1910, it was suggested to Cunard's chairman that their new *Aquitania* should have five funnels. The rationale was that 'if she should be seen alongside one of the new White Star steamers people will believe she is far bigger' and it would 'make an impression' in advertisements!)[43]

Construction was monitored at every stage. Not only were detailed plans and specifications submitted to the Board of Trade by Harland & Wolff, but the government regulator also had a duty to make sure that the ship was constructed according to the plans. Francis Carruthers, who had thirteen years of experience as a seagoing engineer and twelve years of experience with the Board of Trade by 1908, was based at Belfast and had plenty to keep him busy. His work for the board as a ship surveyor 'mainly consisted of superintending the construction of new steamers'. He monitored construction as it progressed,

***

Harland & Wolff noted that both ships had been 'contracted for' on 31 July 1908, when the White Star directors approved the 'Design "D"' concept in principle. Early in August 1908, Alexander Carlisle submitted to the managing directors 'a tracing of the drawing offices showing some proposed additions and alterations which were to be considered as soon as No. 400 is started'. The alterations were 'necessary principally on account of the large amount of decorative and joining and cabinet makers' work involved, for which special plans etc. will be required'. After discussion, the managing directors 'arranged to forward the plan to Lord Pirrie'.[46] The shipyard and engine works were formally 'ordered to proceed, except with [propelling] machinery' on 17 September 1908; the order 'to proceed with boilers' came on 26 February 1909; then they were instructed to 'proceed with remainder of [propelling] machinery' on 20 April 1909.

Plenty of preparation was required assembling the materials and placing orders with subcontractors in advance. The day after the shipyard and engine works had been ordered to proceed, Harland & Wolff ordered 'three rolled steel keel bars [...] made under Lloyds survey' for 'Yard Number 400' from William Beardmore & Co. Ltd. On 26 November 1908, they placed the order for 'Yard Numbers 400 & 401' for 'stern frames and [propeller] boss arms to be of best Siemen's Martin Mild Cast Steel' from the Darlington Forge Co. Ltd. The shipbuilder noted that the specifications needed 'to be strictly adhered to', noting particular structural elements and 'parts coloured red on plans are to be carefully machined'. The finished parts needed 'to be fitted together before despatch, to ensure accuracy'. The immense stern frame and propeller boss arm castings were to be tested and 'passed by Lloyds' to ensure 'quality of material soundness'. The first castings for *Olympic* were to be 'delivered in five to six months' after receipt of the necessary specifications and drawings from Harland & Wolff, 'and the second set in seven to eight months from [the] same date'. The castings were so large that 'in view of the abnormal size of the castings extra precautions are to be taken both as regards material and workmanship to turn out first class job, to the satisfaction of ourselves and owners'.[47]

The same day, Harland & Wolff ordered the rudders for each ship from the same supplier. They specified two rudders of 'solid section' made from the same Siemens Martin steel to the highest

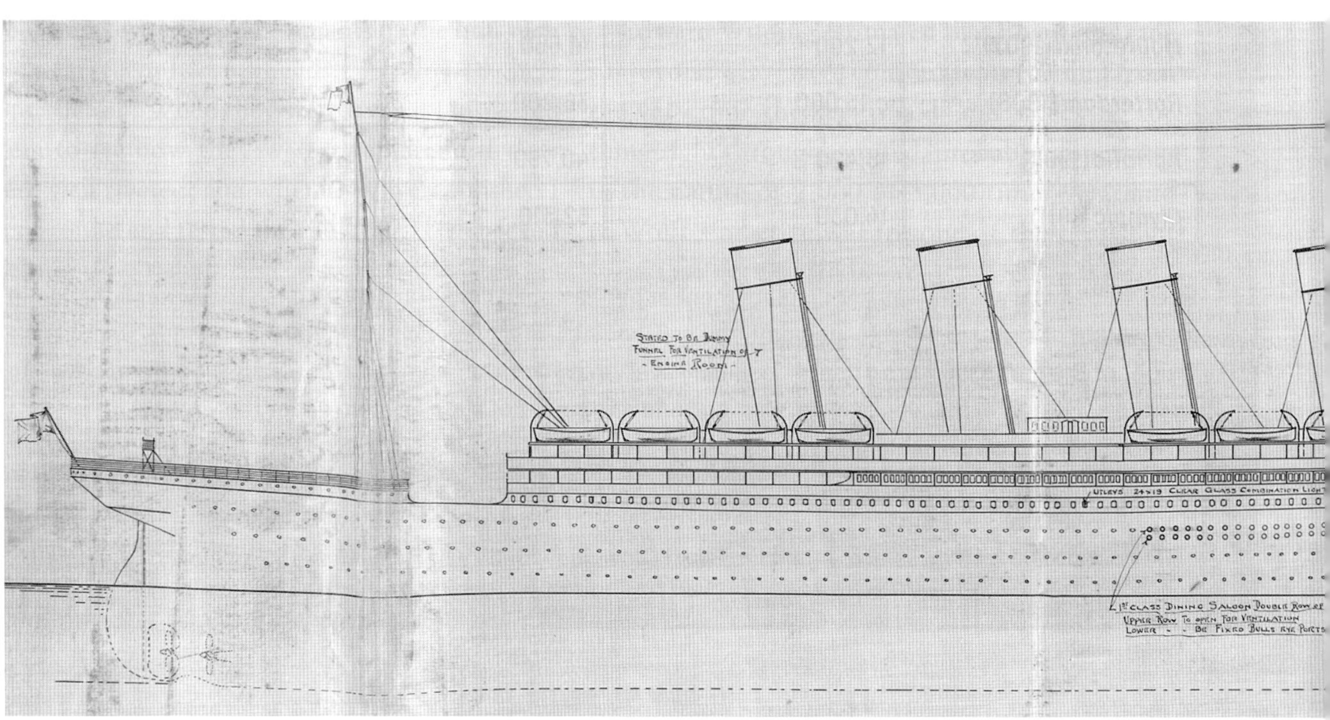

quality, 'each rudder to be put together and proved perfectly lineable to our satisfaction before despatch'. They required 'test pieces to be taken from each separate portion' of the rudders and tested 'in the presence of [a] Lloyds [representative] and our representative if we so require'. The rudder castings needed to have a tensile strength (the ultimate strength the material could take before breaking) of '28 to 35 tons per square inch', with an elastic limit (the greatest stress the material could take without permanently deforming) of 'forty-five per cent' of the tensile strength. The material also needed to be bent 'through an angle of 60 to 80 degrees without fracture' when 'cold'. They required the representative from Lloyd's to certify the tests 'as a further guarantee of tests and soundness of material', with the first rudder due for delivery in January 1910 and the second by July 1910, 'It is understood that castings and forgings are guaranteed for material and workmanship for twelve months from [the] date of each ship starting on her first voyage.'[48]

(The shipbuilder's requirements for *Olympic* and *Titanic*'s rudders were more stringent than those for other ships. The following month, they ordered two rudders from the same supplier for the Royal Mail Steam Packet company's cargo ships, *Berbice* and *Balantia*, both far smaller and under 2,500gross tons. That order specified the same steel and the need to be certified by Lloyd's, but had none of the detailed strength requirements.)

On 29 January 1909, Harland & Wolff placed another order with William Beardmore & Co. Ltd for 'four mild steel plates' for each ship ('3 plates two inches thick' and '1 plate 1½ inches thick'), 'tested and passed by Lloyds'; on 11 March 1909, they ordered 'Nos. 1 and 2 stem bars' for *Olympic* (followed by an identical order for her sister, early in May 1909); and they continued to plan ahead to order whatever was required from the spring and summer of 1909.[49]

Siemens Martin formula steel was 'a natural choice' for the hull plating on *Olympic* and her sister. It was high-quality steel with good elastic qualities and resistance to corrosion, 'ideal for conventional riveting' as well as hydraulic riveting. Harland & Wolff had used it for *Teutonic* and *Majestic* a generation earlier. Their hulls were in 'remarkable condition' twenty years after they entered service. (Harland & Wolff placed an order for '10,000 tons [of] Siemens open hearth mild steel ship plates, tested and passed by Lloyds, Bureau Veritas and/or Germanischer Lloyds when so specified' on 2 June 1909. An order for 15,000 tons of these plates followed on 21 August 1909. The orders were placed with David Colville & Sons' Dalzell Steel & Ironworks at Motherwell, Scotland, one of their regular suppliers.)[50]

The materials that went into all aspects of the ships were subject to extensive testing. Early in March 1909, the shipbuilder submitted the design of *Titanic*'s boilers for separate consideration by the Board of Trade, even though they were simply 'duplicates of those for No. 400'. Two months later, they advised that they had 'ordered the furnaces for these boilers' from John

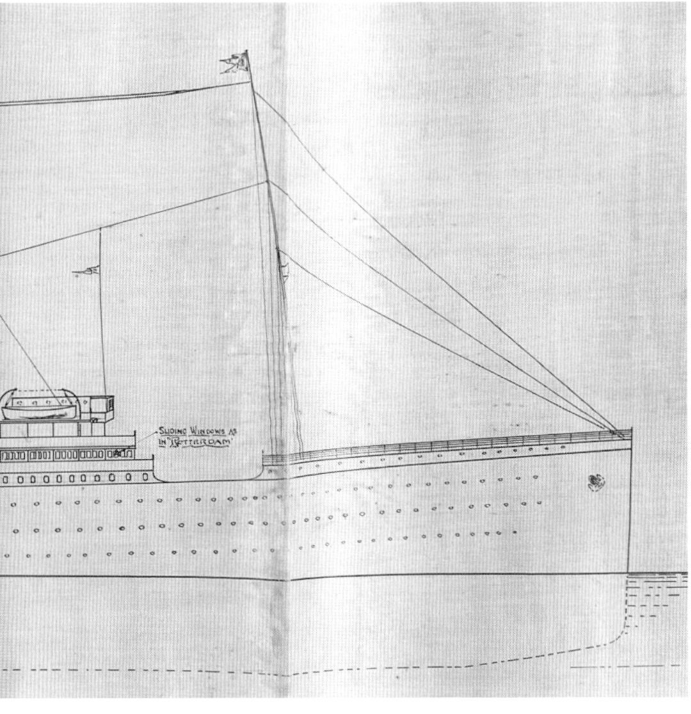

This drawing, dated 4 December 1909, seems to represent Cunard's best guess as to *Olympic*'s design. It notes the B-Deck promenade is enclosed by 'sliding windows as in *Rotterdam*' and the fourth funnel is 'stated to be a dummy funnel, for ventilation of T engine room'. They also knew the first-class dining saloon would have a 'double row of ports', but they placed the saloon too close to the bow. G-Deck is not shown. (Courtesy National Museums Liverpool, Merseyside Maritime Museum)

Brown & Co.'s Sheffield steelworks. By the end of November 1909, Harland & Wolff had advised Francis Carruthers that they had ordered 'the steel boiler plates' and other key boiler materials from David Colville & Sons. The Board of Trade required comprehensive tests which recorded in painstaking detail the 'dimensions of test piece'; the 'actual' and 'per square inch' ultimate stress in tons; elongation; bending; and which part of the boiler the test material was intended for.

Then there were all manner of rivets. The Rivet, Bolt & Nut Company confirmed from their office at 74 York Street, Glasgow, that they had ordered 'a parcel of steel rivets' to be used in the boilers, including 1in-diameter rivets from David Colville & Sons on 29 December 1909. They were sent to Harland & Wolff on 12 May 1910. They made regular submissions to the Board of Trade, documenting steel rivets with bars made from the Lanarkshire Steel Company, Steel Company of Scotland and others.[51]

One problem for the Board of Trade was that 'these two steamers are much larger than any hitherto built'. They looked at the designs for their anchor equipment and had to extrapolate what was required from the existing guideline tables. Harland & Wolff proposed to use wire rope for the centre anchor, which was 'rarely' done and only 'in small vessels'. The port and starboard anchors were 'similar to those put on board the SS *Adriatic*'. In July 1909, the Board of Trade asked Harland & Wolff for more information about the length of the chain-link cables (from 'Messrs. Bullivant & Co. Ltd, Mark Lane, London') and wire rope (from 'The British Machine Made Cable Co., Marine Works, Netherton, Dudley'). Early in August 1909, they thought the length of the cables needed to be increased 'from 300 to 330 fathoms, and the length of the wire rope from 150 to 175 fathoms' for the specifications to be signed off. The weight of the centre anchor was also increased. A sample of the wire rope needed to be tested 'to destruction, or to the maximum power available in the machine used for testing'. They had to call upon Lloyd's Proving House in Cardiff, which had a suitable breaking machine that could apply a strain of up to 350 tons. The wire rope came 'very well out of the tests'.[52]

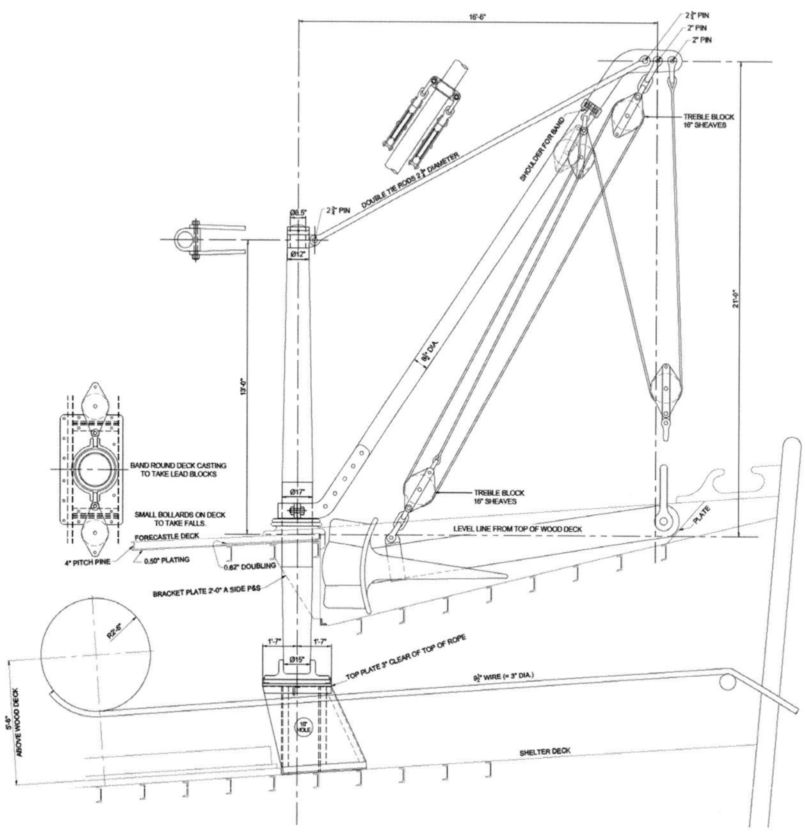

The enormous scale of *Olympic* and *Titanic*'s anchors was matched by the equipment required. This side view illustrates the crane to deploy the centre anchor, which was fitted right at the bow on the forecastle deck. It is based on the 'No. 400 anchor crane' drawing produced by Harland & Wolff. The centre anchor specification, dated 3 May 1910, called for it to be 'at least 15 tons and not exceeding 16 tons'. Francis Carruthers noted on 3 June 1911 that *Olympic*'s centre anchor (a Hall's Patent stockless anchor) weighed 306 cwt, 1 qtr (about 15.25 tons), whereas *Titanic*'s weighed 316 cwt (about 15.8 tons). (© Steven Tuckwood, 2018).

***

Olympic's keel was laid on No. 2 slip on 16 December 1908 and *Titanic*'s followed about three months later on the adjacent No. 3 slip, with the date of her keel-laying recorded as both 22 and 31 March 1909 in Harland & Wolff's records. Harland & Wolff's keel plate plan indicated in red that a number of *Olympic*'s keel bars were 'ex. 391'. That yard number had been assigned originally to HAPAG's *Europa*, planned for delivery by April 1909. The shipyard and engine works were ordered to proceed on 13 September 1906, only for a 'superseding order' to be issued on 25 November 1906 cancelling it. (Oddly enough, she was included as 'new work in hand' in Harland & Wolff's annual report for the year ending 30 June 1907.) From the available evidence, construction never actually started. Her keel bars were the same dimensions as *Olympic*'s, so they were not wasted.[53] On 9 March 1909, *Lloyd's List* noted that *Europa* 'never materialised', suggesting that 'it must [...] be a matter of chagrin for' Albert Ballin 'to see the *Olympic* and the *Titanic* actually in course of construction in the same yard for White Star'.

Progress on both sisters was pleasing. Although it took almost three months to frame the height of *Olympic*'s double bottom, *Titanic*'s was framed in less than two months.

	Olympic	*Titanic*
Framed to height of double bottom	10 March 1909	15 May 1909
Fully framed	20 November 1909	6 April 1910
Plated	15 April 1910	19 October 1910

Early in November 1909, the *Belfast Evening Telegraph* observed that the new ships were 'gradually rising into greater prominence under the huge gantries'. *Olympic*'s 'after body decks up to the upper deck' were plated at the stern, 'amidships and forward the lower deck beams are being placed in position. From the engine room aft the hold columns and deck girders were fitted in conjunction with the framing, to obviate the need of the usual wood shoring aft.' Then the 'shell plating [...] appears above the upper turn of the bilge, the whole so far having been hydraulic rivetted from the keel up'. It was 'interesting to watch the piecing together of the various parts of a gigantic structure, each part demanding special knowledge, experience and skill'.[54] The *Cork Examiner* reported on 22 November 1909 that Harland & Wolff had sent a telegram to White

Looking down on *Olympic*'s double bottom at an early stage of construction, work is progressing steadily. The cellular double bottom was a very strong structure, which contributed to the hull's strength overall. *Titanic*'s slipway is to the right. (Günter Bäbler collection)

Titanic (left) and *Olympic* (right). *Olympic*'s hull was fully plated by 15 April 1910 and *Titanic*'s by 19 October 1910, one day before *Olympic* was launched. (Author's collection)

Star two days earlier, reporting progress on *Olympic* and confirming her hull was completely framed. It was a remarkable accomplishment.

On 25 March 1910, the *Liverpool Daily Post* reported that J. Bruce Ismay had visited Belfast and inspected progress on the two ships, 'The work on both vessels is well advanced, and the launch of the former has been fixed for the third week in October.' The paper 'understood that a conference also took place regarding new steamers' for the White Star Line.[55]

'TECHNICAL SUBJECTS'

Mutual interest drove cooperation between the major shipping lines. One conference, discussing 'Technical Subjects Bearing on Atlantic Shipping', was hosted by the White Star Line at Liverpool's North Western Hotel during August 1910. President Director J.V. Wierdsma, representing the Holland America Line, took the chair. Attendees came from the American Line, Atlantic Transport, Cunard, Dominion, Hamburg America, Holland America, French Line, Leyland, Norddeutscher Lloyd, Red and White Star Lines. They presented papers discussing subjects such as coal bunkering ('results of different systems'); expansion joints ('in connection with creaking in passenger accommodation'); the configuration of walls, floors and disinfectant procedures for provision rooms storing frozen foods; officers' education and crew training; the results of using turbine engines alone verses a combination of reciprocating and turbine engines; electric and oxyacetylene welding; 'distributions of electric light and combination with different dynamos'; corrosion ('the use of zinc to protect stern posts, rudders, etc. The value of J. Dampney & Co.'s "Apexior" [coating]'); the 'supply of fresh and hot water in staterooms' and its discharge 'direct into [the] sea'; and all sorts of regulations and operating procedures, including engineers' working hours.

The Hamburg America Line representative reported the company had ordered a small twin-screw steamer (7,500 tons) with diesel engines and auxiliary machinery 'operated by electricity'. Oil for the round trip would be obtained at Boston or Philadelphia 'and stored in the double bottom and side tanks'. The Red Star Line claimed that their *Lapland* was 'the best ventilated ship on the North Atlantic' and her system included 'an exhaust system extracting the foul air from the top of each compartment'. Cunard officials were interested in the Hotel Metropole ('a good example of the control of temperature by means of thermostats, which automatically cut off the supply of steam to the radiators when the agreed temperature has been reached, and open again when a lower temperature is recorded') and various London buildings, including the 'New Gallery Restaurant (Bond Street)' where 'the conditions existing in the kitchen are somewhat similar to those fitted in the galleys of the *Lusitania* and *Mauretania*'. Unfortunately for the staff, the working conditions were poorer, 'with two fans running, one extracting the hot air by means of suitable trunking leading to the roof, and the other discharging fresh air, the working conditions are much worse'.[56]

The exchanges were mutually beneficial. A report prepared for Cunard's management noted, 'the discussion on the various items was interesting and instructive and one was not pressed to answer any questions, the answer to which might be considered detrimental to the company's interest'. At the end of the second day, attendees inspected White Star's own works at Liverpool, under the guidance of the company's superintendent engineer, Willett Bruce. ('The works appear to be run on excellent lines [...] The works manager is a 6 o'clock man and he sees that the foremen under him are there at the same time.') The visitors were 'struck with the tidiness of the various shops and with the small number of men employed in proportion to the work'.[57] On the third day of the conference, the visitors inspected Cammell Laird & Co.'s works and shipyard on the River Mersey, before being taken to Chester and leaving for the Continent in the afternoon.

On another occasion, Cunard received a request from J. Bruce Ismay for the full-size working drawings of the teak vertical sliding windows on *Mauretania*'s B-Deck. He was impressed with her features. Cunard made sure that the drawings were supplied to Harland & Wolff. Similar windows were fitted for many of the first-class staterooms on *Olympic*'s B-Deck, with clear glass 25½in by 18¾in. A mahogany jalousie and cathedral-glass vertical sliding window were fitted inside to guard the privacy of the stateroom occupants from passengers walking the enclosed promenade outside.

ROYAL MAIL STEAMERS

While their new ships were under construction, White Star consulted the General Post Office. They were being built for service as Royal Mail Steamers, so the postal authorities wanted to make sure the Post Office accommodation was satisfactory. On 8 July 1909, the company wrote to the General Post Office in London to advise them that 'two mail steamers of about 45,000 tons gross register each' were being built, and they enclosed plans 'showing the mail room and post office accommodation proposed'. The postal authorities made a number of suggestions and requests, but some of them were completely unrealistic. They wanted the Post Office placed on the lower promenade deck. White Star explained that 'it is really impossible in a passenger steamer to put accommodation of this nature upon the upper decks, as it would interfere seriously with the earning power of such costly vessels'. They reassured them that the Post Office could be on a lower deck – Harland & Wolff were making 'special provision' for mechanical ventilation of the lower decks and that the air could be 'entirely changed at least once every five minutes'.

The postal authorities worried that the mail room was relatively close to the stern. Only *Caronia*, with a sorting room on a higher deck, had the sorting room closer to the stern, at 33ft (compared to 110ft initially proposed on *Olympic* and *Titanic*). On 22 October 1909, the Post Office Department in Washington wrote to the General Post Office in London:

> In the early days of the sea post service the post offices were located experimentally in the after part of the ship. But it was soon found to be impractical to continue this arrangement because of the greater motion far aft and the extreme vibration caused by the close proximity of the screws. The sea post clerks were unable properly to perform their clerical duties in making out the registry lists, and it also frequently occurred that distributions, which had occupied several hours in completion, were undone by the letters being thrown out of the cases by the violence of the motion experienced in that part of the vessel. As I understand that the length of the *Olympic* and *Titanic* will not be less than 840 feet, the distance of 110 feet from the rudder post is relatively less than it would be in smaller ships [...]
>
> It is of course important that the sea post offices should be close to the hatches in order that the mails may be lowered directly into the mail rooms; and I have noted that the location assigned appears to be close to the after hatch (No. 5) [...] But I also consider it important that the post office rooms should be situated in a part of the ship where motion and vibration shall not be excessive. Without doubt the new steamers will be steadier than ships of less immense proportions, but the vibration from the screws is bound to be very great in the part of the ship where it is proposed to place the post offices, and may not only render satisfactory work extremely difficult but will make eleven hours a day of intricate and exacting duties a hardship for the clerks employed.[58]

Following 'the unsuitableness of the rooms designed for sea post offices in the Cunard steamers *Mauretania* and *Lusitania*', the postal authorities were very concerned with ventilation and lighting arrangements. On 14 June 1910, White Star advised that the Post Office and Mail Room accommodation was being moved to the bow: the Mail Room would be on the starboard side of the bow on the orlop deck, opposite the First and Second-Class Baggage Room; the Post Office and First-Class Baggage would be above on G-Deck. 'It is on the same deck,' they wrote:

> ... but at a slightly higher level above the water, owing to the sheer of the vessel being greater in the fore body, and is located at the forward end of the first class passenger accommodation instead of below the second class accommodation aft, while the facilities for access and for handling the mails, particularly in the matter of the tender lying alongside these large vessels for the receiving or despatch of mail matters, and also the extra space available in the case of exceptionally heavy mails, are we think significantly better than in the previous arrangement.[59]

The General Post Office agreed that the new arrangement was 'a decided advantage, as any possible vibration will be reduced to a minimum'. The 'superficial deck area of the Post Office has been increased from 600 sq. ft. to 640 sq. ft. while the capacity of the mail room is increased from 225 cubic tons measurement to 240 cubic tons measurement,' they noted. 'As far as space is concerned the White Star Line are behaving very handsomely.' They made a number of requests for rack fittings and additional mail cases. By 25 July 1910, White Star were glad that the plans for the postal clerks' staterooms were approved. They explained that concerns about coal dust were unwarranted, 'As respects the method of embarking the mails, there need not be any apprehension with regard to coal dust, as we intend while this ship is in the mail service to coal her by side shoots [*sic*].' A door in the centre bulkhead between the Mail Room and First and Second-Class Baggage Room on the orlop deck would be:

> ... fitted for emergency purposes as suggested, and your request that this door, as well as the one between the [reserve coal] bunker hatch and baggage room, be fitted with bars and padlocks on both sides and for a special lock to be fitted to the door of the sorting room will receive our attention.

In December 1910, they agreed that the postal authorities could inspect *Olympic* when she was completed.

CORONER'S INQUESTS

Progress came at a price. The cost was not only financial but, all too often, in human life. It reflected industrial working conditions of the period. On Thursday, 14 October 1909, 19-year-old John Bell died in an accident on *Olympic*. He was working as an apprentice plater on F-Deck, standing on a plank, when he fell about 32ft down to the tank top below. James Graham was working in the same squad but none of Bell's colleagues appeared to have seen the accident. Another worker below them, James Britty, said that Bell had almost fallen on top of him. The fall was so serious that Bell sustained a fractured skull, which led to a cerebral haemorrhage. He passed away five minutes after admission to hospital.

Two days later, the Old Coroner's Court heard evidence before Dr James Graham, the Belfast City Coroner. The jury were told that the plank Bell had been standing on was steady, even though it 'was not fastened; it was merely laid across the beams' and 'could not be lifted up'. They agreed with the Coroner's view that it had been a 'pure accident' and 'returned a verdict of accidental death, in accordance with the medical testimony'.

Dr Graham said he had known Bell since he was a child and knew his parents, 'The Cregagh district would miss this young man very much, as he was a pioneer amongst his companions in every manly and athletic sport; and the jury would join him, he was sure, in his expression of sympathy.' He noted that there had 'been very few accidents of recent years in either Mesrrs. Harland & Wolff's or Messrs. Workman, Clark & Co.'s [shipyards]'. That was no consolation for Bell's family and friends.

Less than two months later, there was another serious accident at Harland & Wolff. On Saturday, 8 January 1910, James McIlvenny fell about 30ft into a ship's hold. He was admitted to Belfast's Royal Victoria Hospital 'in a very serious condition', suffering from a fractured leg, dislocated arm, scalp wound and a lacerated leg and elbow.[60]

On Tuesday, 8 February 1910, there was an even worse incident early in the evening. Twenty-two-year-old Robert Kilpatrick and 29-year-old John Lynas were 'killed almost instantaneously', while another of their colleagues was 'very seriously injured'. The *Belfast Weekly Telegraph* reported, 'The men were at work on a staging at the side of the big liner *Olympic*,' when 'a number of heavy plates became detached from the chains of the crane and crashed down upon the workers beneath.'

Plater Adam Holmes was one of the witnesses called before the Belfast City Coroner's Inquest. He described six plates on 'double slings with two hooks round the plates in the usual way. When they were fastened he gave a signal to the crane man to heave them up and they seemed to be all right.' The crane then carried them 'about two hundred yards'. The plating was intended for one of the watertight bulkheads, 'The plates seemed to be all right when they were fixed, but they had not reached their destination when they came out of the slings.' The crane operator, Hugh Bell, said the plates were 'swinging slightly' in the wind but did not, to his knowledge, strike the gantry. He saw them fall from 125ft but did not know what had caused them to fall. Assistant Works Manager Thomas Houston said thousands of plates 'were slung in the same way, and by the same crane, and that was the first accident that had occurred'.

Following suggestions that the accident might have been caused by 'short and long plates having been sent up together', the jury added a recommendation that only the same size of plate should be lifted at the same time. The jury foreman said 'it seemed to him that that was a hurried job, and sufficient time was not given to have due care exercised'.[61]

No fewer than 296 accidents were reported for the twelve months up to December 1909 at Harland & Wolff's Belfast shipyard, including three fatalities. Sadly, it rose to 494 accidents for the same period in 1910, including fourteen fatalities. It was partly a reflection of a significant increase in the number of men working at the yard.[62]

Workers had to contend with other problems, too. On 27 August 1910, the *Belfast Evening Telegraph* reported that 'for some time past the workmen in Messrs. Harland & Wolff's shipyard had been losing their food daily, the bread and meat and other things that composed their breakfasts being taken out of their coats while the men were at work'. One morning, thirty breakfasts had been stolen, and Thomas A. Rees was caught 'taking the breakfast of one of the men from the coat pocket'. The police were called and Harbour Constable Running arrived on board *Olympic* at 7.55 a.m. on Friday, 26 August 1910. He found Rees had stolen at least fourteen breakfasts, 'each meal being wrapped up in a separate piece of paper, and the whole bound together in one large parcel'. William Henry Ferguson, working as a joiner on *Olympic*, confirmed one of the breakfasts was his. The thefts had been going on for some time 'and the men had been getting wild about the matter'. Ferguson had never seen Rees before and he did not work for Harland & Wolff.

The case went before the Belfast Custody Court. Rees was visiting from Liverpool. He said he was a fitter and had 'been in Belfast for the past five weeks out of a job. I have pawned everything I possess to obtain food.' Rees admitted stealing, 'I am very hard up, and have been without food for a long time.' He said the rest of the food belonged to him but prosecutor Wellington Young asked Rees how he could have bought the other food if he had no money. The court was satisfied that Rees 'had been systematically robbing the workmen of their food'. The court concluded 'it was quite possible he might have been hard up', but the thefts were repeated and systematic. Rees was sentenced 'to three calendar months' imprisonment with hard labour'.

There were other thefts involving Harland & Wolff's company property. In November 1909, Samuel Easton was 'sentenced to three months' imprisonment' in court in Belfast. He was working for Harland & Wolff when he stole ten paint brushes, '[The] accused [...] had access to the place where the brushes were kept. He got women to pawn them for him.'[63]

***

The process of designing *Olympic* and her sister represented a remarkable effort. It was the culmination of years of experience of shipbuilding and operating in the competitive transatlantic passenger trade. It represented the physical toil of thousands of workers and the collaboration of industry across the United Kingdom. The need to satisfy the regulator that the design met and exceeded requirements, the materials had been tested to the required standards and the ships were being built in line with the plans represented an ongoing and essential process.

Then there was the need to work with the postal authorities and others to make sure they were satisfied with the facilities on board. All of this was complicated enough but, simultaneously, Harland & Wolff and White Star had to consider the ship's safety and watertight compartment design; the precise configuration of the propelling machinery; their choices for the passenger accommodation; and the necessity of financing these ships and making sure the ports, particularly Southampton and New York, would be ready on time.

3

'UNSINKABLE'

By the end of the nineteenth century, shipbuilding had improved to the point that – for many people – it was hard to conceive of a large, modern liner sinking. The use of watertight bulkheads to divide the hull into compartments meant that any flooding from a localised injury to the hull could be contained. The ship would either remain afloat indefinitely or, in a worst-case scenario, only sink after her passengers and crew had all been rescued.

In an article entitled 'Safety on the Ocean', published in the *New York Times* in May 1891, the writer referred to the watertight subdivision of the *City of Paris* and *City of New York*. He claimed, 'These iron walls [bulkheads] will render the latest type of ocean steamers practically unsinkable.'[1] These liners had transverse watertight bulkheads (extending from one side of the ship to the other side) and longitudinal watertight bulkheads (which ran for a distance along the length of the ship):

> No doubt in heavy weather a bad collision, smashing the end of one of the transverse bulkheads and thus opening the compartments to the sea, would sink even one of these splendid ships. The chance is a small one, but its possibility makes the ocean voyage, even in our time, a matter of uncertainty. Fog and ice, worse than all the fabled dangers of Columbus's day, lurk in the path of our ocean steamers, and until science finds some method of baffling them, the voyager will always go off soundings with a little anxiety. In fog lies the danger of collision, and until some intelligible and simple method of communication between vessels invisible to one another is devised, the greatest peril of the deep will still threaten.
>
> Icebergs are less troublesome than ships in a fog, because they can be avoided by steering an extremely southerly course. Unhappily, the travelling public is in a hurry, and captains may be expected to continue taking the chances of cutting 'across lots' along the southerly borders of the Grand Banks.[2]

Three years later, the new *St Louis* was said to have 'a system of bulkheads' that made 'her practically unsinkable'.[3] Following her trials in May 1895, she was the subject of interest for many American shipping observers. Her nine compartments were formed by watertight bulkheads with no watertight doors in them. Edwin Cramp, representing the shipbuilders, explained:

> This feature, completely eliminates the human factor, and, by preventing the leaving open of bulkhead doors in cases of emergency, renders the ship *absolutely unsinkable* [author's emphasis]. More than that, if cut in two, it is not too much to say the separated halves would float independently of each other.[4]

When *St Louis* was several days late coming into port in January 1903, there was speculation that a mechanical problem had led to her breaking down and drifting away from the shipping lanes. Few believed anything more serious could have happened, for 'her watertight compartments render her practically unsinkable, even in the event of a collision and the total derangement of her machinery'.[5] In November 1903, a false report that White Star's *Cedric* had been lost was discounted using very similar language. The same month, there was interest in HAPAG's *Deutschland*. Her watertight doors could be closed automatically in the event of flooding, making her 'practically unsinkable'.[6]

The same confident commentary appeared when *Lusitania* and *Mauretania* arrived on the scene. 'Watertight bulkheads are arranged and fitted with the most stringent regard to making the vessel unsinkable', wrote the *Shipbuilder* after *Lusitania* was launched in 1906.[7] The journal also described *Mauretania*, 'The subdivision by watertight bulkheads and decks is the most elaborate that could be devised for a passenger vessel, the number of separate watertight compartments – 175 – rendering the ship practically unsinkable.'[8]

Similar references appeared in newspapers of the day, including the *New York Times*. An advert in 1906, appearing in the *Shipbuilder*, touted, 'The *Mauretania* is practically unsinkable owing to the watertight bulkhead doors being hydraulically controlled by the Stone-Lloyd system.'[9] The *Belfast Weekly News* said on 4 June 1908 that the new *Rotterdam*'s 'watertight compartments are so arranged as to render sinking after collision a practical impossibility'.

The White Star Line produced a publicity brochure around September 1910, which promoted *Olympic* and *Titanic*, concluding, 'As far as it is possible to do so, these two wonderful vessels are designed to be unsinkable.' The *Shipbuilder*'s Midsummer 1911 'Special Number' explained that their watertight subdivision was 'very complete', noting, 'The captain can, by simply moving an electric switch, instantly close the doors throughout and make the vessel practically unsinkable.'[10]

The use of watertight doors highlighted a compromise between absolute safety and the necessity of designing ships so that they could operate successfully day to day. White Star's first *Oceanic* had been built without a watertight door between the bunker and the boiler room, because it was intended to lift coal from the bunker floor to the top of the watertight bulkhead, but it was not practical. After her first five or six voyages, a watertight door had to be fitted. The *City of Paris* and *City of New York* were also built initially without watertight doors in their bulkheads, but they also had to be retrofitted. Leonard Peskett thought that watertight doors between the engine and boiler rooms were 'necessary for supervision and efficient working of the engines and boilers' on large passenger ships like *Mauretania*. Edward Wilding also thought specifically that a watertight door between the reciprocating engine room and turbine engine room on *Olympic* was required.[11]

***

When Harland & Wolff were designing the watertight subdivision for *Olympic* and her sister, they relied on their own experience and existing practice. Sir Edward Harland had chaired a committee for the Board of Trade looking into 'The Spacing and Construction of Watertight Bulkheads' in merchant ships, which reported in 1891.[12] The committee's highest class of watertight subdivision applied to vessels that were 'subdivided throughout their length, so that they may float in moderate weather with any two adjoining compartments in free communication with the sea'. At the lowest grade were 'vessels subdivided in the fore body, so that they may float in moderate weather with any one of the fore body compartments in free communication with the sea'.[13]

Harland's committee made recommendations concerning watertight bulkhead stiffening; the spacing of bulkheads; access through bulkheads, either by watertight doors or watertight passageways; and how the Board of Trade's surveyors could determine that watertight bulkheads were sufficient. They were concerned about a damaged ship's stability, if she was fitted with both longitudinal and transverse watertight bulkheads:

> A [watertight] compartment subdivided by one or more longitudinal bulkheads should be treated as one compartment only, unless the owners satisfy the Board of Trade that the vessel will not have her stability seriously reduced, or list, so that any part of the bulkhead deck at the side is under water, in the event of the space between a longitudinal bulkhead and the vessel's side, or of two such adjoining spaces on the same side of the vessel, being open to the sea.[14]

Their proposals were never enacted officially. Even so, the Board of Trade recognised them 'as a standard of comparison' and 'referred to that as their standard' for watertight subdivision. In the early 1900s, the committee's highest class of watertight subdivision was taken as the standard for modern passenger liners. The specification book for Cunard's *Carmania*, completed in 1905, stated, 'The vessel must be able to float with any two holds [compartments] bilged [flooded].' A specification book for *Lusitania* and *Mauretania*, dated 6 January 1906, used identical wording, adding, 'The watertight subdivision generally must be as minute as compatible with commercial requirements.' The Holland America liner *Rotterdam* was designed on the same basis.

In December 1911, Harland & Wolff referred to their new *Statendam* 'remaining afloat with any two compartments flooded'.[15] The subdivision rules Harland & Wolff adopted were also 'very similar' to the German regulations of the *See-Berufsgenossenschaft*.

J. Bruce Ismay said that the White Star Line 'were very anxious indeed to have a ship which would float with her two largest watertight compartments full of water'. Harland & Wolff ensured their design met this requirement:

> What we wanted to guard against was any steamer running into the ship and hitting her on a bulkhead, because if the ship ran into her broadside on and happened to hit her right on a bulkhead, that would open up two big compartments.

She 'was designed to float with any two compartments flooded, and to have good freeboard under those conditions', Wilding explained.

Ismay was not concerned with details such as the structure or height of the watertight bulkheads, which he said was 'a technical matter that we are guided in by the shipbuilders'. His concern was that the design objective was achieved. To show him it was, Wilding prepared a 'flooding plan by any two compartments' in 1909. Ismay simply asked if she would float with any two watertight compartments flooded, he 'looked at it and talked it over, and said he was satisfied'.

Harland & Wolff's 'Design "D"' concept, approved in July 1908, included fourteen transverse watertight bulkheads dividing the ship into fifteen watertight compartments. The five single-ended boilers adjacent to the main

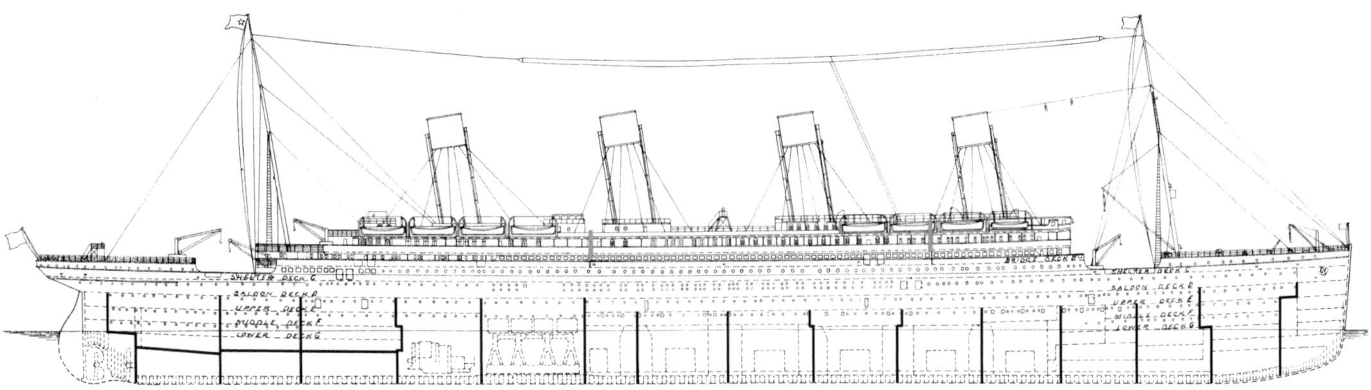

Compared to the original 'Design "D"' concept approved in July 1908, an additional watertight bulkhead was added and another was extended. In the final design, *Olympic* and *Titanic* each had fifteen transverse watertight bulkheads. Most of them extended to the saloon deck, D, and the remainder to the upper deck, E. (Bruce Beveridge collection)

engine room were part of the same boiler room containing another five double-ended boilers. (The arrangement had much in common with *Adriatic*'s. Her aft boiler room had a combination of single- and double-ended boilers in a single watertight compartment.)

By May 1910, plans included an additional transverse watertight bulkhead so that these five single-ended boilers were in their own boiler room and watertight compartment. Although Lloyd's Rules did not yet extend beyond ships that were 680ft long, on a proportional basis it appeared they would have required a total of thirteen watertight bulkheads, less than the fifteen in the final design. The bulkhead plating and stiffening were both in excess of Lloyd's requirements. Harland & Wolff also made sure that the stiffening went beyond the recommendations of the Harland committee (they took a conservative approach, ignoring the brackets attached to the stiffeners when calculating the strength of *Olympic*'s watertight bulkheads). Edward Wilding noted that a head of water 'about 150 feet' deep, from the bottom of the watertight bulkheads would be needed to break the lower part of them, which was many times higher than the highest bulkheads.[16]

In any scenario with two compartments completely flooded, the watertight bulkheads would remain above water by twice the amount recommended by the Harland committee.[17] Harland & Wolff built in such a margin of safety that the design largely met a three-compartment standard, with some exceptions, and even a four-compartment standard in a number of scenarios. All the indications were that the four foremost compartments could be flooded; boiler rooms 4 to 1 could be flooded; boiler rooms 3 to 1 and the main engine room could be flooded; or the four aftermost compartments could be flooded without sinking the ship. (Although the ship would remain afloat, she would nonetheless present an alarming appearance: the flooding of the first four compartments would bring the bow down so far that G and F-Decks would be under water; if the four aftermost compartments were flooded then the stern would be submerged so that the entire rudder would be under water.)

Following their usual practice, Harland & Wolff did not opt for an inner skin inside the hull, longitudinal watertight bulkheads or watertight decks throughout the ship. (The orlop deck at the stern, forming the top of the propeller shaft tunnels, was a limited example of a watertight deck.) Edward Wilding acknowledged that an inner skin offered protection from 'small wounds', but he worried about upkeep, 'By our experience of the upkeep of such very restricted spaces which are very difficult of access, there is considerable risk in our judgement; we believe that such spaces will not be as well looked after as they might be.' Corrosion ('the most subtle enemy that we have to deal with') might even weaken the hull over time.[18]

Teutonic and *Majestic*'s original configuration included a longitudinal watertight bulkhead in the engine room, but Harland & Wolff later pierced them so that they were no longer watertight. The concern was 'from the point of view that if one side of the ship got flooded it might lead to a very serious list, and possibly to a capsize'.[19] Wilding also had grave concerns about the use of watertight decks and felt he 'would rather obtain safety if possible in some other way'. Although it might help to contain flooding in some circumstances, in other situations it might prove fatal. If flooding occurred above the watertight deck, leaving a pocket of air beneath it, then the ship's stability would be affected and she might even capsize, 'as happened in the case of HMS *Victoria* some years ago'.[20] (The watertight subdivision of *Lusitania* and *Mauretania* had more in common with naval vessels. Their design incorporated transverse and longitudinal watertight bulkheads, as well as watertight decks. They were the only merchant ships of their type, whereas the transverse watertight bulkheads on *Olympic* and *Titanic* were commonplace.)[21]

Compared to some earlier liners, the average length of each compartment on *Olympic* and *Titanic* was reduced. It added to the ships' safety, because a smaller proportion of the ship would be flooded for every compartment that was damaged.

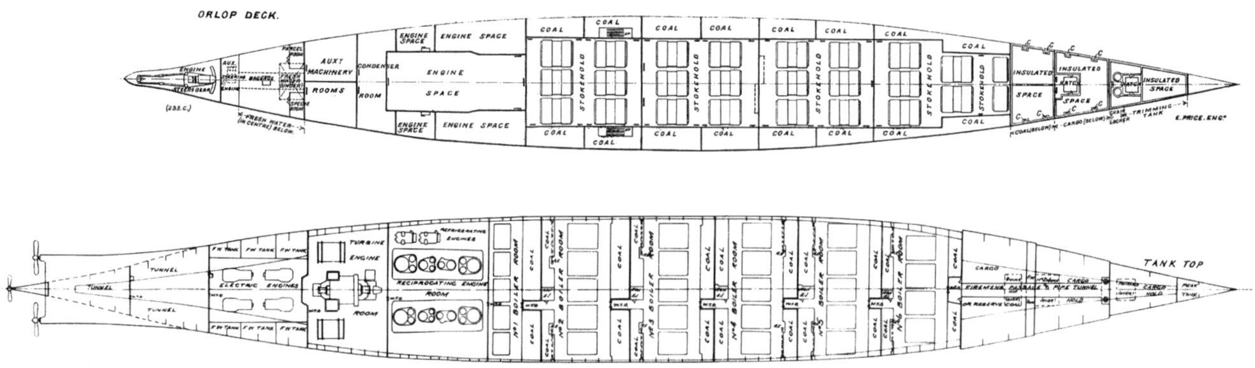

This comparison of *Mauretania* (top) and *Titanic* (bottom) illustrates the different approaches to watertight subdivision. *Mauretania* included a mix of transverse watertight bulkheads (running across the ship) and longitudinal watertight bulkheads (running along part of the ship's length). *Engineering*, 1907, and *The Shipbuilder*, 1911. (J. Kent Layton collection)

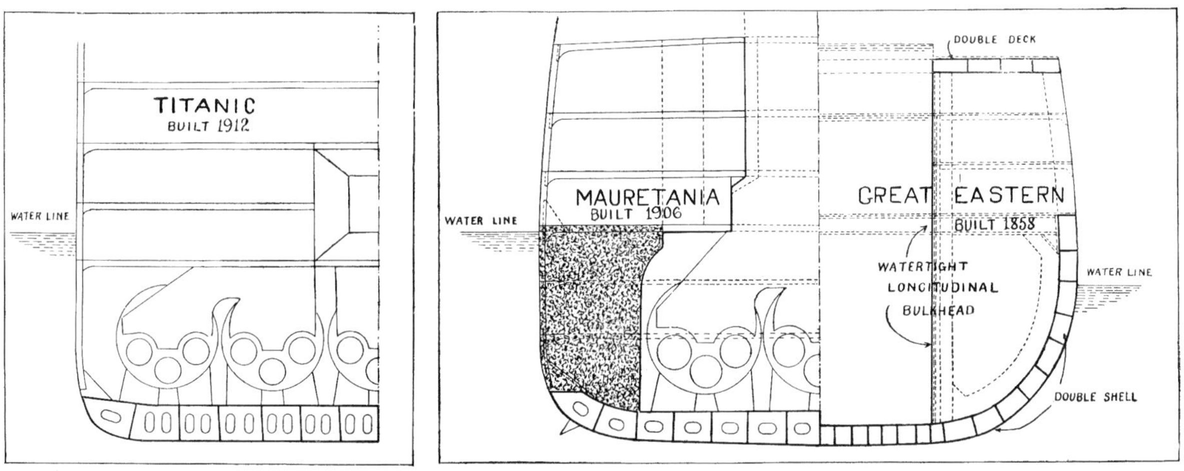

J. Bernard Walker included these comparisons of *Titanic*, *Mauretania* and *Great Eastern* in his article 'The "Unsinkable" Ship' in May 1912. *Great Eastern* proved a commercial failure in her day. Walker advocated longitudinal watertight bulkheads, but naval architects such as Harland & Wolff's Edward Wilding thought they were dangerous. *Scientific American*, 1912. (Author's collection)

	Year	No. of Bulkheads	Average Compartment (% of Length)
Campania	1893	8	11.1
Great Eastern	1859	9	10
Alma	1894	11	8.3
Carmania	1893	12	7.8
Teutonic	1889	12	7.7
Adriatic	1907	12	7.7
Oceanic	1899	13	7.1
George Washington	1909	13	7.1
New York	1888	14	6.7
Olympic	1911	15	6.2
Lusitania	1907	16	5.9

The largest watertight compartment, containing the main engine room, was not much greater in length than the average compartment on *Adriatic*.[22] The four foremost compartments had an average length of 5.7 per cent of the ship's length, meaning that the ship would remain afloat with almost 23 per cent of her length flooded. In the case of the four aftermost compartments, it was about a quarter of the ship's length.

Harland & Wolff worked with the Board of Trade to make sure that the watertight bulkhead arrangements met the regulations. They submitted plans and then discussed any required alterations.[23] One of the changes stemmed from a compromise in relation to the foremost ('collision') bulkhead, following discussions between April and September 1910. It was 'stepped' and did not extend upward in a straight line, which suited the interior layout of the ship but brought part of the bulkhead too close to the bow for the Board of Trade to approve. It also extended to E-Deck rather than D-Deck. Following discussions between Francis Carruthers and Harland & Wolff, including Thomas Andrews, in July 1910, the shipbuilder still did not see the necessity to extend it up to D-Deck. As an alternative, the Board of Trade suggested that the second bulkhead from the bow be extended, as well as trunking up other deck openings so that they were watertight to D-Deck.

Harland & Wolff accepted the suggestion about the second bulkhead, but did not agree with the need for watertight trunking. On 21 July 1910, they wrote:

> If the vessel were damaged forward of this watertight bulkhead, only No. 1 hold and the firemen's passage could become flooded, while the sketch we submitted to you some time ago shows that especially the forward bulkheads of these vessels so arranged that there is a *large* [original emphasis] margin of safety with any two compartments flooded.[24]

They had undertaken extensive flooding calculations, which were checked by the Board of Trade. Alfred J. Daniel wrote for the Principal Ship Surveyor William David Archer on 15 September 1910:

> It appears that if the fore end [of the ship] is damaged, and water floods No. 1 hold and the firemen's passage abaft it, the increase of draft forward will be only 27 inches, and accordingly, I think the surveyors need not now require the openings in the upper deck, leading to the firemen's passage and to the No. 1 hold, to be trunked up watertight to the saloon deck.[25]

For all practical purposes, following the modification the ship had two collision bulkheads.

Another key issue was assigning the ship's freeboard (the height between the waterline and the shelter deck) and load line (approved draught). Not only did that need to be taken into account when confirming the height of the watertight bulkheads, but the ship's approved draught was directly related to her displacement (weight). In turn, confirming the ship's displacement at her load draught was vital to checking the calculations of stresses and ensure she was not overloaded.

Towards the end of May 1910, the Board of Trade were well aware of *Olympic*'s 'unusual dimensions' as the largest ship yet constructed. They wrote to Harland & Wolff. The shipbuilders had been basing their calculations on a draught of 34ft 9in. On 20 June 1910, they noted, 'We are prepared to accept a freeboard slightly greater than Table C, say about 1 inch greater, which will, we estimate, make the load draught about 34 feet 6 inches.' They confirmed all manner of details, including the planned 'distribution of load' and weights within the ship.[26]

Loads for each individual compartment ranged from only 40 tons for the forepeak to 1,610 tons for Boiler Room 2, all included within *Olympic*'s approximate deadweight of 13,200 tons (the sum of her weights of cargo, fuel, provisions, passengers and crew). The Board of Trade looked into the consumption of fuel and stores throughout a crossing to assess the ship's status at the end of a voyage. On 17 May 1911, Francis Carruthers, at Belfast, confirmed to Harland & Wolff that the

'The strength of the bulkheads throughout appears very ample.' This detailed structural assessment of *Olympic* and *Titanic*'s watertight bulkheads looked at details such as the stiffeners and plating, finding them far stronger than what was required. (The National Archives (UK))

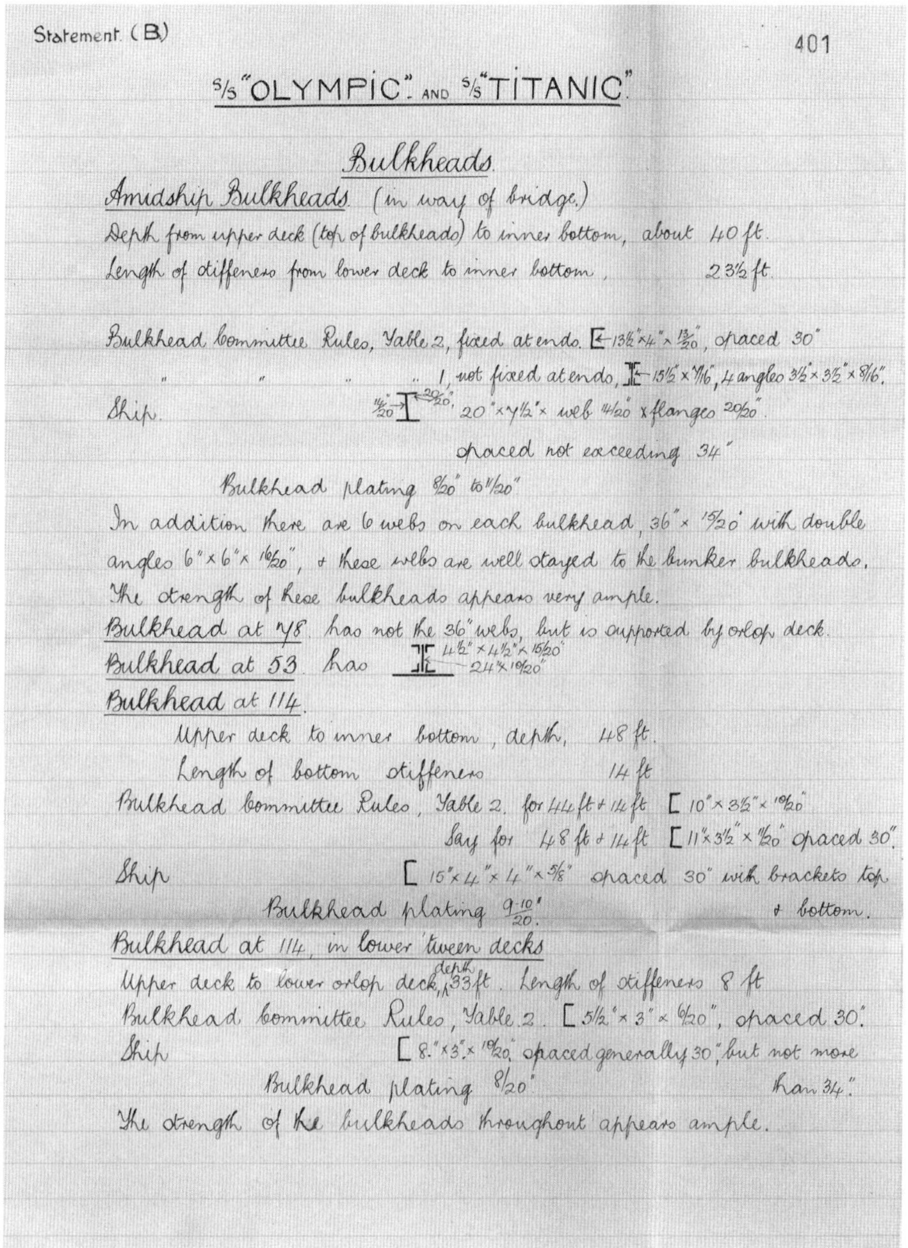

Statement (B) 401

S/S "OLYMPIC" AND S/S "TITANIC"

Bulkheads

Amidship Bulkheads (in way of bridge)

Depth from upper deck (top of bulkheads) to inner bottom, about 40 ft.

Length of stiffeners from lower deck to inner bottom, 23½ ft.

Bulkhead Committee Rules, Table 2, fixed at ends. [13½" × 4" × 13/20", spaced 30"

" " " 1, not fixed at ends, 15½" × 7/16", 4 angles 3½" × 3½" × 8/16".

Ship. 14/20 20/20 20" × 7½" × web 14/20 × flanges 20/20".

spaced not exceeding 34"

Bulkhead plating 8/20" to 11/20"

In addition there are 6 webs on each bulkhead, 36" × 15/20" with double angles 6" × 6" × 16/20", + these webs are well stayed to the bunker bulkheads.

The strength of these bulkheads appears very ample.

Bulkhead at 78 has not the 36" webs, but is supported by orlop deck.

Bulkhead at 53 has 4½" × 4½" × 15/20" 24" × 10/20"

Bulkhead at 114.

Upper deck to inner bottom, depth, 48 ft.

Length of bottom stiffeners 14 ft

Bulkhead Committee Rules, Table 2. for 44 ft + 14 ft [10" × 3½" × 10/20"

Say for 48 ft + 14 ft [11" × 3½" × 11/20" spaced 30".

Ship [15" × 4" × 4" × 5/8" spaced 30" with brackets top + bottom.

Bulkhead plating 9-10/20".

Bulkhead at 114, in lower 'tween decks

Upper deck to lower orlop deck, depth 33 ft. Length of stiffeners 8 ft

Bulkhead Committee Rules, Table 2. [5½" × 3" × 6/20", spaced 30".

Ship [8" × 3" × 10/20" spaced generally 30", but not more than 34".

Bulkhead plating 8/20"

The strength of the bulkheads throughout appears ample.

Board of Trade had approved the freeboard as expected. The exact load draught worked out to 34ft 6½in.[27]

The watertight doors were an important design feature. White Star's older passenger steamers had watertight doors that were either 'screwed down' from the bulkhead deck above or from the stokehold plates at the level of the door. (There was also an option to close the doors from the stokehold. It was a basic door design, used since 1884.)[28]

Harland & Wolff implemented a significant design improvement, which they used for the first time on *Olympic*. The twelve main watertight doors in the boiler and engine room spaces were Harland & Wolff's own latest type, working vertically with a heavy cast-iron door plate. They were held open by a clutch, which was released by a powerful electromagnet controlled from the bridge. Once the clutch was released, they dropped by the force of gravity. The doors could also be closed by hand. Back-up floats were provided beneath the floor level so that, if water entered, they lifted automatically and released the clutches. 'The doors were arranged to drop and close in thirty seconds from the time the electric contact was made from the bridge or when pulling the handle near the door,' wrote Francis Carruthers:

> The remainder of the doors, that is all above the Orlop deck were of the horizontal sliding type [closed manually]. All the doors and frames for the doors were substantial and well made; when completed tests were made of the closing arrangements and the fit after closing of all the doors; all proved satisfactory.[29]

Compared to *Mauretania*'s design, Harland & Wolff preferred the system of watertight doors they adopted for *Olympic*. *Mauretania*'s watertight doors below the waterline were operated largely using the Stone-Lloyd system. The *Shipbuilder* explained that it worked by hydraulic power with 'a powerful hydraulic cylinder attached to each door' with a pipe putting pressure into the closing mechanism (the power needed to be applied until the door was closed). However, one significant weakness in the arrangement was that only the coal bunker doors in Boiler Room 3 were closed on the Stone-Lloyd system. The rest had to be closed manually. If any bunker doors could not be closed in time then the entire boiler room would flood. (One case highlighting issues with watertight doors that had to be closed manually was the sinking of *Republic* in January 1909. When *Florida* struck her 'right in the engine room', the ship's engineers had to 'screw down' the watertight door between the engine room and the boiler room. They managed to close it successfully but the process took 'three or four minutes and by the time it was shut there was 12 feet of water in the engine room and 5 feet of water in the stokehold'.)

Olympic's watertight doors did not need a continuous power supply, because they dropped by their own weight once the clutch was released. She also had a back-up float mechanism to close them, whereas *Mauretania* did not. *Olympic*'s watertight doors remained closed until the officer on the bridge permitted them to be opened, which was potentially inconvenient for the crew below but somewhat better from a safety point of view. (*Mauretania*'s doors could be opened by the crew. They would close again, provided the pressure was still applied.) Although a powered Stone-Lloyd door might be more likely to break through any obstruction, Edward Wilding thought it was always possible for an obstruction to block any watertight door and prevent it from closing. For that reason, he disliked the arrangement on the Cunard ships, because the watertight doors in the longitudinal bulkheads had to be opened regularly to access coal from the bunker. There was no such issue on *Olympic* because the watertight doors were in the transverse bulkheads, allowing access between the boiler and engine room compartments. The doors to the coal bunkers were in non-watertight bulkheads.

Olympic's steam-pumping arrangements were 'more complete than usual', so that if any compartment was flooded or a cock was left open in the flooded compartment then the pumps were not impaired. The eleven steam pumps included two in the turbine engine room, four in the reciprocating engine room and one in Boiler Rooms 2–6.

The end result of all these design choices was that *Olympic*'s watertight compartmentalisation represented a very safe design. Harland & Wolff had gone beyond the highest standards of the time, building in a comfortable margin of safety beyond the design specification.[30]

'THE SAFEST FORM OF TRAVEL IN THE WORLD'

Transatlantic travel between the United Kingdom and the United States and Canada was generally a safe form of transport. The number of lives lost was small compared to the number of travellers. From 1902 to 1911, 4,083,229 passengers embarked on a westbound voyage from the United Kingdom and 1,970,153 passengers were landed from eastbound voyages in the same period. Passenger numbers were growing considerably. The westbound total was almost double the equivalent figure for the decade from 1892 to 1901.[31] Over twenty years up to 30 June 1911, eighty-five of those passengers and 1,570 crew lost their lives on ships registered in the United Kingdom, 'while on voyages between European ports and ports on the east coast of the United

States, Canada and Newfoundland'. Seventeen of the passengers who died were lost on vessels not required to hold a passenger certificate. The remaining sixty-eight were lost in a number of different incidents, including five passengers lost on both *Pomeranian* (February 1893) and *Campania* (October 1905) in stormy conditions when their decks were 'swept by a sea'. Three passengers died when *Republic* sank (January 1909): two were crushed in the collision and one died after being evacuated.

There were, inevitably, incidents which could have been much more serious. *Lloyd's List & Shipping Gazette* carried a report about one particular 'narrow escape' in July 1907:

> New York, July 10.
>
> The *Kronprinz Wilhelm* arrived this morning with her bow looking as if a giant fist had struck it a heavy blow. The point of the prow was pounded off, and seams had started, and there was a large space on the starboard bow just above the waterline. The vessel encountered an iceberg shortly after midnight on Monday, off the [Grand] Banks. The weather was thick and the *Kronprinz Wilhelm* was running at 16 knots, a slower speed than usual on account of the inexperience of the stokers who had replaced the strikers, when the lookout sighted an iceberg 100 feet distant.
>
> The rudder was swung round instantly, and the engines were put full speed astern, whilst the crew were called to close the watertight bulkheads. The steamer lurched with the abrupt change of the screws, and then there was a shock as she struck the iceberg sideways. Her prow, lifting on to the submerged portion of the berg, split off two large fragments, which fell on the deck, narrowly missing a group of sailors. The captain describes the sound of the collision as being like the breaking of a plate-glass window. The passengers were disturbed, but there was no panic, as the danger had been passed and the iceberg was out of sight when they reached the decks.
>
> —*Standard* correspondent.

During the same twenty-year period, 'sea casualties to vessels in the United Kingdom [...] in any part of the world through contact with ice' averaged 42.9 incidents per year. The vast majority of them consisted of damage to the ship and only 1.8 incidents annually were 'casualties involving the total loss of the vessel'. Thirteen ships were 'lost through striking ice' on the North Atlantic routes, including three whaling ships. Only one of them was lost south of latitude 45° north (the cargo ship *Anglo Peruvian* struck an iceberg in fog on 21 April 1906 and foundered three days later). None involved any loss of life. (There were forty-eight ships which simply vanished and so were reported as missing.)[32]

Years earlier, in 1876, Thomas Henry Ismay proposed that passenger lines sailing in and out of New York needed to adopt specific 'steam lanes', as advocated by Lieutenant Maury. He thought that this would 'materially lessen' the 'risks of collision between outward and homeward bound vessels, and with ice'. Unfortunately, it was not possible to reach a general agreement, but 'many of the lines laid down lane routes of their own'. They differed slightly, but 'were on the whole well adhered to'.

By 1891, the major steamship companies had agreed 'on definite Northern and Southern Routes' (Northern from mid-July to mid-January and Southern for the remainder of the year). The Southern Route 'was specially designed to keep the vessels as clear as practicable of the ice and fog which experience had taught were liable to be encountered on any more northerly route at that season'. Seven years later, the routes were adjusted, with the Southern Route moved 'still further to the south' and the return to the Northern Route was postponed to later in the calendar year.

For the first time, a general agreement was reached between the leading British, American and European passenger lines, as well as some of the British cargo lines operating in the North Atlantic. It was 'signed and sealed' at a conference 'summoned by Mr. Ismay' in London in November 1898. The routes remained in force in 1912, although the Southern Route was moved to the south in at least three seasons and, in 1906, the lines returned to using the Northern Route in July rather than August. Each commander routinely reported 'dangers observed on the voyage' to their company, who then ensured other lines were aware. The shipping lines were also in contact with Lloyd's, the Meteorological Office and the press. 'All possible steps' were 'taken on both sides of the Atlantic, and by wireless communication between the ships at sea, to make the warnings received known immediately to all vessels using the routes.'[33]

The number of British-registered ships fitted with wireless apparatus increased rapidly: 116 ships were equipped at the end of 1909; 170 by July 1910; 275 by April 1911 and 410 by April 1912. The majority were foreign-going passenger ships certified by the Board of Trade. Sir E. Sassoon introduced a Bill to the House of Commons in July 1910, which would have made it compulsory for British oceangoing vessels carrying fifty or more people to have wireless apparatus. The British government saw no need to legislate because ship owners were already fitting it.[34]

***

In January 1909, *Republic* became the largest liner yet to be lost at sea. She was rammed in thick fog by the liner *Florida*, but remained afloat for some thirty-eight hours. Ships including White Star's own *Baltic* came to the rescue of her passengers and crew and there was time to evacuate them. In contrast, one reporter wrote that just a 'few years ago a ship [...] would not have floated five minutes'. They said that 'any serious collision twenty-five or thirty years ago meant the immediate foundering of a vessel'. *Republic*'s wireless apparatus had enabled her to summon help.

Few people seemed to grasp how much worse it might have been. The lifeboats had taken several trips to transfer all the passengers. She had not carried sufficient lifeboats to accommodate all her passengers and crew, even though she fully complied with all safety regulations. Although those on board *Republic* were incredibly lucky, her loss served to reinforce a sense of complacency rather than shock the authorities into taking prompt action to improve matters. What

The White Star liner *Republic*. When she foundered in January 1909, she was the largest ship yet to be lost at sea. *Illustrated London News*, 1909. (Author's collection)

might have happened if other vessels had been unable to get to her in time?

Lifeboat regulations had evolved over decades. The Board of Trade's Marine Department was responsible for making sure shipowners complied with the law,[35] but government regulation was relatively minimal. Sir Alfred Chalmers, who was the Board of Trade's long-serving Nautical Adviser while *Olympic* was under construction, viewed the transatlantic trade as 'the safest mode of travel in the world'. He felt that 'it was neither [the] right nor the duty of a state department to impose regulations upon that mode of travel, so long as the record was a clean one'. The regulations dated back to 1894 and followed on from the reports of several government committees from the late 1880s. When the committee examining 'boats, rafts and lifesaving apparatus' met in 1886, the lifeboat regulations then in existence had been laid down more than thirty years before:

> The boat scales in the Merchant Shipping Act, 1854, and Passenger Act, 1855, respectively were framed upwards of thirty years ago to meet a different state of things than now exists. At that time nearly the whole of the passenger traffic was carried on by wooden sailing ships of about 1,200 to 1,500 tons burthen. The boat scales were framed on the basis of the ships' registered tonnage, which in those days of sailing ships gave a very good idea of the vessel's size and of her capabilities for stowing boats, &c., but now the sailing ships have entirely given place to iron or steel steam ships of two or three times their size.[36]

Those 'iron or steel steamships' included Cunard's *Pavonia*. She could carry 1,141 passengers and 155 crew, yet was required to have lifeboat accommodation for just 16.6 per cent of them, which left a shortage of nearly 1,100 people. On average, White Star's fleet was required to have lifeboat accommodation for 17.2 per cent of capacity. They provided 20.4 per cent. (Cunard provided about double the lifeboat accommodation required and went far beyond rivals such as the Allan Line, Inman Line, National Line and Guion Line. The shipping lines all exceeded the requirements but mostly by much narrower margins.)

The committee painted a bleak picture if a sinking ship's passengers had to take to the lifeboats:

> It is obvious that though these boats may save life for the moment, they can but prolong its existence for a time equivalent to the quantity of stores and fresh water they can carry for each person onboard. In the Atlantic, in winter for instance, persons in boats would suffer greatly and probably die of starvation or exposure, unless a ship is lost in the track of other vessels which would, we fear, probably not be the case.[37]

Lifeboat accommodation for every person on board would be of little assistance if they were then left adrift on the open ocean:

> The fullest and most complete equipment of boats and rafts can be of but little avail if a collision or casualty happens to a ship which, owing to the absence of efficient bulkheads, sinks in a very short time and before it is possible to lower the boats. In fact, it appears to us that efficient boat accommodation, though an important thing in itself, is but one of several important things or conditions, which must all be present to make the whole an effectual means of saving life in case of disaster.[38]

They emphasised effective watertight subdivision to keep a damaged vessel afloat, as well as the necessity of trained seamen to operate the lifeboats, and for lifeboats to be well equipped and provisioned. 'Class E' ships ('passenger steam vessels carrying emigrants or others across the Atlantic to ports on the east coast of North America') were 'the most important of any, and we cannot pass over the fact that of late years this traffic has been carried on with remarkable immunity from loss of life'. There were:

> … many liberal and careful shipowners who do all in their power to provide for the safety of their passengers by equipping their vessels with

> boats far in excess of the number required by statute, while at the same time there are others carrying large numbers of emigrants who do no more than they are required to.

They recommended 'that the boats required [...] should be increased 100 per cent, and in addition to them that the owners should be induced to carry sufficient collapsible boats and approved rafts, so that each ship shall have sufficient lifesaving gear for all on board at one time'. The 'room for the extra boats could be found with very little trouble and expense beyond the cost of the boats and their fittings'. Another recommendation was that lifeboats could be carried 'on skids across the beam [breadth] of the ship [...] where they are available for either side'.

In July 1887, the Select Committee examining the broader issues of 'Saving Life at Sea', led by Lord Charles Beresford, issued their own report. Their remit covered lifeboats, lifebuoys and other lifesaving apparatus. They had plenty of expertise at their disposal to inform their recommendations, including Captain Theodore Cook, the commander of Cunard's *Etruria*: 'I am senior officer in the service; I have been in the service now 46 years, since 1841; and I have commanded 24 ships.' In an 'extreme case', with his ship sinking quickly, he argued, 'The great mistake is in abandoning the ship too soon; and it is the rush to the boats that causes the panic.' He was happy with the number of lifeboats: *Etruria*'s could accommodate 800 people compared to the average of 'about 1,000' carried across the Atlantic. They 'could not carry any more boats than we have or larger boats, but we could carry collapsible boats'. He did not think his ship would sink quickly, even if she was badly damaged. 'I think that the compartments that are watertight would keep the ship afloat for several hours, no matter where she was struck; at least that ought to be the case.' The watertight bulkhead doors were 'closed every day throughout the ship by the chief engineer in his department, and by the chief officer in the sailing department', according to Cunard's orders:

> Lord Charles Beresford: So far as your own experience goes, have you no recommendation to make to the committee?
>
> Captain Cook: None whatever. I think everything is done that can be done at the present moment, as far as my own judgement goes, and so far as our own company goes. Of course I am only speaking of the Cunard company.

Cook did not think more standard lifeboats could be carried safely on large passenger ships like his. Cunard did 'rather more' than required by law – 'I think we are in some respects quite exceptional.' He said that collapsible boats could be added to make up the lifeboat capacity and he preferred boats under davits, rather than stored elsewhere and moved for lowering.

Cook suggested the mix of people on board had an impact on the success of an evacuation, particularly if there was a greater proportion of 'first cabin passengers'. 'I should have much less difficulty with them than if I had 1,000 steerage passengers, because they are more intelligent and would obey instructions more readily in case of an accident.'

The committee's report found the existing requirements were inadequate: 'The existing law does not compel any passenger ship over 1,500 tons to carry more than seven boats, while in cargo vessels the boats are often considerably over the number which could possibly be useful to the crew in an emergency.'[39] They suggested 'rafts might be usefully employed in an emergency' and would be 'especially valuable in moderate weather as a temporary means of flotation'. Weather conditions were key:

> Many passenger ships could not, without great inconvenience, carry so many of the ordinary wooden boats as would suffice to carry the whole of the passengers and crew with safety in bad weather. Under such circumstances the crew would not be sufficient to man so many boats; nor could they all be got into the water in sufficient time in the event of very rapid foundering.

> [...]
> Having regard, however, to the fact that accidents occur probably as often in moderate weather as in bad; and having regard also to the fact that the very cause of the accident frequently incapacitates many of the boats, and to the further fact that an insufficiency of boats undoubtedly tends to cause panic, we are of the opinion that all sea-going passenger ships should be compelled by law to carry such boats, and other life-saving apparatus, as would in the aggregate best provide for the safety of all on board in moderate weather.

They left the question of 'the description of boats, and other lifesaving apparatus' for further discussion.

Although watertight subdivision was not in the committee's terms of reference, they argued 'that the proper placing of bulkheads, so as to enable a ship to keep afloat for some length of time after an accident has occurred, is *most important for saving life at sea, and a thing upon which the full efficiency of lifesaving appliances largely depends* [author's emphasis]'. These remarks reflected the evidence they had heard from Sir Digby Murray, the Professional Member of the Marine Department of the Board of Trade. His comments were unambiguous:

> I think all I can say is that the old rules on this subject are entirely obsolete, and we are perfectly well aware at the Board of Trade that they are obsolete. So long ago as 1875 I drafted a fresh schedule for the boats; but the Merchant Shipping Bill which was introduced into the House of Commons was finally given up [...] and the consequence has been that the boat scale has never been amended.[40]

He thought watertight subdivision made ships 'perfectly safe':

> The *Britannic* especially has been saved twice by her bulkheads; once with two compartments full; this last time with one compartment full. And I need not tell you that there is very great risk to life even in smooth weather the moment you have to lower your boats [...] I think you can make ships perfectly safe by subdivision of them; I think the *Britannic* has amply proved that.[41]

'My own impression is, that if you subdivide a ship properly she ought to remain afloat under all circumstances,' he suggested:

> I think the *Britannic* abundantly proved that when she had two compartments full of water, and had a rock through the bottom of her engine room; and she has been in collision lately about 300 miles from New York with the *Celtic*; the *Celtic*'s bows were twisted right round, and the *Britannic* had one compartment filled with water, and still the two ships had no difficulty whatever in getting back to New York. I think that proves that that is the sensible way of providing for safety of life on board ship.[42]

He acknowledged the case of Cunard's *Oregon* the previous year, '*Oregon* foundered after remaining afloat for some time with two compartments pierced, and foundered because they were unable to close a door in one of the bulkheads.' All were evacuated successfully, whereas others had not been so lucky. Cargo ships with ample lifeboat provision had also been lost with all on board because they were unable to summon help. Even though he recommended more lifeboats:

> I do not think you are going to save life by increasing the boats; I think it must be done by the subdivision of the ships. If you really want to save life that is the only way of doing it, and in that case a smaller number of boats would suffice.

The Merchant Shipping Act of 1894 included a table showing the minimum lifeboat requirements. For emigrant ships of 10,000 gross tons and upwards, the Board of Trade required sixteen lifeboats under davits 'with a minimum capacity of 5,500 cubic feet' (about 550 people). If that 'did not furnish sufficient accommodation for all

White Star's first *Celtic* collides with their first *Britannic* in foggy conditions at sea. This sketch is by George Allen Rudd, one of *Britannic*'s passengers. *Illustrated London News*, 1887. (Author's collection)

on board' then there was a further rule to comply with. *Olympic* therefore needed to carry boats or rafts with an additional 75 per cent 'more than the minimum' (4,125 cubic feet), taking her to a minimum of 9,625 cubic feet (about 962 people). In an echo of Sir Digby Murray's comments, the regulations included a 'Rule 12', which provided the potential for a ship that was 'divided into efficient watertight compartments to the satisfaction of the Board of Trade' to carry fewer lifeboats than otherwise required.

The basis of 'efficient watertight compartments' was taken as the highest standard of Sir Edward Harland's committee. It was an incentive of sorts to encourage improved watertight subdivision at the cost of permitting fewer lifeboats, but in practice, shipowners very rarely sought the exception. (Even if a shipowner applied for a Rule 12 exemption and the Board of Trade declined, the Board of Trade did not have the power to enforce a higher standard of watertight subdivision.)

Basing the requirements on the size of ships was problematic, because it meant the legislation would need to be updated as ships grew larger. (John Wilks, President of the Royal Humane Society of Australasia, had spoken to senior Board of Trade officials during his visit in 1882–83. He recommended that the regulations should be based on 'the maximum number of passengers and crew allowed to be carried, and not upon the tonnage of ships'.) The law came into force when Cunard's *Lucania* was already 30 per cent larger than the largest category. By 1907, *Lusitania* and *Mauretania* were three times the size. They were equipped with sixteen lifeboats under davits. Leonard Peskett recalled:

> When the question of boat capacity of such ships as the *Mauretania* and *Lusitania* was brought forward, the special subdivision of those particular vessels was taken into account, and it was considered that owing to the extraordinary precautions which had been taken, the total capacity of boats necessary to be carried would be fully met by the existing Rules.[43]

The Engineer illustrated that *Celtic*'s bow was badly damaged and *Britannic* was struck abaft her engine room. *Scientific American* reported that *Celtic*'s forward compartment and one of *Britannic*'s compartments flooded, but they both remained afloat and stable, 'The White Star company may well be satisfied that [...] nearly two thousand lives have been saved and two magnificent vessels are still safe.' *The Engineer*, 1887. (Author's collection)

THE WELIN QUADRANT DAVIT

It was Alexander Carlisle's turn to consider what was appropriate for White Star's new liners. The 'Design "D"' concept that had been approved at the end of July 1908 included a total of sixteen lifeboats, too. However, many details had not been finalised and were still subject to change. At some stage during the design process, the foremost four lifeboats from the row of eight lifeboats on each side were relocated to the fore end of the Boat Deck. This placed them on the officers' promenade rather than the first-class promenade. Carlisle 'considered that either the Board of Trade or the owners would sooner or later fit more boats to their ships'.[44] He therefore looked at improving the lifeboat-lowering apparatus.

Ernst Axel Martin Welin (known as Axel Welin) was born in Stockholm, Sweden, on 10 November 1862. One of his brothers was the commander of a Swedish freighter and he expressed concerns about the effectiveness of the equipment for launching lifeboats. These concerns prompted Welin to work on a better alternative. By December 1900, he had filed a patent application for 'Improvements in Apparatus for Raising and Lowering Ships' Boats' in the United Kingdom. On 26 August 1901, he amended the specification, adding details and drawings of his 'quadrant davit'. He was awarded the patent on 30 November 1901. (Welin filed similar patent applications in the United States, which were successful.) Welin's design enabled two men to launch a lifeboat, rather than six or eight under the older radial davits.[45]

Shipowners were slow to adopt Welin's system, but he became founder and honorary president of the Swedish Chamber of Commerce in London in 1906. An international exhibition in Italy also gave him the chance to promote the design. The Board of Trade 'formally approved' the design for use on British ships the same year. It was already being manufactured in Germany, Sweden and Austria.

When the British freighter *Star of Japan* foundered in April 1908, her captain reported a distinct contrast. Two of her lifeboats had been fitted under Welin davits but the others were under the old radial davits. The old davits 'proved useless' and the commander attributed 'the saving of every soul on board' to the 'efficiency of the Welin davits'.[46] These observations accelerated adoption of the Welin's design.

In 1909, Welin improved on his design so that he had a 'double-acting quadrant davit', which could service two lifeboats under one set of davits. In an emergency, the outer boat would be swung out, loaded and lowered. Then the davits could swing back inboard for the second, inner boat, so that it could be filled and lowered. Harland & Wolff opted to use these davits on the Union Castle Line's *Balmoral Castle* and *Edinburgh Castle*. (In April 1910, the *Belfast News-Letter* referred to *Edinburgh Castle*'s 'full equipment of Welin davits, which ensures the safe and speedy launching of the boats'.)

Carlisle wanted to go further, with a davit design that could service more than two lifeboats: 'We made a rough design in Belfast ourselves, and then I sent it to him [Welin] to draw the kind of davit he would recommend.'[47] Welin fleshed out a davit design which made it 'possible to double, or even treble, the number of boats, should such increase ultimately prove to be necessary'.[48]

Two key meetings were held in October 1909 and January 1910.[49] Carlisle recalled that he 'came over' from Belfast in October 1909, to attend a meeting with Pirrie, Ismay and Sanderson. Harland & Wolff and White Star both had offices in Cockspur Street, London.[50] Their meeting lasted 'all day' in October 1909 and they 'were talking about two or three thousand things'. They spent 'about four hours together'[51] discussing *Olympic*'s decorative schemes in detail and then spent 'five or ten minutes' on the question of lifeboats in January 1910.[52]

Carlisle explained the advantages of the new davits. There would be 'no expense or trouble in case the Board of Trade called upon them to do something at the last minute'.[53] If the regulations required more lifeboats in future, they could be accommodated by the existing equipment. Ismay agreed that it was a good idea to install the davits.[54] Carlisle was authorised to get 'full plans and designs'.

No decision was taken at that stage about the number of boats that would be supplied for the finished ship. Carlisle recalled, 'The plan showed how the pair of davits held them, but the number of boats was not gone into.'[55] One plan 'showed one pair of davits that you could work four boats on'. 'There was no plan submitted showing sixty-four boats,' he said, 'It was only the set of davits that were shown.' He did not highlight to Ismay or Sanderson 'that as many as 64 lifeboats could be put on'. 'I suppose, if they had gone into it, they could have seen it, but the number was not gone into at the London or Belfast meetings.' (Edward Wilding believed that the Welin davits could 'only handle two' 'full sized lifeboats', 'if the lifeboats were of a collapsible character they could handle three or even four'.)

Carlisle later said that he felt there should be three lifeboats under each set of davits, for a total of forty-eight, but he made no recommendation. Instead, he 'put the plans before the owners and

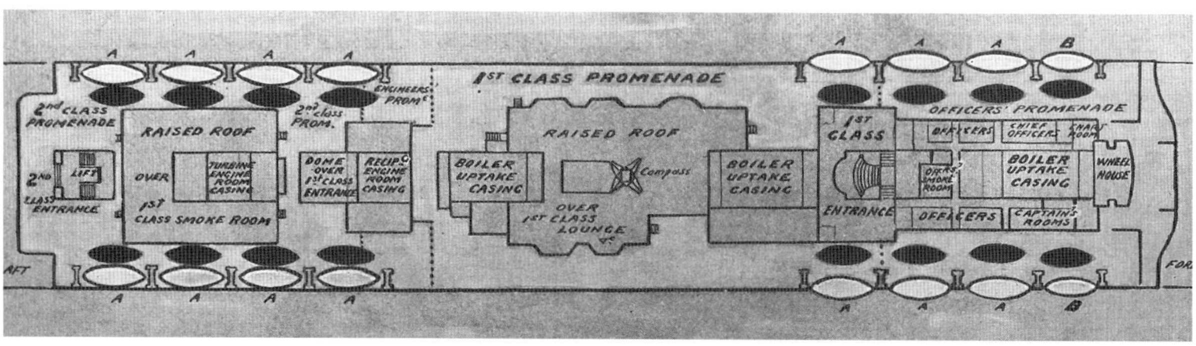

A diagram by W.B. Robinson depicting the layout of *Olympic* and *Titanic*'s Boat Decks. The outline of the lifeboats in black illustrated how the new double-acting Welin davits could accommodate more than one lifeboat for each set, but it would have had little impact on the first-class promenade area. *Illustrated London News*, 1912. (Author's collection)

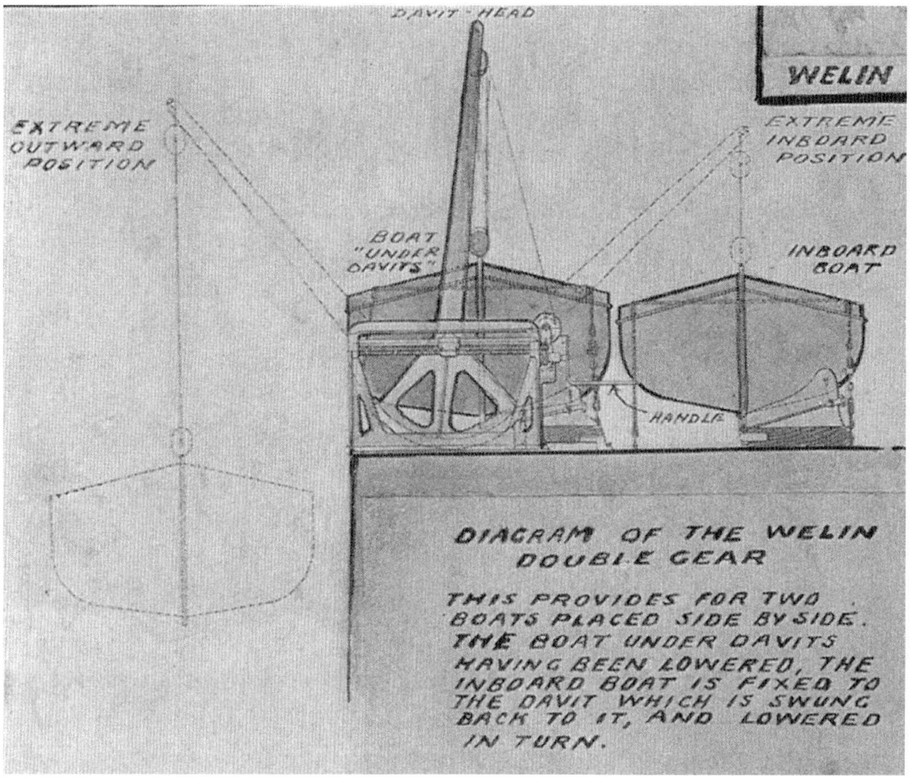

The Welin davit design allowed for one lifeboat to be lowered, then the davits could be swung back inboard to lower a second lifeboat. *Illustrated London News*, 1912. (Author's collection)

let them judge', even though it was the sort of detail they left to the shipbuilders. Carlisle's subsequent comments were also somewhat surprising. The starting point of sixteen lifeboats was in line with other large ships. Why, then, would he have felt that *Olympic* and *Titanic* should carry so many more lifeboats compared to a ship such as *Adriatic*?*

The new davit design needed to be submitted for regulatory approval. By 9 March 1910, Welin had submitted to the Board of Trade his design for double-acting davits which would be used on board *Olympic* and *Titanic*. Seven days later, an official wrote:

* Several years later and in circumstances which were unimaginable in 1909–10, Carlisle was questioned publicly about his views before the British Wreck Commissioner's Court. He explained:

21379. You thought there ought to be three times the number. Did you say so?
I believe I did, but I could not swear.

21380. But it is a very important matter, is it not?
You see I never put my ideas on paper unless I thought they were what should be carried out.

21381. You do not answer my question. You were there apparently discussing this matter. Did you say, as the chairman of the managing directors of Harland and Wolff's, 'I think there ought to be three times as many boats on that deck as we are at present contemplating putting there'?
No, I would not say that I did.

21382. Did you think it?
I thought there ought to have been.

21383. Why did you not say so?
Because I have always been accustomed to put the plans before the owners and let them judge. Unless they asked questions I did not give them an answer.

21384. Do you mean to tell me that on this important matter, having formed the opinion, and the matter being discussed and you being the chairman of the managing directors of the builders, you did not say that?
Certainly not.

> The davits are of the double-acting type and Mr. Welin desires them to be accepted as a standard equivalent to the "A" standard of single-acting davits [...] for boats up to 32 feet [long] by 9 feet [wide]. The davits are fully as strong, or stronger, than those referred to, and [...] might, I think, be approved subject to the various tests suggested [...] being satisfactorily carried out.

A. Boyle noted on 16 March 1910:

> The *Titanic* and *Olympic* are each to be fitted with 32 boats, which are to be carried under 16 sets of double-acting davits, 8 on each side. Sixteen of the boats will be placed inboard of the others, and all the davit frames are to be double ones. Only 8 of the sets of davits onboard therefore can be operated simultaneously. Having regard, however, to the great number of boats and davits to be provided in these two cases, the double frames might be considered as complying with the regulations [...] sixteen boats under davits only are required.[56]

The source of the information that both ships would be fitted with '32 boats' is not clear. It appears to have been a natural assumption under the circumstances, because the ability to carry more boats was the key feature of Welin's davit design. Welin had produced a plan showing how each set of davits could accommodate two or more boats and the inversion of the ships' names in Boyle's comments, with *Titanic* first instead of *Olympic*, mirrored correspondence and plans from Welin's company. (It was not until 27 June 1910 that Boyle actually requested that 'the surveyor [Francis Carruthers] might report how this case stands stating how many boats are to be carried on each of these vessels, and he might obtain a plan showing the deck arrangement'.)

The issue of 'double [davit] frames' caused some concern: the design allowed for the davit frame between two lifeboats to have a davit arm on either side, rather than having to fit two single frames which each had a davit arm. The concern was that with double-framed davits, 'only eight boats can be put out simultaneously'.

'This seems a very small number for such large ships and it is for consideration whether it can be regarded as complying with the spirit of the rules,' another official wrote on 22 March 1910.

A question came up as to whether the inboard lifeboats could 'be considered as being under davits', but it was postponed for later. (The outboard lifeboats already met the regulations for the number of lifeboats under davits on their own.) Welin 'brought a very good model of his double-acting davits' to show to the Board of Trade. The 'means for operating the boat chocks appeared to be very simple and efficient' and Welin's design was approved.

Once Harland & Wolff knew the Board of Trade were happy with Welin's design, they ordered the davits in June 1910.[57] They received the Board of Trade's request for confirmation of how many lifeboats would be carried and a plan of the Boat Deck on 29 June 1910. Harland & Wolff duly submitted a blueprint dated 30 June 1910 (coincidentally, the same day that Carlisle retired). The plan showed the layout of *Olympic*'s Boat Deck, with a total of sixteen boats.

Submitting a plan seems to have been unusual. Cunard's experience was that 'under ordinary circumstances, a boat plan would not be submitted to the Board of Trade'. The lifeboats would simply be fitted 'under davits, and these would be passed by the local surveyor as being in accordance with the regulations'.[58] If they had been using the standard radial davits, then it seems unlikely the Board of Trade would have requested the plan from Harland & Wolff at all.

A decision finalising the number of lifeboats had not been taken before Carlisle left. After Andrews took on Carlisle's duties, Edward Wilding did not recall any 'proposal made that boats should be fitted, or even a plan prepared showing accommodation for everyone onboard'.[59] He saw 'plans prepared for discussion amongst ourselves which did show two boats under each pair of davits' and thought that 'there was a plan submitted to us by Welin showing three boats', rather than one that Harland & Wolff had prepared. On referring to it again, he realised the plan showed 'two drawn and one dotted, giving the impression of three boats'.

One blueprint, which appears to date from around March 1910, showed an 'arrangement of Welin davits and additional boats', including sixteen boats (seven ordinary ones on each side and an emergency cutter), but with an *additional* sixteen (eight on each side) *inboard*. The inboard boats were smaller than their outboard counterparts and had a capacity equating to about fifty people per boat rather than sixty-five. The 'proposed additional boats' would have increased the total capacity in cubic feet by about 82 per cent compared to the original design concept, accommodating about 1,770 people rather than about 970 people.[60]

The ability to carry three boats under each pair of davits was not in question, but Harland & Wolff were 'discussing amongst ourselves at Belfast' whether it was 'a desirable arrangement or not'. They came to the view that 'it was not one which we cared to recommend to the owners', on the basis that a lesser 'number of boats was sufficient for the purposes for which they were most likely to be wanted'. It was one of 'a great many matters of technical detail' left by White Star to Harland & Wolff 'as their experts for decision'.[61]

White Star's approach was to follow what the shipbuilder recommended. Harold Sanderson said:

> I recall that at one of our interviews with Lord Pirrie the question of the number of lifeboats that would be supplied was referred to, but it was only referred to in general terms [...] I cannot recall that any opinion was expressed one way or the other as to the number that Messrs. Harland & Wolff would supply.

Their practice was to exceed the regulatory requirements, 'additional boats were put on, to be on the safe side'.[62] Ismay had 'no recollection of seeing any design which showed the *Titanic* [or *Olympic*] fitted for up to 40 boats'.[63]

Shipowners and builders did not necessarily spend a lot of time thinking about lifeboats. They were just concerned that they met the regulations. For example, the draft specification for the Holland America Line's *Nieuw Amsterdam* dated 23 September 1903 went into substantial detail about the ship's passenger accommodation and equipment and had a single sentence about 'boats' – 'of number and size as may be required to pass the requirements of British and American law'. Later, when the lifeboats were being finalised, they asked Harland & Wolff, 'Would it not be advisable to supply 28 foot 6 inch lifeboats instead of cutters. According to our opinion cutters are not of much use on an ocean steamer.' Harland & Wolff responded:

> The only difference between lifeboats and cutters is, that the latter [recte: former] are provided with buoyancy tanks and other fittings, which of course, makes them heavier and more expensive, and we do not consider it worth the additional cost. It is quite unusual to have all boats fitted as lifeboats.

Ismay believed that lifeboats might be needed 'to pick up crew from another ship'.[64] Sanderson did not think that 'under any conceivable circumstances', all of the passengers and crew would 'be required to be put afloat in boats' or that 'even perhaps 99 times out of a hundred', it could be accomplished successfully:

> The weather conditions in the Atlantic are such that I should look upon it as a very remote contingency, and one to be avoided at all costs. Therefore, in my judgment, I would rather devote myself to [...] make her so safe that we would not have to consider the possibility of putting all these people afloat ...[65]

He was aware of the unusually large number of watertight bulkheads on *Olympic* and her sister. The pumping arrangements were 'exceptional', with 'each boiler compartment having its own equipment, which is quite an unusual thing'.[66]

Sanderson's views on the practical difficulties of evacuating all the ship's passengers and crew were shared by Edward Wilding. Leaving aside the weather, another logistical problem in evacuating a sinking ship was that she would, in all likelihood, be listing to one side. Even a list of a couple of degrees made it that much harder to successfully prepare, fill and lower a lifeboat. A more serious list might put all the lifeboats out of action on one side of the ship. (Leonard Peskett also considered that, in an emergency, it would be easier 'to control the passengers and get them away onto boats from the gangway doors' on a lower deck, rather than 'have a rush of four thousand people on the boat deck, struggling to get into the boats, which would probably be available on the one side only'.)[67] All these concerns supported the view that the best way to keep passengers and crew safe was by trying to make sure they would never need to be evacuated on the open sea.

When *Olympic* was launched, the *Belfast Telegraph* reported:

> The vessels are to be fitted throughout with double-banked boats, operated by the new Welin gear, the first twelve sets of which type were fitted on the new steamers recently ordered by the Union Castle Line [...] The davits are of the double-acting type, each vessel carrying sixteen sets, or enough davits to handle thirty-two lifeboats.[68]

On 19 November 1910, Francis Carruthers reported to his superiors at the Board of Trade:

> There are 14 boats of section A and 2 boats of section D. These I have measured, and the dimensions and capacities are as follows:
>
> A boats; 30' X 9' X 4' = 648 cub. ft.
> Total for 14 boats = 9,072 cub. ft.
>
> 1 D boat 25.2' X 7.1' X 3.15' = 337 cub. ft.
> 1 D boat 25.2' X 7.2' X 3.15' = 343 cub. ft.
>
> The total capacity of the boats is therefore 9,752 cub. ft. The total capacity required by the regulations is 9,625 cub. ft. The davits, which have arrived at the yard but are not yet fitted onboard, are of the Welin double-acting type [...] and the arrangement is such that an additional 16 boats could be carried inboard of those shown on the attached print and operated by means of these davits.

Olympic was almost finished when, on 5 May 1911, Carruthers reported that 'a complete set of double-acting davits' had been fitted on the Boat Deck:

> The davits and gear have been tried, all the boats being swung out and lowered, five into, and the remainder within a few feet of the water. All the gears worked smoothly and well. The time taken to swing a boat out of the chocks and into the water[,] a depth of about 58 feet arranging from 1 minute 40 seconds to 1 minute 55 seconds.

When necessary, any lifeboat could be released at each end simultaneously so that it was free of the falls, by means of Murray's disengaging gear.

Four days later, one of the boats was loaded with enough weights to represent about sixty-five people. Although they were distributed to match real loading conditions, the object was to test the electric boat winches, rather than the strength of the boat. The boat was lowered and then raised no fewer than six times. 'I do not doubt the boats were strong enough to be lowered containing the full number of passengers,' Edward Wilding noted.[69]

On 5 May, more boats were ordered from McAllister & Son in Dumbarton, Scotland.[70] Four 'Engelhardt semi collapsible lifeboats [...] each with outer skin of mahogany' were ordered for *Olympic*. The first two were sent to Belfast on the 24th by a G. & J. Burns steamer from Greenock and then the other two followed the next day.[71] On 2 June, Carruthers confirmed that an additional 'four collapsible boats with a total cubic capacity of 1,584 feet were fitted onboard in such a position that they can readily be put into the water by the forward davits'. The result was that the cubic capacity of all the lifeboats was almost 18 per cent more than the law required.

The lifeboats themselves were typical of the period. Sixteen boats were constructed by Harland & Wolff and designed by Roderick Chisholm.[72] They were wooden with copper buoyancy tanks. Fourteen of the boats were 30ft long. These were numbered from 3 to 16, with the even numbers on the port side and the odd numbers on the

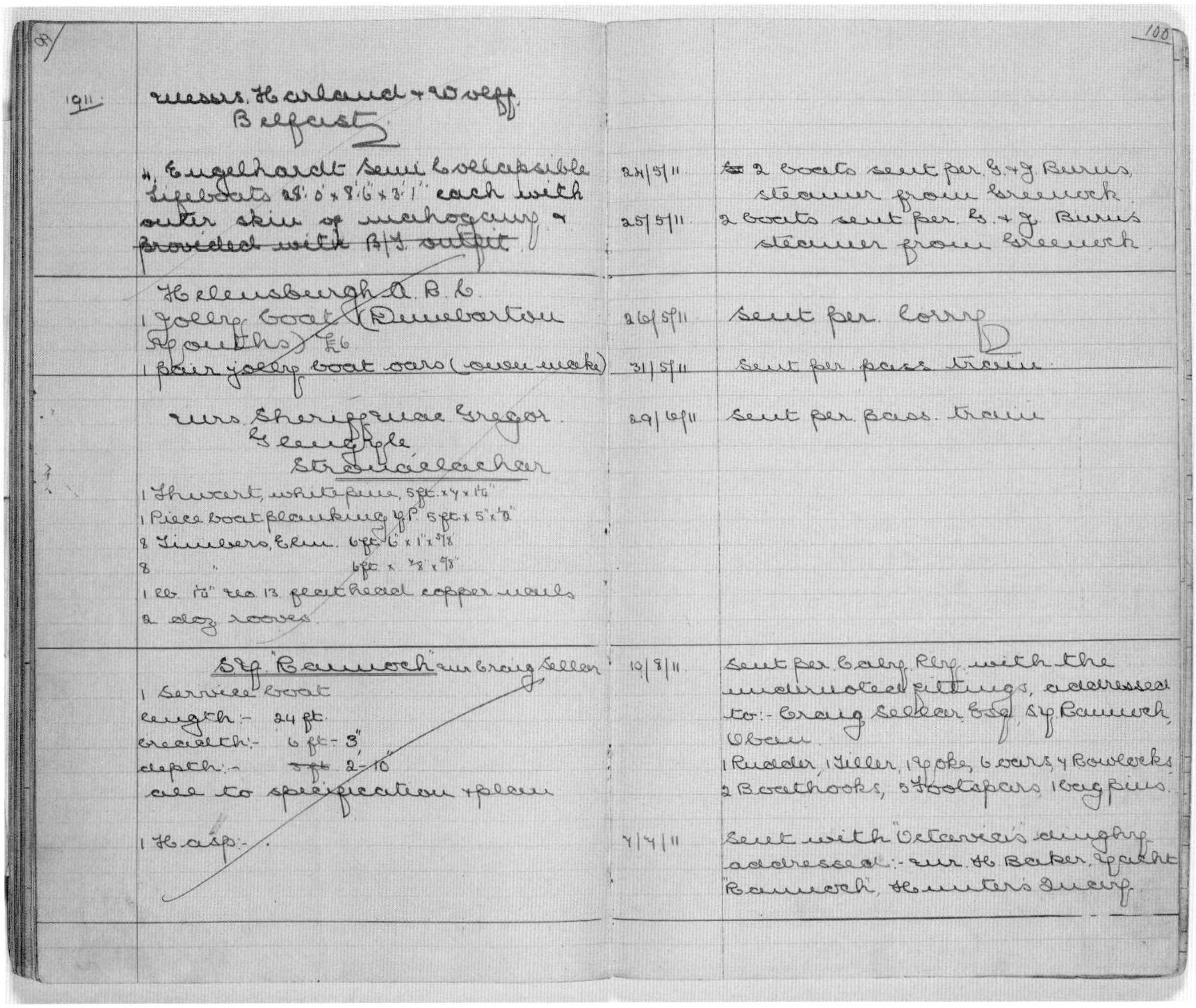

Harland & Wolff placed an order with McAllister & Son of Dumbarton, Scotland, for four 'Engelhardt semi collapsible lifeboats' for *Olympic* early in May 1911. The first two were despatched to Belfast on 24 May and the other two followed a day later. By 2 June, they were placed on board. (Glasgow City Archives)

starboard side. They were arranged in groups, with 4, 6 and 8 situated on the forward port side of the Boat Deck; further aft were 10, 12, 14 and 16. On the starboard side, boats 3, 5 and 7 were situated at the forward end of the Boat Deck, while boats 9, 11, 13 and 15 were installed at the after end. The forward end of the deck was reserved for the officers' promenade and the after end consisted of the second-class deck space and engineers' promenade, meaning that the arrangement of the lifeboats kept the first-class deck space relatively free and open.

Two of the boats were smaller cutters. They served as 'emergency boats'. Numbered 1 and 2, they were positioned immediately behind the bridge on the starboard and port sides respectively. As a rule, they would be kept swung out in the event that they needed to be lowered quickly, such as trying to rescue someone who was lost overboard.

Calculating the number of people that all the lifeboats could carry was a little complicated, because it involved determining their capacity in cubic feet and then assigning a ratio of how many cubic feet to each person. The cubic capacity itself could vary depending on the method of calculation. Under Stirling's Rule, which was used on Board of Trade survey documentation, the lifeboat capacities were:

Boats	Length (ft)	Width (ft)	Depth (ft)	Capacity (cubic feet)	No. Persons
14	30′	9.1′	4′	655.2	65
1	25.2′	7.1′	3′	322.1	40
1	25.2′	7.2′	3′	326.1	40
4	27.5′	8′	3′	376.6	47

These figures differed slightly from Carruthers' 19 November 1910 report but agreed with other Board of Trade documentation. (The lifeboat dimensions themselves also differed depending on the source, either due to rounding or a slightly different method of measurement.) The capacities displayed on the plates fitted to the lifeboats took a conservative approach and rounded down:

Boats	Dimensions (ft)	No. Persons
14	30 x 9 x 4	64
2	25 x 7 x 3	33
4	28 x 8 x 3	40

Using those figures, the lifeboats could accommodate a total of 1,122 people, but their capacity totalled 1,178 people under Stirling's Rule.[73] Compared to the 'Design "D"' concept approved in July 1908, the number of lifeboats had been increased. The number of people the ships were designed to carry was also reduced during the design process. Both these changes effectively raised the lifeboat capacity from about 25 per cent to over 34 per cent of the maximum number of passengers and crew.[74] During an average crossing, the lifeboat capacity would be over 50 per cent of the number of people aboard.

***

There was increasing interest in the issue of lifeboats. On 19 November 1910, the journal *John Bull* carried an article about *Olympic* headlined 'A Serious Insufficiency of Lifeboats'. It drew attention to 'the model of the *Olympic*, a monster steamship of 45,000 tons now building for the White Star Line', which was on display at the Piccadilly Hotel in London. The writer argued she would 'have only sixteen boats', mistakenly saying that her Cunard rivals carried 'forty boats and rafts':

> How, then, can sixteen boats be sufficient for the *Olympic*? The explanation given by the company is that the ship is practically unsinkable, and that for lifesaving purposes boats will not be required. That is a most unsatisfactory explanation. No vessel can be considered unsinkable if it meets with certain concussions [...] What does the Board of Trade say?

The magazine's founder, Liberal MP Horatio Bottomley, was in a good position to raise his concerns. Three days later, he asked about *Olympic*'s lifeboats in the House of Commons. He addressed the President of the Board of Trade, Sydney Buxton, enquiring:

> ... whether his attention had been called to the fact that the new White Star liner, the *Olympic*, now being built, and a model of which was being exhibited in Piccadilly, was provided with fourteen lifeboats only, being less than a quarter of the number furnished to most other vessels of similar tonnage; and whether he would make a representation to the owners on the matter?[75]

Harold Tennant, the Parliamentary Secretary for the Board of Trade, replied on the Board of Trade's behalf:

> I understand that the *Olympic* will be provided with fourteen lifeboats and two ordinary boats, of an aggregate capacity of 9,752 cubic

> feet, which is in excess of the requirements of the statutory rules. I have no information as to any vessel carrying four times this number of boats. The *Lusitania* and *Mauretania* each carry sixteen boats.

John Bull printed another article quoting the answer. It repeated the claim that *Lusitania* and *Mauretania* 'have more boats', toning it down somewhat, but apparently excluding *Olympic*'s two emergency boats from the comparison. They gave some comparisons of other liners, including some smaller ships which carried fewer passengers and had the same number of lifeboats or even more:

> We invite Mr. Buxton to explain how in the face of these facts and figures the Board is satisfied [...] We tell Mr. Buxton, without fear of contradiction from any competent authority, that if he allows the *Olympic* to go to sea with only her present number of boats he will have a weighty responsibility to reckon with, should disaster befall the ship.

Seven days later, they ran another article stating:

> We are glad to be able to say that the owners [White Star] have reconsidered the position, and have decided to install four quadruple armed [*sic*] davits, each of which will swing four boats, making sixty-four lifeboats in all, and in addition there will be the present dinghies. We congratulate the owners.

The source of the information was not made clear and the claim was inaccurate. The davits had been agreed almost a year previously.

On 24 December 1910, *John Bull* printed a letter from an anonymous 'naval architect' (Axel Welin), 'It touches upon a matter which is nothing short of a great scandal.' The commentary preceding Welin's letter noted that the Board of Trade's regulations were 'framed a great many years ago, when there was no steamer in the water of more than 10,000 tons – indeed none approaching that tonnage. This maximum requirement remains today, when there are ships of 50,000 and 55,000 tons being built.' The Aberdeen Line's *Pericles* had foundered 'in a calm sea and fair weather. Her passengers and crew numbered 600. Happily, she had twelve boats, which were more than sufficient to take off every man on board'. If *Olympic*:

> ... with sixteen boats only, had been in the same position, and had had 3,400 souls on board [...] these boats could only have taken off less than 1,000 of the passengers. 2,400 souls must have gone to the bottom of the ocean – assuming, of course, that there was no vessel near enough to render help – in fine weather and a calm sea.

Welin wrote:

> Towards the end of 1909, Mr. A.M. Carlisle, general manager of Harland & Wolff, Ltd., and myself, went very carefully into the question of boats and special davits for large steamers, as we both felt that the time was not far distant when the Board of Trade would feel bound to move and revise their rules, not only as to boats carried, but concerning other special fittings for the safety of the passengers and crew. We accordingly worked out, under my patents, special davits for these ships, each pair capable of carrying three or even four ordinary wood or steel lifeboats. These davits are actually being fitted throughout the vessels in question, and although I have personally no idea of how many boats the builders or owners are likely to fit in these ships prior to their departure, yet the very provision of such gear is the *prima facie* evidence of their desire to be not only up to date, but well ahead of present requirements.[76]

Bottomley asked another question in Parliament on 15 February 1911. He asked if the Board of Trade's President would 'state the date of the last regulations made by the Board in reference to the number of lifeboats necessary to be attached to passenger vessels, and whether, having regard to the increased tonnage of modern ships, he will consider the desirableness of revising such regulations'.[77] Buxton replied, 'Those regulations were last revised in 1894. The question of their further revision is engaging the serious attention of the Board of Trade, and I have decided to refer the matter to the Merchant Shipping Advisory Committee for consideration and advice.'[78]

Sir Alfred Chalmers had considered the issue of lifeboat regulations 'from time to time', but 'it has at no time appeared to me to be urgent'. 'It has always been the case that the boat capacity was equal to only thirty percent of the passengers and crew carried, so the principle is not a new one', he noted.[79]

Chalmers looked at the shipping trade and 'found it was the safest mode of travel in the world'. He thought ships were 'stronger and better ships, both from the point of view of watertight compartments and also absolute strength'. He thought that the maximum number of lifeboats required was already 'the maximum number that could be rapidly dealt with at sea and that could be safely housed without encumbering the vessel's decks unduly'. The routes agreed upon by the shipping lines reduced the risk of collision. Wireless was another consideration. *Lucania* had it fitted in July 1901 and Cunard followed with the rest of their fleet; the Allan Line fitted it in 1902, and by 1904 it was widespread. Ships could warn each other of danger and call for help:

> Another point was the manning. It was quite evident to me that if you went on crowding the ship with boats you would require a crew which were not required otherwise for the safe navigation of the ship, or for the proper upkeep of the ship, but you are providing a crew which would be carried uselessly across the ocean, that never would be required to man the boats. Then the last point [...] was [...] that the voluntary action of the owners was carrying them beyond the requirements of our scale [...] I considered that that scale fitted all sizes of ships that were then afloat, and I did not consider it necessary to increase it.[80]

Nevertheless, three days after Buxton's response in the House of Commons, the Board of Trade asked their principal officers at Liverpool, London and Glasgow to consider lifeboat requirements. Alfred Young, at Liverpool, felt that the lifeboat capacity required should be increased. He prepared a scale for lifeboat capacities, noting:

> ... in any scale [...] that can reasonably be applied to the larger passenger vessels, *there must usually be a large number of persons onboard for whom there can be no boat accommodation* [author's emphasis], and therefore subdivision of the ship into real effective watertight compartments must be, in my opinion, the main factor to be relied on.[81]

Under his scale, ships between 45,000 and 50,000 gross tons – such as *Olympic* – would need to carry twenty-four lifeboats under davits with a cubic capacity of 10,900ft. An additional number of collapsible boats, or rafts, would bring the total capacity up to 19,075 cubic feet. That equated to accommodation for 1,907 people.[82]

Young argued that larger ships might carry fewer passengers than smaller vessels:

> Owing to the enormous expense involved in running such high-speed steamers as the *Lusitania* and *Mauretania* shipowners are arriving at the conviction that new vessels – particularly for the North Atlantic service – should be of less speed and carry more cargo, and it may be, therefore, that vessels of the type and size of the *Olympic* may be of 50,000 tons gross, and more, and yet carry fewer passengers and crew than vessels of 20,000 tons to 30,000 tons.

His comments were borne out by the fact that *Olympic*'s passenger and crew capacity had not increased proportionally to her size. Her passengers were allocated much more space per person. Young also shared Chalmers' concern about the number of crew needed to be able to man the boats:

> Even supposing there were enough sailors to man the boats, assuming for this purpose a proportion of firemen and stewards able to handle an oar, the question still remains – what number of boats would be deemed reasonable,

> and why should any particular number above the present scale be deemed necessary – seeing that the *Lusitania*, say, has been running for some time provided with boats sufficient only for 978 persons, leaving about 2,000 persons unprovided for save by lifebelts? Taking it for granted, however, that a considerable number of persons must be excluded from the boats under any circumstances, what we have to do apparently is to seek for a number of boats which will afford a greater assurance of safety to the travelling public.[83]

Young's counterpart in London largely agreed with his scale, but the officer in Glasgow recommended fewer: eighteen lifeboats under davits, with a capacity of at least 9,700 cubic feet, and additional boats to make the total capacity 16,975 cubic feet. Those would provide accommodation for around 1,700 people.[84]

By early April 1911, the Merchant Shipping Advisory Committee had been instructed by the Board of Trade to look into the matter. True to form, they decided to set up a subcommittee, which Alexander Carlisle was invited to join. They were asked to examine the question of lifeboat capacities for ships 'up to 50,000 tons gross and upwards' and whether to revise 'Rule 12', which took into account a ship's watertight subdivision to see whether it could be exempted from requiring additional lifeboats. They met on 19 and 26 May 1911.

To keep their analysis independent, they were not told about the schemes Young and his colleagues proposed. One question was whether a new scale for lifeboats should be applied retrospectively to existing ships. Sir Norman Hill, chair of the Merchant Shipping Advisory Committee, thought that 'whatever we thought was necessary for the safety of life would have

Another illustration by W.B. Robinson showed several Cunard, White Star, Allan and Union Castle liners and compared their lifeboat provision with the maximum number of passengers and crew. The Allan and Union Castle liners had lifeboat accommodation for between 83 and 94 per cent of the people they could carry, but it was largely a reflection of their lower passenger capacity rather than a higher number of lifeboats. Figures for individual ships vary depending on the source and should be interpreted as an approximation only. *Illustrated London News*, 1912. (Author's collection)

to be enforced on all vessels, whether they were built then or whether they were going to be built'. His colleagues 'generally accepted that view', although Carlisle said 'two or three' times that 'it would neither be reasonable nor fair to extend any new rule to existing vessels'.

Carlisle 'submitted plans for a special davit to carry four boats' for discussion. The group 'as a whole, were not satisfied with them; in fact they preferred the arrangement [...] of stowing boats, such as decked lifeboats, under the ordinary type of lifeboat'. Towards the end, coming up to their report, Hill asked Carlisle if he recommended having four boats on a set of davits. 'No, I think you are right in sticking to the sixteen,' Carlisle responded (eight boats on each side of the ship). The lifeboat scale Carlisle brought to the meeting 'recommended carrying in addition boats readily available for launching' as well as sixteen boats under davits.[85]

One of the points in Hill's mind was the sheer difficulty in 'marshalling the people into the boats' in an orderly way during an emergency. 'You have got a huge hotel, a hotel with thousands of people on board, and how you can marshal those people into the boats is one of the greatest difficulties you can think of.'[86]

The subcommittee reported by the end of May that they were 'strongly of the opinion that every encouragement should be given to secure the provision of vessels which by their construction have been rendered as unsinkable as possible'. They recommended that 'any increase in the number of lifeboats to be carried can probably be best effected by providing for the launching of further boats from the existing davits'. On a ship of 45,000 tons and upwards, their table recommended a minimum of sixteen 'boats to be placed under davits', with a minimum of eight 'additional boats to be readily available for attachment to davits', for a capacity of no less than 8,300 cubic feet. They also proposed 'that vessels divided into efficient watertight compartments to the satisfaction of the Board of Trade should (provided they are fitted with wireless telegraphy apparatus) be exempt from the requirement of additional boats and/or rafts'. (Their proposals 'made a very slight improvement upon the requirements actually in force'. *Olympic* 'carried boats and other lifesaving apparatus in excess of the scale then put forward by the Advisory Committee'.)[87] Carlisle later said that his personal view was that more lifeboats were needed, making it somewhat hard to explain why he did not express his views more fully or forcefully, or why he signed the report. On reflection, he felt he 'was very soft the day I signed that'.[88]

Another recommendation in the report related to the dimensions of lifeboats: 'In future, the depth of lifeboats supplied to a British merchant vessel shall not exceed 44 percent of their breadth.' (The Board of Trade wanted to examine the recommended form and size of lifeboats first, before they made any change to the regulations. They were still looking at it in the spring of 1912.) Further recommendations included 'that all passenger vessels of 10,000 tons gross tonnage and upwards should be required to be fitted with wireless telegraphy apparatus'.

Olympic was now complete. *Titanic* had been launched and other ships were being constructed. Cunard's design estimates dated 26 November 1910 had envisaged their new *Aquitania* with much the same equipment as *Olympic* and *Titanic* – sixteen lifeboats under Welin davits. 'The sum of £1,400 was included for 16 ordinary lifeboats, and £2,100 for Welin's davits, a total of £3,500.'[89]

On 7 August, Leonard Peskett and a representative from John Brown & Co. met with Principal Ship Surveyor William Archer at the Board of Trade's offices in London. Archer told them privately that the Board of Trade were 'formulating a scheme to increase the total capacity of boats on all passenger steamers'. Peskett knew that Harland & Wolff had 'made provision for carrying double the boats required by the rules' by installing Welin's new davits on *Olympic*, even though she was 'considered to be well subdivided'. Archer advised that 'in view of the fact that a change is contemplated', Cunard should submit 'a plan showing [the] position and capacity of boats' proposed for *Aquitania* to the Board of Trade 'for information only'.[90]

By February 1912, Cunard's chairman, Alfred Booth, had concluded that there was no 'halfway house between the current rules and the supply of boats or rafts for every soul onboard'. *Aquitania* would need to carry lifeboats to accommodate everyone on board.[91]

Shipowners were already improving their ships' lifesaving apparatus and adding more lifeboats, but Peskett's view remained that watertight subdivision would 'do more for the safety of passengers and crew than the number of boats that may be carried'. Professionals recognised that the best way to preserve life was to keep a damaged ship afloat. As Harold Sanderson said, the brutal reality of having to evacuate an entire ship's company in the middle of the Atlantic was 'one to be avoided at all costs'.

SELECT CHRONOLOGY: LIFEBOATS

June 1894	Merchant Shipping Act scale for ships of 10,000 gross tons and above requires sixteen lifeboats under davits. If these boats cannot accommodate everyone on board, further boats or rafts are required to provide a minimum cubic capacity of 9,625 cubic feet (about 962 people).
July 1901	*Celtic* enters service, surpassing 20,000 gross tons.
September and November 1907	*Lusitania* and *Mauretania* enter service, surpassing 30,000 gross tons.
July 1908	Harland & Wolff's 'Design "D"' concept shows sixteen lifeboats under davits.
January 1909	White Star liner *Republic* sinks after a collision with *Florida*. The ship's passengers are evacuated before she founders.
October 1909 and January 1910	Anticipating that the lifeboat regulations will be revised in the near future, Harland & Wolff recommend a new type of davit which will enable the number of lifeboats to be increased substantially if needed. White Star authorise going ahead with obtaining new Welin davit designs and using them on *Olympic* and *Titanic*.
March 1910	Welin davit designs submitted to the Board of Trade for approval.
June 1910	Harland & Wolff order the Welin davits.
November 1910	Design estimates for Cunard's *Aquitania* include sixteen lifeboats under Welin davits.
November 1910 to February 1911	Following questions in the House of Commons, from April 1911 the Board of Trade start looking into the matter of lifeboat regulations.
May 1911	Four additional semi-collapsible Engelhardt boats are fitted on *Olympic*, increasing her lifeboat capacity beyond the regulations.
	Subcommittee report recommends a slight increase in lifeboat capacity requirements to the Merchant Shipping Advisory Committee.
July 1911	Merchant Shipping Advisory Committee reports to the Board of Trade and adopts subcommittee's recommendations. Questions about lifeboat form and dimensions delay the Board of Trade taking action to change lifeboat regulations until they can address both points.
February 1912	Cunard's chairman, Alfred Booth, concludes that *Aquitania* will need lifeboat accommodation for all on board.

4

FORGING AHEAD

At a time of rapid progress in marine engineering, Harland & Wolff faced a challenge as they considered how *Olympic* and her sister would be powered. They would be the fastest White Star liners to date. As far back as 26 September 1902, J. Bruce Ismay had been concerned that:

> We must not forget that the IMM Co. has only two steamers that are really capable of doing 20 knots regularly, and it seems to me those responsible for the future management of the company will have to carefully consider if it is not almost necessary to do something to improve their position in this respect.[1]

In 1895, only six of the 'principal steamships employed in the passenger trade of the United Kingdom with the east coast of North America' had a speed above 20 knots. The advent of *Lusitania* and *Mauretania* in 1907 and then *Olympic* in 1911 increased that to nine ships. From 1895 to 1911, the average speed rose from 13.8 knots to 15.2 knots. By contrast, the average gross tonnage roughly doubled to 11,722 gross tons and the total gross tonnage rose from 376,000 to over 1 million tons.[2] *Olympic* and her sister led the trend for increased size, but they would still be among the fastest British ships afloat. Their power requirements meant Harland & Wolff had to go far beyond any ship they had built previously.

***

Teutonic and *Majestic* were the first twin-screw ships built by Harland & Wolff. They were powered by triple-expansion reciprocating engines developing a total of about 17,000 horsepower. This was less than a decade after Harland & Wolff had established its own engine-building capability in Belfast.

When *Oceanic* was completed in 1899, her engine power was increased to over 28,000 horsepower. Her triple-expansion reciprocating engines had four crankshafts as opposed to three on the earlier ships. Arguably, *Oceanic*'s engines represented 'the most outstanding advance, in steam engines, ever made at Queen's Island in a single step' and they represented the starting basis for the design of *Olympic*'s.[3]

The performance of all three of the earlier ships was more than satisfactory. In fact, White Star discussed *Teutonic* and *Majestic*'s performance in 1902. Although both ships needed to be reboilered, their engines were still running well and 'in excellent condition'. On 18 November 1902, J. Bruce Ismay reported to Pirrie that *Teutonic* 'made a good run out, averaging over 20 knots for two days' during a trial of a particular type of coal.[4]

Harland & Wolff's practice was to provide 'ample boiler capacity' to make sure the engines had plenty of steam. *Olympic*'s boiler capacity would be almost double *Oceanic*'s. She needed twenty-nine boilers, including twenty-four double-ended boilers and five single-ended boilers, all 15ft 9in in diameter. The space needed for the boiler rooms would come to no less than 321ft of the ship's length.

They enlarged and refined the design of *Oceanic*'s triple-expansion reciprocating engines. *Oceanic*'s engines each developed 14,000 horsepower while running at 77 revolutions, whereas *Olympic*'s developed about 17,000 horsepower at the same speed. In both cases, the engines were capable of achieving a higher number of revolutions and developing significantly more power. Even so, Harland & Wolff were reaching the limit of what they could achieve with reciprocating engines alone. Once *Olympic* had grown from the smaller initial design concepts to her final size, their calculations seem to have anticipated that about 51,000 horsepower would be required to deliver a service speed of 21 knots.[5] They also wanted to provide ample power in reserve.

While they adopted a tried and tested approach by using triple-expansion reciprocating engines, Harland & Wolff were not against using turbine engines in principle. They turned their thoughts to a combination arrangement. Between the construction of *Oceanic* and the 'Big Four', Harland & Wolff had progressed from using triple-expansion to quadruple-expansion reciprocating engines. (*Adriatic*'s quadruple-expansion reciprocating engines were reportedly the largest built at Belfast.)

The quadruple-expansion engines were more efficient than their triple-expansion counterparts. Steam pressure reduced as it passed through the engines, from the high-pressure cylinder to the first intermediate and second intermediate pressure cylinders, to the final low-pressure cylinder. However, a steam turbine would be able to make use of steam where the steam pressure was too low for reciprocating engines. In effect, they would revert to using triple-expansion reciprocating engines as on *Oceanic* but the fourth stage of expansion would be the low-pressure turbine, making a very efficient quadruple-expansion system.

The concept of using some sort of combination of reciprocating and turbine engines had been around for years. As early as 1894, 'a patent was taken out by Charles Parsons for the "combination" of a reciprocating engine with a steam turbine, whose object was "to increase the power obtainable by the expansion of the steam beyond the limits possible with reciprocating engines"'. In 1902, Lieutenant Soliani of the Italian Navy 'had proposed some combination of engines and turbines'. The same year, the Royal Navy destroyer *Velox*'s quadruple screws were driven by an interesting mix of machinery – 'two small reciprocating engines' exhausting steam into a high-pressure turbine and a low-pressure turbine on each side of the ship.

Around 1906, Sir William White suggested a combination system, 'as applied to the intermediate type of liner of moderately large power and speed'. Various designs were 'prepared for various types of vessels of speeds ranging from 13 to 16 knots'. *Mauretania*'s builders, Swan, Hunter and Wigham Richardson, took 'much interest in this combination of machinery'. Charles Parsons and R.J. Walker jointly authored a paper discussing combinations of reciprocating engines and turbines, which was presented to the Institution of Naval Architects in April 1908.[6]

The first merchant use of combination machinery was *Otaki*. She was completed for the New Zealand Shipping Company by William Denny & Brothers, Dumbarton, in 1908. Her machinery consisted of two triple-expansion reciprocating engines and a low-pressure turbine driving a centre propeller. Her older sister ship, *Orari*, was completed with the conventional two triple-expansion reciprocating engines.

On *Otaki*'s second round trip to New Zealand, the saving in coal consumption was 13 per cent. In August 1909, a Cunard official thought the results for *Otaki* were 'fairly satisfactory', but cautioned that they 'would have been more convincing if the comparison had been made with engines of the quadruple-expansion type, such as are fitted on the *Caronia* [...] which are more economical than the triple-expansion type'.[7] Nonetheless, there could be 'no doubt [...] that compared even with the quadruple type there would be a saving'.

The propelling machinery itself was not the only issue. There was also 'the design of the after body of the ship to receive the three propellers'.

He feared that, 'Unless this is thoroughly considered together with the design and spacing of the propellers, and satisfactory data obtained from tank tests, then the whole of the advantages likely to be obtained by the combination [propelling machinery] might be lost.'

Olympic's earliest known specification already envisaged a triple-screw configuration, but Harland & Wolff opted to test the combination machinery on a smaller scale.

***

In November 1905, Harland & Wolff took out a licence to build Parsons steam turbines, but they lacked the ability to manufacture them. Accordingly, Pirrie approached John Brown & Co. on Clydebank two years later. They forged a working alliance. The Scottish shipbuilders purchased shares in Harland & Wolff and committed to sharing their turbine technology. Pirrie also bought shares in John Brown's.[8]

On 28 January 1907, Harland & Wolff ordered the shipyard and engine works to proceed with Yard Number 394, *Alberta*. The contract was not signed until 30 April 1907 and then a sister was ordered on the same day, Yard Number 399, *Albany*: as a 'duplicate of 394 for Canadian trade'. The two ships were ordered by IMM for the Dominion Line and reassigned later to White Star. They would operate under the 'White Star – Dominion Line' service: *Alberta* was renamed *Laurentic* and *Albany* was renamed *Megantic*. *Laurentic* was laid down on 7 October 1907 on No. 6 slipway and launched on 10 September 1908. *Megantic* followed some months behind on No. 7 slipway: she was laid down on 27 February 1908 and launched on 10 December 1908.

ALL-PURPOSES TWENTY-FOUR-HOUR COAL CONSUMPTION: SELECT PERFORMANCE DATA

Ship	Consumption	Speed
Campania, Voyage 59: Westbound (August 1898)	555 tons	21.5 knots
Oceanic (average 1914)	450 tons	20.5 knots
Baltic, Voyage 237: Westbound (January 1926)	265 tons	15.9 knots
Adriatic, Voyage 202: Eastbound (November 1925)	310 tons	16.4 knots

Coal consumption varied according to the quality of coal, sea conditions and state of loading. However, on any approximate comparison the cost of speed was clear. The previous generation of 'express' North Atlantic liners, such as Cunard's *Campania* and White Star's *Oceanic*, were significantly smaller than *Baltic* and *Adriatic*, but they needed more power and more coal to achieve a higher speed. Although the slower ships were also at sea for longer, *Baltic*'s total coal consumption of 2,432 tons was substantially lower than *Campania*'s 3,427 tons. *Baltic* also had a much higher cargo capacity and space available for passenger accommodation. It was a compelling economic proposition.*

* 'Abstract of Engineer's Log of the Cunard Twin Screw Steam Ship *Campania* from Liverpool to New York', 6–13 August 1898: Merseyside Maritime Museum, B/CUN/4/2/13; 'SS *Oceanic*. Armed Merchant Cruiser', 21 August 1914: The National Archives (UK), MT 23/321; 'Abstract of the Engineer's Log of the SS *Baltic* Voy. 237 from Liverpool towards New York', 23 January to 2 February 1926, and 'Abstract of the Engineer's Log of the SS *Adriatic* Voy. 202 from New York towards Liverpool', 7 November to 17 November 1925. (Author's collection)

Laurentic's overall length of 565ft and breadth of over 67ft, combined with her displacement of around 20,000 tons, made her a substantial vessel by any measure.[9] There was one major design difference between the two sisterships. *Laurentic*'s propelling machinery consisted of two triple-expansion reciprocating engines driving the port and starboard propellers and exhausting their used steam into a low-pressure turbine driving a centre propeller. Her turbine was subcontracted to John Brown & Co. *Megantic*'s arrangement was two quadruple-expansion reciprocating engines driving the port and starboard propellers.

In August 1907, IMM general manager John Lee spoke to the press about White Star's 'two new steamers'. He did not comment directly on their propulsion systems until he was asked specifically:

> When asked if the White Star Company had lodged an order with Harland & Wolff of Belfast to build two turbine steamers to compete with the new Cunarders *Lusitania* and *Mauretania*, Mr. Lee said:
>
> 'We have two new steamers building for the Southampton–Cherbourg[–New York] service of the '*Oceanic*' class, but they will not be any faster. The tendency now for the passenger traffic is to build large steamers with moderate speed.
>
> 'We shall wait, however, and watch the developments of the Cunarders, and if they are a success it may make a difference in the machinery and speed of our new steamers.'
>
> When questioned about the adoption of turbines in the company's steamers, the general manager said that their success had got to be demonstrated with regard to speed, economy in coal and space, absence of vibration, so far as big ocean liners were concerned.[10]

In September 1907, one reporter did not think the combination machinery would be used until it had been trialled successfully on *Laurentic*.[11] Naval constructor William J. Baxter predicted the 'construction of a transatlantic triple screw liner is quite possible', but warned that 'to give it both turbines and reciprocating engines will inevitably result in a failure such as marked that of the *Great Eastern* a half century ago'.[12]

Around 2 p.m. on Friday, 29 January 1909, *Laurentic* left the Abercorn Basin on her 'preliminary trials'. The tugs *Jackal* and *Hercules* aided her and she 'slowly and gracefully' made her way into the Victoria Channel with her 'reciprocating engines going at half speed'. Off Holywood, she worked up towards full speed and the low-pressure turbine was engaged: 'The turbine worked admirably, starting and stopping when called upon with the greatest freedom.' Once she was at full speed, 'there was a remarkable freedom from vibration and in every essential point the machinery more than realised expectations'. Chief Engineer Joseph Bell was a highly regarded White Star veteran. He said that he had never seen 'new machinery start away in so satisfactory a manner'.

The following day, J. Bruce Ismay came on board with officials from White Star and Harland & Wolff. They 'closely followed the experiments that were made'. *Laurentic* made a run to Liverpool and performed even better than she had the day before: 'The most exhaustive and thorough tests were given to the new form of machinery.' That afternoon, a telegram reported her engines were 'highly successful in all respects'.

She returned to Belfast on the Sunday evening and 'was again moored in the Abercorn Basin'. Her builders continued with all the necessary work to get her completed and ready for passenger service. Although seaworthy and with her propelling machinery in working order, she was certainly not finished.[13]

The *Belfast Evening Telegraph* bemoaned the contradictory information leaking to the press, on 12 February 1909, '… between "reliable information", "official confirmation" and "official denial", the public are likely to be bewildered'. Was the combination machinery going to be adopted for *Olympic* and her sister, or not?

In London, *The Times* had written that the success of *Laurentic*'s trial run had 'led the owners to adopt this system of triple screw propulsion for the *Olympic* and *Titanic*'. The *Daily Mail* cited 'official information' from White Star which agreed with *The Times*' report, only for the Press Association to contradict it hours later when they quoted a telegram from Liverpool. It acknowledged the 'gratifying results' of *Laurentic*'s trial run, but said that 'no decision has been arrived at in connection with the machinery' for *Olympic* and her sister.

Laurentic completed her sea trials on 15 April 1909. She sailed from Liverpool on her maiden voyage on 29 April 1909. *Megantic* followed on 17 June 1909. It was 'stated on good authority that on the trials of the *Laurentic* the horsepower was 12,400 or 3,400 in excess of the contract and 2,000 horsepower in excess of that obtained in the *Megantic* with quadruple [expansion] engines, apparently with the same coal consumption'.[14] She exceeded her designed service speed of 16 knots: 'The main engines were designed to run under these conditions at about 83 revolutions, while the turbine made 220 revolutions.'[15]

During a meeting of marine engineering professionals from various leading shipping lines in Liverpool in August 1910, 'the question of turbine verses reciprocating engines verses combined systems received considerable attention. From the discussion it may be taken that the latter type shows a saving of coal equal to 10 to 12 percent, and the White Star appear to be satisfied with the results obtained.'[16] Throughout the course of their first two years in service, performance data showed a 'gain in economy' of 21 per cent, much better than the most optimistic expectation of 14 per cent.[17]

(Later on, Harland & Wolff obtained further valuable data from *Themistocles* and *Demosthenes*, built for George Thompson & Co. and launched

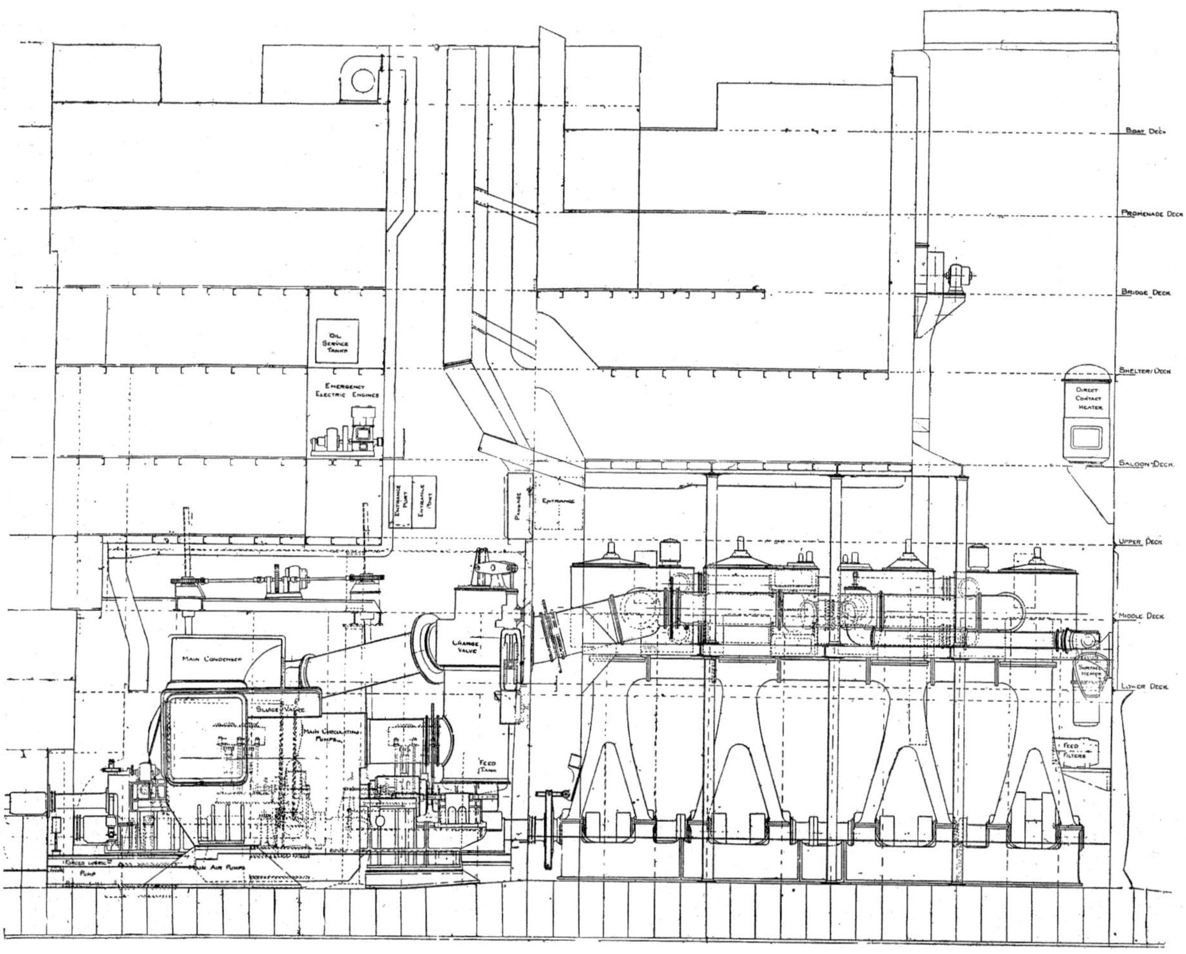

This diagram showing the 'general arrangement of machinery' illustrates the scale of *Olympic*'s reciprocating engines (on the right). The turbine engine (left) was in its own compartment. *The Marine Engineer and Naval Architect*, 1911. (Author's collection)

Olympic's starboard reciprocating engine 'nearly completed in engine works erecting shop' in November 1910 (the view is from the aft end). Edward Wilding wrote, 'I do not think many people who have not been there, realise the enormous power that there is got from the steam pressure in these engines; they move comparatively slowly even when at full power, and the power behind them is, I think I am correct in stating, larger than the power behind the biggest rolling mills in the world. That is, the biggest mills that are used anywhere for the rolling of steel plates, as distinct from the forging of armour plates.' (HOYFM.HW.H1456 © National Museum NI, Harland & Wolff Collection, Ulster Folk & Transport Museum)

'Turbine rotor, showing blading being fitted', *The Marine Engineer and Naval Architect*, 1910. (Author's collection)

in September 1910 and February 1911, respectively. *Themistocles* had quadruple-expansion reciprocating engines driving twin screws; *Demosthenes* had combination machinery. Other than their engines, they were largely identical in design, running on the same route and carrying the same kind of cargo. Data captured in service showed a 20 per cent increase in power 'for the same steam consumption' with the combination machinery; *Demosthenes* enjoyed an extra 0.5 to 0.75 knots of speed for the same fuel consumption. An added bonus was that the 'triple-screw machinery weight was somewhat less than that of the twin-screw reciprocators, in terms of horsepower per ton'.)[18]

Even though the data from *Laurentic* was still limited, the early indications were certainly clear that what worked in theory also worked in practice. The shipyard and engine works had been ordered to 'proceed with boilers' for *Olympic* and *Titanic* on 26 February 1909. Now, they issued orders for work 'to proceed with [the] remainder of [propelling] machinery' on 20 April 1909, five days after *Laurentic*'s sea trials.

The fundamental confidence Harland & Wolff had in the combination machinery is illustrated by how they proceeded with construction of *Olympic* and *Titanic*. When *Lusitania* and *Mauretania* were under construction, their shipbuilders prioritised the forward parts of the hull, whereas progress at the stern was much slower. It was a deliberate decision to allow for flexibility if any alterations needed to be made to the propelling machinery or the design of the stern, because those changes could be made with a minimum of fuss. However, Harland & Wolff proceeded with construction, starting from the stern. If any major changes had been needed to the propelling machinery or the stern of each ship, implementing them would have been very costly, undoing the initial construction work and literally starting again.[19]

All the work involved in designing the propelling machinery included making sure that the propellers were right for the ship. Shipbuilders were constantly trying to find the best propeller configuration, balancing the need for propellers to work efficiently with the need to reduce

The arrangement of the immensely strong stern castings, showing the brackets carrying the port and starboard propeller shafts, the centre propeller aperture, stern post and rudder. *The Shipbuilder*, 1911. (Ioannis Georgiou collection)

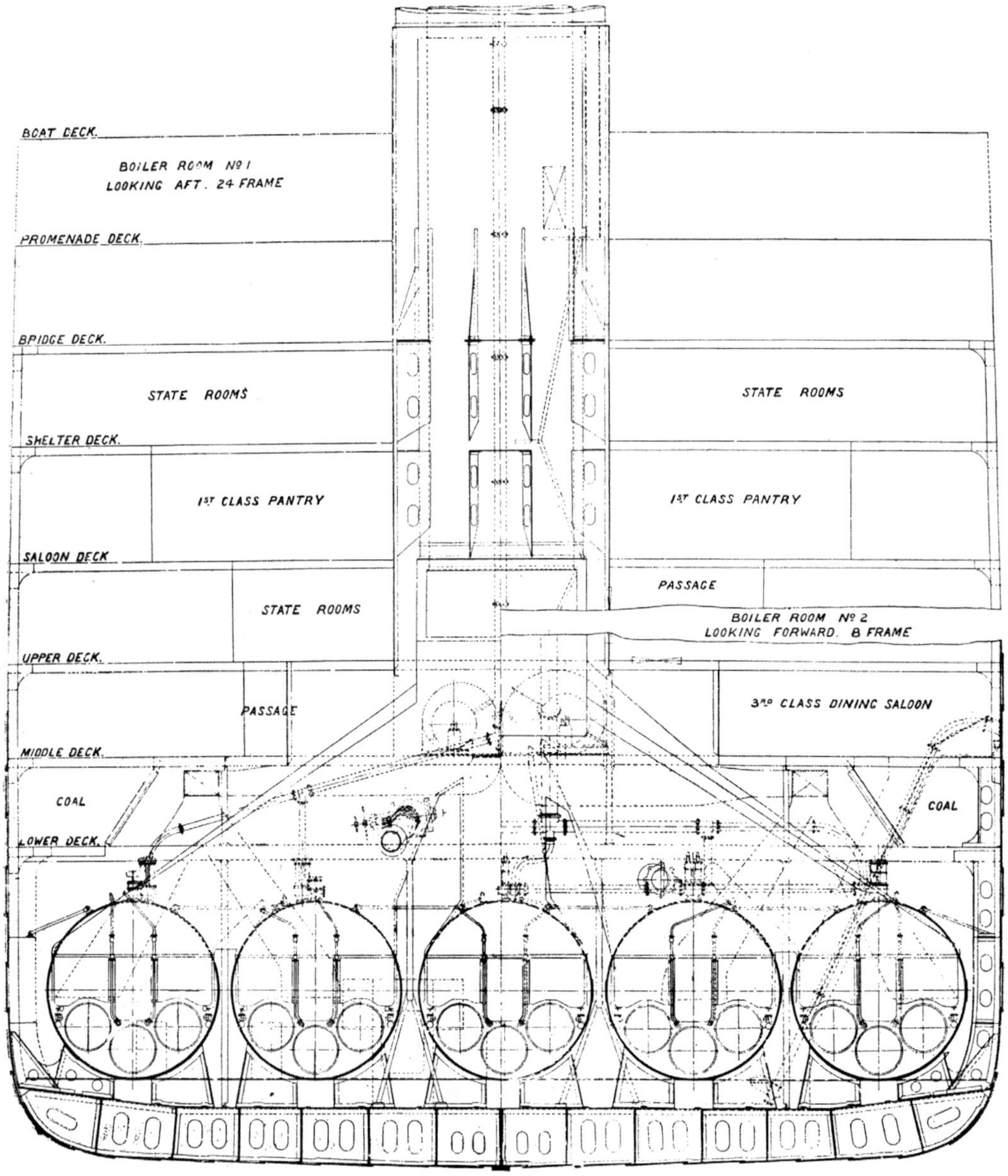

Boiler Rooms 1 to 5 all had five boilers athwartships, whereas Boiler Room 6 only had four because the hull was starting to narrow towards the bow. Boiler Room 1 contained only single-ended boilers, whereas all the other boiler rooms had double-ended boilers. Harland & Wolff's practice was to provide ample steam-generation capacity with plenty in reserve. *The Shipbuilder*, 1911. (Ioannis Georgiou collection)

A set of Kilroy's Patent stoking indicators and regulators was supplied to each stokehold (one for each boiler). The regulator was set to show the rate of firing the engineer ordered. Then the electrical indicators lit up and sounded to alert the firemen when the boiler furnaces needed to be fired. *The Shipbuilder*, 1911. (Ioannis Georgiou collection)

vibration so that the passengers could be comfortable. During their early years of service, Cunard's *Lusitania* and *Mauretania* sported a number of different propeller configurations in order to help determine which one was the best. Fitting different propellers allowed a real comparison to be made to see if a propeller design lived up to its promises. In July 1908, for instance, Cunard's directors were informed 'of the effect produced on the *Lusitania* by the substitution of solid four-bladed wing propellers for the three-bladed propellers previously in use'. Referring to *Mauretania*, they said, 'It would not be desirable to adopt four-bladed propellers of the same design as in the *Lusitania* until a trial has been made of the three-bladed propellers of the special design.'

What Cunard learned as a result of experimenting with different propeller configurations was that propeller design could have a significant impact on reducing vibration and improving passenger comfort. On 13 March 1908, William Watson wrote to their agent Vernon Brown in New York. Various changes had been made to *Mauretania*, including stiffening. He was 'very glad to hear that this ship seems so much improved'. 'We will go on trying to make further improvements. We know the disagreeable action of the side propellers, and we are making an experiment now of a different shaped blade, much smaller at the points, by which we hope to obtain some improvement.' He worried that their ships had developed an unfair reputation in their early months which was proving hard to dispel:

> I have been told that in the drawing rooms of New York and the clubs nothing is spoken about but the vibration of these ships. I myself can testify that in the first class this is absolute rubbish. The rooms that I have been in have been the quietest of any ship I ever travelled in, and you see what Mr. Ismay himself says. But to revert to the beginning of this sentence, I hear these criticisms are made in every house in New York. They are made before passengers who have travelled in the ships, who have related it to us, and they have told the people that it is no such thing.
>
> [...]
>
> The other day Lord Balfour of Burleigh dined in the company of [J.] Pierpont Morgan. He was cracking up the White Star ships when Balfour asked him if he had travelled in our new ships, and he said no, but he heard that the vibration was terrible and they were most uncomfortable – rattling you to bits. Balfour behind his back roared at him, but did not tell him it was nonsense.
>
> [...]
>
> Do you think you could get any press remarks about the exaggeration of the vibration reports? I *know* [original emphasis] individuals contradict it and laugh at the reports – I thought perhaps you might get someone to write a letter to the press speaking especially about it.[20]

Turbine specialist Stephen Pigott was working for John Brown & Co. by summer 1908. His initial intention was to work with them for up to six months, but it proved to be the start of a long and successful career with the firm. The Glasgow shipbuilders were subcontracted to build the low-pressure turbine engines for both *Olympic* and *Titanic*. They looked at the propeller configuration. Pigott's papers contained a notebook with propeller design data and proposals for a number of ships they had built, from *Beagle* (launched in April 1909) to *Yarmouth* (launched in April 1911).

One double-page entry included notes on two 'proposed propeller' configurations for 'Centre Propeller *Olympic* & *Titanic*'. It was anticipated that the low-pressure turbine would be developing 17,000 shaft horsepower and the centre propeller would be making 170 revolutions per minute, on the basis of the ship making 22 knots. They compared two 'proposed propeller' configurations: a three-bladed propeller with a diameter of 17ft and a four-bladed propeller with a diameter of 16ft 6in. (When *Olympic* was completed, her centre propeller was almost identical to the four-bladed configuration in the notebook. *Titanic*'s centre propeller, too, was almost the same as the three-bladed configuration. The exception in both cases was that the pitch of the propeller blades was adjusted and standardised.)[21]

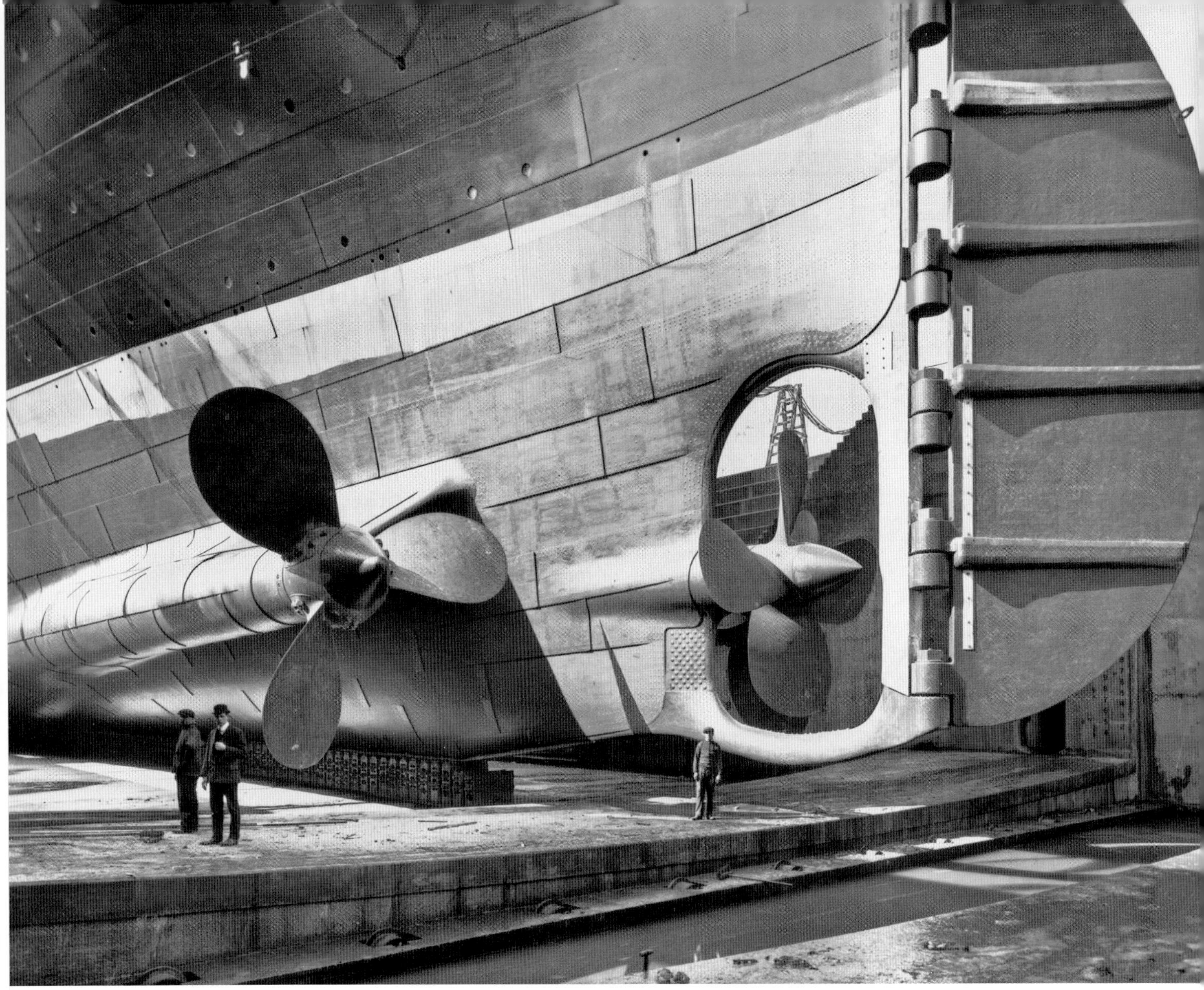

Olympic dry-docked with her propellers fitted in April 1911. Painting is not yet complete. (HOYFM.HW.H1511 © National Museum NI, Harland & Wolff Collection, Ulster Folk & Transport Museum)

LUSITANIA ACTUAL PERFORMANCE 'AT VARIOUS SPEEDS UNDER CONDITIONS PREVAILING ON OFFICIAL TRIALS'

Shaft Horsepower	Speed (Knots)
13,400	15.77
20,500	18
33,000	21
48,000	23
68,850	25.4

Data obtained from *Lusitania* on trials in 1907 demonstrated that a 61 per cent increase in speed from 15.77 to 25.4 knots required a 414 per cent increase in power.

Propelling Machinery	**Coal Consumed (Pounds per Hour per 1 Horsepower)**
Triple-expansion reciprocating	1.54
Quadruple-expansion reciprocating	1.4–1.45
Direct-drive turbine (compound) at high speed (above 23 knots)	1.5
Combination triple-expansion reciprocating & turbine	1.25

The more efficient propelling machinery was not the only benefit of *Olympic*'s design. Not only was it more economical for a given power output; by eschewing any plans to match *Mauretania*'s speed, she did not then need as much power. Figures calculated for Cunard indicated the enormity of the increased power that would have been required for a ship like *Olympic* to achieve 25 knots. To increase the ship's speed from 21 to 25 knots, the effective horsepower needed might increase by as much as 89 per cent.[24] That would have required a similar increase in fuel consumption, leaving aside the space needed to house more powerful machinery, store the coal, accommodate the firemen and trimmers needed to feed the boilers or the enormous loss of earning power by sacrificing space better used for passenger accommodation or cargo. (On a slower intermediate ship like *Celtic*, 65 per cent of the enclosed space in her hull and superstructure was available for passenger accommodation or cargo storage. On *Mauretania*, it was only 40 per cent, because the engine and boiler rooms took up so much space.)

***

Olympic's power requirements were still remarkable. Her propelling machinery took up more than three times the space of *Celtic*'s. It generated about the same horsepower as *Celtic*, *Cedric*, *Baltic* and *Adriatic*'s engines combined. Only *Lusitania* and *Mauretania*'s engines had more power. The number of men – including engineers, firemen, trimmers, and greasers – needed to work *Olympic*'s Engine Department increased substantially. *Celtic* required only 92, which was increased to 130 on *Adriatic*. *Olympic*'s requirements were even greater at 327.[25] That compared to about 364 in *Mauretania*'s Engine Department, but *Olympic* needed slightly more engineers, whereas *Mauretania* needed so many more firemen and trimmers.

***

All the design choices paid off. The combination machinery was a transition propulsion system in many ways, but it provided ample power. In the years to come, there would also be a significant contrast with *Lusitania*, which experienced significant problems with her turbine engines only five years after entering service. By comparison, *Olympic*'s machinery was reliable and dependable. The pleasing results would only become clear once she was completed and saw true service conditions. Harold Sanderson noted, 'She proved to be considerably faster' than her anticipated service speed of '21 to 21½ knots'.[26] She combined a higher speed with a significantly lower fuel consumption. It was a winning combination.

OLYMPIC COAL CONSUMPTION DATA: HALF SPEED AHEAD TO FULL SPEED AHEAD

Reciprocating Engine (rpm)	Speed (Knots)	24-Hour Coal Consumption (Tons)	Hourly Coal Consumption (Tons)
52	16	343	14.3
56	17	391	16.3
60	18	446	18.6
64	19	507	21.1
68	20	575	23.9
72	21	649	27.1
74	21.5	690	28.7
76	22	732	30.5
78	22.5	776	32.3
80	23	822	34.2
82	23.5	870	36.2
84	24	920	38.3

NB: This coal consumption data is 'all purposes' consumption, including coal used to drive the ship as well as providing the power requirements on board. It is based on *Olympic*'s load draught of 34ft 7in. The data is rounded. Engine revolutions and speed through the water also vary somewhat depending on factors such as the sea conditions.*

* This table is based upon a mathematical model developed by Sam Halpern. It took into account the coal-consumption figures provided by Harold Sanderson in January 1916 for *Olympic*, showing her coal consumption at different speeds. (The required power to move a ship through the water is directly proportional to the cube of a vessel's speed. He noted that the power required to move it at a certain speed also depends on the cross-sectional area of the vessel's hull below the waterline. Deeper draught vessels of the same hull form require more power to move them at a certain speed than a shallower draught vessel of the same form and speed. On a ship like *Olympic*, the cross-sectional area below the waterline – approximately a product of the vessel's draught and breadth – will tend to increase in proportion to the draught because its breadth will not change significantly for small changes in draught. Therefore, the amount of power required is approximately proportional to the vessel's draught as well as the cube of the vessel's speed through the water. Fuel consumption is directly proportional to the power required to move the ship.) Further data was obtained from the reported coal consumption on *Olympic*'s maiden voyage in June 1911. When the data from the model was compared with the actual coal consumption as reported, these values were within 1 per cent, providing strong confirmation as to its accuracy.

5

'THE PASSENGERS HAVE NEVER BEEN MORE COMFORTABLE'

When *Oceanic* entered service in September 1899, one observer thought:

> The beauty of her design, as well as her yacht-like lines throughout, called forth much admiration, while a run through her interior enabled one to realise to what a pitch of perfection shipbuilding has been bought in this new transatlantic wonder. Wherever one turned, whether from the stateroom to the second cabin, the steerage berth, or the crew's quarters, there was evidence of all round perfection of the minutest details.[1]

The Engineer explained that the first-class public rooms and entrance had been 'designed by the eminent architect, Mr. R. Norman Shaw, R.A., and executed by Messrs. George Trollope and Sons, and by Messrs. J. Aldam Heaton & Co.'. The decorative features under the dome in the first-class dining saloon were from Messrs Clayton and Bell.

Robert Norman Shaw and the late John Aldam Heaton had worked as a partnership in Aldam Heaton & Co. (In addition to their work on passenger liners, they did interior architectural work for Thomas Henry Ismay's home, Dawpool, at Thurlaston. His impressive residence included a substantial picture room to display items from his art collection.) It was common practice for interior architects to design the most important elements of ships' interiors, such as the first-class entrances, public rooms and even the finer first-class suite or staterooms. The shipbuilders fitted out the rest and panelling or fittings were often similar from ship to ship.

The passenger lines, too, often had an idea of what worked well, and wanted new ships fitted out in a similar way. The Holland America Line instructed Harland & Wolff that *Nieuw Amsterdam*'s passenger accommodation was to be 'similar to that in the *Noordam*, and fitted up in the same style'. First-class public rooms, entrances and passages were 'to be decorated, fitted and furnished complete including chairs, tables, etc. as may be required, by Messrs. Mutters [& Zoon] of the Hague, or other Dutch firm, selected by the owners'.

Cunard employed James Miller to design *Lusitania*'s interiors and Harold Peto to do *Mauretania*'s. They adopted very different styles so that the sister ships each had their own distinct character. *Lusitania*'s interiors were airy and light, whereas *Mauretania*'s included French walnut and mahogany. Peto was tasked with 'designing, drawing and supervising the decoration of all the first class public accommodation', as well as 'about fifty of the best first class staterooms, with the lavatories and bathrooms attached'. The second-class public rooms were designed by the shipbuilder, but they

The first-class Palm Court on board the Holland America Line's *Rotterdam*, completed by Harland & Wolff in 1908. Its cream décor reflected Louis XVI-period styling. The illustrated panels represented 'scenes of bygone centuries' and 'the magnificent and masterful work of the Delft Art Pottery Works'. Furniture supplied by H.P. Mutters en Zoon was similar to that on *Olympic* and *Titanic*. (HOYFM.HW.H1277 © National Museum NI, Harland & Wolff Collection, Ulster Folk & Transport Museum)

had to submit their plans to Peto for approval and work under his supervision.[2]

William Watson wrote to J. Bruce Ismay on 5 December 1908. He explained that Miller:

> … has asked me to give him a letter of introduction to you (which I have done), as [he] would like to see whether he could not get some work from you on your ships. I have great pleasure in doing so, because I think Mr. Miller's work has been very good, and I certainly think you will get good value from him. Personally I prefer the *Lusitania*'s decorations though they did not cost as much as the *Mauretania*'s.[3]

The White Star Line did not take up Miller's services. They used Aldam Heaton & Co. to design the most important first-class public areas and accommodation. Harland & Wolff did the rest.

The sheer size of the new ships was a challenge. On 3 March 1910, Alexander Carlisle chaired a meeting of Harland & Wolff's managing directors. He reported, 'In view of the immense amount of decorating and other work in connection with Nos. 400 and 401, the chairman of principals had authorised the extension of the polishing shop, to be effected by putting a second floor into the platers' shed in the South Yard.'[4] They were impressed with the Dutch firm H.P. Mutters en Zoon, who were contracted to design and outfit twelve of the first-class special staterooms in period styles (six on B-Deck and six on C-Deck). Their designs included both Modern and Old Dutch, Louis XVI, Empire and Queen Anne styles. (It appears likely that the cost of each individual stateroom they designed amounted to several hundred pounds.)[5] Harland & Wolff outfitted a number of other staterooms in a more understated way, either using French oak panelling or white panelling with oak dado beneath.

The shipbuilder tended to prefer lighter-coloured staterooms, based on their experience of what worked well on earlier ships. In November 1903, the Holland America Line wanted the most luxurious first-class staterooms of *Nieuw Amsterdam* to have better-quality ('more elaborate') carpeting and wall panelling, 'The intention has been to have real "Chambres de luxe", as [...] in *Teutonic* etc. viz. "Cabinet finish".' The next month, Harland & Wolff suggested:

> We will fit the luxe rooms equal in all respects to the corresponding rooms in *Celtic* and *Cedric*, but we do not propose to put in polished hardwood bulkheads, nor the elaborate wash basins fitted in the *Teutonic*, as these were not repeated in the following ships, white paint being considered much more suitable for bedrooms.[6]

The first-class suite rooms on the 'Big Four' were relatively plain because no period styles were used. They simply used white panelling, which was little different to other staterooms. The ordinary first- and second-class staterooms on *Olympic* and *Titanic* had a lot in common with what had worked before on other ships, using fairly standard fixtures and panelling but with further improvements for passenger comfort.

The end result was a substantial family resemblance with other ships Harland & Wolff had built. In many ways, the first-class lounge was an enlarged and improved version of *Oceanic*'s; elements of the first-class smoke room were much the same as the design used on *Adriatic*, *Laurentic* and

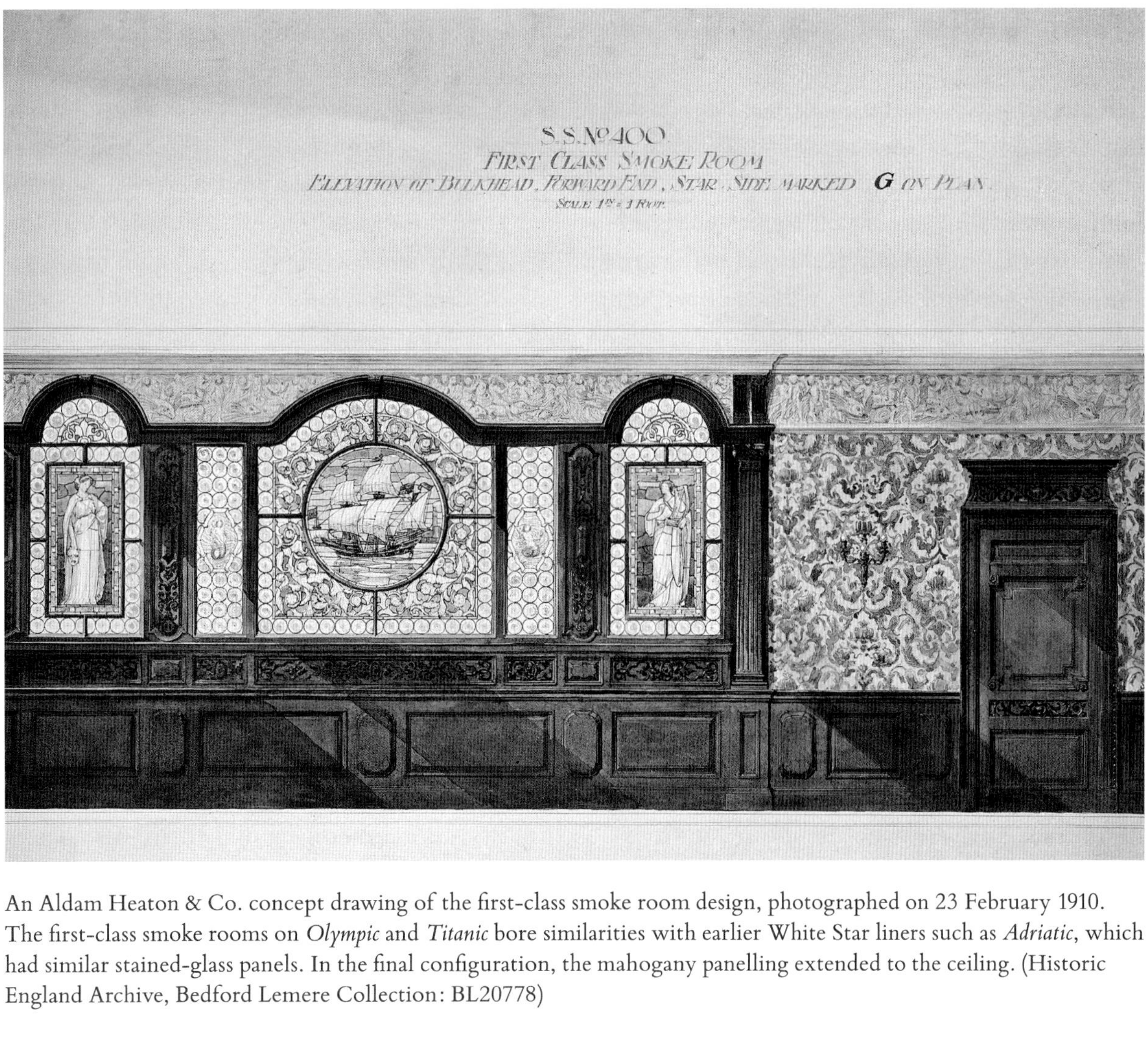

An Aldam Heaton & Co. concept drawing of the first-class smoke room design, photographed on 23 February 1910. The first-class smoke rooms on *Olympic* and *Titanic* bore similarities with earlier White Star liners such as *Adriatic*, which had similar stained-glass panels. In the final configuration, the mahogany panelling extended to the ceiling. (Historic England Archive, Bedford Lemere Collection: BL20778)

Megantic. Regular transatlantic travellers who were comfortable on White Star's earlier ships would feel very much at home on their newest liners.

The relationship between Harland & Wolff and Aldam Heaton & Co. grew closer over time. Harland & Wolff's 'arrangement with this firm [...] proved a distinct advantage to Harland & Wolff and their clients'. Early in February 1912, Pirrie reported 'the purchase of a controlling interest in Aldam Heaton & Co.'.[7] In effect, the shipbuilder acquired their own interior decorator as a subsidiary. (He was kept closely informed of all orders that Harland & Wolff placed with Aldam Heaton & Co. and correspondence between them. It appears his close interest related to the work the firm were doing on *Olympic* and *Titanic*. In October 1911, Pirrie said 'that now that the *Olympic* is finished, in future the practice of sending him copies of all letters [...] might be discontinued'.)[8]

A FLOATING HOTEL

The passenger accommodation depicted in the 'Design "D"' concept from July 1908 was improved considerably in the final design. Minor improvements and alterations continued right up until completion. Frank A. Munsey contrasted the new generation of ships with their predecessors. 'On my first crossing, twenty years ago, I was struck with the lack of consideration by marine architects for the comfort of passengers,' he wrote for *Munsey's Magazine*:

> The boats were small, and their narrow decks were frequently awash. In bright sunny weather, and with a smooth sea, passengers got on very well, but when it was cold and stormy and rough there was practically nowhere to go. There was, to be sure, a dingy little smoking room and a long, plain, barnlike dining saloon. Beyond these two rooms and the pinched little cabins there was nowhere to go but out, and the 'out' was often denied to passengers [...]
>
> The palace hotel liner now gives us dozens of other great public rooms, luxuriously and comfortably and temptingly furnished.[9]

The situation where 'there was practically nowhere to go' had changed beyond recognition. For the White Star Line, *Adriatic* had represented a substantial leap forward with the range of first-class public rooms and facilities available. Now, the 'Design "D"' concept contained similarities and improvements. On the promenade deck, a reading and writing room was situated just ahead of the first-class lounge amidships; the domed first-class smoke room and two veranda cafés were further aft. (*Adriatic* did not have a veranda café at all. Those on *Lusitania* and *Mauretania* were exposed to the elements.) The bridge deck immediately below included extensive first-class suite rooms amidships. The second-class smoke room was about the same size as its first-class counterpart directly above. On the saloon deck, a large reception room where passengers could meet before and after dinner was an innovation. The domed first-class dining saloon could accommodate almost 600 people. A large-second class dining saloon, further aft, could accommodate a similar number. It spanned the width of the ship, aping *Adriatic*. Third class had a general room at the stern. Down on F-Deck, a gymnasium, Turkish bath establishment and swimming pool were available for first-class passengers. The pool was a true swimming pool, whereas *Adriatic*'s 'plunge bath' was tiny by comparison and was used in conjunction with her Turkish and electric baths.[10]

The general arrangement plans dated 31 May 1910 were very close to the finished product.[11] They reflected numerous changes made since July 1908, big and small. The original two lifts (or 'elevators'), extending from the Boat Deck down to F-Deck, were replaced with three lifts serving A-Deck to E-Deck. Both first-class main staircases were changed, with the original straight stairs fore and aft replaced with curved designs. Rather than placing the lifts opposite the forward first-class staircase as the passengers descended, the stairs were relocated so that the lifts were directly behind them. Two staircases in the first-class reception room ('entrance and lounge' on the 'design "D"' concept) were replaced with one, which narrowed going down to E-Deck. In second class, they retained a single

Olympic's forward first-class grand staircase, looking from the starboard side of A-Deck up towards the port side Boat Deck. The figures in the famous clock panel represent 'Honour & Glory Crowning Time'. (Library of Congress Prints & Photographs Division)

lift serving the main staircase from the Boat Deck to F-Deck. (Second-class passengers could not exit on A-Deck because the promenade deck was entirely first class.) They added another large staircase for second-class passengers further aft, serving B-Deck down to F-Deck.

On the promenade deck, the small observation corridor at the fore end was removed. The first-class lounge layout was changed slightly with the number of bay windows increased from two to four. Further aft, the first-class smoke room was altered and four bay windows were added. (The small dome over the smoke room was removed.) Both veranda cafés saw a similar change with new bay windows.

The first-class suite rooms amidships on the bridge deck were fleshed out. Their details and layout were not shown on the earlier concept plan. Now, plans envisaged a spacious 'parlour' suite on each side of the ship with a sitting room ('parlour'), two large bedrooms, private bathroom facilities and wardrobe rooms. Spacious first-class suite rooms with interconnecting doors, private bathroom and wardrobe-room facilities provided flexibility to create suite accommodation for large parties of travellers. A significant change involved moving the second-class smoke room further aft and introducing a new first-class restaurant.

They increased the number of private bathrooms for the first-class suite rooms amidships on C-Deck. The foremost suite rooms on each side amidships became 'parlour' suites similar to those on the deck above. The first-class staterooms and their capacities were marked clearly: 'These outside two berth rooms are fitted with one cot bed 4 feet wide' and another 2ft 6in width, 'or twin beds'. (These beds compared well with others of the period. For example, the first-class berths on board *Nieuw Amsterdam* were 2ft 4in.) They also had a 'sofa seat' in each room, as well as access to their own private bathroom. The ordinary stateroom arrangement was improved to allow more rooms to have access to natural light and ventilation.

The purser's office was moved to the starboard side of the ship abreast the first-class staircase. A maids and valets dining saloon was added near the aft first-class staircase, as well as a Marconi operators' saloon. The second-class lounge was redesignated as a library and the original layout consisting of bench seating all around the room (like that on the 'Big Four') changed. Far aft, the large third-class smoke room spanning the ship's width was replaced with a smoke room and general room.

On the saloon deck, original proposals for the first-class staterooms showed a mixture of three- and single-berth staterooms. They extended three staterooms deep from the side of the hull, a similar arrangement to many staterooms on the 'Big Four', so that only the outermost staterooms had any natural light from portholes. The plans were amended so that the staterooms were only two deep and the inner one had access to natural air and light by a thin corridor extending to the ship's side. (They were typically known as tandem or Bibby staterooms. Pirrie had written on 6 October 1903, 'The arrangement of these rooms in the boats of the White Star and Bibby Lines is, in my opinion, a better one than that adopted in the new P&O boats.') The single-berth staterooms were separated out.

As well as renaming the 'entrance and lounge' as the 'reception room', the gangway doors were moved forward and now led into separate entrance halls. They provided convenient access for passengers in the first-class staterooms on D-Deck or to the lifts to the upper passenger decks. (Thomas Andrews noted an alteration in this area had been 'submitted and approved at [a] meeting held in London 3 June 1910'.) The first-class saloon dome was removed. Further aft, the second-class dining saloon was reduced in size. The uppermost second-class staterooms were improved in a similar way to the first-class staterooms, with a narrow corridor providing access to a porthole and natural light for many of the inner staterooms.

Third-class staterooms forward on the starboard side of E-Deck were removed and replaced with larger first-class stateroom accommodation. Harland & Wolff's revised plans included interchangeable first- and second-class staterooms, many of which now had access to natural light in the same way as the staterooms on D-Deck.

They moved the first-class gymnasium from F-Deck to the Boat Deck, enabling it to have large windows and ample natural light. The Turkish baths, cooling room and electric baths were scaled down and moved to the starboard side. The swimming pool kept its original position. A new feature was a squash court for first-class passengers, extending over two decks forward on F-Deck and G-Deck. (An alteration on 2 August 1910, marked on a separate iron plan, improved the access staircase arrangement.)[12] On G-Deck forward, a compartment with 150 open third-class berths was reduced substantially when the fore part was allocated to first-class baggage storage areas and the post office was relocated from the stern to the bow.

Alterations made after 31 May 1910 were added using tracing paper. The amendments were drawn out and laid over the main general arrangement plan. Typically, they were annotated with a short note explaining that they had been submitted to and approved by the White Star Line on a particular date, then initialled. A table summarising the 'proposed alteration to design' for several sections of second and third-class staterooms noted the changes gained two second-class staterooms ('six passengers'). Third-class lost forty-six open berths and twelve permanent berths, but gained forty-two portable berths.

***

The first-class reception room on D-Deck was an innovation that proved very popular. *Olympic*'s was expanded just months after she entered service and *Titanic*'s was altered similarly. In this early view, a temporary 'D-Deck' sign is visible. (Library of Congress Prints & Photographs Division)

First-class special stateroom C65 was decorated by H.P. Mutters en Zoon in Louis XVI style. The many period styles used for different staterooms reflected different tastes. Nonetheless, the staterooms were very spacious for the period. It was photographed in May 1911. (HOYFM.HW.H1601 © National Museum NI, Harland & Wolff Collection, Ulster Folk & Transport Museum)

The designers of *Olympic* and *Titanic* had the advantage of examining their competitors. It is a safe bet that a great deal of time was spent considering whether certain design features could be adopted or improved upon. Edward Wilding went on board *Lusitania* 'once or twice during construction' and then sailed 'on her sixth voyage', looking over her thoroughly. An unknown observer, who travelled on *Lusitania* shortly after she had entered service, recorded some of the smallest details which contributed to passengers' comfort. (It appears to be someone associated with Harland & Wolff or the White Star Line.)[13] He felt that the jarring of entrance doors and windows made the vibration feel much worse than it was. The binding of the hardwood panelling in some staterooms also had negative consequences:

> During the first two days when we had moderate to fresh westerly gales with rough head seas there was a certain amount of vibration in various parts of the ship, but nothing to fuss about. In smooth weather, such as we have had since, the vibration is very little, except possibly in the Second Saloon quarters right aft, but I do not think it would be very noticeable there were some of the entrance doors and windows properly secured. The constant jarring of these doors makes the vibration seem much worse than it really is; and in my room – B73 – I have noticed nothing at all.
>
> In some of the staterooms which are panelled with hard woods, there is a lot of creaking and noise when she is steaming into a head sea, and I should think that this is attributable entirely to the binding of the woodwork, as I remember quite well we had the same trouble in the *Majestic* and the *Teutonic* years ago, and only overcame it by putting rubber strips between the panels of the woodwork.

There were 'a good many minor matters' that could improve comfort. Electric light switches were:

> … all over the place, sometimes most difficult to find being hidden behind doors, curtains, &c. In my comparatively small room there are no less than four switches in different corners; while in the two parlours of the Regal Suites, there are five, all of which are of course unnecessary.

There were no heaters in the Regal Suite dining rooms. The lavatory door springs were inadequate when the ship was rolling, so the doors:

> [had] to be tied back by towels around the handle of the door and hand rail. These doors, I think should be supplied with hooks, and the best I have seen are those used in the White Star Line which automatically catch the door when pushed back, and merely require lifting to release.

The journey from stateroom B73 to the bathrooms was a long one and the bathroom steward was not always in attendance, 'This morning, for instance, by being away for his breakfast at about half past nine o'clock, when he should of course have been attending to the baths.' Some staterooms, which did have their own private bathroom facilities, were noisy: 'There is a good deal of violent noise, especially noticeable during the night, apparently caused by the sea rushing up the discharge pipe from the bath or basin when the ship rolls. I have heard this noise, and it is quite enough to seriously inconvenience a seasick lady passenger.'

Another related issue was the water supply:

> The first morning we could get no cold salt water in the baths, only red hot, and cans of cold water had to be brought. Sometimes the fresh water in the basins in the bathroom and staterooms has been turned off, so that no fresh water has been available. But these of course are all matters of detail which can be easily attended to.

The toilet bowls were too narrow and shallow: 'I have seen one flush right over, and when the ship is moving in a heavy sea the water in the pans "sluices up".'

A number of the windows needed overhauling:

> The inside windows jamb, and are very difficult to open. In my room I had to get the

> Joiner to work, and in very hot Gulf Stream weather, such as we have been having, it is very disagreeable if one cannot open ones window and get fresh air when required. The same remark applies to the bathroom I have been using where the heat has been terrific because although the outer window would screw down the inner window has been immovable.

The door latches were insufficient, leading to 'the constant swinging and banging of doors, to the damaging of the latter and inconvenience of the people'.

Smokers on the forward promenade deck caused problems. Their ashes blew aft towards other passengers. Second and third-class passengers – some of whom were smoking – were also allowed to use the purser's office on B-Deck to change money and make other arrangements: 'It is not agreeable for first class ladies to be mixed up with them.' The first-class dining saloon arrangements were 'excellent, and the small tables most comfortable'. However, the orchestra tended to leave for the lounge around 8.15 p.m. and play there from 9 p.m. Many passengers did not dine until 7.30 or 8 p.m., so that they missed much of the music. All these little details were important for passenger comfort.

Public expectations were changing, with rising demand for private bathroom facilities. *Mauretania*'s first-class accommodation included ten private bathrooms attached to suites. Two ordinary staterooms had their own private facilities. Norddeutscher Lloyd's *George Washington* (1909) had four suites and thirty-one staterooms with private bathrooms. The Holland America liner *Rotterdam* (1908) set a very high standard: she had eight suites and forty-eight staterooms with private bathrooms. (Standards for the public bathrooms differed by class. Specifications for *Lusitania* and *Mauretania* dated 6 January 1906 called for 'one bath for about every 15 first class passengers' and 'one bath for every 30 persons' in second class.)[14] By the time the design was finalised, *Olympic*'s first-class accommodation included sixteen suites with private bathrooms and twenty-six bathrooms which could be let with staterooms.

White Star felt the need to meet competition for the very finest first-class accommodation. *Oceanic* had two 'special suites of rooms' on the port and starboard side of the upper deck, which each consisted of up to three staterooms and a private bathroom. They did not compare with what was on offer elsewhere. In summer 1907, an Imperial Suite on the promenade deck of Norddeutscher Lloyd's *Kronprinzessin Cecilie* included 'suites of parlour, bedroom, bath and toilet room, private dining room and maid's quarters' and cost $2,000 from Bremen to New York (the rate included two adults and one maid). The two Regal Suites on board *Lusitania* and *Mauretania* were impressive, 'comprising two bedrooms, drawing room, dining room, private bath and toilet'.[15] The dining-room panelling of *Lusitania*'s port-side Regal Suite was Italian walnut. The fireplace in the dining room even had a 'surround of *Fleur-de-Pêche* marble'.[16] There were six sets of 'en suite' rooms ('comprising bedroom and sitting room, with private bath and toilet'), which were renamed as Parlour Suites several years after she entered service. *Rotterdam* had two large suites on the upper promenade deck amidships, including a 'sitting room with open fireplace, two spacious bedrooms, two bath and toilet rooms and room for valet or maid'. The four Parlour Suites on *Olympic* were designed with this competition in mind: each with two large bedrooms, private bath and toilet facilities, wardrobe rooms and a large sitting room or parlour which was well furnished and included a table and chairs. They were very spacious and the bedrooms themselves were each double the size of the first-class staterooms forward on B-Deck.

There was increasing demand for privacy from solo travellers, who preferred single-berth staterooms: *Mauretania* had over forty in first class. *Rotterdam*'s first-class accommodation included 'an unusually large' number of single-berth staterooms, totalling 87. (The 'average first cabin stateroom' had two lower berths and 'the unpopular upper berth' was 'eliminated except in a very few cases'.) *Olympic* had slightly more, provided with beds rather than fixed berths.

The shipping lines tried to strike a balance between meeting public demand and making the best use of the space they had available. 'From verbal replies obtained from a very large number of passengers, it would appear that the majority are in favour of small single berth rooms,' wrote Leonard Peskett in January 1909, 'This class of room may not give the best returns for space.' He also queried the number of private bathrooms on *Rotterdam*: 'Are so many private toilets necessary? Does this arrangement pay?'

In June 1909, *Scientific American* covered the new *George Washington*, 'Some of the innovations [...] are the elimination in the cabins of the first class of upper berths, great number of rooms with [private] baths, children's playroom, two electrically-worked elevators for passengers' and a gymnasium, electric baths, photographers' dark room and twenty dog kennels on the Boat Deck.

White Star were not merely competing against the Cunarders. Standards of comfort were improving all the time on both intermediate and express steamers.[17] Cunard, too, were acutely aware of how their ships measured up. One early worry with *Lusitania* was that staterooms with upper and lower berths had only one long curtain enclosing both, rather than separate ones for each berth, 'Some difficulty might arise with one passenger wishing to have the curtains closed, and the other wishing to have them drawn back.'

Peskett compared the typical second-class staterooms on *Lusitania* and *Mauretania* with *Rotterdam*'s. 'Nearly all' the second-class staterooms on the Cunarders were 8ft long compared to only 6ft 3in on the Dutch ship: 'This difference [...] makes about twenty per cent difference in the numbers of passengers carried in equal areas. Wardrobes can be fitted in eight foot rooms; no wardrobes can be fitted in rooms six feet three inches long.' Passenger comfort competed with profitability: 'The shorter rooms are the most profitable for owners, as the initial cost is less per room, and the [passenger] numbers that can be carried are greater.'[18]

All sorts of little details were noticed by victualling staff or came to the shipping lines' attention from passenger complaints. These were noted down for the designs of their next ships. When they were planning *Lusitania* and *Mauretania*, Cunard specifically requested comments from captains, victualling staff and engineers so that their feedback could be considered. *Umbria*'s Captain Stephens suggested to Cunard in May 1904:

> In as many of the lobbies as practicable, the doors of the staterooms should not be directly opposite, for when the ship is in the gulf stream and very warm, everyone is anxious to have their doors open. I have heard frequent complaints when men have been berthed opposite women. It is embarrassing for both parties.

Etruria's Purser, J. McCubbin thought it was important that the dining saloon was large enough for a single sitting. During high season, passengers complained that they had to wait, 'As to the saloon itself, my idea is that an American prefers light and airy decorations: I do not appreciate the German decorations. I think that Americans have a great liking for Louis XIV [recte: XVI] style – dead white.' He recommended smaller tables and some side tables separated from the middle of the saloon, 'Americans are a high strung people, who value cheerfulness and lightness.' McCubbin pointed out that a passenger might spend a third of their time ('or more if he be inclined to seasickness') in their stateroom, so comfortable stateroom accommodation was 'of primary importance' and 'we may learn something from the German liners'. He thought 'there [*sic*] staterooms are perhaps more comfortable on account of their location over the saloon – in the best part of the ship'. He noticed that demand for single-berth staterooms was 'growing greater every day'. German liners' stateroom fittings were 'very comprehensive, including – amongst other conveniences appealing to ladies – electric heaters for curling tongs etc.'.

Similar comments came from J.F. Banks, Chief Steward on *Ivernia*:

> A number of single berth rooms should be provided with chest of drawers, writing table

> complete, and a small easy chair. I would suggest that every stateroom has a small writing cabinet, that each room has two electric lights, one stationary, and one portable, and that the heating of each room should be separate.

His counterpart on *Lucania* thought staterooms should be larger, because it was 'the one great grievance on board the *Lucania* in the eyes of passengers'. Chief Surgeon B.S. Jones, serving on *Ivernia*, singled out electric lighting in staterooms, arguing that a passenger should be able to see to read in bed or to shave. He suggested 'that there should be a playroom for children. Many passengers do not think it right that that should be disturbed by children playing.' Another recommendation was that Turkish baths and 'a swimming bath should be included [...] in my opinion it would be a great source of amusement and pleasure to passengers'.[19]

Etruria's commander made similar suggestions, advocating a Turkish bath and a gymnasium.[20] *Slavonia*'s Captain J.G. Barr was in favour of a Turkish bath: 'The attendant [needs] to be an expert in massage, manicure, chiropody. It should be a regulation that all stewards, at least those attending on saloon [first-class] passengers should have a bath once a day.'[21]

★★★

Competition helped drive improvements in third-class accommodation. Shipping lines fought for their share of the immigrant trade, but governments were also taking an increasing interest in their welfare. In February 1907, the United States Congress passed an act creating the Immigration Commission. The commission produced forty-one volumes of reports, dealing with all aspects from immigrants' conditions at sea; to immigrants by each industry; immigration and crime; immigrants in cities; immigrants' children in school; and first and second-generation immigrants. Their investigation into conditions at sea in steerage and third-class accommodation included sending agents 'in the guise of immigrants' on twelve transatlantic steamers. Anna Herkner crossed the Atlantic three times as a third-class passenger, and prepared the report into steerage conditions. It was presented to Congress on 13 December 1909, shortly after she had crossed on *Cedric*.[22]

The February 1907 legislation had attempted to improve conditions by increasing the amount of space for each third-class passenger from 1 January 1909 but, ironically, there were issues with the legislation that made matters worse. On 18 January 1908, Philip Franklin, IMM's vice president, accompanied Mr S.C. Neale, counsel for IMM, to a morning meeting of the Immigration Commission in the room of the Senate Committee on Immigration in Washington. Neale explained to the chairman that steamship companies objected to Section 42 of the legislation, which had been intended 'to give greater space to third class passengers than had been accorded them under the old passenger act of 1882', because they thought that 'it would be greatly to their disadvantage' and that Congress had been 'under a misapprehension of the facts in the case'. Thirty-five American and European steamship companies had appointed Franklin 'to appear before the commission and explain to it the exact bearing of section 42'.[23]

Franklin opened by arguing, 'We should like very much to have section 42 of the act amended to conform more closely to the British Board of Trade regulations, which took effect 1 January 1908.' The steamship lines felt the section would 'create very serious hardship upon the various steamship companies, because it would reduce their carrying capacity of passengers from 25 to 35 per cent'. The British regulations defined 'so clearly' where passengers might be carried, 'on any deck which can be properly lighted by side ports and efficiently ventilated mechanically or otherwise', whereas the American legislation allowed 'passengers to be carried on practically the same deck that the British Board of Trade regulations do, but it compels the steamship companies to change the construction of their steamers in order to do that', because there was a dispute as to precisely what the main deck of a ship was – 'no two or three experts can agree'.

Back in 1882, ships might have had three passenger decks, but that had increased to eight or nine passenger decks. Section 42 restricted the third-class passenger accommodation to the main deck and two decks below it.

'We are prohibited from putting the third class passengers into the most desirable part of the ship,' Franklin emphasised:

> We have some ships running that carry only third class passengers, and we carry them up on this deck [he indicated on a ship plan]. It is absolutely contrary to the law, but it is the best deck on the ship [...] It is the deck on which we carry first class passengers, but by your law we cannot use it for third class.

Shipping lines were able to make sure that their ships had fore and aft well decks, thereby making the main deck lower down under the definition, but Franklin felt the British approach was much more sensible:

> The point is this: Every steamer cannot afford to give its upper decks to its third class passengers. They must reserve those for the first and second class. Every steamer is so designed that it can carry passengers in the lowest deck which can be properly lighted by side ports and efficiently ventilated mechanically or otherwise. They arrange the plans so that is accomplished.

Under the British regulations, the authorities took some account of space being used for third-class dining saloons, smoke rooms and lounges. Franklin was asked, 'There is nothing in that law to prevent you doing away with the dining rooms and the smoke rooms?'

He argued, 'We really have to have them from a commercial point of view in order to attract passengers ... It is competition. Everybody is doing the same thing. We would have to provide some. We could not do away with all of them under any circumstances.'

Earlier during the hearing, Franklin had commented:

> I do not think it is quite appreciated how anxious the steamship companies are to give their third class passengers every possible accommodation and every possible comfort that they can. Otherwise they would lose their trade. We are just as anxious to take good care of the third class passengers as you are to have us do so, because our livelihood depends on it.

Third-class passenger traffic had risen substantially. In 1895, 328,246 third-class passengers entered US and Canadian ports, while 154,071 departed; by 1905, westbound traffic reached 1,010,346, but the number of departures had only risen to 246,480. In 1907, about 1.25 million third-class passengers had entered the United States and 562,000 people had left the country as third-class passengers. Comparing the first half of January 1908 with the same period the year previously, westbound third-class passenger numbers had fallen 42 per cent, whereas eastbound traffic had more than trebled. (The onset of an economic recession in the United States was readily apparent.)

The rise in eastbound numbers had been a real help to the shipping lines. Franklin explained:

> We had the *Arabic* sail yesterday from the other side, and she carried only ninety third class passengers. Ordinarily she never thinks of going with less than three hundred. The eastbound traffic is making up for the losses westbound. We have to make it in some way or other, or we would go out of business.

Franklin thought a requirement in Italian law to determine a minimum speed for passenger ships carrying third-class passengers was not necessary. He preferred a longer voyage and discussed the subject with William Stiles Bennet, a Republican Congressman representing New York's Seventeenth District:

> Mr. Franklin: When I travel myself I always go on the slowest ship.
> Mr. Bennet: But you do not travel third class.

> Mr. Franklin: The passengers have never been more comfortable. They do not want to get off. Half of the third class who embark do not know how to eat with their knives and forks. The plan of having their women and children sit down and have a steward bring them three meals a day and supper at night is something they never heard of.

He provided the committee with a bill of fare from the White Star and American lines, two of IMM's constituent companies. For breakfast on a Sunday, third-class passengers on a White Star liner received Quaker oats and milk, smoked herrings and jacket potatoes, boiled eggs, fresh bread, butter, marmalade, Swedish bread, tea and coffee; Sunday 'dinner' (lunch) consisted of vegetable soup, roast pork, sage and onions, green peas, boiled potatoes, cabin biscuits, fresh bread, plum pudding, sweet sauce and oranges; and the evening meal, ragout of beef, potatoes and pickles, apricots, fresh bread, butter, currant buns and tea. Every day, a supper was available, including cabin biscuits and cheese. Fresh fish was 'served as substitute for salt fish' when they had opportunity.

An amendment was adopted in December 1908, superseding the February 1907 legislation. It removed the ambiguous term 'main deck' and instead brought US law into line with the British Board of Trade regulations. The change was welcomed by the shipping lines.

Conditions on new ships were improving dramatically in the first decade of the twentieth century, but there were plenty of older vessels in service. These had 'old type steerage' compartments, 'whose horrors have been so often described' and formed 'a majority of the vessels bringing immigrants to the United States'.

Although passengers had berths, food and air, sleeping quarters consisted of large compartments that could accommodate as many as 300 passengers in each. Their berths consisted of an iron framework with a mattress, pillow, 'or more often a life preserver as a substitute', and a blanket. The blankets were not sufficient to keep warm and 'generally the passenger must retire almost fully dressed' so they were not too cold. To make matters worse, even though passengers had their own bags or bundles, there was no designated baggage space and they had to keep their possessions underneath their berths. Hooks for clothing were not provided and some lines provided eating utensils for each passenger to keep throughout the voyage, but nowhere to store them. The passengers were 'looked upon as so much freight, with mere transportation as their only due'. Open deck space was 'very limited' and there were no separate dining rooms, meaning that the sleeping compartments formed the passengers' 'constant abode'. When storms occurred, the deck could not be used. Passengers only had their own berths and passageways between them to exercise or entertain themselves:

> When to this very limited space and much filth and stench is added inadequate means of ventilation, the result is almost unendurable. Its harmful effects on health and morals scarcely need be indicated [...] In many instances persons, after recovering from seasickness, continue to lie in their berths in a sort of stupor, due to breathing air whose oxygen has mostly replaced by foul gases.

Separate washrooms for men and women were required by US law, but in practice, the requirement for them to be kept 'clean and serviceable' throughout the voyage was not met. Soap and towels were not provided and floors of washrooms and toilets were 'damp and often filthy until the last day of the voyage', when they were inspected by the authorities in port.

Meal times were a challenge:

> If the steerage passengers act like cattle at meals, it is undoubtedly because they are treated as such. The stewards complain that they crowd like swine, but unless each passenger seizes his pail when the bell rings announcing the meal and hurries for his share, he is very likely to be left without food.

One female passenger waited until other passengers had rushed off for breakfast, so that she could dress 'without the presence of men onlookers', but that created another problem – twenty-five minutes later, food was no longer being served. When she explained the reason for her lateness, she was allowed some bread, but then warned 'not to use that excuse again'.

Menus sounded sufficient on paper, but in reality, the quality and preparation of the food was inadequate. Lentil soup, boiled fish, potatoes, gravy and white bread formed a typical lunch; sausage, potatoes, tea, black bread and butter an evening meal. Leftover meats were often used.

Some of the stewards were quite strict in driving men out of the women's quarters, but one steward harassed the female passengers himself – 'not one day passed' without him causing some annoyance. On another occasion, the chief steerage steward took liberties, 'such was the man who was our highest protector and court of appeal'. Crew even married girls from steerage:

> I cannot say that any woman lost her virtue on this passage, but in making free with the women the men of the crew went as far as possible without exposing themselves to the danger of punishment. But this limit is no doubt frequently overstepped. Several of the crew told me that many of them marry girls from the steerage. When I insinuated that they could scarcely become well enough acquainted to marry during the passage, the answer was that the acquaintance had already gone so far that marriage was imperative.

An interpreter, 'who could and should be a friend of the immigrants', only passed through the accommodation twice a day. Getting advice or information from him 'always failed' because he merely replied, 'How in the devil do I know?'

During a twelve-day crossing, Anna Herkner:

> ... lived in a disorder and in surroundings that offended every sense. Only the fresh breeze from the sea overcame the sickening odors. The vile language of the men, the screams of the women defending themselves, the crying of children, wretched because of their surroundings, and practically every sound that reached the ear, irritated beyond endurance.
>
> [...]
>
> There was no sight before which the eye did not prefer to close. Everything was dirty, sticky, and disagreeable to the touch. Every impression was offensive. Worse than this was the general air of immorality. For fifteen hours each day I witnessed all around me this improper, indecent and forced mingling of men and women who were total strangers and often did not understand one word of the same language. People can not live in such surroundings and not be influenced.

Newer steerage accommodation followed 'the plans of the accommodations for second cabin passengers. The one difference is that everything is simpler proportionately to the difference in the cost of passage.' Early in the twentieth century, the newer third-class accommodation was only provided by shipping lines who carried emigrants from northern Europe, who formed 'a small per cent of the total influx'. Depending on the ship, elements of both the older and newer accommodation could be found, such as a basic stateroom which provided relative privacy for up to eight people, but whose bedding was still inadequate.

White Star's 'Big Four' each had a large third-class capacity and set high standards of comfort. Passengers were 'treated with every consideration from the very beginning of his relations with the line'. Tickets were obtained easily and a train brought passengers to the port on the morning of the sailing. A bus conveyed passengers and their luggage to the dockside without any extra charge. Once on board, a doctor examined all of them and then they proceeded to an open deck. Passengers were assigned their rooms and berths, with consideration for their preferences:

> Friends and acquaintances were placed together. The various nationalities were quartered together as much as possible. The few Jewish passengers were assigned staterooms distantly

> removed from all the others. All these proceedings followed a careful plan and were kindly conducted, so that no needless crowding and rough handling resulted. The same consideration that was shown here continued throughout the ten days' journey.

Once the third-class passengers were on board and assigned their accommodation, the ship left the dock. The second- and first-class passengers boarded at the Liverpool landing stage.

Accommodation was plain, simple and 'devoid of all nonessentials', but much cleaner than on older ships. Decks were divided into sections using letters of the alphabet. (The watertight bulkheads closed off one section from another.) The stateroom partitions were wooden, 'painted white, and kept thoroughly clean':

> A current of air was admitted at the base and top of the partitions. The air on these two decks, both in the staterooms and hallways, was remarkably fresh. Berths were arranged in two tiers, and in construction apparently differed in no way from those usually found in the second cabin. Each berth contained a straw-filled mattress and pillow. Both of them were covered with white slips, which, however, were not changed during the ten days. For covering, a pair of heavy grey blankets was provided. These were of ample size and weight to be practically sufficient even on the coldest nights of the journey. At the head of the berths was a drop-shelf that served either as a seat or table, as the occasion demanded. Each room was furnished with a mirror, and hooks to hold clothes were quite abundantly supplied. A lever for turning on the electric light in the room and a bell for summoning [the] steward or matron were within easy reach of both berths. There was plenty of space for hand baggage under the lower berths and also beyond the foot of the berths. The floors everywhere were of plain white boards, and these were kept scrupulously clean. They were scrubbed every day by the stewards on their hands and knees, and were well dried, so as to avoid all unnecessary dampness.

Stewards took great care. At 9 a.m. daily, passengers were asked to vacate their rooms unless they were ill:

> The stewards then went through them all, giving each such attention as it needed – making beds and sweeping or scrubbing floors. At intervals in the hallways were placed cans to receive waste. These were not frequent and convenient enough to be used by all in cases of seasickness, but they did afford a place for other waste. If a criticism might be offered on the staterooms it could only be on the lack of cans to use in case of seasickness. The stewards were untiring in cleaning up the results of the innumerable cases of illness resulting from the rough sea during the first few days [...] Each section was in the distinct charge of a steward, who was held strictly responsible for all order in his section. A steward was always on duty in the hallways, both day and night, and lights burned all night in the passageways.
>
> Men, women and families were assigned each to separate sections [...] Staterooms are provided for 1,600. The two sleeping sections or compartments [...] contained about three hundred berths each. When the steerage is full these two rooms are used as sleeping quarters for men. The berths are constructed and supplied exactly like those in the staterooms.

The many public bathrooms mostly contained five wash basins and five toilets:

> There was always an abundance of soap, and large roller towels were supplied frequently enough to insure the presence of clean ones at all times. The floors in these rooms were of tile. These were practically always in a fit condition for use. Only when an accident occurred, occasioned by a leak in some pipe, was the floor wet, and this was remedied as soon as discovered.

Strict supervision was in place to make sure that 'men were positively kept out of the women's quarters'.

Meals in the dining saloons were taken at 'long tables seating from ten to fourteen persons':

> At meals these were covered with white cloths and each place was set with a thoroughly usable knife, fork and spoon. Bread, salt, pepper and mustard were set all along the centre of the table. Soup and meat were served from the pantry. Vegetables, preserves, pickles and sugar were placed at either end of the table in large dishes and each passenger could serve himself. Each table was in the charge of one steward, who laid the cover, served, and attended to the wants of those there seated. The service and attention were real and all that could be asked. The food was all of a very fair quality and abundant. Absolutely everything served was such as might be eaten without hesitation by anyone. The preserves served with each breakfast and the fresh fruits, apples and oranges given out several times at dinner were of an exceptional quality and would have made endurable meals of a much poorer quality. Coffee, tea, and hot water could be had by women and children at almost all hours of the day from the pantry.

Ginger ale, soda water, stout and 'smokers' necessities' were available from a bar, which opened onto the passageway near the entrance to the men's dining room, 'This received some patronage, but was not particularly popular.' Notices forbade crew from selling provisions, but there:

> ... were some such indirect dealings: passengers appeared on deck continually with apples and oranges when these had not been given out at the table [...] It proved that those who could speak English or give tips to the right persons were abundantly supplied with fruit. Nothing, however, was openly offered for sale except at the bar.

Jewish passengers were looked after by 'a Hebrew who is employed by the company as a cook, and is at the same time appointed [...] as guardian of such passengers':

> He was the first to receive such an appointment. It is his duty to see that all Jewish passengers are assigned sleeping quarters that are as comfortable and good as any; that kosher food is provided and to prepare it. He has done duty on most of the ships of the —— Line. On each he has instituted this system of caring for the Hebrews and then has left it to be looked after by some successor.

An interpreter 'who spoke English, Swedish, Norwegian, and some German was on board', but he was 'evidently not very capable'.

Outside meal times, there were plenty of hours at sea to fill, either on deck or inside:

> A large portion of —deck in the stern of the vessel was set apart for the steerage. This was quite free of machinery and provided an ample open space for games, walking, dancing and other exercises. A part of this deck was partially inclosed [*sic*] and covered, and thus afforded a shelter even during stormy weather. Comfortable benches were placed at frequent intervals [...] The deck was thoroughly scrubbed and hosed off each day [...] At night it was well lighted and might be used until 9 o'clock, when the interpreter announced it was time to go below. On pleasant evenings, when the entertainment was lively, this time was somewhat extended. There were some musical instruments among the passengers, and there was considerable singing and dancing after those first few rough days were over.
>
> For those with whom [...] the open deck did not find favour there were [*sic*] access to the dining rooms. The tables when not prepared for meals were covered with red cloths and could be used for games, writing, or any other purpose. A piano in the women's dining room found an untiring performer in a German student. Notices of the distance travelled each day were also posted in each of the dining rooms. A clock, regulated each day, was also placed there. The men might smoke in their own dining room but not in the other.

Three notices were framed and hung at the entrance to every third-class section: one was a set of regulations for the conduct of third-class passengers; the second was Section 7 of the United States Passenger Act (1882), which regulated 'the carriage of passengers by sea'; and the third informed passengers that valuables could be deposited with the purser's office.

Typically, a day in third class began with the 'wake-up' bell at 6.30 a.m. and the breakfast bell followed half an hour later. After the meal, passengers could go on deck while the stewards cleaned the staterooms and scrubbed the dining room floor. The morning inspection took place and passengers were allowed back to their rooms at 10 a.m. Bouillon was served to female passengers at 11 a.m. and lunch was served at midday. The evening meal or 'supper' was served at 5 p.m. and then 'there was a general withdrawal to the upper deck, where during the evening there was considerable singing, dancing, walking and merrymaking generally. A crowd of Scandinavians frequently played children's ring games.'

On arrival at New York, a customs official came on board with a physician. He passed through the third-class accommodation and the physician examined the passengers. After more than a week at sea, passengers on board the White Star Line's newest ships arrived in a much more comfortable condition than previous generations of migrants. 'It is safe to assume that 99 per cent of the third class passengers have never enjoyed at their own homes the comforts now provided daily as a matter of course on the *Baltic*,' wrote one passenger in an account published in August 1905.[24]

***

When the press reported *Olympic* as a 1,000ft ship in April 1908, William Watson, Cunard's chairman, quipped, 'Of course it has appealed to my own imagination as to where all the passengers are to come from to fill these enormous carriers.' (In January 1909, Leonard Peskett thought that a ship 800ft long and 88ft in breadth, designed for a service speed of 21 knots, could potentially accommodate 740 first, 730 second and 4,000 third-class passengers, for a grand total of 5,470 passengers.) The reality was that accommodating an increased number of passengers was not White Star's primary focus. By building a larger ship, they had significantly more space available to devote to passengers, but that space was used to increase and improve first- and second-class accommodation. That required more space per passenger than if they had focused on increasing third-class capacity and overall passenger numbers. The 'Design "D"' concept had envisaged:

First Class	
190 single berth rooms	190 beds
34 two berth rooms	68 beds
114 three berth rooms	342 beds
338 rooms	600 beds
Second Class	
102 two berth rooms	204 beds
128 four berth rooms	512 beds
230 rooms	716 beds
Third Class	
In enclosed rooms	1,528 beds
In open berths	260 beds
	1,788 beds
Total passengers	**3,104**

The final design capacities, as published in *The Shipbuilder* for *Olympic*, were:

First Class	
96 single berth rooms*	142 beds
106 two berth rooms	212 beds
127 three berth rooms	381 beds
329 rooms (excluding 'parlour' rooms in suites)	735 beds
Second Class	
76 two berth rooms	152 beds
2 three berth room	6 beds
129 four berth rooms	516beds
207 rooms	674 beds
Third Class	
In enclosed rooms	862 beds
In open berths	164 beds
	1,026 beds
Total passengers	**2,435**

* 46 of the 96 single-berth rooms could be converted to two berths if needed. These totals assume they were used as two berth.

Total passenger numbers decreased by 22 per cent. First-class passenger numbers were increased with fewer single-berth staterooms. Second-class passenger numbers fell slightly with a reduced number of two-berth staterooms and those in third class were reduced significantly. Total third-class capacity decreased more than 40 per cent. *Olympic*'s first' and second-class passenger capacities were higher than other liners:

	First Class	Second Class
Mauretania	560	475
Rotterdam	505	580
George Washington	520	377
Olympic	735	674

For White Star, the trend was to increase first- and second-class passenger accommodation, both as an absolute number or as the proportion of passengers carried.[25] This was true of both their 'fast mail' express steamers and intermediate ships such as the 'Big Four'.

		First	Second	Third
Majestic	1890	300	190	1,000
Oceanic	1899	410	300	1,000
Olympic	1911	735	674	1,026
Celtic	1901	347	160	2,350
Cedric	1903	365	160	2,350
Baltic	1904	425	450	2,000
Adriatic	1907	425	500	2,000

Olympic's passengers enjoyed an average of over 50 per cent more space per passenger than their counterparts on board *Mauretania*.[26] The increase

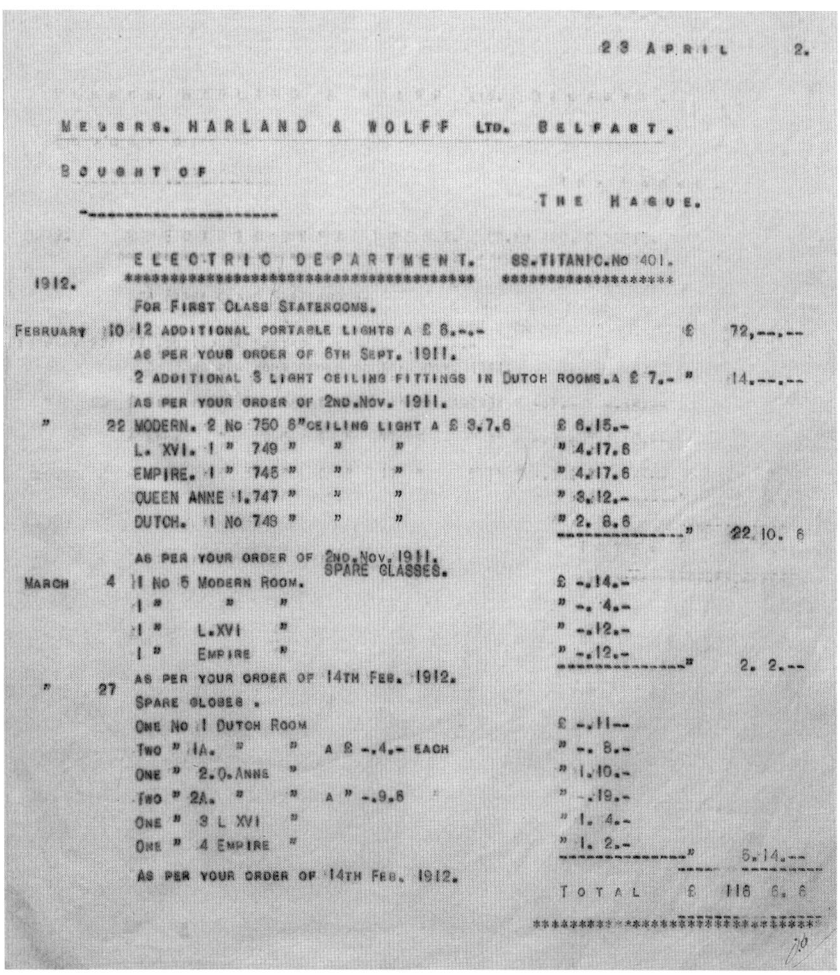
23 APRIL 2.

MESSRS. HARLAND & WOLFF LTD. BELFAST.

BOUGHT OF

THE HAGUE.

ELECTRIC DEPARTMENT. SS.TITANIC.No 401.

1912.

FOR FIRST CLASS STATEROOMS.

FEBRUARY 10 12 ADDITIONAL PORTABLE LIGHTS A £ 6.-.- £ 72,--.--

AS PER YOUR ORDER OF 6TH SEPT. 1911.

2 ADDITIONAL 3 LIGHT CEILING FITTINGS IN DUTCH ROOMS.A £ 7.- " 14.--.--

AS PER YOUR ORDER OF 2ND.NOV. 1911.

" 22 MODERN. 2 No 750 8"CEILING LIGHT A £ 3.7.6 £ 6.15.-

L. XVI. 1 " 749 " " " " 4.17.6

EMPIRE. 1 " 745 " " " " 4.17.6

QUEEN ANNE 1.747 " " " " 3.12.-

DUTCH. 1 No 748 " " " " 2. 8.6 " 22.10. 6

AS PER YOUR ORDER OF 2ND.NOV.1911.

SPARE GLASSES.

MARCH 4 1 No 5 MODERN ROOM. £ -.14.-

1 " " " " -. 4.-

1 " L.XVI " " -.12.-

1 " EMPIRE " " -.12.- " 2. 2.--

AS PER YOUR ORDER OF 14TH FEB. 1912.

" 27 SPARE GLOBES .

ONE No 1 DUTCH ROOM £ -.11--

TWO " 1A. " " A £ -.4.- EACH " -. 8.-

ONE " 2.Q.ANNE " " 1.10.-

TWO " 2A. " " A " -.9.6 " -.19.-

ONE " 3 L XVI " " 1. 4.-

ONE " 4 EMPIRE " " 1. 2.- " 5.14.--

AS PER YOUR ORDER OF 14TH FEB. 1912.

TOTAL £ 118 6. 6

Harland & Wolff placed all manner of smaller orders with H.P. Mutters en Zoon. In July 1911, they needed additional chairs and tables for *Titanic*'s first-class reception room. Orders placed in September and November 1911 and February 1912 included additional portable lights, ceiling lights and other fittings. (1392-01. 1.2.4.30, Haags Gemeentearchief)

in space available for passenger accommodation was used to expand first and second class and allocate more space per passenger. She was simply so much more spacious than other ships.

***

First-class passenger demand drove improvements to their accommodation and catering provision. Cunard felt it keenly, too. William Watson wrote in February 1908:

> I have letters talking about the food and why don't we have a restaurant. One man writes me that he got £5 deducted from his passage money on the German boat and had his meals in the restaurant, which cost, over and above his passage money, rather over £10 including his wine, and there's where the strange thing to me always comes in. People will fight for £1 or £2 and even take worse accommodation, and then will go and spend £5 extra on their wine bill.[27]

Fierce competition for first-class passengers included providing better food and choice for them. Early in 1911, one of his colleagues pointed out:

> Shortly, more competition will be against us, and from what I am told great efforts are being made, particularly through the catering to attract passengers. I am not only referring to the *Olympic*; the Norddeutscher Lloyd are supplying their steamers with everything obtainable in the markets regardless of price; bonuses are given if passengers have been satisfied, and the cost of catering is not mentioned. The Hamburg American Line are not quite so extravagant, but [...] anxious to do all they can to satisfy their patrons. The Holland-American Line are also making special efforts.[28]

Adolphe Smith, who sailed on board several Cunarders in 1908, wrote a letter outlining his views on catering at sea. He noted the differences in taste between passengers of various nationalities. Smith was concerned at what he saw as the inferior food on offer on board some British vessels, which put them at a disadvantage compared to competitors:

> What I feel and feel very strongly is that firstly other nationalities, French, German, Spanish, etc., also travel on British ships, and after all, there is a minority of English people and even of Americans who have travelled and observed with sufficient intelligence to acquire some knowledge of the gastronomic arts. So much is this the case that even in my limited experience, I have not only continental, but British and American friends who prefer going to Havre to cross the Atlantic on the French ships [...] solely because of the superiority of the French cuisine [...]
>
> My point is this, that the advantages [...] from the undoubted superiority of the British seamanship and British shipbuilding should not be in part effaced by the greater skill and art of the foreign cooks.[29]

He had found *Caronia*'s first-class catering rather disappointing, compared to steerage. In steerage, he tasted a dish which was 'aromatic, pungent, odoriferous and really appetising':

> Not only did it contain a liberal supply of paprika but it was well flavoured in other respects. Then the meat had been carefully cooked, it obviously had not been allowed to boil, so it was tender like butter. It was far and away the best made dish I was able to find on the *Caronia*. The purser explained that the art of making a goulash had been acquired during the Mediterranean cruises and was due to the influence of the Hungarian emigrants. I was greatly impressed by this incident. What a picture we have here, the standard of art in cooking is raised to a higher level in the steerage by poor Hungarian emigrants while it remains stagnant amid the wealthy but uncultured saloon [first-class] passengers.

Smith sampled *Lusitania*'s first-class menu on 4 November 1908:

> There was an insipid tasteless *Consommé Valenciennes* and then I ordered, from the menu,

> fillets of *Flounder-Ravigotes*; the flounder was skilfully fried, but the *ravigote* was not reviving at all; it had little or nothing of the 'pick-me-up' about it; chervil, tarragon, vivette, eschalot, capers and other stimulating herbs, etc., which give life to the oil and yellow of egg of a *ravigote*, were more noticeable by their absence than their presence. But after this, I had some jugged hare which was really excellent. The little sweet onions had been carefully browned, the mushrooms were plentiful, there was wine in the sauce which was unctuous and appetising. A real cook had prepared that dish. And then I wound up with a *meringue sicilienne* which was quite a *chef d'oeuvre*. The thick cream was not insipid but absolutely delicious and the meringue paste all that could be desired.

After several days at sea, Smith was introduced to the Austrian chef:

> He at once agreed the dishes were tasteless, but pleaded, as I fully expected he would, that the fault was with the passengers and not with himself. He had been warned when he first cooked for the English and Americans that he must unlearn all he had learnt and refrain from attempting to make things really tasty and appetising. One thing made it almost impossible to cook properly. It is the habit that both English and Americans have of putting a lot of salt and pepper on their food before tasting it. Of course this is enough to spoil any delicate sauce. In France a person who helps himself to salt and pepper without first tasting is considered rude and badly brought up. The cook is supposed to flavour to the taste of the majority; and, only when he had failed to do so, or when a person has an altogether abnormal or exceptional taste, is it allowable to add salt or pepper to the dish. But as the English and Americans have not, in this respect, at least, been taught good manners, the chef on board the *Lusitania* naturally found it very difficult to know how to proceed. He had to flavour a sauce not as it should be flavoured, but in such a manner as to render it less liable to be utterly spoilt by the consumer.

The chef 'determined to show me that, in spite of the restrictions of the menu, he could cook'. At dinner with his table companion, Charles Mortimer Dally, Smith enjoyed a *chateubrian bordelaise* 'that was so resplendent with the aroma of eschalot that it [caused] quite a sensation, for, as the lid was removed, people sitting near sniffed and wondered from whence came so appetising a perfume [...] It was very good indeed; and people would not complain of the tasteless, unappetising cooking on English ships if such dishes as those which were specially made for me were more generally served.'

Smith thought that dishes should be prepared and flavoured continental style. The option could be made clear on the menu, accompanied by information about the history of a particular dish or sauce:

> The chef will offer us one or two dishes accommodated with a béchamel sauce [...] King [...] Louis XIV, employed Louis de Bechamel, Marquis de Nointel as Lord High Steward of his household. It is true that this nobleman has been described as a prodigy of meanness and a gigantic parasite, but he was nevertheless recognised to have been, though an idiot, yet an inspired idiot. In one instant he did what professional cooks had striven for years to accomplish, he immortalised his name by mixing velvet down sauce with an equal amount of cream and then created the sauce 'Bechamel'.

He concluded:

> The dish would be ordered out of curiosity, and if really well done would rebound to the credit of the ship. But the description given should serve to warn off the plain roast and boil people who might be incapable of appreciating the ingredients, the high flavouring and art of the chef.
>
> In any case, it does seem to me that better use should be made of the talents of the chef. Something should be done to render it more evident that the refinements of continental cooking can be obtained, even on a British ship. The time has gone when we can be content

with saying 'they manage these things better on the continent'. We have got to manage them better ourselves.

HAPAG's *Amerika* had introduced the first extra-tariff à la carte restaurant in 1905, in addition to the first-class dining saloon. The general manager of London's Carlton Hotel, who examined *Amerika*'s restaurant during her maiden voyage, thought its popularity had 'surpassed all expectations':

> Men and women in evening dress dined in the restaurant, decorated with lilies of the valley, roses, and orchids, and listened to the Hungarian band as closely as if they had been in the Carlton or the Waldorf [Astoria Hotel, in New York] [...]
>
> The refrigerating machinery worked perfectly, enabling us to have fresh supplies, peaches and grapes, as well as poultry and game, throughout the voyage. The ventilation of the restaurant surpassed anything I have seen on land.
>
> There are no fixed hours for meals. Passengers can promenade the deck and come in when they feel hungry. The restaurant is open from 7 o'clock in the morning until midnight.[30]

White Star decided that a restaurant was essential on *Olympic*. It was not included on the July 1908 plans, but J. Bruce Ismay wrote, 'It has been decided to install an restaurant,' on 25 January 1910. The standard of food would be as high as anything ashore.[31] One writer, commenting about the improvements in catering at sea in June 1912, explained:

> Some people eat to live, but the majority live to eat. Exemplification of this fact will be found on any one of the great ocean steamers, where luxurious living is enjoyed as it never was before. Those who have not taken an ocean voyage have little comprehension of the superabundance of luxuries that all the great steamers provide, some of them in greater degree than others [...] Now, on any of the larger steamers, the passengers snap their fingers at the waves. They are as much at home as they are in a hotel on land.[32]

The majority of *Olympic*'s passengers were first or second class, who expected a high standard of service. Consequently, the increase in the victualling staff was considerable. Intermediate ships such as *Celtic* carried about 179 victualling staff (one to every sixteen passengers), reflecting that so much of her capacity was third class. However, on *Mauretania*, the number of victualling staff was about 323 (one to every 6.7 passengers), reflecting that she carried more first- and second-class passengers. *George Washington* carried about 283 victualling staff (one to every 10.4 passengers), which was again influenced by her large third-class capacity. On *Olympic*, the number of people in the Victualling Department increased to nearly 500 (one to every 4.9 passengers). The à la carte restaurant and other new features contributed to this as well as a higher general standard of service. All of this had to be planned for when arranging the crew accommodation and it was yet another factor why the ship had to be so large.

***

The working and crew areas were practical and utilitarian compared to the passenger accommodation. This is a 'dishwasher' aboard *Oceanic* about a decade earlier. *The Lady's Magazine*, *c.* 1900. (Author's collection)

The Shipbuilder devoted a 'special number' issue to *Olympic* and *Titanic*, describing the 'unrivalled extent and magnificence' of the passenger accommodation. White Star and Harland & Wolff created what was, in many ways, an enlarged and improved version of their preceding ships that had proved so successful. They chose to provide an unprecedented range of facilities for first-class passengers: Turkish and electric baths, the first proper swimming pool afloat, a gymnasium and even a squash court. Innovations such as the reception room and the first extra-tariff restaurant on a British ship highlighted that a transition was taking place, where an Atlantic crossing was genuinely more like spending time in a luxurious hotel ashore.

The range of spacious first-class 'parlour' suite and staterooms on B and C-Decks, in particular, was remarkable. Ordinary first-class staterooms frequently used standardised fittings similar to other ships, with incremental improvements for passenger comfort, but the simple provision of much more spacious accommodation was beneficial. They provided more first-class lifts (elevators) than any other ship and introduced the first lift in second class. Second-class passengers enjoyed a standard of comfort that was arguably as good as first class a generation earlier. (Simply having a large enclosed promenade deck area was something that first-class passengers had not enjoyed until *Rotterdam* entered service in 1908, but now second class had theirs too.) Their spacious public rooms were scarcely comparable to the much smaller accommodation on earlier ships. Third class continued the higher standard of accommodation seen on ships such as the 'Big Four'. Passengers' experiences across all three classes bore scant resemblance to that on so many older passenger ships.

6

'WE HAVE REACHED THE LIMIT ...'

While *Olympic* and *Titanic* progressed from design to physical reality, there were serious practical obstacles to overcome. Trading conditions were poor in 1908, then they began to recover in 1909 and 1910; the White Star Line had to raise the necessary finance and the facilities at New York and Southampton needed improvements to accommodate them. Lord Pirrie reportedly said 'he was quite prepared to build a thousand foot liner of 50,000 tons. It was merely a question of money – and harbour accommodation.'[1]

Writing for the *New York Times* in early January 1909, Philip Franklin lamented the tough times facing the shipping lines.[2] Passenger traffic in 1908 reflected a 'tremendous depression', which had set in during September and October 1907. It was 'the worst and most unprofitable year that the transatlantic lines experienced in many years'. He spoke about the 'promise of a return to prosperity' in 1909 and noted that they were planning for that future:

> The most important announcement of new tonnage contracted for the Atlantic trade was made by the White Star Line, who stated that they have ordered from the celebrated yards of Messrs. Harland & Wolff, Belfast, two steamers that when completed will be the largest and most luxurious vessels ever constructed, having a gross registry of about 45,000 [tons] each. These mammoth vessels which are to be named the *Olympic* and *Titanic*, have been particularly designed for the New York–Southampton service of the White Star Line, with a speed of not less than 21 knots.

President J. Bruce Ismay noted in IMM's annual report that a surplus of $4,033,780 in 1907 had turned into a deficit of $1,729,982 in 1908. He explained to shareholders that the deficit:

> … was chiefly due to the business depression […] which cut down Westbound freights heavily in the first half of 1908, reducing the tonnage below that for the same period for a number of years and making it necessary for a number of the lines to lay up several steamers.[3]

(Competition on the Atlantic was intense. Albert Ballin's plans for a common agreement on sailing schedules between the major lines were contentious.) Ismay said, 'You can confidently look for a material improvement in results.' Their new ships would be ready to take advantage of the upturn. White Star laid the groundwork for raising the necessary finance.

***

On 1 August 1907, Harland & Wolff signed off an agreed schedule of 'instalments due' for Yard Number 400. The same would apply for her sister. Payments from the White Star Line would fall due to the shipbuilder at five stages of construction:

when the keel was laid; when the ship was framed to the height of the double bottom; when the ship's hull plating was completed; when the ship was launched; and when the ship was delivered. The payments schedule was the same arrangement used for the majority of ships completed by Harland & Wolff for many different shipping companies in the 1890s and 1900s. White Star had to invest a huge sum of money in building the ships before they would see any profit from them.[4]

J. Bruce Ismay, Harold Sanderson, Lord Pirrie and the remaining IMM directors considered the best way to finance the project. From 1903 to 1908 inclusive, IMM paid about 72 per cent of its cumulative net earnings just to pay the interest due on its debt. In 1904 and 1908 IMM reported overall losses. By contrast, White Star was profitable throughout, partially offsetting losses elsewhere in IMM. The result was that IMM's debt (bonds traded on the stock market) was trading at distressed prices. They were discounted by about 30 per cent relative to their issue price.[5] The interest rate IMM paid to its debt holders was 4.5 per cent on the redemption price, but an investor buying the bonds at the lower market price got a yield of over 6 per cent. If IMM borrowed any more money, the interest rate would have to match the market price, driving up its borrowing costs and making the situation even worse.

Unsurprisingly, J. Bruce Ismay and his colleagues concluded that White Star, which on its own was much more creditworthy than the IMM combine as a whole, should borrow the money They explained in IMM's annual report in 1908, 'This was considered by your directors a much more satisfactory method of financing than an attempt to sell the International Mercantile Marine Company's Four and One-Half Per cent Bonds at their present low market price.'[6] White Star's borrowing costs were much lower.

The money White Star borrowed would benefit IMM in another way. They decided that the funds raised could 'only be used in payment for new property and the total amount of debentures [bonds] issued cannot exceed three-fourths of the actual value (not exceeding cost) of the new property [ships]'. The consequence was that 'at least twenty-five per cent of the value of the new property acquired' would provide additional security underlying IMM's own bonds 'without increasing the outstanding amount of such bonds'.[7] Following an assessment of value of the White Star Line's fleet, on 22 September 1908, Messrs C.W. Kellock & Co. reported:

> The vessels are twin screws, and aggregate 311,403 tons gross register. They were all constructed by Messrs. Harland & Wolff, Limited, Belfast, with the exception of the *Cretic*, which was built by Messrs. R. & W. Hawthorn, Leslie & Co., Limited, Newcastle, and they are specially adapted for the particular trades in which they are employed.
>
> After careful consideration we estimate the value of the above-mentioned steamers to be £4,850,000.

On 5 October 1908, White Star issued a prospectus to potential investors, setting out why they were borrowing the money and why. Their existing fleet provided security against the money being borrowed.[8] They explained that the company was building 'two steamships for a Canadian service, and has recently placed with [Harland & Wolff] an order for two steamships of large size and of the highest class for its Southampton and New York Royal Mail Service. The cost of these four vessels is estimated at £3,600,000' (the vast majority of that cost was accounted for by *Olympic* and *Titanic*, at about £1.5 million each). During the preceding ten years, 1898 to 1907, the company's profits had been rising steadily: between 1901 and 1907 they increased from £454,024 to £848,486. Eighty per cent of the fleet had been built since 1898, so it was modern and profitable. The new ships would generate more profit, ensuring the investors would be paid back.

They were authorised to issue £2.5 million in debenture bonds, but they only issued half:

> Messrs. Glyn, Mills, Currie & Co., 67 Lombard Street, London, as Bankers of the Company, are instructed to offer, on behalf of the Company, for subscription:

£1,250,000 4½% First Mortgage Debentures

The First Mortgage Debentures form part of an authorised issue of £2,500,000, which will be secured by a Trust Deed creating a specific First Charge upon twenty-three vessels of the Company's fleet, valued at £4,850,000 and upon four additional vessels to be built for the Company by Messrs. Harland & Wolff, Limited, at the estimated cost of £3,600,000.[9]

Investors paid an issue price of £97 10*s* for each bond ('£5 on application, £22 10 shillings on allotment, £35 on 16 November 1908, £35 on 31 December 1908'), giving them a yield of 4.6 per cent on their initial investment or 4.5 per cent based on the redemption price of £100. IMM's annual report for 1909 noted:

The debenture bonds of constituent companies held by the public [...] include £1,250,000 debentures issued by the Oceanic Steam Navigation Company, Limited, in 1908, to provide for the construction of new tonnage, and from the proceeds of which payments are now being made on account of the new steamships *Olympic* and *Titanic*, as they become due.[10]

The bonds White Star issued were held almost entirely in the United Kingdom. Lord Pirrie, writing to Lord Dufferin in June 1899, argued, 'I look upon steam ship debentures [bonds] as one of the most secure of all investments.'[11] Many private individuals felt similarly. It appears that J. Bruce Ismay himself made a personal investment. His monthly office accounts for November 1908 included the initial instalment on an 'allotment of £84,700' of the bonds.[12]

The Company has at present building in the yard of Messrs. Harland & Wolff, Limited, of Belfast, two Steamships for a Canadian Service, and has recently placed with the same firm an order for two Steamships of large size and of the highest class for its Southampton and New York Royal Mail Service. The cost of these four vessels is estimated at **£3,600,000**.

The net earnings of the Company, before charging interest and providing for depreciation, have been certified for the past 10 years by Messrs. Price, Waterhouse & Co. as under :—

30th September, 1908.

To the Directors of The Oceanic Steam Navigation Company, Limited.

We beg to report that we have examined the books of The Oceanic Steam Navigation Company, Limited, for the ten years ending 31st December, 1907. We have regularly audited the books for the last five of these ten years and we have made certain adjustments as required in the profits of the earlier years, in order that the accounts of the whole ten years may be prepared upon the present basis.

We certify that the profits, after charging all outgoings, including repairs and renewals and all expenses of management, but before charging interest on borrowed money and Income Tax, and before providing for depreciation on steamships and other property, have been as under :—

		£	s.	d.
1898	...	523,838	9	0
1899	...	444,590	2	4
1900	...	780,900	19	0
1901	...	454,024	11	0
1902	11 months ...	451,670	8	9
1903	13 months ...	651,622	14	8
1904	...	502,730	8	10
1905	...	712,679	3	4
1906	...	805,309	9	3
1907	...	848,486	7	2
Total of 10 years	...	£6,175,852	13	4
Average of 10 years	...	£617,585	5	4

(Signed) Price, Waterhouse & Co.

Messrs Price, Waterhouse & Co. certified the Oceanic Steam Navigation Co. Ltd (White Star Line) net profits for 1898 to 1907 inclusive. It was important information for the investors who were purchasing the bonds because it indicated the company was profitable and would have the ability to pay them back. (The National Archives (UK))

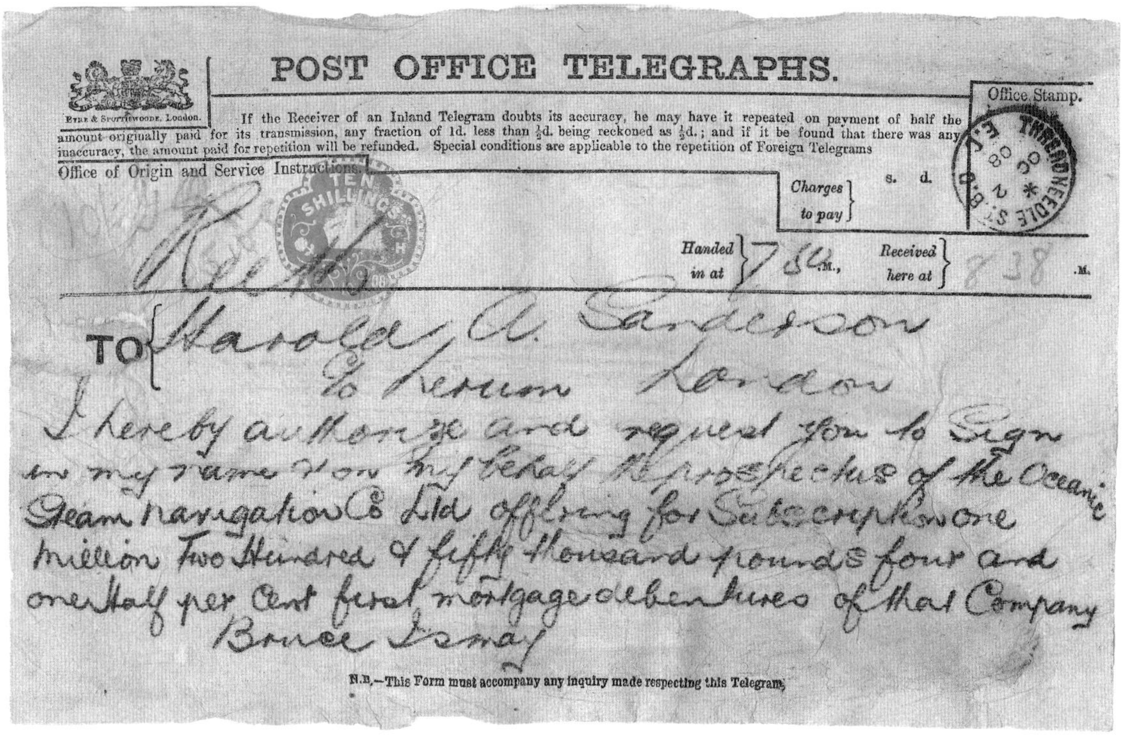
POST OFFICE TELEGRAPHS.

If the Receiver of an Inland Telegram doubts its accuracy, he may have it repeated on payment of half the amount originally paid for its transmission, any fraction of 1d. less than ½d. being reckoned as ½d.; and if it be found that there was any inaccuracy, the amount paid for repetition will be refunded. Special conditions are applicable to the repetition of Foreign Telegrams

Office Stamp.

Office of Origin and Service Instructions.

Charges to pay s. d.

Handed in at 7.54 M., Received here at 8.38 .M.

TO Harold A. Sanderson London

I hereby authorize and request you to sign in my name & on my behalf the prospectus of the Oceanic Steam Navigation Co Ltd offering for Subscription one million Two Hundred & fifty thousand pounds four and one half per Cent first mortgage debentures of that Company
Bruce Ismay

N.B.—This Form must accompany any inquiry made respecting this Telegram.

J. Bruce Ismay sent a message to his fellow director, general manager Harold Sanderson, authorising him to sign off the prospectus on his behalf. (The National Archives (UK))

There was quite a bit of comment in the shipping press about the amount of capital the major shipping lines were having to raise. HAPAG argued that *Lusitania* and *Mauretania* were having 'a most disastrous influence upon the shipping trade'. The German lines were under pressure to compete. There was 'a natural desire to build equally large or larger vessels' and 'the necessity to construct giant vessels has precipitated an era of most over-burdensome expense'. *Lloyd's List* had little sympathy, writing that Albert Ballin had 'by his own actions done much to stimulate Atlantic luxury'. He was 'forcing the pace all round'.

***

IMM's directors were 'confident' that *Olympic* and her sister would 'at once establish themselves in the favour of the travelling public'. Prosperity was returning:

> The business during the first half of 1910 as compared with 1909 has shown a distinct improvement, though still falling somewhat below the record year of 1907, and it is expected that the results for this year [1910] will compare very favourably with those of the previous two years.

White Star's own profits fell 65 per cent in 1908. They recovered to £1,057,519 in 1910, representing a seven-figure sum for the first time.

It was White Star's financial strength that helped support IMM. In 1903, IMM's management had argued that the combine's existing fleet was capable of generating net earnings of $9.5 million each year. They calculated it on the 'average rate earned per ton' during 1897 to 1902 and did not account for 'any savings' they might make by reducing 'unnecessary duplication of expenditure'. Unfortunately, net earnings only reached $8,004,035 in 1906 (they would not exceed $9.5 million until 1913). The anticipated savings did not materialise. In 1910, White Star's own reported net earnings equated to the majority of the reported net earnings for IMM as a whole.[13] Poorer returns from other lines in the combine were a drag on its overall results.

White Star declared substantial dividends – payments which, of course, were distributed to IMM as the controlling shareholder – but IMM itself remained unable to pay dividends, even on its preferred shares.[14] The combine needed the funds to pay its own debts. By 1912, IMM's net earnings adjusted for the size of their fleet stood at less than half the equivalent for Cunard, Norddeutscher Lloyd or HAPAG.[15] IMM had more debt than HAPAG, Norddeutscher Lloyd or Cunard, whereas White Star, considered on its own, did not. IMM's combination of lower net earnings and higher debt was not good.[16]

The combine's profit-sharing agreement with the major German lines was not successful. HAPAG and Norddeutscher Lloyd received payment from IMM regardless of whether the conditions for them to pay IMM were met. The healthy subsidies anticipated from the United States government did not materialise. Then, the failure to stymie competition contributed to reductions in ticket prices amid 'rate wars'. An agreement 'to fix minimum rates' for first-class passengers between the major lines, effective from 31 March 1902, had included twelve companies before Cunard withdrew in 1905. Cunard reduced rates on the new *Lusitania* in August 1907, then the French Line undercut them on eastbound fares. IMM followed, reducing eastbound rates on the American Line ships and White Star's *Oceanic*, *Majestic* and *Teutonic*. Rate cuts followed on westbound traffic and then the seasonal winter fares. Competing lines responded quickly to reductions.

Cunard's chairman, William Watson, bumped into J. Bruce Ismay and Harold Sanderson on 22 October 1907. They were concerned about HAPAG. He suggested, 'Why don't we ignore Ballin altogether, and restore rates if everybody else wants to do it, instead of foolishly chucking away money.' By mutual agreement, Ismay raised it with Norddeutscher Lloyd, 'Would it not be possible for the lines to revert to the agreed first and second class rates' and ignore HAPAG 'entirely'? He argued that approach 'would be an object lesson to Mr. Ballin, in so far that it would show him that it was not the intention of all those interested in the North Atlantic trade to be controlled and dictated to by him'.

On 10 January 1908, *Lloyd's List* reported White Star had cut its westbound second- and third-class rates, only for Cunard to follow suit 'with a corresponding reduction'. White Star were unhappy that passengers did not pay more for *Lusitania* or *Mauretania*. Watson referred to it as 'this little war which has arisen between the White Star and ourselves'. He had argued privately, 'If rates are reduced these ships will attract all the passengers, because people will willingly pay £23, £24 and £25 for such accommodation, comfort and speed as we can give them, and no one else can.'

On 31 January 1908, the leading shipping lines met in London to discuss first- and second-class rate agreements and a 'general pooling scheme' for third-class business. Their discussions went on for days. Cunard was alarmed at a proposal for a minimum first-class fare of £30 on *Lusitania* and *Mauretania*, because it thought it was far too high. On the other hand, the continental lines thought Cunard's suggestion of £24 was 'far too low' because 'the lines wanted to make money, and were not existing for philanthropic purposes'. Harold Sanderson and the Anchor Line's Richard Henderson supported Cunard and the rate was reduced to under £26. White Star secured a minimum first-class rate of under £23 for *Adriatic*, under £22 for *Oceanic* and under £19 for *Teutonic* and *Majestic*.

The discussions reflected all the complexities of shipping lines' day to day operations. It was 'unreasonable to expect lines with high rated ships – such as the *Lusitania* and the *Kronprinzessin Cecilie* to charge passengers who wished to take servants, the high minimum rate for their servants'. They agreed to a compromise for lower rates, which 'must not be below second cabin rates'. Then there was the issue of motor cars, which were often carried as cargo free of charge: 'They should be carried as freight – and charged for.'

Cunard advocated 'fair and equitable terms' for any pool of third-class business but pointed out 'they had more tonnage and they had more capacity' than in earlier years, which needed to earn them a higher share. They all came to an agreement to divide the third-class business between shipping lines according to an agreed percentage, with penalties for companies who did not comply. It covered about 90 per cent of the third-class passenger traffic to the United States. The new rates under the agreement were issued on 10 February 1908, binding for three years.[17]

SOUTHAMPTON AND NEW YORK

Olympic and her sister were taking shape at Belfast, but the unfortunate fact was that the port facilities at Southampton and New York were inadequate. During a general discussion about the increasing size of steamships early in 1907, Emile Boas, general manager of HAPAG, explained some of these concerns:

> Larger steamships mean a greater depth of channel, longer piers, and more space between them. Practically, the size of the future steamship depends on the depth of the channel [...] We have at present at this port [New York] a depth of 34 feet at high water in the channel. This allows us to load to a depth of 31 feet and a half. We must have at least two feet and a half under the keel. It is expected that the Ambrose Channel will have a depth of thirty-five feet, and this will allow for a slightly deeper draught. [...]
>
> The length of steamships has now reached about 800 feet, and there is no reason, from a technical point of view, that we should not have the thousand foot steamship. We will be able to save in working expenses in the large size steamships. The question, however, has many sides. The present piers are 800 feet long. They would have to be longer with longer vessels, and then, again, with greater length comes greater beam, and as the present space between the piers is only 250 feet, it would mean another change. We have reached the limit, I believe, with our present facilities.[18]

Boas' views were born out by comments in *Lloyd's List & Shipping Gazette* on 30 May 1907. Existing ships were encountering issues in New York:

> The grounding of the *Baltic* in the lower Bay of New York this week is precisely what has happened to other large vessels. Sunday afternoon a week ago it was the *Minneapolis* that stuck in the mud. On the last day of April it was the *Celtic* that blocked the channel. Other liners which have recently discovered the uncertainties of channel navigation in the Sandy Hook vicinity are the *Caronia*, *Etruria*, and the *Deutschland*. These grounding troubles are due primarily to the fact that the harbour is trying to do business on the basis of the shipping requirements of twenty years ago. Ocean carriers have become veritable giants within a decade. It is only now, when the task should have been completed, that the work is really in progress of enlarging the old channels to meet the new demands.[19]

The answer lay in improving the port facilities, further dredging and extending the piers. White Star were acutely aware of the situation. New York did not have an encouraging history: an application by the Dock Commissioner to extend the piers of the North (Hudson) River was refused in December 1903, another was declined four months later and a final appeal was declined in July 1908. White Star made their first application on 5 January 1910, which was refused on 3 March 1910.

On 31 May 1910, Charles P. Sumner, as the manager of Cunard's New York office, wrote to Alfred Booth, Cunard's chairman, about a short talk he had with Philip Franklin, 'He was not very communicative, but I judge at [the] present time that he is very much in the clouds as to what he will be able to do.' Sumner continued:

> I indicated to him that I had been informed he had been to Washington on the subject and been turned down, and from his reply I judge such is the case – that he has abandoned all hope of making any progress with the present officials,

> but that he hopes to make a renewed and more successful effort either late in the winter or early in 1911.

'It is a good deal of a conundrum to know what docking facilities the White Star Line will be able to secure for their *Olympic* and *Titanic*,' Sumner reported on 18 July 1910:

> What concerns us more closely is what docking facilities we could get should we build a steamer about 850 to 870 feet over all. I think it would be well should we be able to ascertain the tonnage of the suggested new Hamburg American steamer, to exceed their figures by a few tons – sufficient to be able to advertise the largest steamer – I think we need have no fear about the fact that we will own the fastest.

Cunard already had plans to build a liner larger than *Lusitania* and *Mauretania*:

> I do not suppose either the New York Dock Department or the War Department at Washington would raise any question of our [pier] extension into the river – there would always be the danger of some ferryboat or tug running into us and at times a heavy flow of ice during the winter months might make it difficult to secure the steamer.

Although lengthening the piers was 'an unknown quantity' and there were practical risks without those improvements, Sumner thought that 'if any extra length is necessary [...] in order to secure a tonnage in excess' of new German tonnage, 'I would rather take this risk than to be shut off from the privilege of advertising the largest steamer in the world.'

By 14 October 1910, Sumner had 'not been able to learn of any change of plan' by White Star. He understood that Cunard's rival had considered a plan 'of mooring these big floats or pontoons to their steamers', but thought that plan would not work. He suggested 'a pontoon of sufficient width to cover our [ship's] overhang moored strongly to the dock'. Alfred Booth wrote, 'We now know definitely that the length of the *Olympic* and *Titanic* is 850 feet between perpendiculars, so the problem of solving the pier difficulty will rest with them [White Star], and I have very little doubt that a practical solution will be found.' Booth added that a set of plans for Cunard's proposed new liner would be sent to the New York office for their criticism and suggestions, 'You will have to be very careful that no information gets out about these plans, and that no one in your office sees them except those you can absolutely trust.' Sumner noted that White Star 'have interested Mr. J.P. Morgan in bringing all the influence he has to bear upon the government at Washington to permit some increase in the length of their piers' on 17 December 1910.[20]

Early in November 1910, J. Bruce Ismay arrived in New York aboard *Mauretania*. He told reporters, '*Olympic* must go somewhere when she arrives here next June.' When they informed him of the harbour authorities' objection to extending the necessary piers, Ismay recalled the same issue when *Oceanic* was nearing completion over a decade earlier, 'I cannot see what difference another 100 feet would make to the tides.' He hoped 'that the War Department will relent and change its attitude toward the extension of the Chelsea piers before that time'. (There was keen speculation about who would command the two new ships. Ismay was quoted telling reporters that *Adriatic*'s Captain Smith 'would probably have' charge of *Olympic* and *Oceanic*'s Captain Haddock 'would have the *Titanic*'.)[21]

On 5 December 1910, Franklin wrote to Jacob Dickinson, Secretary of State for War, whose consent was required. The letter represented an appeal against the decision to refuse permission for pier extensions. He argued that the Ambrose Channel had already been dredged to a depth of 40ft at 'immense' cost, for the very purpose of making sure that 'the very largest steamers possible' could enter and dock in New York. 'This expenditure, without any provision for adequate terminal facilities, would be a needless waste of money,' Franklin explained. 'It seems unreasonable for the government to prevent the City of New York from developing its facilities.'

Morgan 'took up the matter' with Dickinson but 'little progress' was reported. He promised 'to act without delay' on the application, but pointed out 'that President Taft, when Secretary of War, had once refused to approve of lengthening the docks along the North River, and Secretary Root had twice made a similar refusal'. Dickinson also referred to a collision between a German and a French liner on 21 November 1910, 'He could not consent to endanger shipping by approving any lengthening of docks along the river.'

Following reports that IMM were considering building piers at Montauk Point for the new liners, which would take two years and involve new railway facilities, Franklin was quoted in the *New York Times* on 9 December 1910:

> If they [the War Department] persist in this attitude we must look for facilities elsewhere. The piers in South Brooklyn are long enough, but there is not enough water there for big ships at present. We have Boston, Staten Island, Montauk, and other points under consideration.

Early in January 1911, matters were looking up. Dickinson had referred the application to the Harbour Line Board, who would hold a public hearing to consider it. Franklin explained to a reporter:

> At the present time the navigable space in the North River at that point is 2,750 feet, and the proposed extension, would still leave 2,650 feet which would be ample [...] Our idea is to have a wooden pile extension composed of three or four tiers of piles, so that the water could flow freely underneath without interfering with the river currents in any way.

He had written to Mayor Gaynor and Dock Commissioner Tomkins, who were both 'heartily in favour [...] and agree with me that it would be a great pity if the *Olympic* and *Titanic* had to dock elsewhere'.

Colonel William T. Rossell presided over the hearing on 25 January 1911, assisted by Colonel William Black and Colonel S.W. Roessler. They considered simultaneously a proposal for extension of the piers used by the German and Dutch lines. Although Captain Smith could not attend the public hearing, he gave his own testimony before leaving in command of *Adriatic*. He thought 'a 2,600 foot channel would be ample room for two large ships to get about without any risk to either'. Asked whether there would be enough room for two large liners on opposite sides to back out simultaneously, Smith 'replied that no sane man in command of a ship would attempt to do such a thing'.

On 29 January, the *New York Times* asked, 'Is New York [...] to lose much of her trade because of the refusal of the Federal Government to allow the extension of her North River pier line to accommodate the large steamships now being built?' Franklin thought the port would 'suffer in its prestige in the eyes of the world if these vessels can be accommodated in a small port like Southampton, England, and not be able to find pier room in New York'. Colonel Black said 'that there had never been any fixed plan for the harbour of New York formulated by the city government'. He felt that legislation should be passed to fix the pier-head lines, 'In his opinion the future development lay in constructing a quay in the Riverside section, where the largest ships could be berthed comfortably. This would take time to carry out.' While the work was being done, a 'revocable permit' might be issued 'to allow the city to extend the piers temporarily'.

A fortnight later, one of the *New York Times* reporters had been informed 'on good authority that the report of the Harbour Line Board would be favourable to the steamship lines, and that it would recommend to Secretary of War Dickinson the granting of a revocable permit'. The permit would allow an extension of 100ft, provided that the tide could flow freely underneath and that there were no structures above. On 6 March, Dickinson authorised the extension on a temporary basis, three months before *Olympic* was expected to arrive on her maiden voyage.[22]

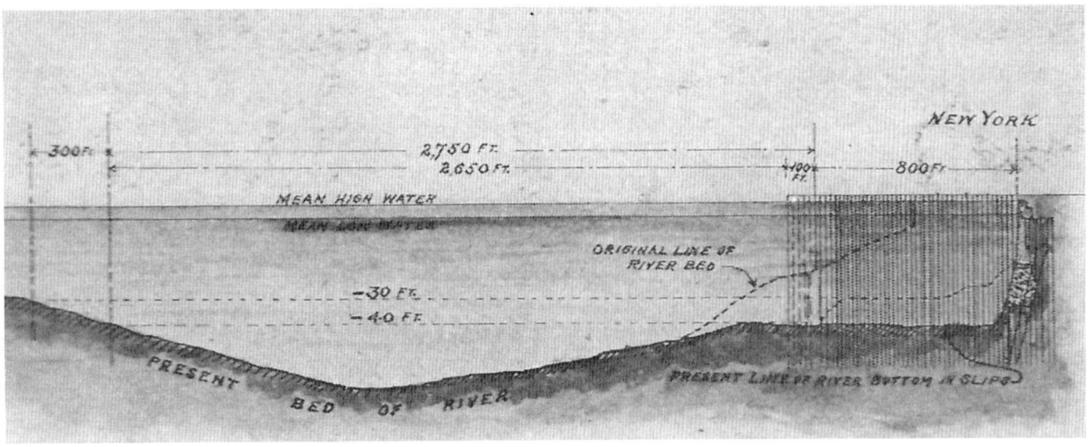

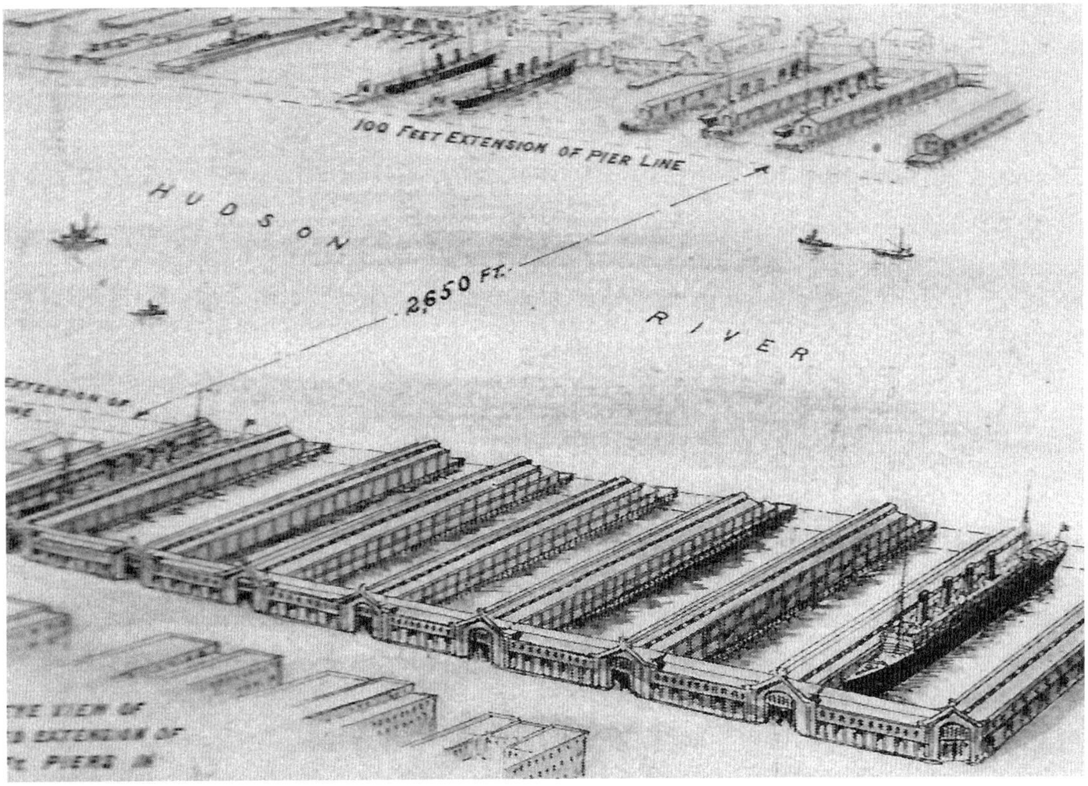

Top to Bottom: 'Cross section of Hudson river at Chelsea District, showing increased width of river channel due to dredging', *Scientific American*, 1911. (Author's collection); An illustration of the 'Chelsea piers' compared to the Hudson River for scale (top) and an enlargement of the piers with *Olympic* pictured docked in the foreground. *Scientific American*, 1911. (Author's collection); *Olympic* 'at an 800 foot pier, showing [her] stern 90 feet beyond pier ahead' (left) and a 'plan of 800 foot piers with 250 foot slip showing berth room occupied with coal barges' (right). The size of the 'new Cunarder' (*Aquitania*) was somewhat exaggerated. *Scientific American*, 1911. (Author's collection)

***

Southampton also had its challenges. Concerns about the depth of water were not new. On 30 April 1907, the White Star Line had instructed their New York office that *Adriatic* would need to be loaded only to a 'light' draught for maiden eastbound crossing to Southampton. They pressed for the necessary dredging work to be completed but it was a constant struggle. One issue at Southampton was frequent disputes between the Southampton Harbour Board and the London & South Western Railway as to who was responsible for what. Philip Curry, White Star's manager at Southampton, wrote to the Southampton Harbour Board on 18 December 1908:

> It is with pleasure we have to inform you that with the object of bringing their Southampton–New York service up to the highest standard of efficiency, and of developing to the fullest extent the traffic via Southampton, the management of the Oceanic Steam Navigation Company, Ltd. (White Star Line) are now having built at Messrs. Harland & Wolff's shipbuilding yards at Belfast two steamships of the highest class, which will provide every attraction and comfort for travellers by the line, and which will much exceed in tonnage any vessel now afloat or under construction.
>
> It will be obvious to your Board that if the port of Southampton is to derive full benefit from the advent of these steamers, it must be safeguarded against any possibility of reproach, or reflection upon its facilities, such as would arise were detention to be caused to the vessels in consequence of any deficiency in the draught of water in the channels and approaches to the port and docks.
>
> We therefore deem it our duty to inform you that the builders estimate that these two vessels, when ready for sea in the Southampton service, will draw approximately 35 feet of water, and it is expected that on their arrival at this port from New York their draught will be in the neighbourhood of 32 feet 6 inches.
>
> The steamers may be expected to take their places in the service in the spring of 1912, and we confidently hope to receive an assurance from your board that whatever dredging may be necessary to enable them to sail promptly, and to land their passengers without detention, will be undertaken and completed before they are delivered.[23]

Commissioner A.J. Day argued:

> As Mr. Curry states in his letter that these new vessels of the White Star Line will not be coming to Southampton before the spring of 1912, there can be no great urgency in dealing with this most important question of further dredging, and many things may possibly happen between now and then [...] When the time does arrive for the Board to deal with this question, I presume both the Railway Company and the International Mercantile Marine Company will be asked to guarantee that these new vessels of the White Star Line will not only come to Southampton but will remain here for a term of years.[24]

Estimates submitted to the Works committee on 5 March 1909 put the cost of the necessary dredging at £28,650. On 1 April 1909, the subcommittee opted to 'recommend that consideration of the matter be deferred for the present'. They also recommended 'that a letter be sent to the Cunard company, enquiring if they propose to use Southampton as a port of call'.

On 4 May 1909, Commissioner H. Wilding wrote a letter to Commissioner W. Bone, 'in order to save time at the next meeting':

> There have been suggestions as to making the channel more than 600 feet wide, and making the embayment of Fawley where large steamers could safely be moored and lie for some time. No doubt these suggestions have arisen from a desire to make Southampton attractive to the Cunard company – a very excellent object, and one with which I am in entire sympathy – but it does not arise out of the request put forward by

> the White Star Line, and the expense of doing this should not be mixed up.[25]

Almost ten months had passed since Curry's original letter of 18 December 1908 and, on 8 October 1909, a letter from the White Star Line was read to the Works committee asking for a response. A special meeting of the Works committee was called for 22 October 1909. It was only then that they resolved it was 'advisable to dredge the channel to 35 feet from the Thorn Knoll [buoy] up to within 600 feet of the docks, provided that a satisfactory financial arrangement can be made to ensure a proper return on the necessary capital outlay'.

Another special meeting was called for 5 November 1909, which referred the matter to the Finance, Parliamentary and Works (joint) committees 'for consideration'. On 29 November 1909, they recommended 'that the further consideration of this matter be deferred for six months'.

The bureaucratic machine ground its way ahead. On 21 December 1909, a subcommittee was appointed 'to confer with the White Star Line, and to report'. The special subcommittee resolved to ask the White Star Line:

> 1. How often do you consider it probable that these two vessels would arrive at dead low water.
> 2. Should they be arriving at dead low water, as this would be known long before arrival here, say at or about Plymouth or Cherbourg, would there be any hardship in slowing down sufficiently to come in with the rising tide.
> 3. As it could not be considered a business proposition to spend this large amount of money for the sake of two vessels only without securing a return to pay the interest to the stockholders, and provide for the sinking fund, we shall be glad to hear what financial arrangement you would be prepared to enter into if the Board would provide 3 feet more water, the loan would be for 55 years.[26]

The next day, a delegation from White Star met with the committee: Harold Sanderson, director and general manager; Philip Curry, local manager at Southampton; E.W. Bond, assistant local manager at Southampton; Captain Steele, Marine Superintendent at Southampton; Captain Haddock, commander of *Oceanic*; and George Bowyer, who piloted the company's large vessels. Councillor R.G. Oakley chaired:

> I may say that the Board as a whole hesitates to incur such a heavy expense – some £100,000 for the benefit of two vessels which may or may not require more water than the 32 feet we have now; we hope therefore to hear from our friends of the White Star Line their expression of opinion with reference to the matter, and to discuss in a friendly manner one or two questions which I shall ask from the chair. Mr. Sanderson will answer these questions in his remarks.

Sanderson opened by assuring them that 'we may all look upon the conference which we have this afternoon in a friendly way, and not binding to either party':

> I am sure that the first is practically an impossible question to answer in a definite manner, but to the best of my ability I will answer it.
>
> Having regard to the necessity we would be under of making best water sailings this side, the ships leaving New York the same date will arrive at worst tides. We have two routes in the Atlantic trade – long and short; the long from about the middle of January to the middle of August. We have to sail the ships from New York according to tidal conditions and cannot always adhere to a fixed hour, not because of insufficient water but owing to strength of tide we have to sail the ships from the piers when the tide is dead.
>
> On the long route we expect to make Plymouth [on] Tuesday, Cherbourg Wednesday morning, and expect to reach Southampton about noon Wednesday and at low water, we should have to wait off Cowes Roads for water to Southampton approaches, and this may happen very frequently in January to August.
>
> With regard to the second question – that is a matter of opinion: in these days people look

> upon any delay as a hardship. Our one aim and object is to satisfy these people, and detention will be looked upon as a source of complaint, and there will be complaint.[27]

He was surprised by the third question, 'that we should be asked to provide improvements at the ports the ships will frequent':

> I really think that such a novel proposal would not have been made by any other port in Great Britain.
>
> We are spending £1,500,000 on each ship, and as you are aware although you have estimates, you usually find extras.
>
> In bringing two ships' trade to Southampton, we feel we are doing all that might be asked, and cannot hold out any hope that your suggestions as to assistance can be entertained. Purchases for the ships would be made at Southampton.
>
> With regard to the question of what these ships would draw, we do not think it possible to tell what the maximum working draft might be at some future time. We have not realised our expectations here as to cargo, possibly owing to the bad time since we came down here, and feel that we can for the present limit our ideas of weight of cargo to the quantity our ships secure at present but we must not count upon less than 35 feet.
>
> As a fact the very least the ships must have at the bottom of two feet clear, and some think that it is not prudent to have so little. 37 feet of water in the channel is the minimum for safe navigation of those ships.
>
> We find when looking at it broadly, in 1909, if we had sailed these two ships the *Olympic* and *Titanic* each week, on 80 occasions in the year they could not have sailed to time.
>
> On the Wednesdays in 1909, there have been 11 occasions on which the ships could not have sailed at noon. It will be to the point to consider that argument, which is unanswerable. Out of these 11 occasions, there would have been one for which we could not have found a convenient sailing time at all.
>
> That opens up a question – the point which I should like to mention is that years before we came to Southampton, it was the boast of Southampton that ships could arrive and sail at Southampton without having to wait for water, as in other ports.
>
> If you gentlemen come to the conclusion that this boast is one which you do not agree to maintain, the value of our argument vanishes [...] The question for us to consider is whether there are attractions sufficient for us to remain, and whether Southampton's conditions will induce further business here [...]
>
> If Liverpool could hold out the natural advantages of your port to the whole of the shipping trade, so that they could come and go at any time without waiting for water, by the expenditure of £100,000, it would make them the happiest place in Great Britain, or five times, or even ten times, the amount they would decide instantly to spend.
>
> It is for your gentlemen to decide, not for me. If adequate water is not provided it would be impossible for Southampton to continue her fortunate position.[28]

He argued that 'in Liverpool the people are not willing to put up with the delay. The drawback is very serious there, and we have to land passengers at Holyhead.' If the company had to advise eastbound passengers to disembark at Plymouth then it would 'be a reflection upon Southampton':

> You might not think that any serious drawback but do you not see any revenue to come from this to meet your expenditure, purchases are made at the port of Southampton. That, I think, is a great point to bear in mind. The Atlantic business is one of the greatest importance in Great Britain for everyone. On the Fishguard Station; at Belfast; at New York, large sums of money have been spent; in New York no end of money in making 40 feet depth and 1,000 feet width, as they think it wise to make their ports available at all times, and for larger tonnage than at present exists.
>
> If Southampton desires to continue to be a port of great consequence, it is only fair to look upon the matter from an all round point of view with an eye to the future.

> These ships pay large sums of money in wages, and it is not very hard in my mind to see that Southampton derives some benefit in the large sum of money directly or indirectly brought to the town by the crews and their dependents and the men who work the ships in port.
>
> I do not think it is fair that we should be looked harshly at for being the first people to bring large ships to Southampton. We were not the first to produce large ships, and we shall be followed in a very short time by others. The Germans will follow us in a very short time.
>
> How far you attach importance to shipping I do not know. It is an important matter to Southampton. If there happens to be a vessel leaving Hamburg drawing 35 feet of water, what can they do at Southampton to accommodate her? I should think it only reasonable that the directors of the Hamburg company may desire to call at Southampton, but would not think they should have to wait, and would look somewhere else for a calling port.
>
> I feel gentlemen that the blame should not be upon us, and it is hardly fair. To find that we are merely the forerunners with these big ships will only realise our expectations.[29]

Curry explained the basis of the estimate that the ships could not have sailed on eleven occasions on Wednesdays in 1909:

> We have only allowed sufficient margin to pass through the channel on the falling tide, before the water goes away, and arriving in Cowes Roads one hour to 1½ hours before low water, or on account of fog we might find ourselves unable to get up for two or three hours.
>
> The channel which we are talking about is only 600 feet wide and even supposing we could get away by time, in the event of fog we should have to push on, we could not stop or go back. We are working on a very narrow margin. Southampton really would be wise to provide deeper water, and apart from the White Star Line, I think Southampton as a port would be wise to remove this drawback.[30]

The questions began.

Councillor Hamilton assumed, 'It is not likely the company would consider the question of contribution toward the capital outlay.'

On harbour dues, Sanderson protested, 'I do not think because we are more enterprising we should be penalised and called upon to pay additional harbour dues.'

Councillor Beavis remarked, 'The complaint as to passengers might be good if passengers did not now get off at Plymouth.'

'From a passengers' point of view we have to consider whether the attractions are attractions to them; they object to the long railway journey to London. These are the people we should have complaints from,' Sanderson responded.

'You speak about German ships, and of two large steamers that cannot come into Southampton. At present, some ships lay in the Cowes Roads, and land their passengers, and we get nothing out of them,' Beavis countered.

'I am advised that [the] ships you refer to, do what you say, but I suggest that if there were facilities for ships coming up here, they would probably come.'

'They don't very often come up.'

'Three out of four,' Councillor Bowyer pointed out.

'When we came to Southampton three years ago,' Sanderson explained, 'I then made a statement – it was our intention coming to Southampton [that] we would do everything in our power to increase the trade, and to make the line a very successful and excellent one. After this, we produced these two ships, and there seems to be in the minds of some gentlemen resentment for having done it.'

'No. No,' protested several committee members.

'We do think we are entitled to sympathetic feeling at your port. We are indeed doing our best for the port. Whether you give us water, or not, do not look upon us as a nuisance, for we are doing what we think is best.'

'You mention about coming to Southampton that you would do all in the interests of Southampton, but you came in the interest of the White Star Line,' Beavis replied.

'At the same time'. Sanderson might have been forgiven if he was frustrated. 'If these ships cost £1,500,000, they will run very frequently – 12 or 13 voyages a year, unless accident.'

'Supposing in ten years' time the ships go elsewhere; where should we stand?' asked Councillor W.T. Hurst.

'That is not a reasonable thing to look at. Where will you be if you don't improve the port?'

'So far, so good, but is it really necessary to do the present work?'

'I hope you will think so.'

'There is no proof at present.'

Sanderson argued, 'We started with a little ship called the *Oceanic* 400 feet long, and we increased ten feet at a time, each time feeling our way; undoubtedly a success. The last step was a big one, as you see we were willing to invest three million of money.'

'Do these ships coal at America for in and out?' Beavis wondered.

'No. The *Adriatic* does, but these two new vessels have not the accommodation. They will coal at America and Southampton.'

The meeting came to an end. 'I am very grateful for your kind reception,' Sanderson concluded, 'and I hope that anything we want to discuss in future we shall exchange views in the same friendly way as today. I should like you, however, to bear in mind that I say what I think is only right and true. You will be wrong if you consider this matter in the light of two ships only. They are the forerunners of other ships of a like capacity.'[31]

On 2 February 1910, the subcommittee 'resolved unanimously':

> That this committee recommends to the Harbour Board the deepening of the channel to 35 feet, conditionally that the White Star company contribute the sum of £2,000 per annum in excess of their dues for ten years, and that the London & South Western Railway company dredge the 600 feet outside their dock premises to a similar depth, within a reasonable time, and also assist the Board in financing the dredging required on the same terms that they assisted the Board in the dredging to 32 feet.

The Harbour Board approved the recommendation thirteen days later, but other concerns were raised throughout 1910. On 14 November 1910, the London & South Western Railway 'were anxious that the Board should bear part of the expense of dredging a new swinging ground off the entrance to the new docks'. Unfortunately, the Harbour Board 'considered their liability ended in providing deep water up to within 600 feet of the dock head (south-east corner) […] but they agreed that should the railway company undertake the dredging, the Board was not at all likely to raise any objection'.[32]

On 14 January 1911, Mr T.M. Williams, manager of the Docks and Marine Department for London & South Western Railway, wrote to ask whether the company could 'undertake the dredging of the ground in question, without prejudice to the point as to liability for the cost'. He also drew to the Harbour Board's attention to 'the fact that unless the work be promptly commenced, it will be impossible for the new ships of the White Star Line to make use of the docks in June next, as contemplated'. The subcommittee did not object, on 'the clear understanding that the whole of the cost be borne by the company'.[33] By the time *Olympic* was ready, the necessary dredging was complete.

***

The dredging was not the end of it. A new dock needed to be built. The contract for the new 'White Star' dock was agreed in October 1907, forming 'the last link in the chain of dock-extensions which were undertaken by the London & South Western Railway Company when they purchased the docks from the Southampton Dock Company in 1892'. The contractors had to employ two additional dredgers to expedite the work. It was only 'by working their plant day and night' that they got the dock ready in time, dredged to 40ft and with the turning basin outside to 35ft. While all was ready for *Olympic* to use Berth 44 in June 1911, the adjoining quays 'were not finished and equipped' until afterwards. It was an impressive piece of engineering.

Francis Ernest Wentworth-Shields, speaking to the Institution of Civil Engineers, thought it was fair to claim that 'the White Star dock would meet all requirements for many years'.

The dock was shaped like a parallelogram. Its 1,700ft length and 400ft width provided a water area of 15½ acres. It was dredged to provide 40ft of water at the low spring tide. Southampton's docks were tidal (as opposed to being closed with lock gates) and it had two natural advantages compared to many other ports. The range of the tide was quite moderate – about 13ft in the spring – and the tidal water did not have much silt in it. By one estimate, the new dock itself, its equipment and the associated dredging of the channel 'must have cost about £750,000'.

One observer, Frederick Palmer, wondered whether the two berths were 'too short' for two present-day vessels almost 900ft long or had the designer 'been bold enough to anticipate the coming of the 1,500-foot vessel'? The investment showed that 'provision for those leviathans must, of course, be made, and in incurring that heavy liability the London & South Western Railway Company have shown that the passenger traffic between England and the United States should not be diverted from Southampton for want of adequate facilities'.[34]

The 'White Star' dock or 'New Deep Water Dock' included construction of 'four large steel cargo sheds'. One was finished by June 1911, 'It is 700 feet long and 120 feet wide [...] Waiting rooms and lavatories have been provided at the southern end with cages for bonded cargo, and offices for the medical officer, foremen and clerks.' A smaller annexe included storehouses for baggage and dock gear. Then there were the passenger gantries:

> Owing to the great height of the *Olympic* and her sister ship, the *Titanic*, the passenger gantries are of the two-decker type, in order that they may be available for passengers at different states of the tide. There the passengers alight from the gantries, pass down into the shed by broad stairways, and proceed to the trains. The examination of baggage will take place in the shed, so that passengers will be under cover the whole time when passing between the ship and the train.[35]

'View of the world's largest dock looking from the entrance.' This view helps to illustrate the scale of the works and the construction challenges faced by the engineers. *The Sphere*, 1910. (David E. Olivera collection)

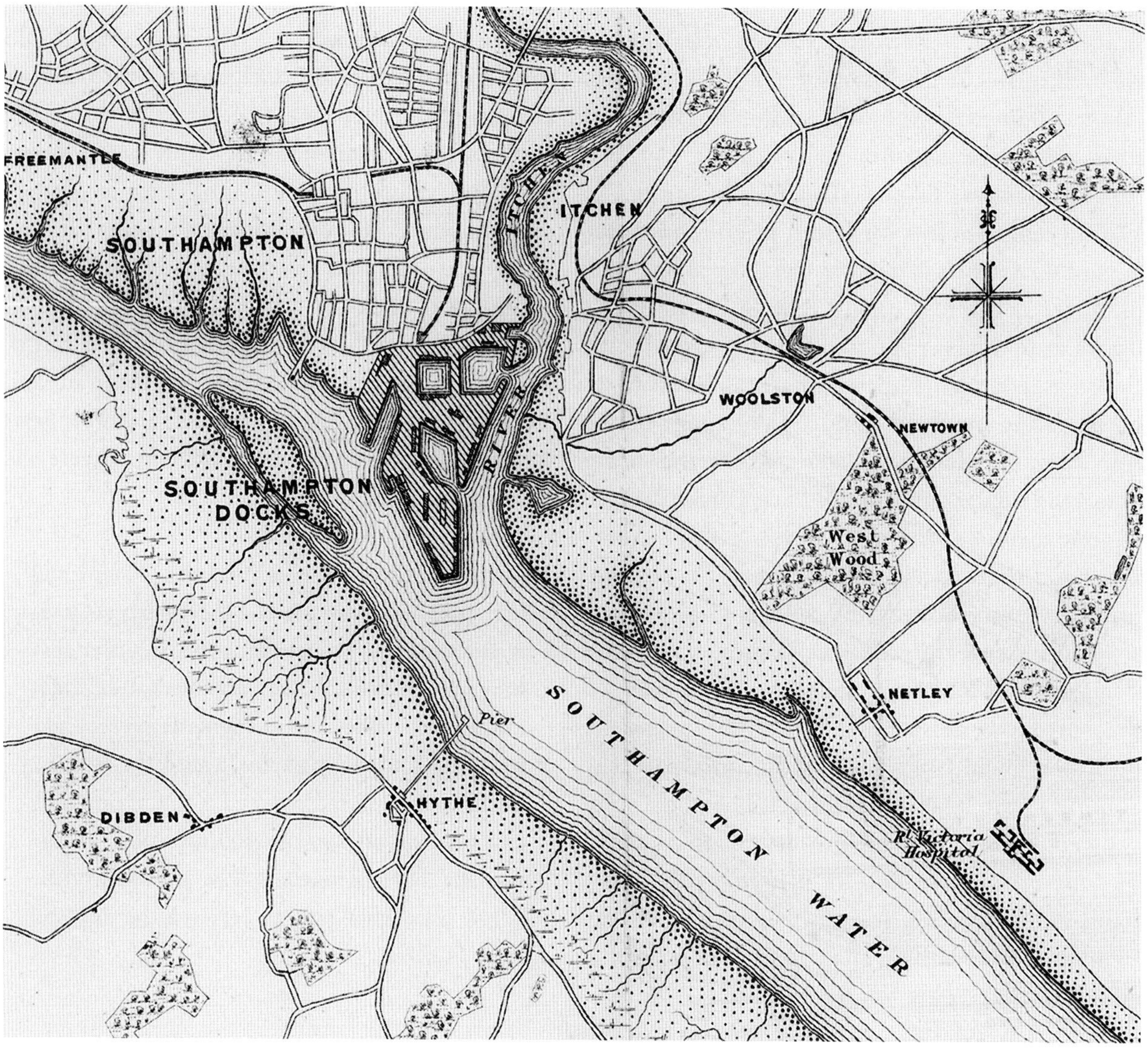

Southampton lies at the confluence of the River Test (left) and River Itchen (right). Ships leaving the docks headed down Southampton Water towards the open sea. *Proceedings of the Institution of Civil Engineers*, 1914. (David E. Olivera collection)

'Special trains from and to [London's] Waterloo Station run directly to and from alongside the mail or passenger steamers on days of arrival and departure,' the London & South Western Railway Company explained, 'The journey occupies about an hour and three quarters.'

The development of Southampton Docks over nineteen years was remarkable. The total length of all the quays increased from 12,054ft in 1892 to 23,250ft in 1911; the cargo and passenger sheds expanded from 554,718 square feet to 1,064,946 square feet; and the length of railway sidings more than doubled, from 15 miles to 37 miles.[36]

Southampton's existing Trafalgar Dry Dock, which was only opened in 1905, was already 'too small for vessels of the *Olympic* type'. *The Mariner* noted:

> Its length is to be increased from 875 feet to 897 feet, a gain of twenty-two feet; its width at [the] entrance is to jump from 90 feet to 100 feet; and its depth over sill [...] will be increased from 33 feet to 35 feet. The widening is being effected by cutting away the faces of both walls and setting back the altars. A sliding caisson is to take the place of the entrance gates.[37]

Opposite: Shortly before *Olympic*'s launch in 1910, *The Sphere* illustrated how she would appear berthed in the new dock. The illustration contained several inaccuracies, including placing the first-class gymnasium on F-Deck (it had been moved to the Boat Deck by June 1910). (Jerry N.J. Vondeling collection)

NEW FOREST
SOUTHAMPTON WATER
SEA
DRY DOCK
MAIL VAN
MAIL VAN
PASSENGER LUGGAGE VANS
FIRST CLASS LOUNGE
FIRST CLASS DINING SALOON
SECOND CLASS LIBRARY AND THIRD CLASS PUBLIC ROOMS
SECOND CLASS DINING SALOON
FIRST SECOND AND THIRD CLASS AND STEWARDS
THIRD CLASS DINING ROOM AND GYMNASIUM
SWIMMING POOL, BAGGAGE, RACQUET COURT AND THIRD CLASS
ION OF
WALL
DEPTH OF WATER AT LOW TIDE 35 FEET
MASS CONCRETE RETAINING WALL
FLOOR OF DOCK
CLAY OR MUD
DEPTH FROM TOP OF WALL 73 FEET
G. Terrance Stephenson 1910

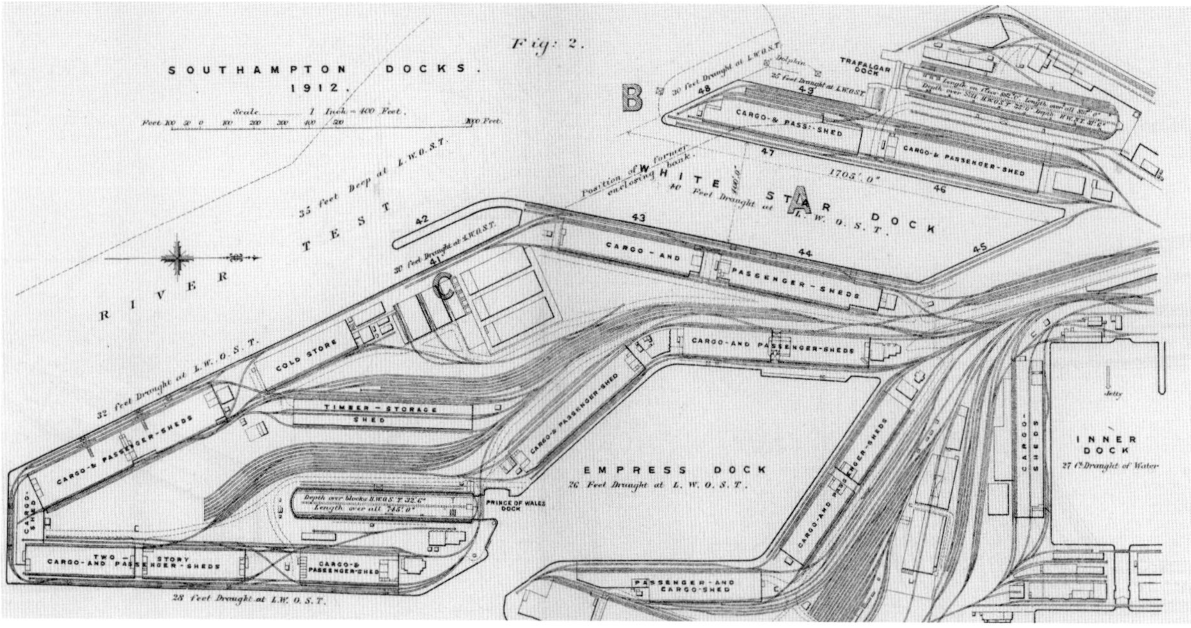

The layout of the docks in 1912. At the top right, the Trafalgar Dry Dock is visible. Enlargement works enabling it to accommodate *Olympic* were not completed until 1913. The new 'White Star' dock is marked by the letter 'A'. Their ships passed Berth 42 as they left the dock and proceeded down the River Test. *Proceedings of the Institution of Civil Engineers*, 1914. (David E. Olivera collection)

All the work would not be finished until 1913, forcing *Olympic* to use the dry dock at Harland & Wolff in Belfast in the meantime.

***

The changes at Southampton and New York were matched by smaller-scale preparations at Cherbourg, Queenstown and Plymouth. At Cherbourg, about 85 miles south of Southampton, the port was 'well protected from the heavy gales of wind and high seas often experienced in the English Channel during the winter months, by a semi circular breakwater extending from Greves Point on the east to Fort Querqueville on the west, a distance of about 4½ miles'. Two openings were available for ships to enter the harbour and the one on the western end, about half a mile wide, was used by passenger liners; vessels proceeded inside the breakwater for about a mile and then anchored. There was ample water for merchant ships to swing at anchor 'inside the harbour in the space laid down by the authorities for merchant ships having a minimum depth of 42 feet [during the] ordinary spring tide'. No dry dock was available for vessels of *Olympic*'s size and the only engineering repair shop was the French navy's, but as she was only calling at the port, there was no problem.

A single railway ran along the quay wall from the railway station ('which is situated about half a mile in a direct line from the Custom Shed, where all baggage is examined on landing'), where the special train started out for Paris. Passengers were able to walk around the town once their baggage had been cleared through customs and there was 'a large waiting room fitted with tables and writing materials' as well as a 'refreshment bar attached to the Custom Shed'. In 1908, four tenders were in service (two for Norddeutscher Lloyd, one for the HAPAG and one for White Star and other IMM companies).[38]

The White Star Line ordered two new tenders. On 22 June 1910, they contracted Harland & Wolff to build 'Yard Number 422' – *Nomadic* (a 'passenger tender'). The order for 'Yard Number 423' – *Traffic* (a 'baggage tender') followed within a month. *Nomadic*'s length between perpendiculars of 220ft was about half that of the first *Oceanic* and her moulded breadth of 37ft made her nearly as wide. *Traffic* was smaller, with a length of 175ft and a breadth of 35ft. Progress was swift:

	Nomadic	***Traffic***
Contracted for	22 June 1910	14 July 1910
Shipyard and engine works ordered to proceed	25 June 1910	19 July 1910
Keel laid	22 December 1910 (Number 1 Slip)	22 December 1910 (Number 7 Slip)
Fully framed	18 January 1911	25 January 1911
Fully plated	25 March 1911	7 April 1911
Launched	25 April 1911	27 April 1911
Delivered	31 May 1911	31 May 1911

Both would be ready to serve *Olympic* when she made her first appearance at the French port.

Queenstown had no significant docking facilities. In 1903, the longest Cunarders in service were *Campania* and *Lucania*, but *Lusitania* and *Mauretania* were substantially larger. Before they entered service, Cunard and the Harbour Commissioners arranged the dredging of the inner harbour so that there would be enough deep water to accommodate them. Unfortunately, bad weather conditions could make it problematic to enter the harbour.

On 29 December 1907, *Lusitania* grounded lightly when she was entering Queenstown to take on mail. The Harbour Commissioners thought the incident was 'entirely due to the pilot, who committed an error of judgement', which could not 'be attributed to bad weather

Cherbourg – 'La Gare Maritime'. *Nomadic* (left) and *Traffic* (right) pictured some years after they entered service. (Author's collection)

or any fault of the harbour', but Cunard were worried that the incident happened 'despite the precautions which were exercised by the pilot and captain'.

In 1909, Cunard omitted the call at Queenstown on *Lusitania* and *Mauretania*'s eastbound crossings. Several years later, on 23 February 1913, *Mauretania* arrived at Queenstown and had to enter the inner harbour because a heavy gale blowing from the south-south-east made it 'impossible' to take on mails outside the harbour entrance. She anchored safely and loaded the mails, but the severe weather conditions prevented her leaving 'until the flood tide had swung her head to the southward'. It left her in a 'precarious position': the strong wind on one side of the ship and ebb tide on the other put her lying 'athwart the tide with her anchor cable strained at right angles'. She was loaded to a draught of about 35ft but the depth of water fore and aft was only around 38ft. Cunard worried that 'if the anchor had dragged or if the cable had parted, the damage incurred would have put the vessel out of commission for a long period'. Cunard decided that their two largest liners 'could not be allowed to enter Queenstown Harbour again when the weather was such that the mails could not be taken on board in the outer roads'.[39]

White Star could rely on the existing tenders *America* and *Ireland* to convey passengers and mail from the quay. The harbour entrance itself opened to the south and was about 1 mile wide, but the Harbour Rock shoal separated it into two channels to the west and east. The eastern channel had a minimum width of 380ft compared to the wider western channel, but the harbour pilots preferred to use it for large ships because it was straighter and better defined than the western one. The channel depths of 38 or 42ft were relatively shallow. It left 'practically no margin' in the case of *Lusitania*, according to Captain J.T.W. Charles, 'you always have to allow even going at moderate speed for a set down of 2 feet [...] we certainly do not like going in with less than [...] 42 feet'. (Depending on the conditions, large ships anchored outside the harbour itself, off Roche's Point.)

The same situation applied at Plymouth. Large ships could not dock but were served by tenders including *Cheshire*, *Sir Francis Drake* and *Smeaton*. They were available to service liners calling in on the eastbound crossing (*Sir Francis Drake* was well suited to carry first- and second-class passengers, including their baggage; *Smeaton* could carry third-class passengers and baggage; *Cheshire* conveyed mail, parcels and other cargo). The outer anchorage at Plymouth had a limiting depth of 36ft of water at low spring tides, although the channel had a minimum width of 1,500ft and did not need dredging to maintain it.[40] *Olympic* and her sister had reached the practical limit of size for the time being, but all of the planning and effort by the White Star Line ensured that the port facilities were available for them.

'THE GREATEST SHIP IN THE WORLD'

As [*Olympic*] lay abreast the stage yesterday the general opinion among the many thousands of sightseers who thronged both banks of the river was that the Mersey had never borne a ship of more beautiful proportions or more graceful lines. So exceedingly well balanced are the proportions of the vessel that at first glance her vast bulk is hardly realised; indeed it is impossible for the eye to form any complete vision of the tremendous size of the vessel without comparison; that is until some other vessel becomes dwarfed by her vicinity.

The Journal of Commerce, 2 June 1911

Olympic caused a sensation when she first appeared in Southampton [...] She looked colossal, and even 'uncanny' as she towered high above the waterline, with her tiers of spacious decks and her four huge funnels, dwarfing all other craft within sight.

Dr John C.H. Beaumont, *Ships – and People*, 1926

Her immense yet harmonious proportions, the gracefulness of her lines, her great towering height, and above all the idea of strength, stability, and enormous power which a first view of her conveys, touch the imagination with a sense of wonderment which only increases as one has the opportunity of studying her in more detail. She is the highest achievement of the shipbuilder's craft, the finest production of his genius, and with her sister ship, the *Titanic*, she stands unapproached in the race for leviathan liners.

The Cork Examiner, 16 June 1911

> castings made. A large number of the boilers have been water tested, and work has been started on one of the funnels, of which there will be four in the vessel.

Alexander Carlisle was 'reported as saying that the *Olympic* and *Titanic* will far surpass all other vessels' except in speed. The reporter added, 'Their combined tonnage far exceeds that of all the vessels engaged on both sides of the Spanish Armada.'[2]

Harland & Wolff were planning ahead to make sure *Olympic* could be launched successfully. The Belfast Harbour Commissioners' engineer had informed Harland & Wolff that he 'did not consider the work' necessary 'at the Victoria Wharf' could 'be carried out in less than seven months'. Alexander Carlisle spoke to the engineer on the telephone and advised his colleagues on 5 May 1910 that the engineer had been 'unable to say how long it would take'.

Harland & Wolff were keen to get the work done so they could meet the agreed launch date of 20 October 1910. They wrote to the Harbour Commissioners and proposed that they would employ 'Messrs. Henry to do what was necessary, under the supervision of the Harbour Engineer and to his plans'.[3] Later that month, Harland & Wolff also issued a strict general notice to their workers concerning all ship launches:

> Any of the staff or their friends wishing to witness a launch must send in an application on a question & reply form the day prior to the launch to his superior, unless he has an actual duty to perform in connection with said launch and has instructions either from his superior or from one of the managing directors.
>
> Any violation of this rule will mean instant dismissal.[4]

Early in August 1910, the managing directors 'discussed and approved' the launching arrangements, 'subject to some slight alterations proposed by Lord Pirrie'. They confirmed that, 'as arranged with Mr. Ismay', nobody, 'with the exception of Mr. Ismay and his own party (consisting of a limited number), the press and the company's [Harland & Wolff's] own officials, including the foremen and their wives or daughters, would be admitted to the works' for the launch. Plans 'of suitable stands for the accommodation of Mr. Ismay and his party, the press and the company's officials should be prepared and submitted to Lord Pirrie for his approval'.

By 11 October 1910, the launch was little more than a week away. They approved the arrangements for the guests. Lord Pirrie and Charles Payne decided to use the services of four Liverpool tugs as well as Harland & Wolff's *Hercules* and *Jackal*. There was 'no necessity to have a salvage pump on board', as Payne explained, because *Olympic*'s watertight bulkheads would already be made watertight 'to the height of the lower deck [G]' and 'that even with the fore hold flooded the vessel would not sink'. They decided 'it would be more economical to fit up the necessary workshops and stores' on board ship for 'carrying on the work after launching', rather than putting up 'special sheds at the Deep Water Wharf for this purpose'. They would set up in the third cargo hold 'a temporary Smithy [blacksmith works] for engine works, the necessary fitters benches etc., also joiners', painters', plumbers', fitters' and electricians' stores in other convenient parts of the vessel, from which the men working on board will be able to draw the materials required for carrying on work'.

They were considering what to have for the 'arrangement of flags on Arrols gantry' and decided 'that these should, if possible, have a meaning, such as "God Speed" or "Good Luck"'. They approved the proposals for platforms 'for photographing and cinematographing purposes'.[5]

One problem was that *Olympic* would block some of the lights in the harbour after she was launched, because she was so large. W.A. Currie, as secretary to the Belfast Harbour Commissioners, issued a 'Notice to Mariners' that from 20 October 1910:

> … and until further notice, the *two vertical green lights* at the South end of the *East Twin Island* will be obscured by the SS *Olympic* lying at the Deep

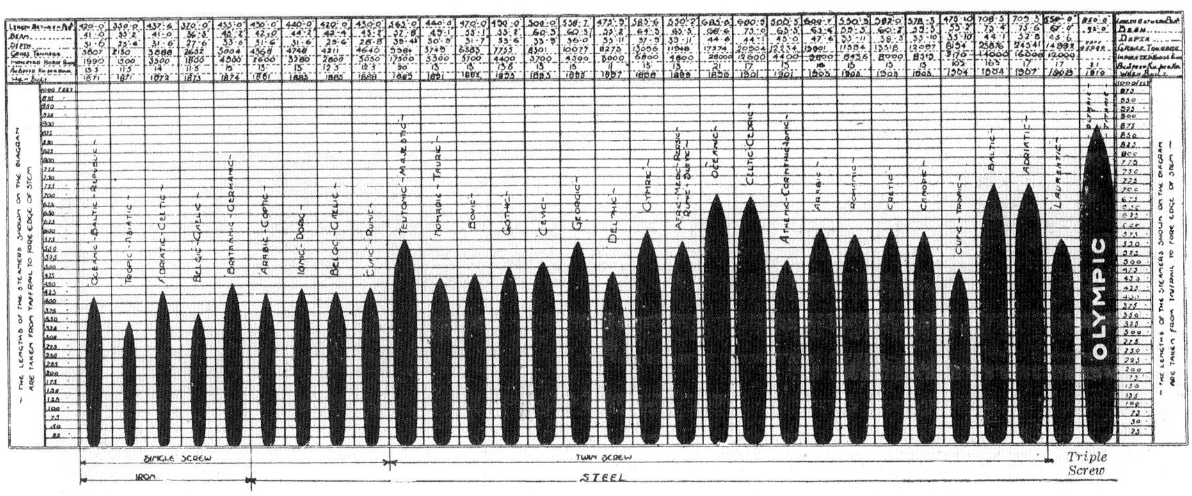

Top to Bottom: An illustration of *Olympic*'s size compared her to Piccadilly in London. The caption stated that 'four Savoy Hotels' would be required to accommodate her 2,500 passengers. *Illustrated London News*, 1910. (Author's collection)

Another comparison showed the progress made over seven decades, from *Britannia* to *Olympic*. *Illustrated London News*, 1910. (Author's collection)

'A diagram showing comparative sizes of various types of White Star steamers from 1871 to the present time.' The company had progressed from single-screw iron ships to twin-screw steel ships, and now a triple-screw steel ship. *The Marine Engineer and Naval Architect*, 1911. (Author's collection)

> Water Wharf, North of the New Graving Dock, to Vessels passing Northwards along the Victoria Channel until they are nearly opposite.

To protect *Olympic* herself, she would 'lie with her stern towards the North, and will exhibit a *bright white light* on her *quarter*'.[6]

There were particular issues to consider with *Olympic*, because of her huge size. However, standard procedures were followed for every launch. Harland & Wolff's 'Memo for Launches' outlined the many tasks to complete, including finding out the 'harbour soundings [depth of water] in line of launch'; removing any buoys 'in line of launch'; checking 'all mooring ropes are on board'; arranging 'for flag pole on stern' and bringing flags 'aboard in good time'; advising the Harbour Master and Harbour Police Superintendent of the arrangements; testing the ship's tanks; ensuring the collision and B bulkheads were watertight; completing a 'launching check' of the chains and wire ropes in place; examining the ship's rudder and seeing it was secured in position; making sure donkey boilers 'are properly lashed'; arrange 'for tugs, if any required, and men to go in same'; and sending a note '("Round Robin") to foremen, captain of tug, etc. etc. fixing hour for meeting to discuss launching arrangements'.[7]

Officials looked at the practical preparations, including the various gates and entrances to 'the harbour that might need to be guarded against crowds rushing in'. The Belfast City Commissioner agreed to use 300 policemen, split equally between each side of the river, 'to guard against the inrushes of the crowd on all the dangerous points of the harbour estate'. The Antrim Line (as 'iron ore miners, coal importers and steamship owners') were concerned about their yard. Three days before the launch, they wrote to W.A. Currie at the Harbour Office in Belfast, requesting 'some police protection for our Coal Yard at the lower end of Albert Quay […] during the launch of the *Olympic*'. They explained:

> On occasions of this kind large numbers of youths, and also full grown men, invade the premises and clamber unto the roof, which is highly dangerous, and they also climb over the coals and reduce them to powder. We would be glad if you would afford the premises police protection on this occasion.

***

On the morning of Thursday, 20 October 1910, the weather in Belfast was 'fairly bright'.[8] One observer worried that the wind appeared 'to be blowing with a velocity which might interfere with the success of the launch', from the north-north-east. It eased off as the launching hour approached. The River Lagan was 'comparatively placid, and the conditions could scarcely have been more fortunate'. Despite the time of year, the 'sun shone with a warmth and power reminiscent of spring'. It was reported that trams were crowded 'from an early hour' as spectators made their way to Harland & Wolff:

> The scene at the yard was one of great interest and animation, and the presence of many distinguished visitors helped to make the launch ever-memorable. Commodious stands had been erected for the accommodation of the principal department heads and their friends, specially invited guests, the newspaper representatives, and the builders and owners of the ship. Lord Pirrie was early on the scene, and although he looked very nautical in his yachting cap, as well as self-possessed and calm, one could not help but think that it was a trying time for him.

The *Irish Times* wrote that there 'was practically a general holiday […] in many of the principal workshops and factories of Belfast'. Queen's Road was busy. As people reached the yard and *Olympic* came into view, they 'were gazing with undisguised awe at the leviathan which reared her gigantic hull far above their heads', according to the *Belfast Weekly News*.

Lord Aberdeen, the Lord Lieutenant of Ireland, arrived with Lady Aberdeen 'in good time for the launch'. They had travelled up from Dublin the previous day, arriving by special train at the Great Northern Railway terminal in Belfast's Great Victoria Street Station. Lord and Lady Pirrie had welcomed them and shown them all over Harland & Wolff's Queen's Island works. They enjoyed a 'special stand, covered with white and crimson cloth', alongside other 'distinguished guests', including Prime Minister Asquith's daughter; the lord mayor and lady mayoress; and the High Sheriff.

The White Star Line's guests came over from Britain on the *Duke of Cumberland*, one of the Fleetwood, Lancashire, steamers that the White Star Line had chartered specially for the occasion. They included 'nearly one hundred pressmen from various parts of the United Kingdom, many of them representing American and Continental newspapers and shipping trade journals'. White Star's party included J. Bruce Ismay, Harold Sanderson, E.C. Grenfell, Philip Curry, Captain P.D. Murray and Captain C.F.D. Beresford.

The chairman of the Belfast Harbour Board, Robert Thompson, was on the River Lagan aboard *Musgrave*. He 'enjoyed a capital view of the launch'. Meanwhile, across the river, Workman, Clark & Co.'s North Yard had a special stand prepared for that shipbuilder's own guests, including members of the Engineering and Scientific Association of Ireland. They had come to Belfast specifically to enjoy the launch.

'Beneath the great ship swarmed [...] some hundreds of busy labourers removing the last supports of the gigantic structure above their heads,' noted the *Daily Mail*. *Olympic*'s hull 'overhung the cradle on the sliding ways about 90 feet at the after end and 50 feet forward', according to *The Shipbuilder*. Her bow rested on an 'elaborate and substantial' launching cradle with 'the height of the standing ways above ground' about 14ft. The 'trigger gear' was relatively simple, similar to what was used when *Oceanic* was launched eleven years earlier, 'About 300 feet from the bow the usual timbers supporting the standing ways gave place to two solid steel castings, each weighing about 15 tons, in the body of which was fitted a cast iron trigger.' The trigger was kept in place by a ram exerting a hydraulic pressure of 500lb per square inch. Once the 'keel blocks and shores' were removed by the army of workmen, all would be ready and the pressure released. Her passage would be eased by 15 tons of tallow, 5 tons of tallow and train oil all mixed together and 3 tons of soap.

The Union Jack and the Stars and Stripes flew 'high on the great gantry' and 'midway between the two national emblems fluttered a code flag which spelled "Good Luck"'. According to the *Belfast News-Letter*, the sunshine was 'glowing' and 'took the autumnal sting out of the atmosphere'. On each side of the river, the 'vast grounds' waited in eager anticipation.

At 10.50 a.m., two rockets were fired to signal the launch was imminent. Then, ten minutes later, Lord Pirrie issued his command from the stand: 'Now.'

In response, Charles Payne released the hydraulic pressure. *Olympic*'s motion was almost imperceptible, then she gathered speed. 'It was scarcely conceivable that nearly 30,000 tons of steel and iron should move along so easily.' Yet she did. The *Daily Mail* recorded:

> Then, indeed, does the pent-up excitement burst forth. The silence is shattered with roar upon roar of cheering, wildest and highest from those who are saying to each other, or themselves, 'I helped to build her'.
>
> And through it all in perfect majesty the mightiest vessel in the world goes gliding on into her element as easily as if she were but a tug of a hundred tons.

She reached a maximum speed as she was moving down the slip of 12½ knots. Sixty-two seconds elapsed from the first sign of movement until she was fully waterborne, then 'about another forty-five seconds sufficed to pull her up almost within her own length':

> In the bed of the river were placed three heavy anchors on each side of the ship, each anchor being connected by a 7-inch steel wire hawser to eyeplates riveted to the shell plating. There were also placed in the bed of the river two piles of cable drags, each weighing over eighty tons,

> connected in a similar manner with an 8-inch steel wire hawser. These were all so arranged that when the vessel was nicely clear of the end of the slip the drags and anchors acted simultaneously in bringing the ship to a standstill. As a further precaution the ship's own bow anchors were stowed in the hawse pipes, ready for letting go in case of emergency.

Now *Olympic* was afloat safely, 'cheers rent the air'.

One of Harland & Wolff's officials, Mr Barker, had seen over 200 ships launched at the yard. At the launching of those ships he had 'thought nothing more of them than of a bucket being thrown into the water', but *Olympic* had moved his 'stony old heart': 'When I saw that vessel sliding into the waves [...] I actually took off my hat and waved it with the rest.' Keen spectators from the crowd 'scrambled about across the greasy ways in order to get a view of the *Olympic*'.

Olympic's stern makes contact with the water as she starts to move down the slipway. (Author's collection)

'After the launch': *Olympic*'s elegant lines belied her enormous size. Her appearance harked back to *Oceanic*'s launch, eleven years earlier. *The Marine Engineer and Naval Architect*, 1910. (Author's collection)

> Her sharp bow, beautiful lines, and clean run from stem to stern suggest speed, her tremendous weight, rigidity, and strength indicate stability, and like the *Oceanic* eleven years ago she may proudly claim the title of 'Queen of the Seas'. Alongside her ponderous and giant frame big steamers look like an ordinary steam tug. A fine Bibby liner, the Belgian steamer, *Albertville*, and other ships under construction or being fitted out and repaired seem puny and insignificant when compared with the last produce of the Queen's Island.

She came to rest with her bow about 500ft from the end of the ways. The draught at the stern was 20ft 5in, whereas it was 15ft 8in at the bow. Overall, she was riding light with a draught of about 18ft, 'corresponding to a displacement of 24,600 tons'.[9]

The successful launch vindicated her builder's detailed calculations. Harland & Wolff had the benefit of data and experience collected over decades of launching ships to inform their analysis. Their 'launching curves' included scenarios with the 'height of tide' at 15ft 3½in and 15ft 9½in on the Hamilton Dock sill. They had calculated that, as she went down the ways, *Olympic*'s stern would begin to lift after she had run between about 685 and 666ft and the pressure on the 'fore poppets' or launching cradle was between 4,830 and 5,650 tons (this was the point when the moment of buoyancy equalled the moment of weight about the fore poppets).

She was brought to a standstill as anticipated. The tugs began their work of moving her carefully towards the fitting-out berth. A strong gust of wind brought *Olympic* into contact with one of the buttresses, slightly denting her hull plating, but it was a routine repair.

The Lord Lieutenant of Ireland prepared to catch his special train for Dublin, scheduled to leave at about 2.15 p.m. That afternoon, the press went to Belfast's Grand Central Hotel for lunch at the invitation of Messrs Ismay, Imrie & Co. on behalf of the White Star Line. The reporters 'much appreciated' the 'generous hospitality'. Frank Bullen toasted the builders and owners, 'The largest ship in the world has been launched without the slightest bustle or clamour of ceremony. It is particularly British. The people who do great things are not the people who talk about it.' The response on behalf of Harland & Wolff hoped that 'the development of transatlantic traffic' was 'a factor making for peace'.

'We in Ireland today are proud of Belfast and of Belfast men,' wrote the *Irish Times*. 'The launch of the *Olympic* restores to this country the credit of having built the largest ship afloat.' *Olympic* would 'exceed the big Cunarders by a tonnage which is greater than that of the largest merchantman in existence twenty years ago'. If other ships were 'floating hotels', then '*Olympic* will be a floating city, with a population exceeding that of many Irish towns'. Their reporter would 'hardly repeat the stories of the almost preposterous magnificence of this ship which have been current during the past few days', but credited her with a non-existent nursery, ballroom, skating rink and roof garden.

Reporters used statistics to add to the superlative descriptions of *Olympic*'s size, noting she was 175ft tall from the keel to the top of her funnels; there were 'over 2,000 sidelights [portholes] and windows in the ship'; the rudder weighed over 100 tons; the port and starboard propellers weighed 38 tons and the centre propeller, 'which is of solid bronze, 22 tons'. *The Shipbuilder* noted that *Olympic*'s launch weight, 'which is officially stated to have been 27,000 tons inclusive of ways and cradle, exceeded by as much as 12,000 tons that of the heaviest ship previously floated at Belfast' (*Adriatic* in September 1906). She took over the record from *Mauretania* as 'the heaviest steamship previously sent off the ways anywhere'.

'There is something awe-inspiring in the proportions of the *Olympic*,' wrote the *Belfast News-Letter*:

> She bulks largely in one's imagination, and suggests marvellous developments in future transmarine operations. Moderate terms and comparison between this gigantic creation of the shipbuilder's genius and her predecessors fail of their purpose when applied to such a ship,

> and the mind is almost staggered by her wonderful size, her general dimensions internally, and the luxury and completeness of her fittings and appointments.

The reporter concluded, 'She was not built to fail; a higher destiny awaits her.'

***

After the dignitaries had gone home, work continued to make *Olympic* ready for sea. The *Irish Times* reported on 14 November 1910:

> The new White Star liner *Olympic* is at present lying alongside the fitting out wharf, and the gigantic boilers are being transferred from land to the vessel by means of the huge floating crane, capable of lifting 150 tons, which is berthed between the vessel and the wharf.

On Christmas Eve, it reported the official announcement that her maiden voyage would take place on 14 June 1911. There was no room for delay.

While the two ships were on the stocks 'side by side', interest was divided between them, but now *Olympic* commanded attention:

> Since the launch on the 20th October the work on board has been proceeding very rapidly, thanks to the adequate provision made for its execution. In addition to the gigantic floating crane [...], the special electric plant arranged for the necessary temporary lighting has been an important factor [...] to enable the work to be carried on satisfactorily, in addition to which current has to be provided for a large number of electric cranes, drills and other special portable appliances. All the boilers, refrigerating and electric engines, masts, etc., have been placed on board, and the main engines are now all practically in the vessel. The plating and riveting of the houses on [the] promenade and boat decks, and other iron work, including that in connection with stairways and electric hoists, the laying and caulking of the promenade deck, the drilling of the boat deck, and the fitting up of staterooms on all decks; the insulation of provision and cargo chambers, the fitting of sanitary and water service pipes, scuppers, etc., throughout the ship, are some of the items of the work in hands since the launch, and it is interesting to observe that a start has been made with the fitting of the side framing and ceiling panelling of the first class [dining] saloon.[10]

The Scotsman hailed 'substantial progress made with the carpentry and furnishing, with the result that she now lies many inches deeper in the water than when launched, and had yesterday to be moved to a new berth'. She moved from the New Wharf back to the Alexandra Wharf, where she had been immediately after launch, but would return to the New Wharf early in January.

The pressures Harland & Wolff were facing were illustrated on 17 January 1911. Their management discussed 'the dates for the departure of No. 400 and launch of No. 401'. George Cuming pointed out that adhering to *Olympic*'s departure date 'would in all probability strain the resources of the engine works to the utmost'. The same was true of other departments, but they had got a date, which had already been announced. They fully intended to meet it.[11]

White Star transferred *Laurentic*'s chief engineer, Joseph Bell, to Belfast while *Olympic* was under construction. Bell had entered the mercantile marine in the early 1880s, serving Lamport & Holt. In 1885, he joined White Star and served on many ships, including *Oceanic* and *Adriatic*. On 5 November 1908, he 'crossed to Belfast' and remained with *Laurentic* as she neared completion. After serving for almost two years and gaining experience with her propelling machinery, he transferred to Belfast on 13 September 1910.

Bell celebrated his fiftieth birthday in March 1911. He was rated as 'a most capable officer' with 'a large circle of relatives and friends'. It seems likely White Star thought his experience with *Laurentic*'s combination propelling machinery would be invaluable. A 'well known engineer' of the time argued that 'the engine room of a

A mere six months after launching, the world's largest ship is in the world's largest graving dock. (It was only named officially as the Thompson Dock, after the Chairman of the Belfast Harbour Commissioners, in 1915.) *Olympic*'s funnels are in place and workmen are busy painting. The temporary structures off the port-side forecastle deck are workmen's toilets. (Public Record Office of Northern Ireland)

seagoing vessel provides one of the best opportunities that exists for the training of a mechanical engineer' and Bell certainly had decades of experience. Ismay recalled that Bell 'remained in Belfast during the whole of the time she was being built, superintending her construction, making any suggestions which he thought would lead to improvements'.[12]

On 2 March 1911, Harland & Wolff's engine works management were preparing for *Olympic*'s basin trials, 'It would greatly assist us in our preparations if the boilers could be filled before docking.' They asked colleagues in the Design Department to 'please look into the matter and say if there is any objection'.[13] Any issues or defects could be identified and addressed prior to her sea trials.

***

On 1 April 1911, the world's largest liner entered the largest dry dock in the world, which was now operational.[14] Thousands of people gathered all around Belfast Harbour to witness the sight that Saturday. A large number of guests were on board the tug *Musgrave* to witness the operation. When she left at 10.30 a.m., the party on board included Robert Thompson, chairman of the Belfast Harbour Commissioners and a Member of Parliament. At the dry dock itself, Lord and Lady Pirrie were present, alongside managers from Harland & Wolff.

'From an early hour everything was in readiness on board the liner for her transfer from the fitting out jetty, a short distance to the south of the dock entrance.' Under Thomas Andrews' and head manager J. Dickinson's direction, 'the liner was got under way, and with the assistance of the tugs *Hercules* and *Jackal*, was soon in the centre of the river and facing the new dock entrance':

> The approach of the *Olympic* towards the dock was necessarily a slow process, but gradually the vessel was hauled in by the powerful hydraulic capstans at the dock head, and at 10.55 exactly the bow crossed the entrance of the dock. The caisson, or entrance gate, had two hours previously been hauled round to a berth alongside the west quay wall. There was an exciting moment after the bow had safely passed the entrance [...] there was very little space on each side, but so well was the vessel under control, and so accurately had the distances been gauged, that it passed in without touching the wall. Not so much as a piece of paint was scraped from the side of the *Olympic* during the whole dry-docking operation. It speaks volumes for the experience of the men who were assigned the difficult task of transferring the vessel that this important duty was accomplished without the slightest hitch, and in a comparatively short space of time. So admirable were the arrangements that the docking was completed inside fifty minutes, and at twenty minutes to 12 o'clock, just at high tide, the *Olympic* was in her new quarters. It was a wonderful spectacle to stand on the breast of the quay and see the huge vessel towering so high above. It was only at this stage that one got an adequate conception of its colossal size. The three hydraulic capstans, which are such a feature of the dock's accessories, were largely instrumental in having the vessel docked to quickly.[15]

Olympic had only 1ft 6in on either side of her at the dry dock entrance. The divers reported that the bottom of the dock entrance was free from any obstructions and the floating caisson (dock gate) was placed in position, the dock being closed at 12.30 p.m. A little over half an hour later, the powerful pumping engines were brought into operation. The water level dropped slowly.

> It was not long until the *Olympic* was securely resting on the 350 keel blocks, and then the work of supporting and steadying the vessel by means of large pieces of wood or 'shores', extending from the ship's side to the quay wall, was entered upon. This operation occupied a considerable time, and it was not until late in the afternoon that the dock was emptied of water, and the entire hull of the vessel was visible. The *Olympic* will remain in the dock three

> weeks, and during that time the propellers, each of which consists of three six-foot blades [*sic*], made entirely of brass, will be affixed to the liner.

***

On 12 April 1911, Harland & Wolff's managing directors discussed the Nottingham Engineers' Society's imminent visit to Belfast. They wanted to look over *Olympic*, but they 'should be informed that it was against the rules and would interfere with the progress of the work on the vessel'. Any of the shipbuilder's managing directors were 'of course, at perfect liberty to personally take onboard the *Olympic* any friend of their own' but, even in that case, 'it was desirable to, as far as possible, avoid doing so until after say the 20th May'.

Work would be continuing on board during the Easter holidays. Lord Pirrie reported that he had met with the Works committee of the Belfast Harbour Commissioners a few days earlier and arranged for 'the necessary dredging at the Deep Water Wharf to enable the *Olympic* to lie afloat at all states of the tide; this work to be carried out while the vessel is in dry dock'. The Victoria Channel would be dredged to allow a depth of 32ft 'at ordinary high water, spring tides [...] prior to the departure of the *Olympic* on the 29th May' and further dredging was needed prior to *Titanic*'s launch. The Harbour Commissioners 'should do their utmost to have the turning basin ready for the *Olympic* to turn in on the 28th May'.

The next day was the 13th and it certainly proved unlucky for one labourer. Thirty-one-year-old Peter Ward was working on *Olympic* with other men, lowering a steel beam. He was sat across the beam, trying to release one of the tacklings holding it, but 'both came away, and the beam fell'. He fractured both his legs and was admitted to Royal Victoria Hospital, passing away a few weeks later.[16]

In the middle of April 1911, *Olympic* was 'being rapidly fitted out':

> The fittings and upholsterings for the saloon and other interior portions of the huge vessel arrive every hour, and the scene on board is an animated one. A large number of visitors have viewed the liner since she was moved into the dry dock, and though no one, except those on business, is allowed on board, a good idea of the proportions [...] can be had by the observer from the dock side.[17]

Olympic's bottom had been painted and water 19ft deep was allowed into the dock:

> During the past few days the propellers have been fixed in position, and she will be removed from the dry dock on Saturday, 22nd April. Last week a number of White Star Line officials viewed the vessel. A tour of the ship was mapped out for them, and when this route was afterwards measured it was found to be only four yards short of three miles.[18]

On 2 May 1911, '*Olympic* underwent her basin trials', which consisted of 'the engines being turned while the vessel was securely moored in the deep-water fitting out basin, and this preliminary test proved entirely satisfactory'.[19]

Seven days later, Harland & Wolff conducted a test of the electrical boat winches with one of *Olympic*'s lifeboats. They made sure there were weights distributed aboard to represent a load of sixty-five people, then raised and lowered the boat six times. After the weights were removed, the boat was lowered into the water. Edward Wilding noted, 'There was nothing the matter with her; she was watertight.'[20] The ship's emergency circuit was intended to make sure power would be available whenever needed, connecting 500 lamps 'throughout all passenger, crew and machinery compartments, at the end of passages, and near stairways, also on the boat deck, to enable anyone to find their way from one part of the ship to another'. The four electrical boat winches were also connected to the emergency supply by means of changeover switches, as well as other essential equipment such as the Marconi wireless apparatus; lights on the bridge, including the captain's navigating and chart rooms, wheel house, telegraphs and Morse signal lanterns; and cargo and gangway lanterns. All had to be tested to make sure they were working correctly.

Deliveries continued to arrive. On 17 May 1911, 250 tons of 'large Caprington screen steam coal' was delivered from Scotland and 'trimmed into bunker of No. 1 stokehold', from John Kelly Ltd. Two days later, 51 tons of the slightly more expensive 'best Caprington house coal' was delivered for the saloon galley ('21 tons at port side and 30 tons at starboard side').

Francis Carruthers worked to complete all the necessary documentation for the Board of Trade in London. He reported all the ship's relevant structural details as a steel ship 'with the Rules of Lloyd's Register for 1885'. Under his 'general remarks', he wrote, 'The workmanship is of the highest class, the vessel is new and has been built in accordance with the plans.' Another official initialled Carruthers' comments in blue pencil. *Olympic*'s Survey for Freeboard was stamped '13 May 1911'.

Carruthers referred back to the correspondence about the collision bulkhead specification the year previously and the Board of Trade's decision to approve her assigned load draught. His duty was to report that the completed ship now complied with what the regulator had laid down while she was under construction. By 2 June 1911, he was able to report, 'This vessel is now completed, and all the requirements of the Board contained in this minute have been satisfactorily carried out.' All the myriad calculations made during construction were checked again and reaffirmed.

On 23 May 1911, Harland & Wolff decided that free tickets to inspect the completed *Olympic* would be a worthwhile gesture to their staff: submanagers were allocated four each and 'two additional if applied for'; heads of office staff were given four each; general office staff 'except juniors' had three and juniors had two; head foremen and pupils had four; and 'leading hands and

'The lifeboats are in place and she is a hive of activity.' (Public Record Office of Northern Ireland)

Olympic's lifeboat davits are in place. This photo shows the view from the port-side bridge wing, looking aft along the Boat Deck. The deck in the foreground is the officers' promenade and the officers' quarters are visible to the left. The lifeboat davit arm is marked 'Welin Quadrant Davit Patent', but the Engelhardt collapsible boats have not yet been placed on board. (BELUM.Y18721 © National Museum NI, Harland & Wolff Collection, Ulster Folk & Transport Museum)

chargemen' had two tickets each. They arranged for 'two commissionaires for the purpose of taking the tickets at the gangway'.[21]

Two days later, the White Star Line held their annual general meeting. The directors submitted their reports for the calendar year 1910. Lord Pirrie retired by rotation and put himself up for re-election as a director. The *Syren and Shipping* looked at that year's results for the White Star Line, Union Castle Line and Royal Mail Steam Packet Company: 'Between them these undertakings own considerably over half a million tons of the very highest type of shipping, and they are representative of the four great trade routes of the world.' It thought that 'perhaps the most striking feature in the whole business is the enormous increase in the profits of the White Star Line'. The improvement in 1910 compared to 1909 was 'little short of phenomenal' and 'the directors are certainly most modest about it all [...] But, after all, it is chiefly when things go wrong that boards of directors launch out into explanations [...] The money may be safely left to do the talking.' The publication wondered 'how the current year is shaping' and suggested:

The local and national press carried many adverts placed by the White Star Line for 'the largest steamers in the world'. (Roy Derouin collection)

Olympic presented an impressive sight to the many visitors queuing up to see her on 27 May 1911. (Author's collection)

> It should eclipse even 1910, for in addition to the extra traffic which is certain to result from the Coronation there will be the advent of the *Olympic* to be reckoned with. That marine palace is to sail on her maiden voyage on June 14, and she will thus play a leading part in the passenger trade of the coming summer.[22]

The same day, the company registered *Olympic* at Liverpool. Her official number was assigned as 131,346. The detailed paperwork included her dimensions and engine specifications, as well as all the tonnage calculations. The register entry noted specifically that the enclosed promenade on each side of B-Deck ('open spaces port & starb'd on bridge deck in way of windows'), was not included. It was over a length of 357ft and would have added 751.68 tons to the figures for *Olympic*'s gross and net tonnage. All sixty-four shares in the ship were held by White Star's legal entity, the Oceanic Steam Navigation Company, with Harold Arthur Sanderson ('30 James Street, Liverpool') as the designated manager.

On Saturday, 27 May, she was opened for public inspection. '*Olympic* seems to be – perhaps on account of her name – a favourite of the gods, or, at least, of those of the fabled deities who have anything to do with the control of the weather,' wrote the *Belfast News-Letter*:

> Her launch took place on the 20th October last in the mellow sunshine of an ideal autumn day; delightful spring weather [...] prevailed on the 1st April, when she was transferred from the fitting out wharf to the new graving dock, the largest of its kind in the world; and, on Saturday last, when the huge liner was thrown open for public inspection, the glory of early summer made the day an entrancing one.

She was open from 2 to 7 p.m. For the first two hours, admission was charged at 5*s*, then, after 4 p.m., 2*s*. The majority of the visitors were thought to be Belfast residents and the proceeds of the ticket sales went to four city hospitals: the Royal Victoria; the Mater Infirmorum; the Hospital for Sick Children, Queen Street; and the Ulster Hospital for Women and Children in Templemore Avenue.

An hour beforehand, newspaper reporters were given the first chance to look over the ship and Harland & Wolff officials 'pointed out and explained some of the notable features'. They saw:

> ... the most luxurious apartments; indeed, in wandering through the lengthy corridors or the spacious and brilliantly-lighted apartments and glancing into the splendid staterooms, furnished in various styles, it is hard to believe that one is actually on board ship, and not in a modern palatial hotel in some of our leading cities.

A strict route was in place. Two gangways – one forward, one aft – 'stretched from the wharf to the upper deck, and entering by one of these the sight-seers emerged by the other'. There was already a long queue when the first public visitors were admitted 'and every minute the throng was reinforced as visitors arrived in motor cars, by tramcar and on foot, so that a substantial sum must have been raised for the benefit of the hospitals'.

The *Scientific American* wrote a detailed account of inspecting *Olympic*'s engineering spaces before she left Belfast, 'The smallest steamship ever built by Messrs. Harland & Wolff [...] could easily be accommodated on the poop deck.' It noted, 'The whole space devoted to the machinery is simply enormous [...] certainly the impression gained in all three engine rooms is one of vastness – vastness not so much of the machinery, though that is big enough, a cylinder column alone, for instance, weighing 21 tons.' The electric engine room 'is bigger than many a big ship's main engine room'. It admired the Kilroy's stoking indicators in the boiler rooms ('one for each boiler'):

> ... very neat little electrically operated signal disks with loud gongs, which give visible and audible intimation to the firemen as to the exact moment when each furnace is to be fired. The instruments are so arranged that there shall be the minimum number of doors open at the same time, and no two opposite doors in the same double-end boiler shall be open together.

> We propose to arrange for the steamer *Slieve Bearnagh* making a special trip on Tuesday night, 30th May, from Queen's Bridge Jetty, Belfast, at 9.15 p.m., to enable passengers to view the *Olympic* illuminated in the Lough, the steamer to return to Queen's Bridge Jetty at about 11 p.m.; also for the steamer *Slieve Bearnagh* leaving Queen's Bridge Jetty on Wednesday 31st May at 11.15 a.m. for a cruise down the Lough, returning afterwards to the harbour in time to permit passengers witnessing from the deck of the steamer the launch of the SS *Titanic*, the steamer to lie opposite the Alexander Graving dock, or as far up the river as you can permit, returning to Queen's Bridge Jetty after the launch.[28]

The press turned some of their attention to the launch of '*Olympic*'s sister ship'. *Titanic*'s launch was set for 12.15 p.m. 'The Harbour Commissioners and the tenants of Albert Quay have very kindly permitted the committees of the two local children's hospitals to enclose the Albert Quay and make a charge for admission in aid of their funds,' noted the *Belfast News-Letter*. 'The enclosure is directly opposite the launching slip, and is practically the only available spot from which the public will have a clear view of the launch from start to finish.' They recommended that 'no one should miss this exceptional chance'. Tickets were available from 'Messrs. Thornton's; Leahy, Kelly & Leahy's; Jordan's, and several music shops in the city at a reduction, if bought on or before tomorrow [Tuesday, 30 May]'. There was 'no fear of overcrowding', since the space was so large, and 'an ample service of tramcars' would be running:

> In addition to the interest attaching to the launch itself, it will doubtless be a considerable inducement to many that the small amount they pay for seeing this interesting night in comfort will be devoted to two of the most deserving charities in the city. Nothing can appeal to the sympathies of the charitably disposed more than the relief and care of the children of the poor in time of sickness, and any person who has visited these institutions and witnessed the tender manner in which poor little patients are treated in their pain and suffering must have come away with a feeling of thankfulness that our city is fortunate in having two establishments which so admirably carry out this beneficent purpose. The Belfast Hospital in Queen Street and the Ulster Hospital at Mountpottinger serve the two halves of the city, but unfortunately in neither case does the regular income nearly approach the expenditure, although the greatest economy is used and no money is spent unnecessarily.

'*Titanic* will slip from the cradle on which she has been created in Harland & Wolff's Belfast yard, and her arrival in her proper element will be saluted by her wonderful twin sister,' wrote the *Daily Telegraph*'s correspondent. *Olympic* 'is now full grown, and is about to justify her claim to the title of Queen of the Seas'. The *Northern Whig* said:

> To launch in the morning the largest steamer in the world and despatch in the afternoon on her first trip the most gigantic liner that has ever dared the seas is a 'double event' that [Harland & Wolff's] best equipped rivals are not likely to best for many a day.

The paper called it 'a perfect day for the launch', with 'sweltering May sunshine tempered by a southerly breeze that crinkled the waters of the channel and set the flags of the shipping tugging joyously at their halliards'.

Harland & Wolff repeated the arrangements that had worked so well when *Olympic* was launched only seven months earlier, including the flags flying the Union Jack and Stars and Stripes with the code signal 'Success'. One newsman thought 'decorations are superfluous' with 'a boat like the *Titanic*', 'Towering up under the cage of the gantry she caught and held the eye to the exclusions of everything else.' *Olympic*, 'with her gleaming white sides, made perhaps a braver show' when she was launched, whereas *Titanic*'s 'dark red and black, edged with a narrow line of white on the upper deck [*sic*], gave a more notable impression of sheer irresistible strength'.

One of the workmen called *Titanic* 'a rare big upstanding brute', arguing that she did not look like 'a mass of dead iron, but a live thing straining at the leash'. The preparations seemed to have been completed particularly quickly that morning:

> The gangs had been hard at work all morning, and by the time the stands began to fill and the crowds to edge down as far as they could go on either side of the vessel very little remained to be done. The forest of shores under the bilges had dwindled to a few stray clusters, and these were being knocked away one by one.[29]

There was not long to go:

J.P. Morgan (left) and Lord Pirrie (right). The photo was captioned as both men attending *Olympic*'s launch but the occasion seems to be *Titanic*'s. *Leslie's Magazine*, 1911. (Author's collection)

> Lord Pirrie [...] descended the gangway and consulted with Mr. Charles A. Payne, who had charge of the launching apparatus. He examined the dial face, peered at the cylinder half hidden in the timber of the ways, the piston rods of which hold the triggers in position, and gave an order. At the word all the foremen's whistles blew shrilly [...] a quiver ran through the crowd, and away in mid channel one could see the yellow funnel tops of the tugs moving fussily into position like terriers nosing round a rat hole.
>
> Then began the real thrill of the launch, the tense packed moments of waiting for the final signal. Through the dense mass of spectators one could feel the ripples of excitement gathering force, half, eager intentness for the arrival of the great moment, half unconscious fear that something would go wrong. The workmen, however, toiled on solidly and steadily, hammering and hauling with the greatest unconcern [...] as each of the remaining shores was knocked away a painter was always at hand to coat over with a dripping brush the rusty circle it had left on the vessel's hull [...]
>
> By now the needle on the pressure gauge had crept round nearly half a circle. Already it was far above the 435 tons which it touched when the *Olympic* was launched, and was close up to the five hundred mark. The engineers were scanning it more intently than ever, tapping the brass case and flinging an anxious eye at the knots of men still swinging their sledges in the gloom under the bilges. The vast majority of the gangs has scrambled clear by this, and lined along inside the barriers were mopping faces ashine with sweat, and glancing expectantly at their lagging comrades. Suddenly the note of a whistle blown furiously cut through the air, and with it came a cry taken up all along the line, 'Stand clear'. On the word a swarm of men came darting out of the holes in the launchways, while the batch of painters, with one eye on the launching gear and the other on their brushes, dabbed hastily at the marks left by the last of the shores. Again the warning cry rang out, on this occasion with an emphatic 'Hurry on' tacked to it, and as a few stopped to pick up their tools the command came with a hint of anxiety, 'Never mind the hammers, come as you are!' It was still a few minutes to launching time, but apparently those in charge were taking no chances. Very speedily the foremen mustered and counted their squads, the words 'All right' were passed

> along [...] and amid a dead silence at thirteen minutes past twelve – two minutes before the time fixed – at a signal from Lord Pirrie, Mr. Charles A. Payne gave the fateful order.[30]

Titanic's hull began to move. She 'entered the water without the slightest hitch, amid the hooting of syrens and the enthusiastic cheers of the vast multitude who had assembled on both sides of the river', reported the *Belfast Telegraph*. She reached a maximum speed of 12 knots, taking sixty-two seconds down the slipway – virtually identical to her sister. Her displacement of 24,466 tons was slightly less than *Olympic*'s, meaning that her older sister retained the record of being the heaviest ship launched to date.[31] *Titanic* was brought to a stop, 'then the tugs fastened on to her and in a marvellously short time dragged her down to the deep water berth so lately vacated by her sister'.

Lord and Lady Pirrie 'entertained a number of guests, including Mr. J. Pierpont Morgan and Mr. J. Bruce Ismay'. J.W. Kempster 'presided at a luncheon to a large number of visitors and officials of the Queen's Island' at Belfast's Grand Central Hotel, on Royal Avenue, about half a mile away from Belfast City Hall.

One family would remember the 31st as a sad day. Forty-three year old James Dobbin suffered an accident which cost him his life. He was one of the many workers knocking away the timbers beneath *Titanic*'s hull, but was struck by one and pinned to the ground. When he reached hospital, 'he was suffering from shock and contusions', according to evidence before the City Coroner, Dr James Graham. Although he underwent an operation to try to save his life, the doctors' best efforts were unsuccessful and he died some days later. The court returned a verdict of 'accidental death' and expressed 'sympathy with the widow and family'.[32]

OLYMPIC'S DELIVERY TRIP

Olympic's delivery crew, who would 'take the vessel to Liverpool and Southampton', signed on the same day. There were some 278 crew who signed on initially, but about twenty of them left the ship by consent before leaving Belfast. The crew included almost sixty in the Sailing or Deck Department and about 222 men, including engineers, firemen and trimmers to operate the propelling machinery. The arrangements varied depending on their role. Captain Smith and his officers, as well as the chief engineer and senior engineering staff, were effectively starting a long period of employment with the ship. The others were men who signed on just for the delivery voyage, 'Each man to receive £4 for the run to Liverpool and Southampton and third class rail and steamboat ticket back to Belfast.' Many of them were young local men from Belfast, Carrickfergus or surrounding areas, in their twenties and thirties.[33]

The weather was still 'beautiful'. Around 2.30 p.m., *Nomadic* left the quay with 'a small party of representatives of the owners and builders on board' to take them out to *Olympic*, including J. Bruce Ismay, Lord Pirrie, J.P. Morgan, Harold Sanderson and Henry Concanon.

Two hours later, *Olympic* sailed for Liverpool. *The Shipbuilder* noted approvingly that she had 'been completed and handed over to her owners twenty-nine months after the laying of the keel, and seven months and eleven days after her launch'. It was an impressive record and mirrored *Oceanic*'s construction at the turn of the century, yet *Olympic* was far larger. *The Northern Whig* said, 'A big crowd of holidaymakers, who had spent the morning in witnessing the launch of the *Titanic*, decided to make a day of it by seeing the *Olympic* off.' Trains to Whitehead and Bangor 'were crammed'.

Opposite: J. Bruce Ismay (standing in between two shipyard workers on the platform, top right) inspects the launching gear beneath *Titanic*'s mighty hull. Lord Pirrie descends the stairs (bottom left). (Davison Photo, 'Kempster Album')

The people of Liverpool waited eagerly for *Olympic*. Plenty of their city's companies had fulfilled orders for the ship. Elkington & Co. Ltd's premises at 27 and 29 Lord Street displayed 'specimens of the Service of Electro Plate expressly designed and manufactured by them';[34] Hocking & Co. Ltd, another Liverpool firm, 'patentees and manufacturers of evaporators', advertised their '"Climax" Fresh Water Distillers (as supplied to the *Olympic* and *Titanic*)'.[35] Messrs Chadburn's Ship Telegraph Co. Ltd had acquired their works at Cyprus Road, Bootle, in 1904, which were being enlarged and modernised in 1911. The firm manufactured the Willett Bruce Automatic Whistle Control, 'as used on the *Olympic* and *Titanic*'.[36] The Liverpool Engineering and Condenser Co. Ltd supplied the Quiggin-type evaporators for the turbine engine room; Railton, Campbell & Crawford manufactured 'the famous Railton, Campbell and Crawford's Patent Silent Ash Hoists' at their Washington Foundry in Cherry Lane, Walton; and Thomas Utley had supplied Utley 'sidelights [portholes] and patent windows'.[37]

The Journal of Commerce argued in May 1911:

> Amongst the many radical – one might say revolutionary – changes that have taken place in the method of construction of passenger steamships of recent years nothing perhaps has been more striking or important than the method of natural lighting, and both the *Olympic* and *Titanic* will from this point of view prove unrivalled.[38]

They reported that Harland & Wolff had asked Thomas Utley to do special new designs for portholes, which 'would not become obsolete in twenty years' time'. They needed to 'be of immense size' and the shipbuilder also made clear that they needed to be round – 'under no circumstances' would the shipbuilder contemplate cutting rectangular holes in the ship's hull (doing so would compromise the structural strength).

J. Bruce Ismay had taken a personal interest in all the arrangements. He was undoubtedly pleased with the Utley Patent Combination Light, a porthole design which provided ample natural light. On the Boat Deck, A-Deck and B-Deck it was possible to use rectangular windows, because these decks were superstructure and did not contribute to the strength of the hull. The A-Deck arrangement included fifty Utley Patent French Windows ('the latest idea in marine lighting [...] fitted in front of the d[eck] house and down each side in the principle staterooms'). The Utley Patent Sliding Windows on B-Deck were regarded 'as the cleverest example of efficient simplicity ever placed on the market. This class of light was brought out in 1909 by Mr. Utley, having been specially designed for the two liners [...] and they have since become the most popular standard of sliding window in use on shipboard.'

When *Olympic* arrived in the River Mersey on 1 June, she was set for a busy day. Ensuring the public inspection went smoothly was a logistical challenge. Harland & Wolff had already drawn up plans, marking out the route that visitors would take, by 19 May. They would board using the first-class entrance on the starboard side of D-Deck, then cross the reception room, go through the first-class dining saloon, the galleys and second-class dining saloon, and take the aft second-class staircase up to C-Deck. From there, they would cross the second-class library and proceed forward through the first-class accommodation, with barriers placed at the 'lower half of doors' to the first-class staterooms amidships. They then went up to B-Deck and walked aft using the starboard promenade, entering the second-class smoke room and then proceeding through the second-class entrance on the port side before heading forward through the first-class accommodation to proceed up to A-Deck. Their route then took them aft, past the reading and writing room, through the first-class lounge,

Opposite: Amid excited cheers, *Titanic* thunders down the slipway. (Davison Photo, 'Kempster Album')

Olympic at Liverpool. The *Journal of Commerce* reported, 'As she lay abreast the stage yesterday the general opinion among the many thousands of sightseers who thronged both banks of the river was that the Mersey had never borne a ship of more beautiful proportions or more graceful lines. So exceedingly well balanced are the proportions of the vessel that at first glance her vast bulk is hardly realised; indeed it is impossible for the eye to form any complete vision of the tremendous size of the vessel without comparison; that is until some other vessel becomes dwarfed by her vicinity.' (Author's collection)

first-class smoke room and port-side veranda café, then along the port promenade deck and back to the forward grand staircase. Then, they headed aft on the Boat Deck's starboard side and into the second-class entrance. That took them down to E-Deck, where they could disembark from the aft gangway on the starboard side. The tour took in a significant number of the first and second-class public rooms, as well as the most luxurious first-class stateroom and suite accommodation, but third-class areas were not inspected at all.

Cunard were keen to check out the new ship. Alfred Booth, who had succeeded William Watson as chairman the previous year, A.D. Mearns, the company's general manager, and Leonard Peskett, their naval architect, spent several hours on board. Peskett authored a memo that was shown to the executive committee: 'It was the general impression that taking the first class accommodation as a whole (including the public rooms) a better effect and general result might be obtained at a much less cost.'[39] Judging *Olympic* as a 'commercial asset', he thought his plans for *Aquitania* were better. They included more first- and second-class staterooms and more third-class passengers.

The two ships were going to be similar in size. He thought the combination propelling machinery would 'show a small percentage of saving in coal consumption', but 'it necessitates two engine rooms, two sets of casings, etc., none of which contribute to revenue'. He did not like the arrangement of the boiler rooms either, which were 'primitive and badly thought out'. The boilers, fitted across the breadth of the ship, did not allow for longitudinal watertight bulkheads; 'the uptakes from wing boilers take away deck space'; and 'the great width of the ship demands interior longitudinal support'. The coal bunkers spanned the breadth of the ship, making the coaling process 'tedious and costly'. (Peskett's comments reflected his professional views on longitudinal watertight bulkheads, which were quite different to Harland & Wolff's.)

He also disliked the position of the electric engines, low down in the ship in their own engine room, 'Any water getting into the lower part of the ship aft, would soon put them out of action.' He did, however, like *Olympic*'s emergency dynamo arrangement. The two emergency dynamo engines were fitted level with E-Deck and the steam pipe supplying them could be supplied from any of the boiler rooms. His plan for *Aquitania* had a petrol-driven emergency dynamo. Later, he reflected that it would 'perhaps be better to consider a similar arrangement' to *Olympic* instead.

Olympic's 'general outline' was good, with 'a dummy funnel being fitted to obtain balance of proportions'. Her sides were essentially vertical, whereas other ships had a 'tumble home' because their sides gradually curved in, narrowing the hull on the higher decks. The lack of a tumble home on *Olympic* gave 'maximum deck area for cabins and promenades':

> The boat and A decks overhang beyond the beam of the ship to get width of promenade and allow [life]boats when being lowered to clear sides. The promenades and spaces for games are spacious. The reception room and first class saloon are well planned, but the other first class public rooms are not equal in planning or decoration to the corresponding rooms on [the] Cunard company's latest steamers.

The 'large number of fans' grouped around the base of the fourth funnel were too loud and 'the hum from these fans when they are working is very objectionable in the first [class] smoke room below'. Along the B-Deck promenade, the screen windows were different from some other ships. Cunard's new *Franconia* used the Laycocks patent, enabling passengers to open and close the windows with ease, but on *Olympic*, 'the screen windows are made in a metal frame and are opened or closed with a portable crank handle which works a rack and pinion, and is under control of the deck attendant only'.[40]

Cunard's superintendent engineers, J. Currie and A. Galbraith, were on board the same day. They expressed their gratitude to 'the kindness of' W.J. Willett Bruce in allowing them to see 'the propelling and auxiliary machinery, boilers, etc'. To their great surprise, *Olympic*'s engine room was 'so spacious', 'everything is well arranged and readily accessible, and appliances and room for overhauling have received special attention'.[41] Their own impression of the ship's engine and machinery spaces was much more positive than Peskett's:

> On the bottom platform [in the main engine room], sufficient width has been left between the columns of the reciprocating engines to allow two specially designed electrically driven winches to be fitted. During overhauling of heavy parts, this will enable the services of a large number of labourers to be dispensed with, and consequently will materially reduce working expenses.
>
> The machinery spaces have evidently been planned with a view to enabling the engineer on watch in the engine room to have everything more or less under his eye. For example, the refrigerating machinery is placed in the port wing on the lower platform of the main engine room, while the electrical installation is situated in the tunnel, aft of the turbine engine room, and consequently is cool and thoroughly ventilated. This is important when it is remembered that the efficiency of electrical generators rapidly decreases if they are not kept cool. From this point of view, the proposed position of the plant on the *Aquitania* is not ideal.
>
> A small auxiliary lighting set is also carried, being placed high up in the after funnel.

Fournier long-distance temperature-recording instruments enabled the engineers in the engine room to check the temperatures 'of the tanks on the top deck for supplying both hot salt and hot fresh water to the baths, etc'. The engineers' workshop impressed them, measuring 'some fifty feet long' and 'equipped with modern machine tools for turning, shaping and milling, etc. Thus it will be possible to carry out a large number of operations on work which otherwise would require to be sent ashore.' This 'step in the right direction' helped improve efficiency.

They thought the boiler rooms were 'well arranged':

> There are four exits from each boiler room, where the coal heavers can go in and out when in port, and none of these men are allowed in the engine room. There are a number of doors out in the cross bunkers, and this should reduce the number of trimmers required.

The spiral staircase leading from the firemen's accommodation in the bow down to the firemen's tunnel and Boiler Room 6 was 'a great mistake, as only one man can ascend or descend at a time, and a broad stairway with landings would have been much better'.

During their tour, Chief Engineer Joseph Bell accompanied them and 'readily gave us all information asked for. It is only by having opportunities of this kind that one can see how other people have improved on our practise, and also how we may improve on theirs.' They had recommended a number of points 'in connection with the *Aquitania*, and their adoption on the *Olympic* strengthens our recommendation'.

While *Olympic* was at Liverpool, she generated a number of charges for the White Star Line to settle: 'Arrived new, from Belfast on the 1st June. Pilotage 32 at 4*s*'; harbour rates were quoted at 'half Southampton dues'; 'Thirteen tug trips with visitors at 5*s*' generated a charge of £3 5*s*. She 'sailed 9 p.m. same day'.[42]

Trinity House Pilot George Bowyer and the inward Isle of Wight pilot, Wallace Caws, joined *Olympic* at Liverpool for the run to Southampton. Four years earlier, Bowyer had treated rumours that White Star was going to 'build two large ships of double the tonnage of *Adriatic*' with scepticism – 'The jump from 25,000 to 50,000 tons appeared to be almost impossible.' Now, when the two men arrived at Liverpool's landing stage and saw *Olympic* anchored in the Mersey, he thought 'she looked immense'. They took the tender and looked up when they drew alongside, 'We could hardly believe our eyes that there was such a ship.' When they went up to meet Captain Smith in his quarters, they remarked how large she was.

'Yes,' Smith responded, 'but after you have been on board for some time her size will wear off.'

During the run to Southampton, they 'had a good look round the ship, and went the whole length of her'. Unlike Smith's experience, Bowyer recalled one of them saying, 'I do not know what you think of her, but I don't see that she is getting any smaller.' They both agreed.[43]

8

'THE GREATEST SHIP IN THE WORLD'

Olympic benefited from an early morning high tide when she arrived at Southampton. The 'Movements of Liners' reported 'from Lloyd's and Special Telegrams' on Saturday, 3 June 1911, noted her arrival at '11.35 p.m. yesterday'.[1] She was met by five tugs, who helped her into the new dock bow first. She passed *Oceanic* at the head of the docks, the White Star Line's old flagship looked small in comparison, but 'no difficulty was experienced in manoeuvring the huge ship, which was safely berthed between three and four o'clock'. Pilot George Bowyer noted that work was still under way to get the dock completed and only Berth 44 was ready.

Olympic's 'four immense funnels, easily the biggest ever seen at Southampton, were eloquent of the enormous growth that has lately taken place', wrote the *Southampton Times*:

> Everything in the vicinity was made to appear dwarfish. Her funnels were almost as high as the largest building in the docks, whilst her towering masts were easily the tallest things in the vicinity. Her hull was a thing of wonder and awe, and the visitors who thronged to the dock from the early hours of the morning were greatly impressed with the biggest ship the world has ever known.

She had had a good run from Liverpool to Southampton and her officers told the press that they 'were well pleased with the manner in which she performed the passage'. *Olympic* 'maintained a good average speed, and arrived in the Solent well up to her advertised time'. Southampton's mayor regretted that the time of *Olympic*'s arrival prevented the city from welcoming her in the same way they had for *Adriatic*, four years earlier. Nonetheless, she was greeted by the signal 'Port Admiral Welcomes *Olympic*' on arrival. Mayor Colonel E. Bance visited the new liner at 11 a.m., accompanied by his deputy, Alderman C.J. Sharp, the Sheriff, Alderman E.R. Ensor, and Town Clerk, R. Linthorne. Captain E.J. Smith received them.

The mayor remarked that he was there as Southampton's mayor, as Admiral of the Port, and as Chairman of the Southampton Harbour Board. He was interested in the ongoing dredging work, to make sure that Southampton could accommodate large ships such as *Olympic* 'at all states of the tide [...] Southampton had always prided itself on the fact that the port was ready to receive any ship, whatever her size might be, and he thought they would all do their best to maintain the reputation of the port.' The mayor's comments belied the White Star Line's experience over the preceding three years.

The following Thursday, 8 June, plenty of visitors came to admire the new ship. 'On the quayside this morning nothing could be seen but an over towering mass of black,' wrote a reporter for the *Aberdeen Daily Journal*:

An early view of *Olympic* taking on cargo and supplies at Southampton. She is docked port side to the quay, which was not the case on her maiden voyage. (Jens Ostrowski collection)

> Labourers [...] were shovelling bucket on bucket of coal from barges which looked little more than wooden tubs. One by one we climbed up the gangway and disappeared through a small hole in the *Olympic*'s side. It was really a full sized doorway but it seemed a mere keyhole by comparison.

The 'fine view' from the Boat Deck 'was like standing at the top of the Aberdeen Townhouse tower surveying the countryside for miles around. Nothing around was so high as to interrupt the view.' They looked down:

> The coal heavers filling their buckets seemed mere specks, mere ants industriously crawling over the dust heaps. At the other side [of the dock] lay the *Carisbrook Castle*, the South Africa liner which some ten or twelve years ago was regarded as a leviathan. Today it appeared to be but a yacht's dinghy.

Olympic's interiors represented 'the acme of luxury':

> Sitting in the reception room I found it hard to believe I was not in the Ritz or any of the other sumptuous West End hotels of London. Low set easy chairs in green upholstery are scattered over the dark, richly coloured carpet. A blaze of bronze ceiling lights set a-glittering the gilt decoration of the white enamelled walls.

The grand staircase was 'magnificent'. He wrote his account at a small table in the reading and writing room, 'The sun streams into the apartment, and rests upon the pure white walls.' First-class staterooms were 'large and well furnished. The old-fashioned bunk is displaced by a brass or oak cot bed. Most of these beds are four feet wide.'

White Star held a luncheon on board for local dignitaries and invited guests. Many of them travelled in a large party from London on a special train, as guests of Messrs Ismay, Imrie & Co., White Star's management company. Harold Sanderson presided. Attendees included Mayor Colonel E. Bance, Lord Pirrie, Hugh Drummond, as chairman of the London & South Western Railway, representatives of the port and commercial authorities of Southampton and newspaper representatives. The *Southampton Times* reported that Mayor Colonel E. Bance proposed a toast to the new ship:

> He had seen many vessels, many hotels, many palaces in various parts of Europe and America, but he knew of nothing that could combine the perfection of hotel life, first class residential life, and he could add 'palace life' in the same degree as could be found on board that magnificent steamer.

When he had seen her the preceding Saturday, 'he was wonderfully struck with the details of every part of it'. There 'was nothing which they required in the shape of magnificence, luxury, or size in connection with the journey across the Atlantic which was not provided'.

Harold Sanderson extended his welcome to the guests and said his only regret was that J. Bruce Ismay was unable to be there. *Olympic* represented White Star's 'endeavour to produce a ship which would satisfy all requirements'. He recalled meeting some of the same gentlemen four years earlier when *Adriatic* arrived in Southampton. He had commented to them that White Star intended to make its express service from Southampton successful and he hoped that they understood that White Star was true to its word – she 'was only the forerunner to two such vessels and the *Titanic* would shortly make her appearance in Southampton'. They had 'endeavoured to provide everything in the form of luxury and comfort which the most fastidious traveller could justify, or at least reasonably require':

> In the third class they had provided very fine accommodation indeed, and the second class, he ventured to say, was quite equal to the first class of a few years ago. Some of those present who had read descriptions of the vessel in some of the papers might have been disappointed at not finding the motor racing track, the ice skating rink, the botanical gardens, and the

> zoological gardens, but even they would not be so disappointed as the motorists, the professional skaters, the head gardeners, and the lion tamers who had applied to the White Star for appointments which they had not been able to give them.

Lord Pirrie was 'very heartily received'. He 'spoke of the progress of shipbuilding and urged the importance of dock authorities keeping abreast of the times':

> He mentioned, in connection with the *Britannic* [1874], that it was necessary to cut five feet away from her, as there was no dock long enough. The *Great Eastern* would have been a success if it had not been for the absence of dock accommodation. A ship a foot longer than the *Olympic* could not be got into the graving dock at Belfast, which was only ready just in time for that vessel. Till the dock authorities were prepared to go further in dock accommodation it would be foolish for his firm to advise their clients to go in for larger ships than the *Olympic*.[2]

By the time Harold Sanderson closed proceedings, the White Star Line could be pleased with a very positive reception in Southampton. One newsman heard a visitor describe her that day as 'one of the wonders of the world' and say that 'you can't take her in all at once'. 'That is quite true, sir,' responded one of the crew, 'and it would take you a week to know your way about her.'

On 16 June, the *Western Gazette* described its editor's visit over a week earlier. He said she 'literally dwarfed the other liners berthed nearby, and seemed to appal and almost take the breath away'. Entering 'by one of the large gangways you could, without stretch of imagination, believe yourself to be in a huge, palatial hotel, furnished and fitted with every conceivable luxury'.

Hundreds of miles to the north, an invited guest known only as 'W.B.' was on a train heading south on 'a fine, if a trifle warm, run to London'.[3] They got there on time and settled in to Cranston's Waverley Hotel, Southampton Row. The following morning, Friday, 9 June, they joined a 'special train' which left Waterloo Station in London around 10.30 a.m. They had a 'non-stop run to Southampton to inspect the *Olympic* and generally to have a very good time on board'. When their guide was showing the group round, he lost his way and had to apologise, 'Well, it is like this. I have only been on board four days, and I admit I do get mixed up a bit in this town!' W.B. noted for the *Motherwell Times* that there were 'two electric winches in the engine room, the product of the skill of Messrs, Chambers, Scott and Co., Engineers, Motherwell', 'Is it to be wondered if I had a little extra interest and pardonable pride in knowing that our good town contributed to this noble ship?'

W.B. thought the first-class accommodation was 'magnificent, and you have the greatest difficulty in remembering you are on board a ship'. The only reminder were 'the comparatively low ceilings. They are not low, but in a palace or grand hotel, they would of course be much higher.' Second class was 'so good that we thought the second was really the first class, and certainly the second class on this ship is equal to the first class of a few years ago'. The 'ingenious arrangement' of second-class staterooms made sure that 'many, if not all, the second class inside cabins have a porthole which means a very great deal to travellers when the weather is warm, and gives natural light'. Third class was 'compact and comfortable'. Lunch was 'smartly served' on tables of two to eight people. 'Seated at a table for two, I had as companion an agent from Guernsey, one of the Channel Islands, and he was interested to meet an agent from Scotland.' They lamented that 'the hours on board passed all too quickly, and stepping ashore, we joined our special [train] lying alongside the vessel, and had another non-stop run to London'.

On Tuesday, 13 June, Mr Davis (from the interior architectural firm Mewes & Davis) and Mr Saunders (Selby & Saunders, Quantity Surveyors) toured *Olympic* so that they could report back to Cunard. They liked the first-class grand staircase, writing that 'the general planning [...] was considered very good; that the position of the elevators was very suitable'. The swimming pool was 'quite a good bath', but 'it would have looked much better in marble; the cost of which would not have been any greater'. The electric-bath arrangement was 'an improvement upon the German ships'. Davis preferred grouping the gymnasium, squash court, Turkish and electric baths and swimming pool all together, 'in a similar manner to the Royal Automobile club'. They liked the lighting in the first-class dining saloon but criticised the lack of a 'small well and balcony' in the centre.

Davis did not like the 'unsuitable' styles of many of the first-class special staterooms ('in most instances the detail being course and heavy'). He thought many had too much furniture in them. On the positive side, the wardrobe rooms 'are a great addition'.

The restaurant's décor was 'considered the best upon the vessel', but the space for the orchestra 'was in too close proximity to the passengers'. The reading room's windows and lighting were 'excellent – particularly the lights [windows] above the boat deck'. They liked the 'very good proportions' of the first-class lounge, but the décor was 'considerably overdone'. 'One of the greatest objections is that passengers must be continually crossing the room to get to the smoke room or palm court [veranda café].' Davis 'was particularly interested in the corridor between the smoke room and lounge, and would like to obtain one on the *Aquitania* as wide as possible so as to form a gallery'.

Dr Beaumont recalled that '*Olympic* caused a sensation when she first appeared in Southampton [...] She looked colossal, and even "uncanny" as she towered high above the waterline, with her tiers of spacious decks and her four huge funnels, dwarfing all other craft within sight.' Visitors had to be 'led around by guides, who themselves were by no means sure of the direct route'. He thought:

> Everything is on a big scale, but put together with good judgement – very 'roomy' but in excellent order – with an absence of awkward spaces or corners and overcrowding. The interior wood work, decorations and fittings, are very handsome, and in excellent taste, with an absence of anything tawdry or 'loud'. Experts have told me that her engines are marvellous in design and structure.[4]

All the preparations needed to make sure *Olympic* was ready for her maiden voyage kept Beaumont and his colleagues busy.

LABOUR DISPUTES

Labour unrest was causing concern for the shipping lines. The day before *Olympic* sailed, Joe Cotter, Secretary of the National Union of Ships' Stewards, Cooks, Butchers and Bakers, announced at Liverpool 'that everything was ready for the international strike, and that the signal would be given this week'. The *Irish Independent* reported that a Belgian union had held a delegates meeting and that 'the international strike would begin'. On 15 June, the *Dundee Evening Telegraph* said, 'No serious difficulty has been experienced with outward bound Liverpool liners,' except for issues with *Baltic*'s and *Teutonic*'s crews striking for higher wages.

Seamen's representatives were due to meet with officials from the Cunard and Booth Lines. White Star was in negotiation, 'and hopes are entertained in Liverpool that the strike will soon terminate'.

A writer for *Justice*, the newspaper of the Social Democratic Federation, which became the British Socialist Party in 1911, spoke to a 'comrade' from the 'Southampton branch', who was among *Olympic*'s crew.[5] He was not happy:

> The crew were engaged at Southampton, and left here on Saturday, May 27. On reaching Liverpool we were put aboard for Belfast with nothing whatever in the way of sleeping accommodation and no food was provided. On arriving at Belfast on Sunday morning we found we could not get aboard the *Olympic* until the afternoon. From seven o'clock in the morning till between three and four o'clock in the afternoon the only food we got was given to us

> in a shed under disgraceful conditions; there was practically nothing in the way of cooking apparatus, and we couldn't make a drop of tea until seven o'clock that evening. In fact, all the time we were aboard the *Olympic*, the food was bad and disgracefully served.

They confirmed that there had been 'a dispute in connection with the firemen'. Most were Liverpool men, although 'there were a few from Southampton'. They had experienced such 'long hours' during sea trials 'that they refused to work any longer'. He said that 'boilermakers' mates from Harland & Wolff's yard' had been engaged to make sure *Olympic* sailed and alleged they 'were locked in a shed before being transferred to the ship, whilst others were told to go on board, and did not know until they got there what they were wanted for. Not more than two or three of them had ever been to sea before.' There 'was a great deal of discontent during the whole run [to Liverpool], but the crew, being Southampton men, could not very well leave her so far away from their homes'. They 'were expected to work any hours we were told to work, and overtime was not taken into account [...] Our hours of work were from the first thing in the morning until the last thing at night.' During the public inspection at Liverpool, the men 'were kept fully occupied all day long till nearly eight o'clock at night attending to visitors on the ship, and then had to get her ready to leave at midnight'.

On Monday, 5 June, he reported the 'coal porters refused to work on the *Olympic*', demanding a wage increase from 3*s* to 5*s* 'for fixing up their gear' because 'the height of the decks' was 'greater than those of other steamers'. While White Star were considering that dispute, the coal porters also demanded that their payment per ton of coal be increased from 1½*d* to 2*d*. White Star were 'willing to concede a portion of the increase' for fixing up the coaling gear, but not for the rate per ton of coal. He alleged 'men have been drafted from the Tyne, the Tees and the Humber in order to get the coal on board [...] to enable her to get away by the fourteenth'. The writer concluded, 'It is not at all unlikely that these incidents may be the spark that will kindle the fire' of a general strike. 'The shipping companies, while catering in the most luxurious fashion for the wishes and tastes of their wealthy patrons, will grant decent conditions to those who work in these marine hotels only when forced to do so by organised effort and legislative measures.'

On Saturday, 10 June, the Seamen's Union sent a deputation to Russell Rea, the recently elected Liberal Member of Parliament for South Shields. They voiced their concerns that 'Rea's firm [R. & J.H. Rea] was importing non-union men' to coal *Olympic*, but Rea said that 'was altogether wrong'. His firm 'paid a higher wage to the coalporters than any similar firm, and the men had made a demand equal to four shillings a day, which was impossible to comply with'. White Star 'had been obliged to get men where they could, as they must have the ship coaled'. His firm had offered to meet with men from the Seamen's Union and his son had arranged to see the union representative, 'who had refused to consult him'.[6]

Philip Curry, White Star's manager at Southampton, told the *Southern Daily Echo* that the company were 'absolutely committed' to making sure *Olympic* sailed on schedule. They made concessions to make sure she did, granting increases to the sailors and 'engine room staff'. He was also responsible for the American liners and said *New York* would be laid up. He was looking at laying up other ships, 'as circumstances required, which meant driving away trade'.[7]

PROVISIONING 'A SMALL TOWN'

All manner of supplies were coming on board. Catering for everyone aboard *Olympic* was a challenge, but it was a benefit that she sailed on a comparatively short voyage. In December 1911, *Railway & Travel Monthly* calculated that a ship such as the Aberdeen Line's *Demosthenes*, completed by Harland & Wolff earlier the same year, had to carry the equivalent of 'sixty thousand days' food for one person', because there were 1,500 passengers and crew on a forty-day voyage to Australia (while she replenished fruit and fresh vegetable stores when she stopped at various ports *en route*, 'the bulk of the supplies for both the out and home voyage' were 'shipped at the same time'). *Olympic* carried more people, but on a much shorter voyage, so the total amount of food required was substantially less.[8] The *Dundee Evening Telegraph* noted that catering for her passengers and crew was 'equivalent to that of a small town':

> It may be assumed that a ship's company of this size is not wholly composed of teetotallers, so that all the requirements of a cosmopolitan community have to be met. The American will call for his cocktail; the Englishman for his small Bass, Guinness or Scotch and 'polly'; the Frenchman or Italian for his favourite vermouth.

Fifteen thousand bottles of ale and stout were required; 500 bottles of spirits; 1,000 bottles of dry and sparkling wines; and 7,500 cigars would 'be wafted across the wide expanse of the Atlantic'. She would be stocked with a 'truly colossal list of things essential to the pleasures of an ocean trip', including:

> 75,000 lbs fresh meat;
> 10,000 lbs 'wet' fish;
> 7,500 lbs dried fish;
> 25,000 lbs poultry;
> 7,000 lbs bacon and ham;
> 35,000 eggs;
> 40 tons potatoes;
> 200 boxes oranges;
> 150 boxes apples;
> 2,000 lbs hothouse grapes;
> 250 barrels of floor;
> 10,000 lbs cereals ('oatmeal, rice, peas, etc.');
> 3,500 lbs jam;
> 1,500 gallons fresh milk;
> 1,200 quarts fresh cream;
> 5,500 lbs butter;
> 1,500 lbs coffee;
> 1,000 lbs tea ('choicest blends');
> 10,000 lbs sugar;
> 1,500 quarts ice cream

Olympic had 'to be equipped with as much glass, cutlery, plate, chinaware and linen as would stock half a dozen decent sized shops'. These included 7,000 pieces of glass, 25,000 pieces of china, and 21,000 pieces of earthenware 'in the shape of plates and dishes'.

Exploring *Olympic*'s Boat Deck. This area on the aft starboard side was allocated to second-class passengers. The second-class staircase deckhouse is visible on the left and the raised roof over the first-class smoke room in the middle. (Author's collection)

Linen stores included over 1,000 items 'including sheets, pillow slips, serviettes, and tablecloths'. The American Line and White Star Line had their own combined 'laundry on shore, capable of dealing with 20,000 articles a day, and installed with all the latest hygienic appliances. Night as well as day an army of workers is constantly employed ministering to the requirements of "those who go down to the sea in ships".' The *Southampton and District Pictorial* called it a 'department of activity which rarely occurs to the mind even of the Atlantic traveller, much less to the general public'. Passengers might notice the 'spotless napery' of a dining saloon table, 'never paying heed to the size of the company's "linen cupboard" or to the highly organised and specialised methods by which it is plenished':

> Picture an *Olympic* with three thousand souls onboard; reckon up their weekly requirements in the way of linen, serviettes, sheets, bedspreads, towels, and the like; then add to the sum the requirements of the steamers necessary to maintain a weekly service for two companies, and you will appreciate the necessity for a laundry plant capable of turning out and with a high pressure capacity twenty-five per cent in excess of that figure.[9]

Sitting inside 'a great corrugated iron building in a corner of dockland, employing nearly fifty hands', the laundry was on the 'spacious top floor' and had 'the most up to date laundry machinery engineers have yet devised. Some of the machinery [...] is the biggest and most speedy in the world.' The equipment was all 'electrically driven' and included ironing and washing machines:

> The washing machines are huge but delicate [...] which swallow linen as fast as it can be served; the hydros, shaped like huge vats, extract the water until the cleansed articles are half-dry when they come out. The ironing machines practically do the rest for sheets, serviettes, and the like [...] Blankets, from their nature, have

A June 1911 advertisement for Ross's Belfast Dry Ginger Ale pointed out proudly that it was available on *Olympic*. (Jerry N.J. Vondeling collection)

This stylish illustration exaggerated the size of the first-class à la carte restaurant. 'Colman's D.S.F. Mustard is being exclusively used on this magnificent floating palace. It is here, as always, the finishing touch to the perfectly served meal.' *Illustrated London News*, July 1911. (Jerry N.J. Vondeling collection)

> to be dried on a different principle, being placed on special drying horses fitted with steam coils, which extract all the moisture, the dampness being expelled in the form of steam through the roof. The temperature of the blanket drying receptacles is kept at summer heat [...] so that the articles are dried under natural conditions. The hand irons, for starched articles, stewards' jackets, etc. are electrically heated.

There was a water-softening plant, which reduced the amount of soap needed, while the water was reused a second time after 'the soapsuds resulting from the first washing are discharged'. The 'side lines' included a 'wonderful carpet cleaning machine'. '"Padded arms", electrically operated, are set in motion by a lever, and the beating commences. One only has to see the whirling arms hurtling round to appreciate that very little dust remains in a carpet after a few moments "thrashing".' Another useful piece of equipment was the disinfectant machine 'in an adjoining compartment of the red-brick boiler house, which lies at the foot of the great shed'.

Normally, the vast quantity of laundry arrived on a Saturday evening. During *Olympic*'s first year of service, she would go on to set a record for 'up to 75,000 pieces' in a single week. After unloading, they would be taken up to the top-floor laundry on the elevator and separated out:

> The washing machines and the hydros clean and dry the plain articles in a few minutes, and they are ready for the great ironers. Each of them has a capacity of 1,848 serviettes per hour, or 78 12-foot tablecloths per hour. When it is remembered that on the *Olympic* there are 44 varieties of articles, ranging from blankets to serviettes, and that there are 23,000 of the latter in the *Olympic*'s stock, it will be seen that the 'wash' of the world's largest liner is a considerable business, requiring, primarily, a supreme talent for organisation. By Tuesday evening, the various processes are complete, and the clean linen is sent through another 'shoot' and placed on the liner.
>
> On the following morning the linen of the American liner in port is received, representing approximately 22,000 to 23,000 articles. The processes are repeated, and by the following morning the work is done, and the linen cupboard of the vessel sailing on the following Saturday is replenished.

Plenty of other items needed to be supplied. One Southampton firm, F.G. Bealing & Son, were proud that they never needed to advertise their services. F.G. Bealing founded his nursery and horticultural florist in 1890 and was 'the driving force' growing the business. Through connections with Oakley & Watling, the White Star Line's fruit suppliers, they got a contract to supply all the company's Southampton ships. They loaded 'flowers, palms and plants' onto carts to take them down to the dockside. Typically, they would be taken aboard and then distributed throughout the ship: *Olympic* required 300–400 plants.[10]

THE COMMANDER AND HIS CREW

Captain Smith certainly had a lot to oversee at Southampton. He was well compensated. If contemporary press reports were accurate, he earned £1,250 annually with a potential annual bonus of £200 if there were no accidents to the ship under his command. (By way of comparison, when Captain Turner assumed command of *Mauretania* in January 1910, Cunard increased his annual earnings from £700 to £800.)[11]

Smith certainly earned far more than the rest of the Sailing Department. His reported annual earnings were the equivalent of over £100 a month. The chief officer earned £25 a month; the first officer, £17 10*s*; the second officer, £14; the third officer, £9 10*s*; the fourth officer, £9, and the fifth and sixth officers, £8 10*s*. *Olympic*'s chief surgeon, William O'Loughlin, received £10 each month; the master at arms, £5, and a lamp trimmer, storekeeper or quartermaster, £5.

In the Engine Department, there was a similar pattern, where wages fell steadily with rank: the chief engineer rated £35 a month, second only to the captain; the senior second engineer, £22; the junior second engineers, £18; and the most junior engineer, £8 10*s*. A fireman received £5 10*s* for their hard physical exertions. One anomaly was the pay of the restaurant staff. The manager, Antonio Pietro 'Luigi' Gatti, earned about £550 a year and a number of positions in the restaurant staff were more highly paid than the ship's senior officers.[12]

Smith obtained his Master's Certificate in Liverpool shortly after he turned 25 years old, then served one voyage each on *Arzilla* and *Senator Weber*. He commanded *Lizzie Fennell*, which was just over 1,000 gross tons, from 1876 to 1879, then joined White Star as the fourth officer of *Celtic* in March 1880. Eight years later, he had his first command of a White Star liner with *Baltic*. He was in charge of a number of ships before he joined *Majestic* in July 1895. He spent almost nine years with her until June 1904, except for a couple of voyages on *Germanic* in 1902 and 1903.

From then on, he was at the peak of his career, taking command of White Star's newest and largest vessels, *Baltic*, *Adriatic* and then *Olympic*, for their maiden voyages.[13] He had joined her at Belfast on 15 May 1911, so he had been in command almost a month by the time she sailed from Southampton for the first time. Smith's ship was 'authorised to wear [fly] the Blue Ensign' because he was in the Royal Naval Reserve.

Contrary to popular belief, he was not in command of *Germanic* when she capsized in freezing conditions at New York in February 1899. J. Bruce Ismay thought highly of Smith, 'I should think very few commanders crossing the Atlantic have as good a record as Captain Smith had […] I think he had an exceptionally clear record.' Harold Sanderson said Smith was 'a man in whom we had special confidence; otherwise he would not have been in the position'.[14]

Smith was affectionately known as 'E.J.' and 'quite a character in the shipping world'. *Oceanic*'s first officer in 1911, Charles Lightoller, said he had 'a pleasant quiet voice and invariable smile', 'a voice he rarely raised above a conversational tone – not to say he couldn't; in fact, I have often heard him bark an order that made a man come to himself with a bump'. Smith was 'a great favourite, and a man any officer would give his ears to sail under'. During years sailing with him on *Majestic*, Lightoller found it an education to see Smith:

> … con his own ship through the intricate channels entering New York at full speed. One particularly bad corner, known as the Southwest Spit, used to make us fairly flush with pride as he swung her round, judging his distances to a nicety; she heeling over to the helm with only a matter of feet to spare between each end of the ship and the banks.

Ship handling was far from the only aspect of the commander's job. He was, in effect, the authority over a floating city and all manner of issues might arise which he was, ultimately, responsible for. Smith had dealt with plenty of unusual incidents over the years, as illustrated by one of *Adriatic*'s round trips from Southampton to New York a little over a year previously.

Adriatic left Southampton on the morning of 23 February 1910 after a number of crew either deserted, 'failed to join' or 'left by consent'. Smith's ship also took on a stewardess, 'owing to an unexpected increase in passengers', meaning he had to certify that 'there was no opportunity of taking her before the proper authorities'. The same day, a trimmer was reported 'off duty owing to an injury to his right great toe, caused by a piece of coal falling on the foot'. Another trimmer suffered an 'inflamed finger due to injury received during performance of his duties', which kept him off work for a few days. Five days into the westbound crossing, a third trimmer had to be fined 'for disobedience to lawful commands'. He admitted to 'refusing to obey orders', pleading that 'the duty took him to the engine room, and that he signed articles to work in the stokehold only'.

Shortly before reaching New York, the ship was searched 'for dutiable articles' and nothing was 'found concealed by any member of the ship's company'. *Adriatic* arrived on 4 March 1910. While she was in port, some crewmen transferred to *Cretic*. Sixth Engineer Arthur Ward was admitted to hospital in New York due to 'suspected appendicitis', meaning that he would miss the eastbound crossing.

There were a few further desertions at New York, so additional crew had to be taken on for departure on 12 March 1910. Then, while *Adriatic* was leaving her pier, one of the ship's stewards heard a revolver shot. It came from a second-class stateroom, which was locked from the inside. The ship's crew forced it open and found a passenger 'lying on the deck with a bullet wound in the right temple'. Captain Smith reported, 'The revolver was found lying close to the man's right hand. The ship's surgeon was called and pronounced life extinct.'

Edward Ettridge (who used the stage name 'Ed Beppo' for his English music-hall performances) had shot himself in Alfred Burgess' stateroom. He was mistaken briefly for Burgess. One of the ship's officers had to call for a tug to take the body off the ship, an action which did not escape the notice of the newspaper reporters.[15]

Smith signed the entry in the ship's log that Ettridge had died of a 'bullet wound in right temple', countersigned by Purser McElroy and Chief Surgeon William O'Loughlin.[16]

***

Fifty-three year old Joseph Evans signed on as *Olympic*'s chief officer. The Welshman 'was a very unassuming personality' with 'a kind and tender disposition'.[17] He had joined the White Star Line as a junior officer in September 1891. Evans had been a chief officer since August 1900, serving on *Teutonic* from April 1906 to January 1910, then *Adriatic* from January 1910 to May 1911.

First Officer William Murdoch was fifteen years younger and transferred with Evans from *Adriatic*. Murdoch had been with the company since June 1899, climbing through the ranks until his promotion to *Celtic*'s first officer in May 1904.

They were joined by Second Officer Robert Hume, completing the senior officers in Smith's team. He served on *Oceanic*, *Cedric* and then *Adriatic*. Hume was another White Star veteran, joining in January 1900. He had served on *Majestic* from April 1905 until November 1910, punctuated by a few voyages on other ships such as *Arabic*, then joined *Oceanic* in December 1910.[18] (Unlike the other senior officers, he had not served with Captain Smith before.)

Murdoch and Hume were both Scottish; born in the same year, 1873; and from Dalbeattie and Kirkcudbright, respectively, which were about 15 miles apart. Both men were in their late thirties. They appeared to have promising careers ahead of them.

Thirty-two year old Henry Cater joined as third officer. He had been with White Star since May 1904 and knew Hume from their time aboard *Oceanic*. They both retained their rank when they transferred, joining *Olympic* on 23 May 1911.

Twenty-seven year old David Alexander joined as fourth officer, less than four years after his first appointment with White Star. He had served with Captain Smith on *Adriatic*.

Twenty-nine year old Alphonse Tulloch joined as fifth officer and 30-year-old Harold Holehouse served as sixth officer. It was quite striking that Holehouse had been with White Star less than a year. He had served as fourth officer on board *Cymric* and *Afric*, stepping back to sixth officer on *Olympic*. It was not uncommon for officers to move back in rank when they joined a more prestigious ship. He was joining the largest ship in the world and serving for the first time with Captain Smith.[19]

Smith and his officers all recorded their previous ship as the 'same', reflecting that they had joined *Olympic* at Belfast for what was a voyage in the 'home waters' of the United Kingdom to Liverpool and then Southampton. Now that she was preparing to sail to New York, she was a 'foreign going ship' and they were officially signing on again. *The Southampton Times* said, 'Nearly all the principal positions on the new ship have been allocated to officers of the *Adriatic*,' naming Chief Officer Joseph Evans, Purser Hugh McElroy and Chief Surgeon William O'Loughlin. It reported that White Star intended to fill the vacant positions on *Adriatic* with Liverpool men in anticipation of her transfer to the Liverpool to New York service later in the year. While *Olympic*'s crew were drawn from a variety of other ships, those from *Adriatic* were certainly well represented in senior positions.

The crew were arranged into three main areas of work: the Sailing Department, including Captain Smith and his officers, seamen and lookouts; the Engine Department, including engineers, firemen and greasers; and the Victualling Department, including the purser and his staff, stewards, cooks and roles such as attendants for the Turkish baths. *Railway & Travel Monthly* noted:

> The old time title of 'victualling' is still applied by steamship companies to the provisioning of modern liners, otherwise this word is now rarely used save amongst agricultural labourers in some districts, who still call their food 'vittals', a slurred pronunciation of victuals. The legal definition of a certain class of innkeepers is 'licensed victualler'.

There were a number of positions new to *Olympic*, as far as White Star was concerned. In the Sailing Department, White Star hired seamen with the rank of quartermaster, whereas on smaller ships the helmsman continued to be one of the able-bodied seamen. Two quartermasters were stationed on the bridge, either to steer the ship or run messages, while a third stood watch on the after docking bridge. In the Victualling Department, a racquet steward was needed to attend to the squash court (23-year-old Frederick Wright).

Although *Adriatic* had a single lift for first class, *Olympic* had three – as well as the new second-class lift, all of which required lift attendants. There were forty-five à la carte restaurant staff, reflecting the premium service being offered. They were internationally diverse. Nineteen of them were Italian, including the majority of the waiters, but the restaurant chef, Pierre Rousseau, and most of the galley personnel were French. These jobs included a sauce cook, fish cook, roast cook, soup cook, pastry cooks and assistants to support them. There was even a coffee cook, as well as a 'kitchen porter', plate steward, carver and others. British, German and Swiss staff added to the diversity. (The youngest was 17-year-old Herbert Vine from London, employed as a 'page boy'.)

In the Engine Department, it appears that an additional fifth engineer was carried to attend to the low-pressure turbine, following an arrangement on *Laurentic*.[20]

The Turkish bath attendants and stewards, as well as the gymnasium instructor, were roles already in existence on *Adriatic* (29-year-old Maud Slocombe transferred from *Adriatic* to *Olympic* as a Turkish bath assistant.) The man who initially signed on as 'gymnasium steward' was one of a number of crew who failed to join on sailing day.

John H. Thearle signed on as senior second engineer. He had already been with *Olympic* since joining her at Belfast on 15 March 1911. Born in Liverpool in 1869, he served an apprenticeship

A rare view of *Olympic* docked at Southampton prior to her maiden voyage. (Jens Ostrowski collection)

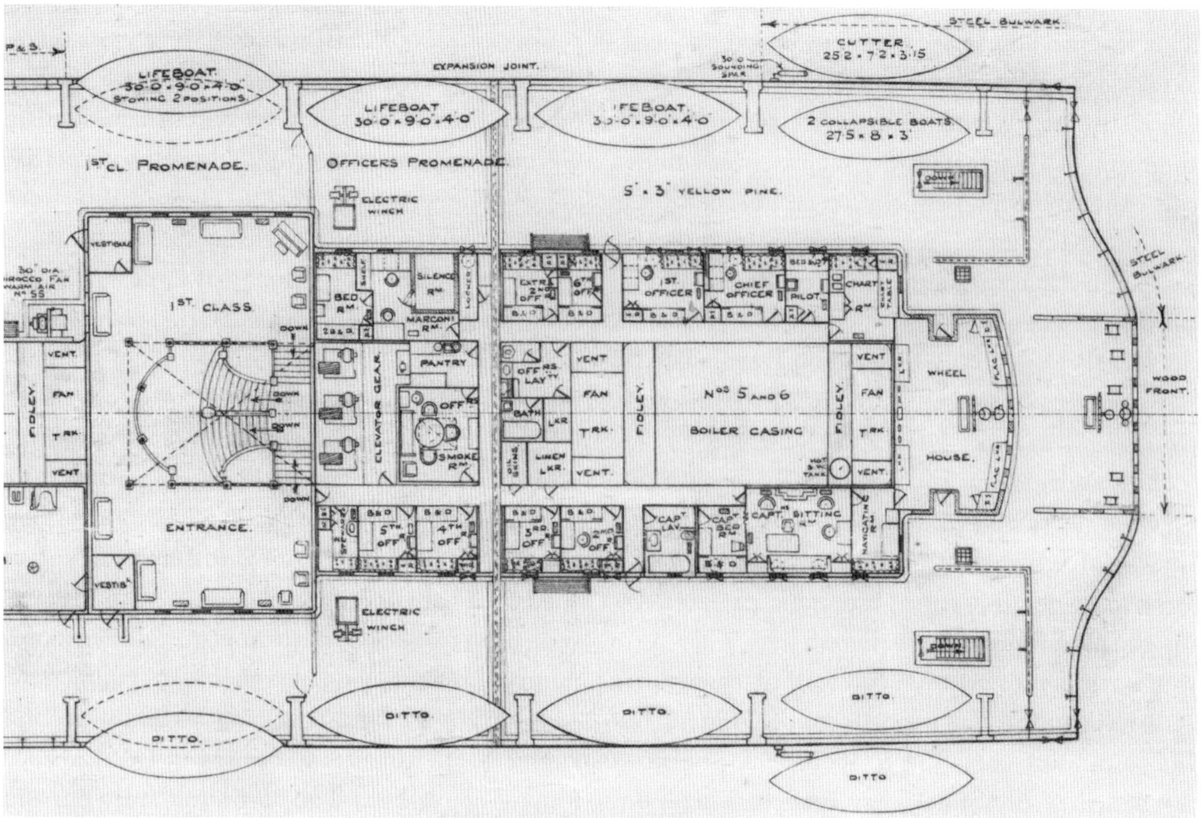

Olympic's officers' quarters and bridge when she entered service in 1911. Captain Smith's accommodation comprised a private bathroom, bedroom and sitting room on the starboard side, behind the navigating room and wheelhouse. The Marconi operators' bedroom, Marconi Room and Silent Room were on the port side, further aft. (HOYFM.2010.73.187 © National Museum NI, Harland & Wolff Collection, Ulster Folk & Transport Museum)

at the engineering works of David Rollo on the Mersey from 1885 to 1890, working for both the Lamport & Holt and Hall Lines. He then joined the Dominion Line, transferring to White Star in 1903. After five years as a second engineer on *Republic*, he transferred to *Oceanic* in the year before *Republic*'s loss in January 1909. He was 'specially selected' for his position on *Olympic*'s maiden voyage and highly qualified, holding his first-class engineer's certificate for eighteen years.[21]

Thearle was joined by 35-year-old Charles William McKimm as senior third engineer. (Both men were starting on a lengthy service with *Olympic* that would see them both through to retirement.) McKimm was a Belfast native.[22] Born on 6 January 1876 in the Curator's Lodge of the Royal Belfast Botanic Gardens, he went on to do an apprenticeship with Harland & Wolff. He had held his first-class engineer's certificate for the twelve years that he had been with White Star.

Thirty-two-year-old John H. Hesketh signed on as junior second engineer. He had served with Chief Engineer Bell on *Laurentic*. Hesketh was a 'level headed man', who had been with White Star since he was a 14-year-old apprentice. Following service on numerous White Star ships, including *Majestic*, where he was in charge of the refrigerating machinery and 'was instrumental in securing a greatly improved quality of cargoes', it was natural for him to progress.

Twenty-two-year-old Arthur Ward joined *Olympic* as fifth engineer from *Adriatic*. Born in Hampshire, he had served apprenticeships with Mitchell Bros, Romsey, and John L. Thornycroft & Co., Southampton, before joining White Star in 1908. He served on *Teutonic*, *Adriatic* and now *Olympic* in quick succession. *Olympic*'s engineers were invariably from other White Star ships, with some exceptions, such as the American Line's *St Paul*.[23]

Thirty-one year old John Propent transferred from *Oceanic* as chief electrician.

A significant number of firemen had been recruited in Liverpool and signed on there. These included men who had worked on White Star ships such as *Teutonic*, *Cedric*, *Arabic*, *Adriatic* and *Cymric*, but others came from Cunarders, including *Mauretania* and *Campania*. *Olympic* provided plenty of employment opportunities.

Violet Jessop was pleased to join *Olympic* as a first-class stewardess. 'The mecca of better times was the prospect of a "new ship",' she wrote:

> As is always the case at sea, rumours and exaggerations about her were rife, but no one denied she would be a paragon [...] She came up to all our expectations and I sensed at once she was going to be a kindly ship, for all ships have a character of their own which some people, more sensitive than others, are quick to realise. That is why there are happy ships, while others have a definitely depressing atmosphere.

The Victualling Department presented Thomas Andrews with 'a magnificent walking stick' in appreciation of his efforts to improve the crew accommodation, 'a consideration which was customarily of secondary importance'. Violet thought Andrews 'one of the finest and kindliest of men it has been my privilege to meet'. She thought *Olympic* 'was the first ship to have proper bathrooms installed for stewards. The fact that they were soon afterwards taken over by the second steward as storerooms, forcing the men to revert to the inevitable bucket for a washdown, is neither here nor there!'[24]

There were roles which were anomalies, either by the department they were in or the general circumstances. *Olympic*'s two Marconi wireless operators ('telegraphists') were both 25. John Clayborough Hawkhead (transferred from *Minnewaska*) and Frank Beatson (transferred from *Afric*), received a weekly wage from the Marconi Company and a monthly rate of hire from the White Star Line. They were listed under the Victualling Department but, perhaps more appropriately, were considered as part of the Sailing Department. Messages for the ship's commander and officers took priority over passenger communications. (Passengers wishing to send messages could do so from the Purser's Enquiry Office. The purser's staff took payment according to the length of the message and then passed it to the operators through a pneumatic tube.)[25]

Scarcely had they had time to adjust to the new ship when they experienced a significant change. In June 1911, *The Marconigraph* reported, 'The greatest event during the month of May was the equipment of the largest modern liner afloat, the SS *Olympic*.' It was 'natural' that she 'should be equipped with a high-power plant' and it was anticipated that she would be able to communicate 'at a very long range'. Her wireless apparatus, installed at Belfast, was already being upgraded:

> For the trip from Belfast to Liverpool, thence to Southampton, this vessel was only fitted with a standard 1½ kilowatt Ship Set, but a 5 kilowatt plant is now being installed and will be completed before she sails on her maiden voyage [...]
>
> Although similar power plants have been installed on other big liners now afloat, the greater range which it is anticipated will be attained by the *Olympic* will be due to the greater span between the masts as compared with that of other liners.

The next month, they reported:

> She was fitted with a 5 kilowatt motor generator set, and the latest type of valve receiver was installed as a standby. The installation is designed to provide, when employed with a suitable aerial having a mean height of 170 feet, a working range of 250 nautical miles over winter, and a maximum range considerably exceeding the above figure.

Olympic's two masts 'stepped 600 feet apart, are used for the support of the aerial'. The electrical supply was a continuous current supplied at 110 volts. The plant itself 'consists of a motor generator set, comprising a continuous current motor mounted on a common bed with and direct coupled to an alternating current generator'. The Morse Key was 'provided with telephone short circuiting contacts to prevent the operator's ear being rendered insensitive by noise in the telephone during transmission'. The equipment aboard ship included 'a tool box containing an electrician's complete repairing kit, together with all special tools and spare parts for the different machines'.[26]

***

The Sea Post Clerks were unusual in that the British and American governments had agreed that they would not sign on as members of the crew, but be subject to ordinary discipline as passengers. Some concerns had arisen about the arrangement in January 1909:

> The Sea Post Clerks are in uniform which very clearly resembles that worn by the ship's officers, and to allow the men the same latitude as is allowed to passengers may bring discredit upon the officers of the ship. By way of illustration, passengers are allowed to drink and play cards in certain parts of the public portion of the ship, but it might produce an uneasy feeling of insecurity if a passenger, not certain of the identity of the Sea Post Clerks, saw one or more of them drinking and playing cards in the smoke room or elsewhere, and this one passenger on a ship where gossip is so prevalent might be capable of a considerable amount of mischief before the actual facts became known. True, the Sea Post Officers may be drinking a glass of lemonade, but it is sometimes impossible to distinguish aerated water from spirituous liquor at a distance, and the smoke room is sometimes visited by passengers whose thoughts do not frequently turn to aerated water.[27]

Another issue was that the shipping lines forbade them from smoking on deck, but in effect, they usually did, to the disapproval of captains. The Controller of the London Postal Service wrote:

> I am informed that on one occasion a commander peremptorily ordered an American officer to remove his cigar. I understand that all the commanders object to the practice, and Commander Haddock of the *Oceanic*, of whom the Sea Post men speak in the highest terms, spoke strongly on the subject.

Phillip Franklin suggested discussing it with Phillip Curry, White Star's manager at Southampton, so that a compromise could be reached. Curry and his staff objected to the clerks drinking and playing cards in the smoke room or other public areas, but did not object to smoking in the smoke room or having a drink during their meals or in their staterooms. The authorities decided to address it when they revised the rules and, in the meantime, they decided to ask White Star whether the Sea Post Clerks could use the Boat Deck for recreation as well as the promenade deck, 'It would do away to some extent with the mingling of the Sea Post Officers and the passengers, which the companies so much object to [...] if the Sea Post Officers were admitted to the same deck as the ship's officers.'

By 1911, the new breed of ships represented by *Olympic* led to other discussions. The British and American Post Offices paid jointly for accommodation for the postal sorters and the agreement stipulated that they could have first-class accommodation. In practice, White Star and other companies allocated the men accommodation near the sorting room, but they could use the first-class dining saloon. The British authorities followed the insistence of their American colleagues in using first class, but they thought:

> The first class accommodation [...] entitles them to use all the public apartments in the ship which are provided for first class passengers, and it is probable that the presence of a sorting clerk in the luxurious lounges and smoking rooms of such ships as the *Olympic* would offend the [White Star] company. There is in fact an instance on record of one of the sorters being ordered not to smoke on deck, such an order would never have been given to a bona fide passenger, and it may be inferred that the company do not really regard these men as entitled to all the privileges of first cabin passengers.

On the other hand, 'the men themselves might assert that the department regards them as entitled to first class accommodation and the disciplinary authority of the captain might be called into question'. One official wrote:

> Our men do not belong to the class of persons who would, if travelling at their own expense, take first class tickets even on the smallest liners on the Atlantic; and they must, I think, find themselves very much out of place in the first class accommodation, especially on such luxurious steamers as the *Olympic* and *Titanic*.
>
> On other ships, so as to avoid difficulties with the passengers, the White Star and American companies arranged for the sorters to dine in the first class saloon, but after the ordinary dinner hour.[28]

In the event, no immediate action was taken, but arrangements were being made to get *Olympic*'s Post Office ready. On 7 June, Commander Foakes and Mr Parsons from the General Post Office, London, visited Southampton to inspect the Post Office arrangements on board *Oceanic* and *Olympic*. They examined *Oceanic* before she left for New York, then *Olympic*:

> The [mail and sorting] rooms are spacious, light and well ventilated, and are superior in every way to any other steamer of the White Star Line. The whole of the accommodation is self contained and after the mails are on board there does not appear to be any reason why any person not employed in the Post Office should have access to it.

There was 'ample room' for the storage of ordinary mail (2,200 bags) in the mail room 'and sufficient space left over for bag opening and newspaper sorting. The baggage room which is connected by a door with the mail room will form a useful annexe for any surplus bags.' Thomas Andrews explained to them that the bag racks for the sorting room were being constructed at Harland & Wolff's works in Southampton and would be placed on board before sailing. Unfortunately, the racks for the mail room were not going to be put on until New York. Andrews reminded them that it had been explained in previous correspondence. Andrews took the measurements for fitting 'revolving fillets' for the names of towns and cities to be fitted on the sorting cases, and he made arrangements for the work to be completed before sailing. 'On the whole the Post Office accommodation is satisfactory and its unusually spacious dimensions will no doubt conduce to more efficient work on the part of the sorters,' they concluded.[29]

***

When signing on, the crew were officially engaged:

> ... on a voyage from Southampton to New York (via Cherbourg and Queenstown) and/or if required, to any port of ports within the North Atlantic and South Atlantic oceans, trading as may be required until the ship returns to a final port of discharge in the United Kingdom for any period not exceeding twelve months.

The agreement and account of crew for a foreign-going ship confirmed that *Olympic* had certified accommodation for '850 seamen'. They agreed 'to conduct themselves in an orderly, faithful, honest, and sober manner, and to be at all times diligent in their respective duties, and to be obedient to the lawful commands of the said master, or of any person who shall lawfully succeed him'. Any 'embezzlement or wilful or negligent destruction of any part of the ship's cargo or stores' would be 'made good to the owner out of the wages' of the offending crewman. All members of the crew agreed 'to perform any duties that are required of them during "stand byes"'.

Blankets supplied to the crew that were not returned to the chief steward at the end of the voyage would 'be charged for' at 5*d* each. Firemen and trimmers would be 'selected by engineer to run electric light in New York, from 6 p.m. to midnight, and not to work during the day', 'no overtime to be paid'.

Captain Smith had twenty-two 'instructions to masters' to obey. Among them, he was bound 'to have a legible copy (omitting the signatures)' of the crew agreement 'placed in an accessible part of the ship'. In the agreement, the term 'seaman' included 'every person, except masters, pilots, and apprentices (duly indentured and registered), employed or engaged in any capacity onboard any ship'. Any captain 'fraudulently altering, or making any false entry in, or delivery a false copy of any agreement, or being party to such an act' would be punished by the infliction of a financial penalty 'or by imprisonment with or without hard labour for any period not exceeding six months'.

The Merchant Shipping Act (1906) specified the scale of crew provisions. These foodstuffs were 'in addition to the lime and lemon juice, and sugar, or other anti-scorbutics required by the Merchant Shipping Acts'. Each crewman's weekly entitlement was:

- water (28 qts.);
- soft bread (3 lbs.);
- biscuits (4 lbs.);
- salt beef (3 lbs.);
- salt pork (2 lbs.);
- preserved meat (2¼ lbs.);
- fish (three quarters of a pound);
- potatoes (6 lbs.);
- 'dried or compressed vegetables' (half a pound);
- 'peas, split' (two-thirds of a pint);
- 'peas, green' (one third of a pint);
- 'calavances or haricot beans' (one third of a pint);
- flour (2 lbs.);
- rice (half a pound);
- oatmeal (eight ounces);
- tea (1¾ ounces);
- coffee (four ounces);
- sugar (1¼ lbs.);
- butter (half a pound);
- 'marmalade or jam' (one pound);
- 'syrup or molasses' (one-third of a pound);
- suet (four ounces);
- pickles (one-third of a pint);
- dried fruit (five ounces);
- fine salt (two ounces);
- mustard, penny and curry powder (a quarter of an ounce each); and
- onions (three ounces).

There were numerous 'conditions and exceptions'. Soft bread was not necessary 'if rough weather renders the making of the bread impracticable'; 'fresh vegetables, or vegetables preserved in tins, may at any time be substituted for dried or compressed vegetables in the proportion of half a pound of fresh vegetables, or vegetables preserved in tins, to one ounce of dried or compressed vegetables'; and 'the dried fruit [...] must be raisins, sultanas, currants, figs, or prunes'. A list of 'substitutes and equivalents' was laid down ('not to be used without reasonable cause').

Any member of the crew 'striking or assaulting any person on board or belonging to the ship (if not otherwise prosecuted)' was liable to be fined. 'Bringing or having onboard spirituous liquors' rated a 5*s* fine, as did a first offence of drunkenness (the fine rose to 10*s* on a second offence). Banned possessions included 'any fire arms, knuckle-duster, loaded cane, slung shot, sword-stick, bowie knife, dagger, or any other offensive weapon or offensive instrument, without the concurrence of the master'.

The firemen who signed on at Liverpool were instructed to be on board *Olympic* by 3.30 p.m. on 13 June rather than 6 a.m. on the 14th for other crew. A number of trimmers, who had signed on at either Liverpool or Southampton, had been promoted to firemen by sailing day. There were plenty of complications, with some men failing to join or deserting; others left by consent. By the time she was ready to sail, her crew numbered approximately sixty-three in the Sailing Department; 327 in the Engine Department and 462 in the Victualling Department, for a total of about 852 when leaving Southampton.

Compared to *Adriatic*, the number of jobs in her Engine Department increased two-and-a-half-fold; her Victualling Department doubled; and the Sailing Department showed a slight increase. The total number of crew was roughly double. In effect, she represented

WHITE STAR LINE

ROYAL AND UNITED STATES MAIL STEAMERS.

TRIPLE-SCREW R.M.S. "OLYMPIC," 45,000 TONS, THE LARGEST AND FINEST STEAMER IN THE WORLD, AND TRIPLE-SCREW S.S. "TITANIC," 45,000 TONS (BUILDING).

REGULAR SAILINGS BETWEEN

SOUTHAMPTON—CHERBOURG—QUEENSTOWN—NEW YORK.

Calling at QUEENSTOWN (Westbound) and PLYMOUTH (Eastbound).

SECOND CLASS RATES AND PLANS.

ISMAY, IMRIE & Co.
30 JAMES STREET. LIVERPOOL.
1, COCKSPUR ST. S.W., 38, LEADENHALL ST E.C., LONDON.
CANUTE ROAD, SOUTHAMPTON.

MARCH, 1911.

Bookings were being taken well in advance of *Olympic*'s maiden voyage. This second-class brochure for all of White Star's Southampton–New York express steamers was printed in March 1911. (Daniel Klistorner collection)

Some Statistics about the "Olympic"

Tonnage, registered	*45,324 Tons*
Tonnage, displacement	*66,000 Tons*
Length over all	*882 Feet, 9 Inches*
Breadth over all	*92 Feet, 6 Inches*
Breadth over boat deck	*94 Feet, 0 Inches*
Height from bottom of keel to boat deck	*97 Feet, 4 Inches*
Height from bottom of keel to top of Captain's house	*105 Feet, 7 Inches*
Height of funnels above casing	*72 Feet, 0 Inches*
Height of funnels above boat deck	*81 Feet, 6 Inches*
Distance from top of funnels to keel	*175 Feet, 0 Inches*
Number of steel decks	*11*
Number of watertight bulkheads	*15*
Passengers carried	*2,500*
Crew	*860*
Rudder	*100 Tons*
Anchors (each)	*15½ Tons*
"Wing" Propellors (each)	*38 Tons*
Turbine Propellor	*22 Tons*
Windows and sidelights, over	*2,000*
Links in anchor chains (each)	*175 Pounds*

A brochure for the White Star Line's 'new triple-screw steamer *Olympic*' included some of her key statistics. The ship's displacement (total weight) was mistakenly given as 66,000 tons (an error also found on a postcard issued to mark her launch the year before). She actually weighed 52,310 tons at her designed load draught. (Jens Ostrowski collection)

two *Adriatics*, reinforcing Harold Sanderson's old argument with the Southampton authorities about the economic benefits she would bring to the port. *Titanic*'s arrival the following year would only increase the opportunities for Southampton's seafarers.

***

On Sunday 11 June, Thomas Andrews wrote a four-page letter to his sister using the South Western Hotel's own headed paper. Only a few days before the maiden voyage, his time was taken up by final preparations – 'every minute of the day appears occupied'. He disapproved of the White Star Line's handling of the strikes, noting, 'Our big ship has caused much confusion in the labour market and strike after strike is the order of the day and night. First the coal men then the firemen and seamen, now the stewards, cooks and bakers.' He thought the passenger list was reasonable and he anticipated that the ship was practically full for the return crossing eastbound.

It is easy to imagine the sense of excitement and optimism for the maiden voyage, combined with the inevitable anxiety that everything should go well.[30] The days, then the hours, then the minutes, ticked by as sailing day approached. She was scheduled to depart on Wednesday, 14 June.

9

'ONE CANNOT HELP BEING IMPRESSED'

'Every point of vantage was occupied by visitors who were anxious to get a glimpse of the vessel,' according to the *Southampton Times*. Pilot George Bowyer was in charge on the bridge. Five tugs supplied by the Isle of Wight Company assisted the operation as *Olympic* 'got under weigh without the slightest difficulty'. A keen observer spotted that she was drawing 34ft 1in of water at the stern. 'She came out of the new dock stern first, with the *Hector* and *Hercules* leading, and when the vessel was off the dock head she was slowly and safely turned – a performance which excited the greatest interest.' 'On sailing day we had to go out stern first,' Bowyer recalled. 'Then on down to the entrance to the [River] Itchen, where at that time was the only swinging ground then dredged.'[1] The tugs departed at 12.45 p.m.

'Perhaps the most striking thing about her departure was the practical comparison which it allowed between the size of the *Olympic* and that of what may now almost be called the last generation of liners,' wrote one observer:

> The Red Star ship *Kroonland* and the *New York*, of the American Line, were moored hard by. The *Kroonland* is about 15,000 tons and the *New York* over 10,000, yet from the highest deck of the *Olympic* they seemed to be dwarfed into bewildering insignificance. Bewildering, too, is the interior of the new vessel with its corridors, halls, public rooms, street-like decks, and room-like cabins.[2]

Many passengers were on deck to watch the departure, but their attention turned to getting settled in. Stewardess Violet Jessop recalled the problems the crew faced in getting acquainted with their new surroundings. The challenge they faced was that '98 per cent of the passengers expect the same prompt service they would get on an old ship'. Ten minutes after the boat train arrived, she 'answered the impatient summonses' of 'a certain well-known society woman' with 'haughty, gimlet eyes'. 'She looked over me,' Violet recalled, 'while she minutely detailed her orders for the entire week – to commence immediately.' She let her stewardess know that she was a 'friend of the president of this line'. (During her service for White Star, Violet had encountered thousands of passengers claiming to be Ismay's friend.)

There were compensations:

> I got a fresh thrill every time I went through *Olympic*'s beautiful staterooms, the Adams room, the Regency room, the Dutch, Georgian and so on, with their exquisite woodwork and sumptuous silk furnishings. I have always maintained that never before or since have such materials of so perfect a quality been used to fit out any ship.[3]

Olympic's elegant lines and great size impressed spectators. On the reverse of another contemporary postcard, someone has written, 'Isn't this a beautiful boat?' (Jens Ostrowski collection)

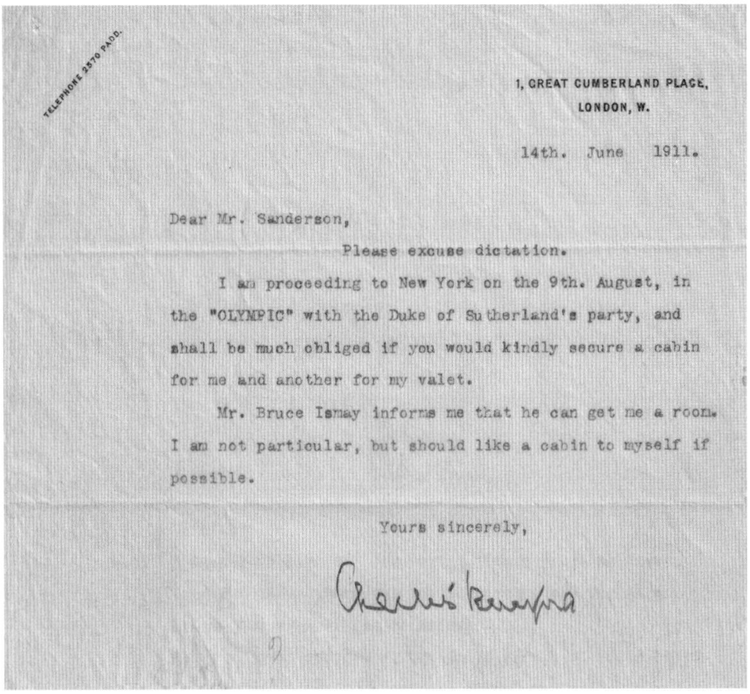

TELEPHONE 2270 PADD.

1, GREAT CUMBERLAND PLACE,
LONDON, W.

14th. June 1911.

Dear Mr. Sanderson,

Please excuse dictation.

I am proceeding to New York on the 9th. August, in the "OLYMPIC" with the Duke of Sutherland's party, and shall be much obliged if you would kindly secure a cabin for me and another for my valet.

Mr. Bruce Ismay informs me that he can get me a room. I am not particular, but should like a cabin to myself if possible.

Yours sincerely,

On the same day *Olympic* left Southampton on her maiden voyage, an enquiry was made to Harold Sanderson about first-class accommodation later in the season. (Jens Ostrowski collection)

The relatively short run across the English Channel towards the French port of Cherbourg was 'no test' in 'perfect weather'. Her expected arrival time was reported as around 4 to 5 p.m., but she steamed at a leisurely pace. *Olympic* arrived at Cherbourg around 7 p.m. and the *Cork Examiner* reported she entered the harbour itself about half an hour later. *Le Petit Journal* reported that local dignitaries, including the Mayor of Cherbourg, went on board. They were received at the gangway by Captain Smith and his officers, while the ship's orchestra played '*La Marseillaise*'. They were given a tour of the public rooms and passenger decks.

One unfortunate problem was that the intended procedure for operating the tenders at Cherbourg was not followed. White Star had 'very carefully gone into' the matter at a meeting in Southampton on 3 May 1911, to make sure everything ran smoothly. J. Bruce Ismay observed that Captain Beresford left *Olympic*'s side 'in the *Nomadic* long before the baggage was transferred from the *Traffic*' and wondered why. 'It occurs to me that it is part of the duties of a marine superintendent to remain by the ship until all the baggage has been transferred.' Thomas Andrews had drawn up a memo, 'which clearly states the manner in which the passengers and baggage should be embarked', and Ismay made sure it was passed on to Captain Beresford.[4]

He also became aware of visitors to the engine room during the passage from Southampton to Cherbourg. Chief Engineer Bell confirmed that 'three gentleman have been in the engine room during the passage from Southampton all by Mr. Blake's instruction', including two gentleman connected with a French steamship company. Ismay was surprised and resolved to get an explanation about why Blake had thought it justified.

Olympic left Cherbourg about two hours after she had entered the harbour, around 9.30 p.m., for the run westwards towards Queenstown. Thomas Andrews kept busy, including writing notes and recommendations right from the start of the crossing. He thought the first-class gangway at Southampton should have landed the passengers at 'the first class entrance on D deck instead of on B deck thus providing a level gangway', acknowledging that 'the position of these gangways will always vary due to condition of tide'. Another improvement would be to always berth *Olympic* 'with the port side towards the quay thus providing direct communication through special (shipside) engine door' and eliminating the need to pass materials 'through the present second class entrance on E deck'. His recommendation

An early view of *Olympic* at Cherbourg. (Author's collection)

meant *Olympic* would depart bow-first, rather than have to back out of the dock stern-first, as she did the first time she sailed. (In the event that she did dock or undock stern-first, he thought the pilot should always be 'on the docking bridge aft' for Southampton, as was already the practice for White Star liners at Liverpool.)

Andrews was also concerned for passenger safety when docking or undocking: 'Third class passengers should not be allowed on the poop or forecastle decks, as in the event of a rope carrying away someone would be sure to be killed.' It was not an issue for first or second-class passengers as their decks were far out of the way of any docking operation work. The propeller notice boards, warning that *Olympic* had triple propellers and to keep clear of the blades, needed to be fitted permanently 'on outside of ship's side rails in way of after docking bridge', in the same way as on other White Star ships. Otherwise, Andrews worried that placing and stowing away 'these boards every voyage' would cause damage 'to the painting and printing'. Fitting them permanently would avoid the notice boards looking unkempt and save the ship's crew from the hassle of constantly fixing and removing them.[5]

The *Cork Examiner* reported that, on the run from Cherbourg to Queenstown, *Olympic* had a 'strong south-easterly wind and heavy sea, which had no more effect on the liner than it would on the Eddystone Lighthouse'. She was originally expected off Cork Harbour at around 10 a.m. on Thursday, 15 June, with the passenger and mail tenders leaving the White Star wharf two hours earlier, but they did not leave until 1 p.m. The weather was 'overcast during the morning, with a freshening breeze', but 'gradually became more threatening'. The tenders came out to meet her 'about a mile outside Roche's Point, but inasmuch as there was a fresh wind and a rather choppy sea, her captain deemed it advisable to come into the harbour proper for the transference of passengers and mails'. Pilot John Whelan guided *Olympic* in an 'admirable manner':

> She steamed in with a majesty and grace of movement which was the admiration of all who beheld her, and came to a standstill with a certainty and security which almost inspired one with the notion that she was endowed with some kind of volition of her own.

A stylish illustration of *Olympic*, intended to convey her power, size and strength. The large liners were frequently pictured with smaller boats in publicity material to emphasise their size. (Author's collection)

Another common approach was to contrast *Olympic* with famous large buildings. (Author's collection)

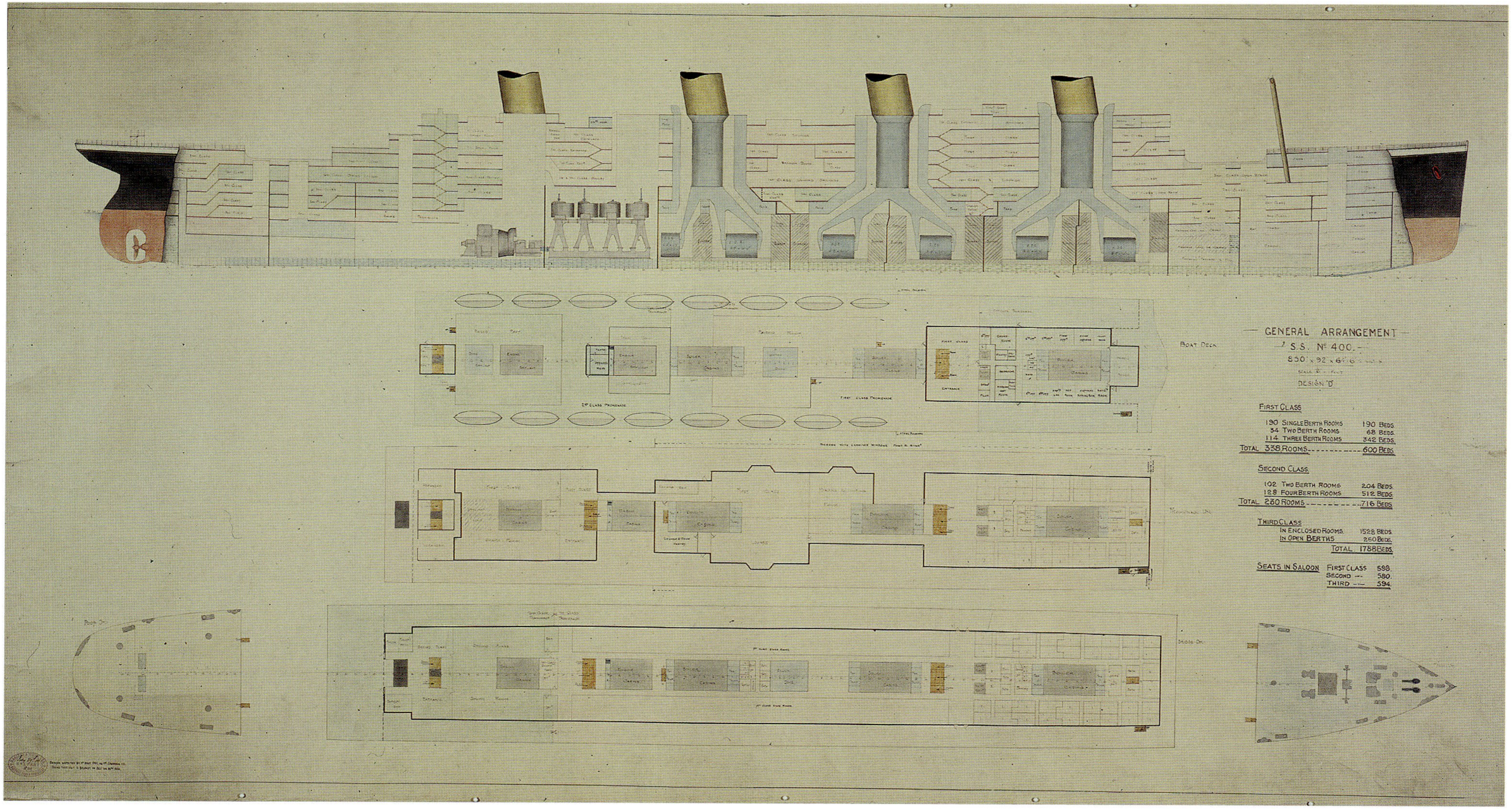

The 'Design "D"' concept, which J. Bruce Ismay and Harold Sanderson approved on 31 July 1908, represented the key fundamentals of the new ships' design, but was subject to many further revisions over the following two years (see pages 118–20). (HOYFM.2010.73.188, 189, 195 and 196 © National Museum NI, Harland & Wolff Collection, Ulster Folk & Transport Museum)

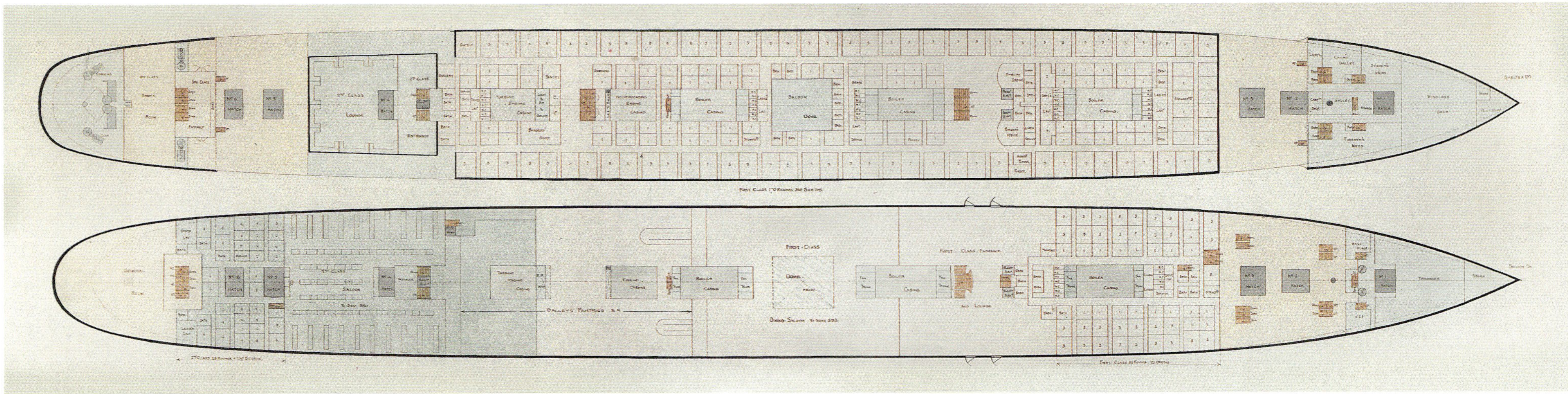

The concept showed the original plans for the layout of the shelter deck, C, and saloon deck, D. (HOYFM.2010.73.188, 189, 195 and 196 © National Museum NI Collection Harland & Wolff, Ulster Folk & Transport Museum)

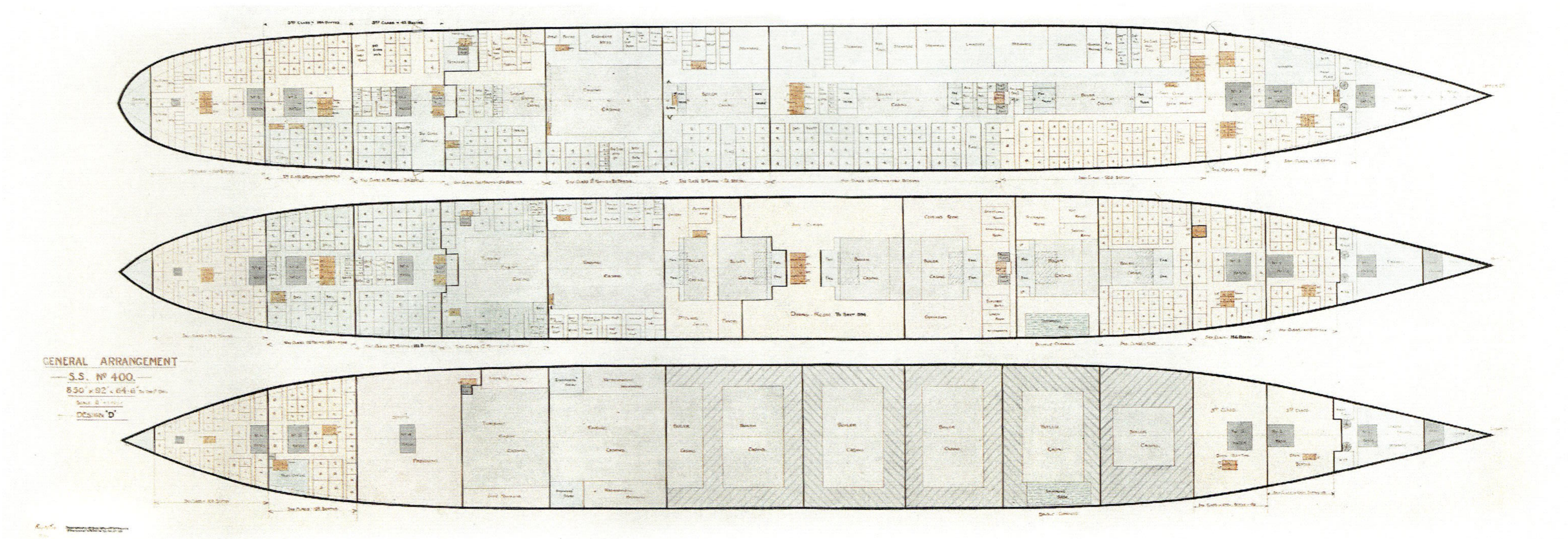

The upper deck, E; middle deck, F, and lower deck G were depicted in detail. Note that an additional watertight bulkhead was added in the final design, dividing the single ended boilers into their own boiler room (see pages 75–6). (HOYFM.2010.73.188, 189, 195 and 196 © National Museum NI Collection Harland & Wolff, Ulster Folk & Transport Museum)

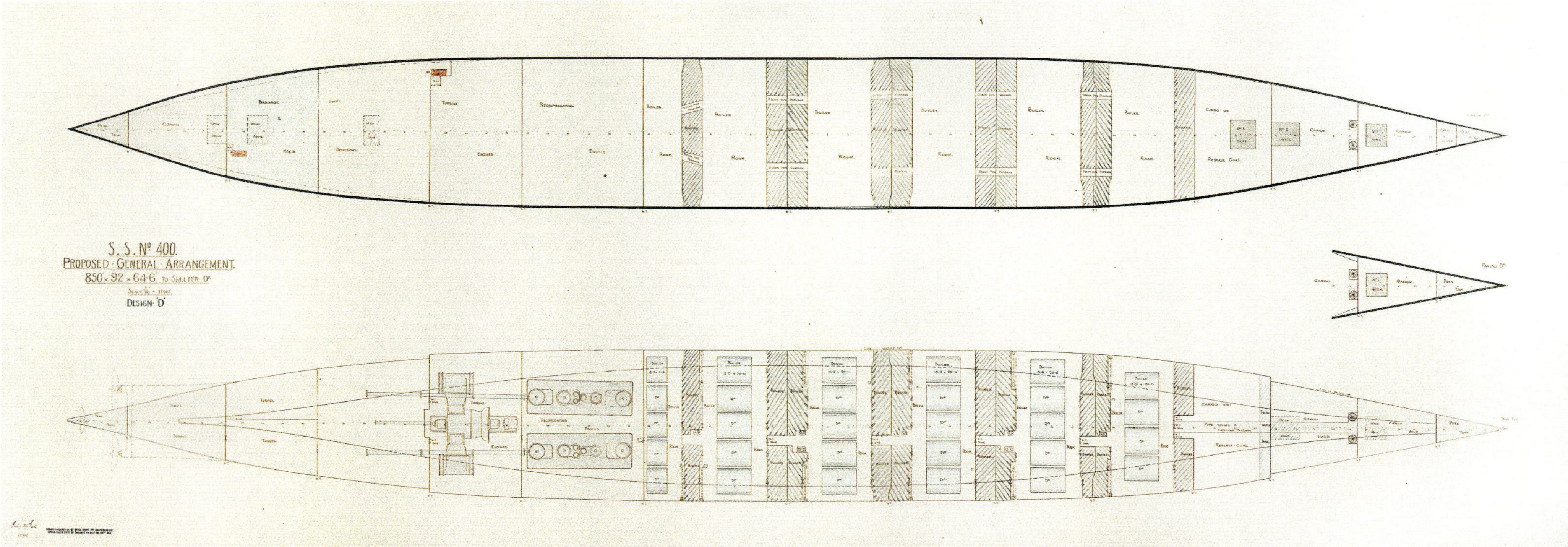

The Orlop Deck, Lower Orlop Deck and Tank Top. (HOYFM.2010.73.188, 189, 195 and 196 © National Museum NI Collection Harland & Wolff, Ulster Folk & Transport Museum)

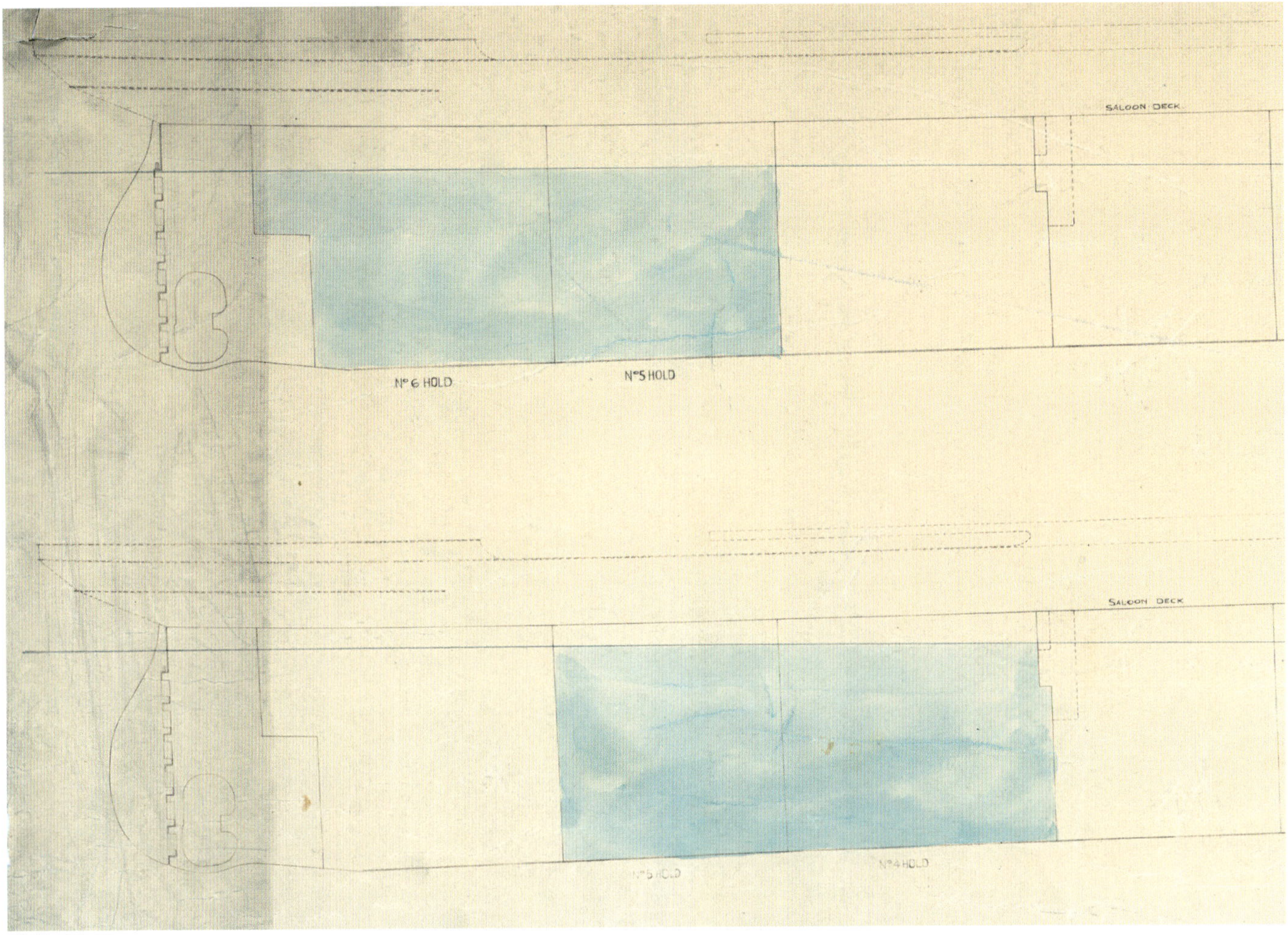

Extracts from a plan entitled 'SS *Olympic* and *Titanic*. Nos. 400–01. Plan Giving Waterlines with any 2 Adjoining Compartments Flooded', stamped by Harland & Wolff Ltd and dated 1 May 1912. Both ships could float with any two compartments flooded and were designed with ample margin to spare, so that they were almost 'three-compartment' ships and could float in several scenarios with four compartments flooded. They would remain afloat with two of the large after-cargo holds flooded, nonetheless, they would settle deep in the water at the stern.

After the *Titanic* disaster, the White Star Line directed Harland & Wolff to make changes to *Olympic* and *Britannic* so that they could float with a significantly greater number of watertight compartments flooded (see pages 359–60). (We know from Edward Wilding's recollection that a key reason behind these changes was to use them for publicity purposes and reassure the travelling public.) However, most passenger liners continued to be built in a similar way to *Titanic*. Harland & Wolff adopted a practice where they would design 'very large vessels' (over 20,000 gross tons) to remain afloat with any three watertight compartments flooded, which was a modest improvement. A generation later, Cunard's *Queen Mary* (completed in 1936) was also built as a three compartment ship by John Brown & Company. Decades later, in December 1996, Chris Hackett and John G. Bedford presented their paper 'The Sinking of SS *Titanic* – Investigated by Modern Techniques' to a Joint Meeting of the Royal Institution of Naval Architects (RINA) and the Institution of Engineers and Shipbuilders in Scotland. They argued that 'the design of the *Olympic* and *Titanic* is configured to comply with today's regulations with only a few exceptions … which says much for the care taken at the time they were planned', concluding: 'In regard to both [watertight] subdivision and [life]boat capacity, the builders went well beyond what was required at the time … The same cannot be said of the majority of passenger vessels today where the minimum requirements are normally treated as the maximum …' Whereas Harland & Wolff's design would 'almost certainly meet present day requirements … with minimal amendment, the *Lusitania* and *Mauretania* design 'could not due to the position of the longitudinal bulkheads'. (HOYFM.2010.73.191 © National Museum NI Collection Harland & Wolff, Ulster Folk & Transport Museum)

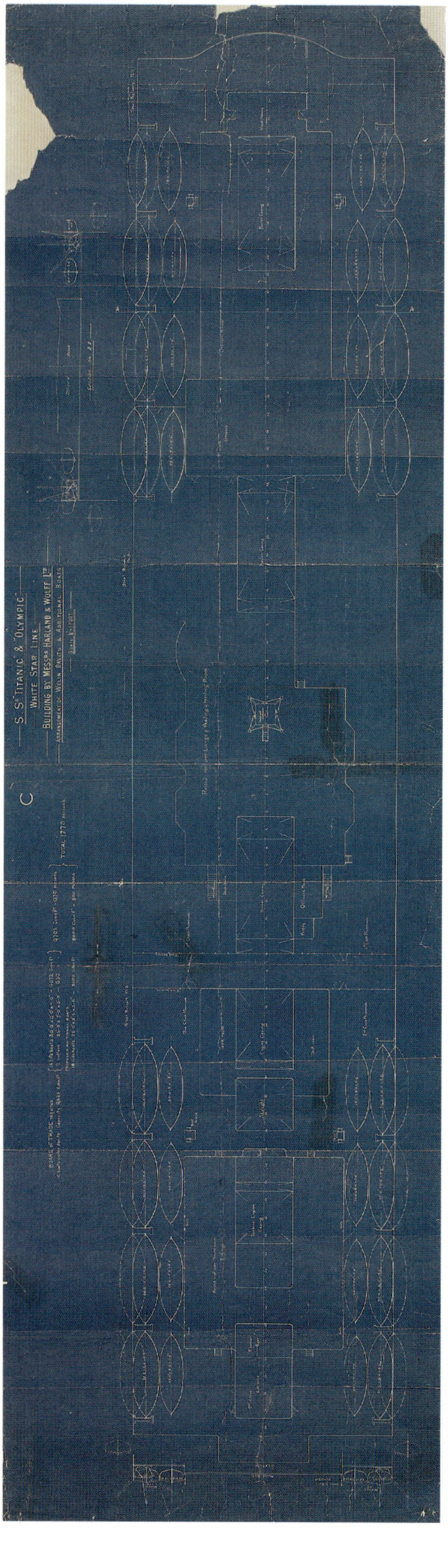

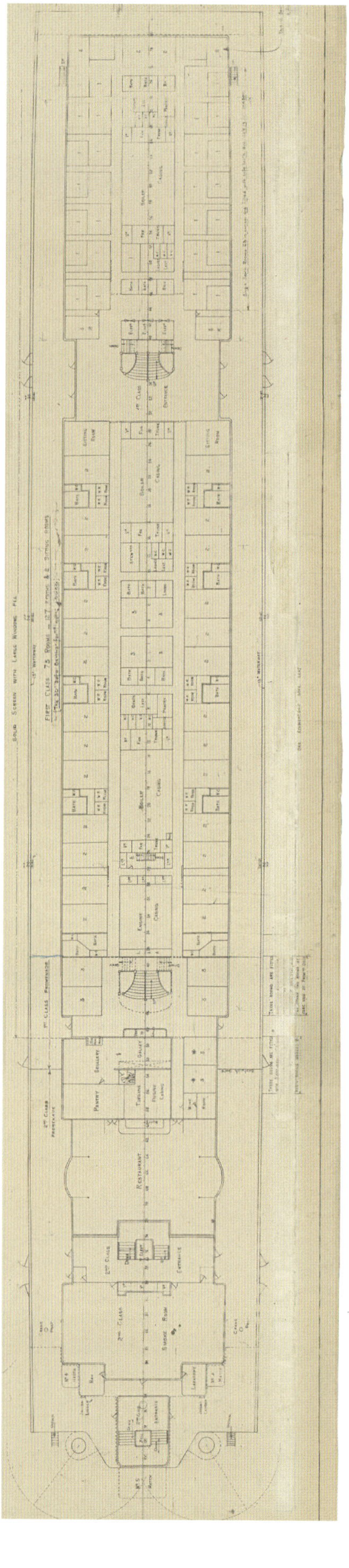

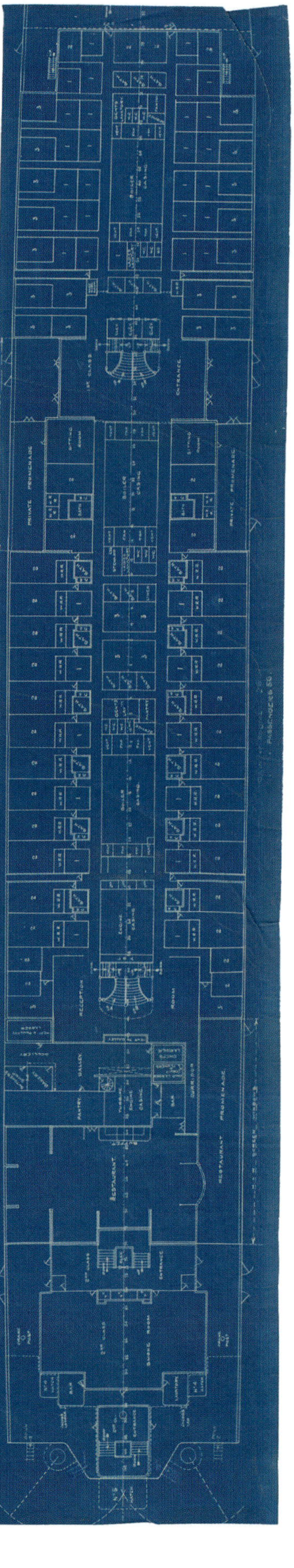

Opposite page:
Top: This blueprint appears to have been produced by the Welin Davit Company to show how their new davit design could accommodate a smaller, additional lifeboat inboard of each of the intended sixteen. The new davit design could not be used on *Olympic* and *Titanic* until it had been approved by the Board of Trade (see pages 89–90). (Courtesy National Museums Liverpool, Merseyside Maritime Museum, B/CUN/8/1912/2)

Middle and Bottom: Plans produced by Harland & Wolff showing *Olympic*'s original B-Deck configuration when she entered service in 1911 (top) and a blueprint showing the expansion of the first-class accommodation on *Titanic* (bottom). It notes a gain of twenty-six rooms and fifty-six passengers. (HOYFM.2010.73.162 © National Museum NI Collection Harland & Wolff, Ulster Folk & Transport Museum)

This page:
Right: Another blueprint showing the 'quadrant arm detail' of the double-acting Welin Quadrant Davits, submitted to the Board of Trade for approval. (The National Archives (UK))

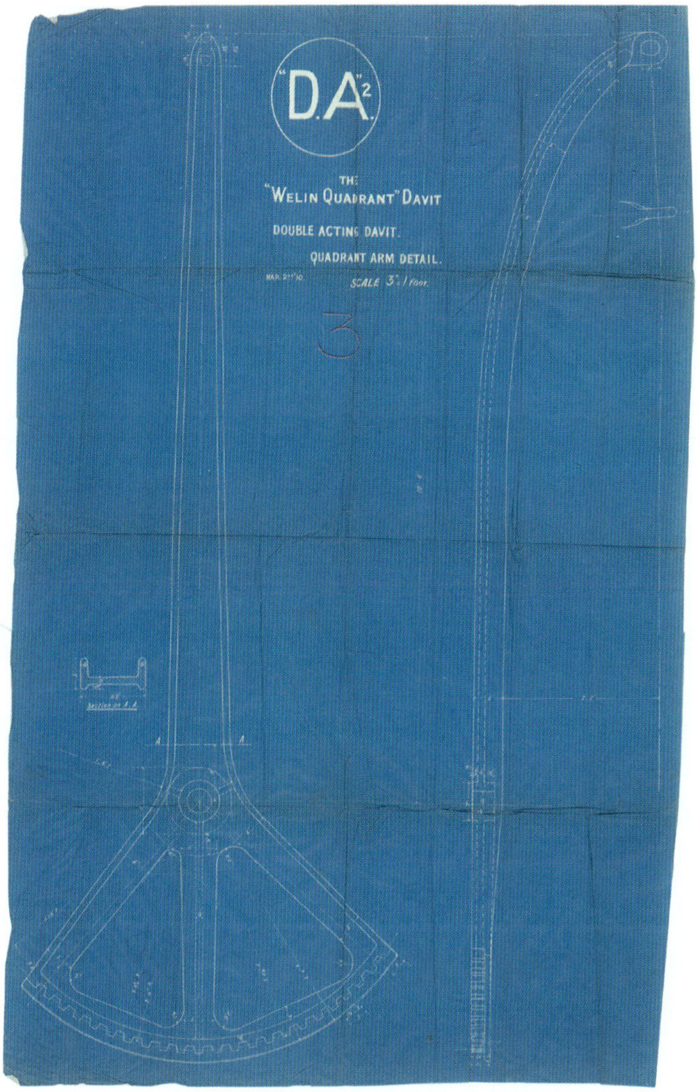

Below: Harland & Wolff maintained a series of volumes documenting the key technical details of each ship as completed. *Olympic* (400) and *Titanic* (401) can be seen near the bottom. Each ship's details ran across two double pages. This is the second of the double-page entries (the left-hand page is cut slightly narrower so that each ship's yard number is still visible). All the engine, bearing and pump details for *Titanic* were the 'same as No. 400', but her propeller specifications were different. An alteration to the pitch *of Olympic*'s port and starboard propeller blades in November 1911 is footnoted, as well as an increase in their pitch on *Titanic* to 35ft. (Courtesy of Harland & Wolff and the Deputy Keeper of the Records, Public Record Office of Northern Ireland [Reference No. D/2805/SHIP/8A-E])

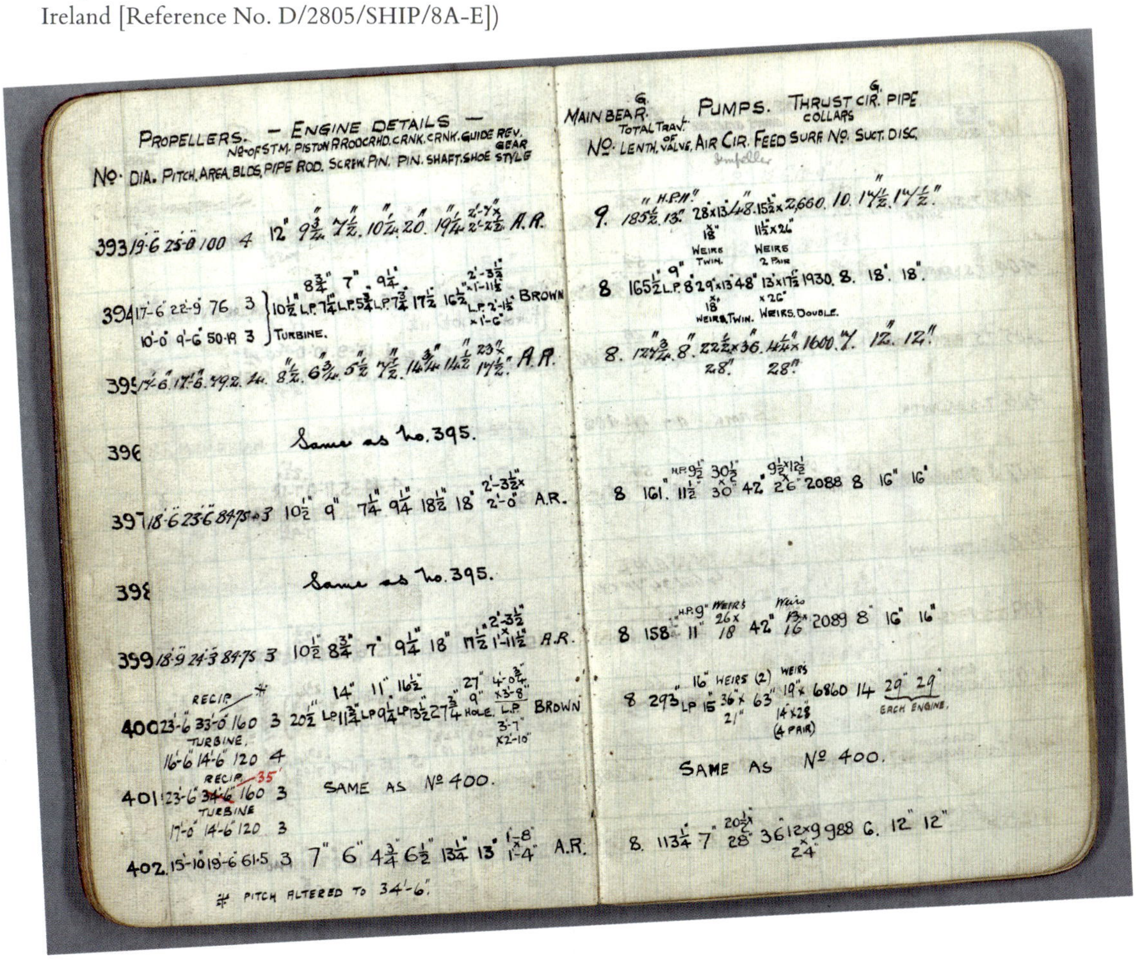

A digital rendition of *Titanic*'s first-class grand staircase, looking from the starboard side of A-Deck forward and up to the port-side Boat Deck landing. (*Titanic: Honor and Glory*, Vintage Digital Revival)

The Regency-style 'parlour' of *Titanic*'s starboard parlour suite on C-Deck. It is possible that this was one of the rooms that Leonard Peskett had in mind when he wrote that the decorative detail was 'overdone'. (*Titanic: Honor and Glory*, Vintage Digital Revival)

One of the styles used in *Titanic*'s first-class special staterooms was Modern Dutch … (*Titanic: Honor and Glory*, Vintage Digital Revival)

… another was Louis XV style. (Giovanni Castro)

The luxurious surroundings of *Titanic*'s first-class à la carte restaurant. In 1911, its counterpart on *Olympic* won recognition from some observers as being one of the best-decorated rooms on the ship. (*Titanic: Honor and Glory*, Vintage Digital Revival)

Looking forward on the starboard side, this view of *Olympic*'s first-class reception room on D-Deck depicts its original layout. The grand staircase is out of sight to the left; the wood-panelled elevator foyer is visible behind the archway to the left; and the entrance vestibule for the D-Deck gangways is behind the double doors on the right. This area was modified on *Olympic* in October and November 1911, when the entrance vestibule was reduced in size to enlarge the reception room, and the wood-panelled area was replaced by white panelling. The same change was made on *Titanic*. (Giovanni Castro)

Olympic's Turkish bath cooling room ... (Giovanni Castro)

... and its counterpart on *Titanic*, which was rearranged based on experience with *Olympic*. (*Titanic: Honor and Glory*, Vintage Digital Revival)

The forward second-class staircase at Boat Deck level, showing the lift (elevator) doors. It was a new feature in second class. Kate Buss remembered being impressed with the second-class staircase on *Titanic*. (*Titanic: Honor and Glory*, Vintage Digital Revival)

Titanic's second-class smoke room on B-Deck was spacious and much larger than its counterpart on earlier White Star liners. (*Titanic: Honor and Glory*, Vintage Digital Revival)

Titanic's second-class library on C-Deck was large and comfortable. (*Titanic: Honor and Glory*, Vintage Digital Revival)

Titanic's second-class dining saloon on D-Deck. (*Titanic: Honor and Glory*, Vintage Digital Revival)

Titanic's third-class smoke room on C-Deck was panelled and framed in oak. (*Titanic: Honor and Glory*, Vintage Digital Revival)

Titanic's third-class general room on C-Deck was finished in white enamel. (*Titanic: Honor and Glory*, Vintage Digital Revival)

(It is probable that she came into the harbour through the eastern channel. Pilot John Cotter, who was employed as the licensed pilot for decades and knew Queenstown very well, went in to take *Olympic* in and out of the harbour 'a good many times'. He always used the eastern channel, which was preferred by the pilots of large ships.)

Captain Smith decided to get *Olympic* 'into the roads between Fort Carlisle and Spike [island]', where the inner harbour widened out. She dropped anchor around 2.30 p.m. *The Cork Examiner* recorded:

> She came right into the inner harbour of Queenstown today, and notwithstanding her great depth of water and enormous size, she manoeuvred in the harbour and at Dognose under her triple screws with as much ease as a motor launch.
>
> On inspection today it was found very hard to give full expression to the thoughts created by the splendid accommodation provided for passengers, particularly the spaciousness which was found everywhere. No one class suffers at the expense of the other on board this fine vessel, for a visit to both the second and third class departments proved this, and it is safe to say that the accommodation provided for the second class passengers is as good, if not better, than that which was provided ten or fifteen years ago for the first class passengers.

Crowds gathered to admire *Olympic* at Church Bay and Grab-all Bay, to the west of Roche's Point at the harbour entrance. Her 'graceful outlines and leviathan proportions […] excited great admiration' and plenty of amateur photographers were there to try and get a good picture. Again, the *Cork Examiner*'s reporter gushed with praise, calling her 'a marvel of shipbuilding':

> Her immense yet harmonious proportions, the gracefulness of her lines, her great towering height, and above all the idea of strength, stability, and enormous power which a first view of her conveys, touch the imagination with a sense of wonderment which only increases as one has the opportunity of studying her in more detail. She is the highest achievement of the shipbuilder's craft, the finest production of his genius, and with her sister ship, the *Titanic*, she stands unapproached in the race for leviathan liners.

Queenstown's residents gained 'some idea of the monster ship as she stood out to sea, with her four great black and yellow funnels appearing over the edge of Spike'. Those on the tender found a ship towering above them as they drew near:

> It was only when the tender drew up under her side, when one had perforce to crane one's neck to catch a view of the people crowded at the rails on the upper decks – which by the way, extend considerably over the line of the bulwarks – that the proportions of the *Olympic* were borne home to the minds of the onlookers.
>
> [She] seemed as if she might be able to brace the fiercest storms and act the worst fury of the deep at defiance. In her side view the outline curved gracefully in the distance, and in front the great mass of her hull seemed the paragon of power and stability.

Olympic was 'a testimony to the attainments of those who designed her, to the efficacy of her builders and the enterprise of her owners'. Her name was 'fittingly reserved for a vessel that is the result of the highest intelligence and skill'.

Mails and passengers came aboard 'under the personal supervision of the local firm of agents, Mr. J.W. Scott, H.C., assisted by the chief clerk, Mr. George Harris'. Everything went with 'the utmost smoothness'. White Star's Marine Superintendent at Queenstown 'exercised his charge with unfailing kindness and consideration'. Another account of the maiden voyage recalled:

> … the familiar visits of the ladies with the soft Dublin accents soothingly asserting bargains in Irish lace, and ready with a witty rejoinder to the individuals – of whom there are always some – who make a point of driving a bargain to the bitter end.

> One such was an extremely homely little man – the prize pug of the ship's company – who had made a show of investing and was ready with an offer of £3 for a collar for which the vendor had demanded £3 10s.
>
> 'All roight, me darlint', said the woman quickly, 'I'll take off the shillin's because you're so very handsome'. The sale was not made.[6]

A number of passengers from Southampton and Cherbourg also disembarked at Queenstown. The *Cork Examiner* reported, 'They were all unstinted in their appreciation of the qualities of the great liner for smoothness and ease of running, and entire absence of vibration.' They found it difficult to realise 'that one was on board a ship at all, so smooth was the motion, and this, too, when the vessel was going at a high speed'.

The *Irish Times* discovered 'from the engine room staff' that 'during the run from Southampton to Cherbourg and Queenstown the engines worked most satisfactorily, and gave promise of a rapid passage by the liner across the Atlantic'. Passengers reported that 'although there was rather a big easterly swell on in the Channel, it had no effect whatever on the ship, which remained perfectly free from any unpleasant motion'.

One person was embarking at Queenstown. A 31-year-old Swedish man, giving his name as 'L. Wilson', joined the ship there. He gave his address as the local sailor's home and was engaged as a substitute able-bodied seaman.

Perhaps some people failed to notice when *Olympic* was on her way again. After a delay of several hours, she left Queenstown. (Captain James Tobin, who commanded mail tenders at Queenstown over four decades, once saw *Olympic* 'leave the harbour at dead low water, and she came out along the Turbot Bank like a small yacht would, and answered her helm just like a yacht'.)[7] *The Irish Times* reported that 'close on 1,300 bags of mails and one hundred passengers were embarked, and at 3.50 [p.m.] the *Olympic* sailed for New York'.

Olympic had 489 first, 263 second and 564 third-class passengers for a total of 1,316, plus about 853 crew. She reported passing Daunt's Rock Light Vessel at 4.22 p.m., then steamed off the southern coast of Ireland towards Fastnet Light, a lighthouse on a rock several miles off the south-western coast of Ireland. From Fastnet, *Olympic*'s course was set to follow the 'great circle route' across the Atlantic, taking her to a point called 'The Corner' at latitude 42°N, longitude 47°W several days later. From that point, she would have to change course to head more directly to New York, steaming to a point south of the Nantucket Lightship, then straight towards the Ambrose Channel Light Vessel, which marked her arrival at New York's harbour.

***

'Once away from Queenstown, with its iridescent sea – a blend of greens and blues and purples so lovely that the pen of a Lafcadio Hearn alone could put its glowing shimmer into words – the voyage may be said to have fairly begun,' wrote one passenger, who recorded the voyage in detail for the *New York Times*:

> It is the middle of the afternoon, trunks and bags have been opened and their contents strewed about the staterooms, the purser has answered the same series of questions a thousand times, and is still answering them (if you see him first) and the ordinary routine of ship life has begun.

Three days into the crossing:

> … you will have decided that you know every face onboard, especially as there haven't been the usual absentees from table on a ship which is nearer like a big London or New York hostelry than anything else you can imagine. But after all, five hundred [first class passengers] is a good number. The last day out you may meet fifty or seventy-five people who seem to have come on overnight.

One 'prominent Russian woman' made an appearance 'late on deck' on the third night out. The ladies had retired for the evening but a number of men remained on the promenade deck:

> Suddenly a figure all in white seemed to be wafting toward us. As she passed we had the gleam of a pair of wonderful black eyes, a suggestion of tawny hair woven into a tight coil at the nape of the neck, a gown of pure white, so closely fitting that every line of the figure was displayed.

She was not seen again.

He used the swimming pool every morning, 'The pool is a delightful supplement to half an hour spent in the gymnasium where, after punching the bag and indulging in various exercises that lots of us neglect at home, you may, if you have the nerve, mount a horse and take a ride'. The mechanical horse was:

> ... an ingenious arrangement, or a devilish contraption, according to your mood, which gives you all the sensations of a swift canter on a bronco, and correspondingly taut sore muscles the next day, if you have not been inclined to athletics previously or are not in the pink of physical condition.

The squash court attendant was 'skilled in hitting the ball into places where you are least expecting it'.

Olympic's public rooms were impressive, including 'a commodious smoking room, and all furnished comfortably and in excellent taste'. The two veranda cafés were 'charming rooms done in trellis work, with climbing vines, cool hospitable retreats, where you may sit and have your coffee and liqueurs, smoke if you like, the meantime enjoying the ocean breezes without a suggestion of discomfort'. Next to the dining saloon, the reception room 'proved to be the most popular resting place on the ship':

> Here a very good orchestra plays before and after meals, and tables and chairs were always at a premium. Not the least delightful feature of the main dining room is its arrangement of windows on all sides, these being paned with opalescent glass, backed with electric lights, so as to produce an effect of perpetual sunshine. We had one afternoon of gloomy weather, with rain and fog outside, that gathered shortly after the luncheon hour had started. To all appearances, as judged from within the dining saloon, there were fair skies overhead and the sun had never ceased to shine.
>
> It is here, too [...] that the amazing steadiness of the ship was a frequent source of comment. For two or three days – in fact until the words 'what a wonderful ship' had passed into the final bromidic stage – the most frequent inquiry was 'Have the engines stopped'? But the engines had not stopped, the screws were still plowing up the foam.
>
> After covering the distances between decks and staterooms a few times the elevators [lifts] – there are four – are welcome additions to the comfort. But it will take more than one voyage to acquaint the crew – several voyages, perhaps, to inform the average passenger of the exact spot he is seeking. It is all so big. There are so many decks, so many corridors. There is so much ground to cover that even the most experienced stewards find themselves guessing on occasion. You may carry the fact in your head that the barber shop is situated thus and so, but the strong chances are that while doing it you will have overlooked the exact geographical relations of the smoking room and the main saloons [...]
>
> With so much space to spare the possibilities for enjoyment are practically unlimited. A day of sports, admirably arranged by one passenger who especially enjoys that sort of thing, brought the familiar quota of deck athletes, but under conditions that made them seem truly like a little ocean marathon. In half an hour £30 had been collected to pay for prizes. There was, for example, a relay race over a mile course, with some creditable shows of speed, and a succession of pillow fighting, with the contestants balanced on a boom. It brought hard knocks to the contestants and no end of fun to the onlookers. Have you ever tried to 'eye the whale', the shipboard equivalent of the drawing room pastime 'putting the tail on the donkey?' And if so, can you explain why the blind-folded contestants invariably proceed outward towards the ship's rail, and almost without exception display an amazing ignorance of piscatorial anatomy by

> placing the eye as near as the extreme southern end of the fish as the space will allow? Have you ever tried, for instance, to walk across a room holding an egg in a spoon, the handle of which is placed firmly between your teeth? If you have, you will do honour to the six year old who made the more hazardous experiment, and succeeded, on the deck of this ocean mammoth, with contrary winds titillating the fragile hen-fruit at every step she took.

He wrote about 'the opportunities for individuality and exclusiveness in entertaining' on board such a large ship:

> You will have an appointment one afternoon to meet Mrs. Smith and a few friends for tea in the veranda café, or the following afternoon Mrs. Jones, to add the novel touch, will entertain you and other friends at bridge in her own suite, the tea served there at a proper moment, having all the suggestion of a private and personal function, with the ship's steward acting as general provider of refreshment.

Life on *Olympic* was 'very charming, very interesting, and very restful, and in effect it makes of the ocean a site for man's dwelling whereon [...] he may take his ease, may satisfy his hobbies, and have his comforts as if he were on land and at home'.

The indications were that White Star had been conservative in their estimates of *Olympic*'s speed. When he booked first-class tickets at White Star's London offices, he had been told that she would dock in New York around 3 p.m. on the Wednesday afternoon, 'And it had become apparent soon after we made the final pause at Cherbourg [*sic*] that no special effort was being made to force her. It was common gossip that the fires under five of her boilers had not been lighted at all.' Shipboard gossip was that she lacked a 'full supply of coal, the strike of dock workers at Southampton having necessitated the employment of a number of green hands whose ability at storing fuel was considerably below the average of the regular workers'.

***

Olympic performed well from the start. Chief Engineer Bell reported 'the engines and boilers worked very well during the passage from Southampton'. He regulated the speed throughout the crossing 'according to instructions received and a good command of steam was maintained for the various speeds required, the boilers being connected up as necessary when speed was accelerated to order until the 24 double ended boilers were all in use and the five single-ended boilers in reserve'. The boilers 'worked without trouble'. The stoking controller and electric telegraph, enabling communication between the engine room and boiler rooms 'worked satisfactorily and enabled the engineer in charge to control the work in [the] boiler rooms as desired'. Bell thought this equipment was 'an absolute necessity for such an extensive installation of boilers'. He found the reciprocating engines 'worked exceptionally well', with 'the speed being increased gradually to order'. The low-pressure turbine, too, did not give 'the slightest trouble'.[8] (J. Bruce Ismay noted that the reciprocating engines never reached their full potential during the crossing, with 'the highest number of revolutions being 79 port and 81 starboard, at which speed the ship was practically quiet'.)

By noon on Friday, 16 June, she had covered 428 miles at an average speed of 20.5 knots. *The Scotsman* published a telegram commenting on her progress:

> On Board the *Olympic*, Saturday [17 June]. The first day's performance of the *Olympic* in the Atlantic was satisfactory. She had made at midday 428 nautical miles since passing Daunt's Rock at four o'clock [*sic*] yesterday afternoon. The sea was rough to moderate, and the vessel so far has shown herself distinctly steady.

It was common for ship's passengers to place bets on each day's run. They would wait keenly for it to be posted in the smoking room or companionway.

Noon on Saturday represented her first complete day at sea and her run of 534 miles confirmed that she was averaging 21.5 knots. There were 'light to strong' south-east to north-east winds and a 'rough sea'. By the Sunday, with sea conditions moderating, her speed had increased. She reported a further run of 542 miles at an average of 21.9 knots.

It appears that it was later on Sunday that *Olympic* encountered fog. Captain Smith slowed his ship down. J. Bruce Ismay reported that the fog caused a delay of one and a half hours, while a report in the *New York Times* said *Olympic* had slowed for four hours. If she had steamed at a reduced speed for four hours, causing a one-and-a-half-hour delay in total, then she possibly reduced speed from around 22 knots to about 14 knots (or about 'half ahead'). When the day's run was posted as usual at noon on Monday, 19 June, it showed 525 miles: an average of 21.2 knots.

Her final complete day at sea saw a 'moderate westerly wind' accompanied by a 'smooth sea' and it reflected in her performance. When the run was posted, it showed she had covered 548 miles, which was her highest yet. Her average speed increased to 22.1 knots. Heading westbound, the current was typically against her, meaning that it reduced her rate of progress 'over the ground' compared to her speed through the water. From that point, she only had a further 317 miles left to cover to reach the Ambrose Channel Light Vessel.[9]

***

News of her progress was relayed ashore. *The Daily Telegraph*'s New York correspondent reported on 19 June that *Olympic* 'came within wireless touch of Cape Race, Newfoundland, today'. The telegram was paraphrased by the *Irish Times*, which noted, 'The absolute steadiness of the great vessel seems assured. A glass of water, almost filled, which was placed by a passenger on the table of a stateroom overnight was found unspilt the following morning.' (The report echoed an anecdote from one of *Cedric*'s smoke-room stewards back in February 1903. On that occasion, it was a 'small wine glass filled with champagne' that was left on a sideboard throughout the crossing and 'not a drop of wine' was spilled.)

The next day, the *Irish Times* reported, 'The *Olympic* is steady as a rock and on Monday we had a series of *Olympic* games, followed by a dance.' Although she had been delayed slightly by fog, the average speed she was capable of was 'magnificent for a vessel that does not seek speed honours so much as a reputation for the maximum of luxury, comfort and health afloat'.[10]

A 'Marconi wireless to the *New York Times*' from 20 June reported that the maiden voyage was 'exceeding expectations' and 'the passengers are unanimous in praise of the big ship', 'The public rooms and the staterooms are huge, and the ventilation is wonderful [...] The weather has been beautiful and the ship very steady, without vibration.'

***

Despite the gushing press reports, Thomas Andrews kept busy examining all sorts of details which could be improved. He noted some instances where the plans had not been followed or where experience had highlighted that particular changes were needed. He thought that the after part of the officers' quarters deckhouse on the Boat Deck could be converted and used as first-class stateroom accommodation in future. The hat and cloak room on the port side of the first-class staircase on C-Deck, opposite the Purser's Enquiry Office, was 'not sufficiently used to warrant this loss of earning power space'. Instead, there was room for first-class staterooms which would bring in revenues. (One of the plans Andrews was considering was 'fitting hooks for hats and cloaks across the bulkhead facing elevators on D-deck' instead.)

Elsewhere, he proposed 'dispensing with' first-class staterooms C144 and C146 in the aftermost section on C-Deck so that the chief steward's stateroom could be placed there and the assistant doctor's room enlarged. (His initial proposal changed so that C142 became the chief steward's stateroom and C144 became the restaurant manager's stateroom. Stateroom C146, which had originally been shown as the restaurant manager's room on February 1911 accommodation plans, became the assistant doctor's stateroom.) Taken as a whole, his suggested changes increased the first-class stateroom accommodation.

The à la carte restaurant was 'short of table accommodation', so he proposed 'fitting eleven additional four seated tables'. He thought that a privacy screen could be fitted 'on the side of the restaurant/buffet supt. [superintendent] entrance door so as to prevent passengers from seeing behind the buffet'. In the first-class dining saloon, he thought that the two single serving doors on the port and starboard sides which connected the pantry with the saloon 'do not appear to be necessary for service and are better kept shut owing to noise'. 'The space in way of these doors will provide accommodation for one additional four seated table P & S [on the port and starboard sides] in the saloon.' The first-class reception room proved a victim of its own success and was 'the most popular room' in first class. Andrews noted it was 'more or less crowded after lunch and dinner' as well as 'for afternoon tea' at 4.30 p.m. 'It was found necessary to bring up all the spare caine [*sic*] chairs from the baggage room to provide temporary seating accommodation, would strongly recommend additional permanent caine [*sic*] furniture being ordered.' Smoking was permitted 'at all times' in the reception room, so he recommended 'an exhaust fan'.

Andrews spoke with Captain Smith, who 'strongly recommends protective windows with round bullseye lights to be fitted in way of square windows on centre shelter [of the] navigating bridge as in *Adriatic*'. The captain's sitting room had a 'full room carpet', so Andrews wrote that linoleum tiles were 'not to be fitted'.

On the A-Deck promenade, there was an issue with the deckchairs. They were not numbered in line with the plans supplied to the workmen, so they needed renumbering, 'H&W to enquire if any change was made on the numbering of the approved plan to which they were supposed to work.' The photographer's darkroom on the starboard side of the promenade was 'not used'. He suggested a photographer could be employed to 'sell photographs to passengers' and develop their films.

He had many details to address to improve passenger comfort and security – 'bolts should be fitted on the inside of the third class stateroom

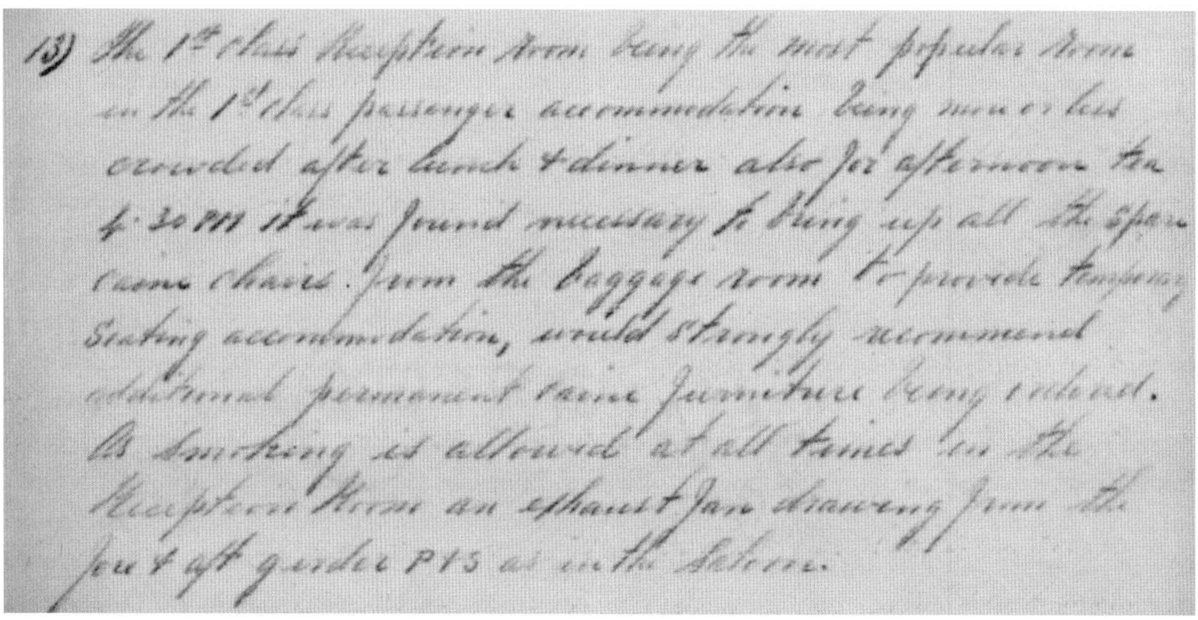
13) The 1st class Reception Room being the most popular room
in the 1st class passenger accommodation being more or less
crowded after lunch & dinner also for afternoon tea
4.30 PM it was found necessary to bring up all the spare
caine chairs from the baggage room to provide temporary
Seating accommodation, would strongly recommend
additional permanent caine furniture being ordered.
As Smoking is allowed at all times in the
Reception Room an exhaust fan drawing from the
fore & aft girder P & S as in the Saloon.

Thomas Andrews' Note 13 refers to the first-class reception room as being the 'most popular room' in the first-class accommodation. He noted spare chairs had been brought in to provide temporary seating accommodation, but additional permanent furniture was required. (Author's collection)

Olympic's bridge, photographed in May 1912. The protective shutters with the round bullseye lights are visible beneath the windows. (Günter Bäbler collection)

doors to give protection to such rooms occupied by women'. It was not uncommon for female third-class passengers to be subject to unwanted attention from crewmen or fellow passengers. The ability to lock their doors from the inside gave additional peace of mind.

Most of his recommendations related to first-class suite and staterooms. Sponge holders had been omitted from the private bathrooms on B- and C-Decks; cigarette holders had been omitted from the gentleman's lavatories in first and second class; and hat and coat hooks had been omitted from some of the first-class suites. All of them needed fitting. The suite staterooms on B- and C-Deck needed 'back plates for electric reading lamps' fitted 'over beds etc.', as with the ordinary staterooms.

Then there was a privacy problem. 'The mirrors in the wardrobe doors adjoining the entrance doors into inside staterooms on C-deck to be dispensed with as when the entrance doors are left open on the deck anyone in the passage can see the occupant of the room dressing or undressing.' Another issue was that the doors to first-class staterooms B32 and B33 opened directly towards the grand staircase. Andrews' first thought was that the doors' location should be altered to provide the stateroom occupants with more privacy. (These stateroom doors were shown in their original location on first-class accommodation plans in 1912 and 1913. One possibility, based on practice on other ships including *Titanic*, is that curtains were fitted instead to provide privacy.)

There was an 'excessive draught in the steward's stairway leading from first class pantry to [the] working passage on E-deck', so Andrews suggested 'a fore and aft wood bulkhead with sliding doors be fitted at foot of stairs as shown on plan'.

The printer's room could be enlarged by removing the sleeping berths, which were not necessary because berths were provided with the steward accommodation. In the third-class baker's shop, he suggested 'an additional trough' should be fitted in place of a dresser which was being removed. A drop table needed to be 'fitted in fore and aft passage outside third class galley sufficiently high to enable barrels of flour to be stored underneath'. The pantry near the Turkish baths was 'not used', so he suggested 'converting this room into a room for the female Turkish bath attendants'. He also thought an electric bell needed to be fitted outside the Turkish bath cooling room door because 'this door has to be kept locked during the hours the bath is being used by ladies'. A potato peeler would make the crew's galley more efficient. He specified that 'notices denoting the lettering of the various decks to be fitted in the second class accommodation' – no small matter, given that the forward second-class staircase extended over seven decks.

Andrews was in regular contact with Chief Engineer Bell and the other engineering staff. He wrote that 'no excessive vibration was experienced' during the crossing, but there was a 'considerable vertical pulsation motion at certain speeds of the main engines which was particularly bad when both engines were running'.

Bell, himself, kept a detailed record of *Olympic*'s performance and the operation of all her systems and machinery. He said that the ten See's ash ejectors (two in each large boiler room) 'worked well'. The electrical machinery produced 'all the current required for light and power […] without approaching the maximum load for which [the] dynamos are designed' and the emergency dynamos 'worked efficiently and was seen on required circuits from dusk until daylight'. Likewise, the refrigerating machinery 'worked very satisfactorily', maintaining the required temperature, while the ice-making machines worked well.

The engineers did have a slight problem with 'regulating the brine [salt water] necessary for [the] water coolers', which was needed to prevent them 'freezing up'. It was a common problem with new installations.

The boiler room telephones needed 'improved insulating from the ship's structure in engine room', but the engine room telegraphs 'worked very well' and Bell found them 'remarkably easy in working' with no adjustments necessary. The watertight doors 'always closed readily and promptly when required for testing' but Bell identified a problem in that there was no indicator on the bridge, 'I consider that a "tell tale" indicating on the bridge when doors are closed would be an improvement.' The hot-water temperature indicators were 'a great convenience and have worked very satisfactorily'. The 'steering gear' which operated the ship's rudder was 'an exceptionally good fitting' and it was 'practically noiseless in working', while the telemotors 'worked very satisfactorily throughout' the voyage.

All the lifts in the passenger accommodation and the storage areas were 'very satisfactory' and 'much used'. The ventilation fans worked satisfactorily but Bell said that additional ones were required. Some additional insulation was required in the passenger areas, but it was already 'being attended to'.

Bell's comments reflected well on Harland & Wolff, who had produced a fundamentally well-functioning ship.[11]

***

Olympic 'was reported 433 miles east of the Ambrose Channel Lightship at 6.58 o'clock yesterday morning [Tuesday, 20 June 1911]', according to the *New York Times*. Captain Smith then sent a telegram noting he expected his ship to reach Quarantine at around 3 a.m. on the 21st. The *Chicago Daily Tribune* reported that she was 'several hours ahead of schedule' and she had passed Fire Island.

She reached the Ambrose Channel Light Vessel, which marked the end of the westbound crossing, at 2.24 a.m. It was an impressive performance, given that she had been delayed leaving Queenstown and then lost one and a half hours due to fog. Without those delays, she would have arrived on Tuesday evening.

Pilot J.J. Adler came aboard around 2.30 a.m. The *Syracuse Herald* reported *Olympic* then arrived off Quarantine at 3.15 a.m. 'There was but little delay in the examination by Dr. Alvah H. Doty, the health officer of the port, wrote the *New York Herald*, 'but the *Olympic* got under way after the quarantine tugboat left, and proceeded as far as Stapleton, where she again anchored, awaiting the return of the tugboat to take off the health officer and his assistants.'

The crossing seemed to be over all too quickly for the victualling staff who had to prepare for the arrival in New York. 'Any docking day is nerve-racking apart from the work, with the sheer enthusiasm expended everywhere,' wrote

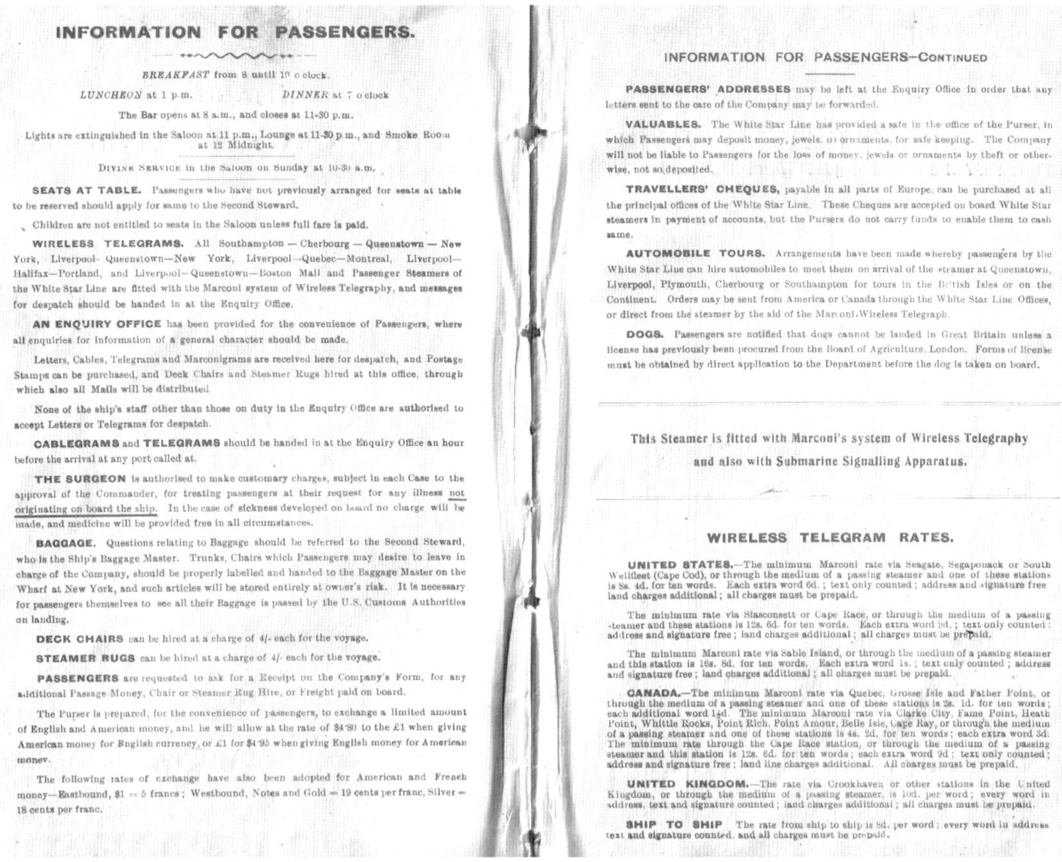

INFORMATION FOR PASSENGERS.

BREAKFAST from 8 until 10 o'clock.

LUNCHEON at 1 p.m. *DINNER* at 7 o'clock

The Bar opens at 8 a.m., and closes at 11-30 p.m.

Lights are extinguished in the Saloon at 11 p.m., Lounge at 11-30 p.m., and Smoke Room at 12 Midnight.

DIVINE SERVICE in the Saloon on Sunday at 10-30 a.m.

SEATS AT TABLE. Passengers who have not previously arranged for seats at table to be reserved should apply for same to the Second Steward.

Children are not entitled to seats in the Saloon unless full fare is paid.

WIRELESS TELEGRAMS. All Southampton — Cherbourg — Queenstown — New York, Liverpool—Queenstown—New York, Liverpool—Quebec—Montreal, Liverpool—Halifax—Portland, and Liverpool—Queenstown—Boston Mail and Passenger Steamers of the White Star Line are fitted with the Marconi system of Wireless Telegraphy, and messages for despatch should be handed in at the Enquiry Office.

AN ENQUIRY OFFICE has been provided for the convenience of Passengers, where all enquiries for information of a general character should be made.

Letters, Cables, Telegrams and Marconigrams are received here for despatch, and Postage Stamps can be purchased, and Deck Chairs and Steamer Rugs hired at this office, through which also all Mails will be distributed

None of the ship's staff other than those on duty in the Enquiry Office are authorised to accept Letters or Telegrams for despatch.

CABLEGRAMS and **TELEGRAMS** should be handed in at the Enquiry Office an hour before the arrival at any port called at.

THE SURGEON is authorised to make customary charges, subject in each Case to the approval of the Commander, for treating passengers at their request for any illness not originating on board the ship. In the case of sickness developed on board no charge will be made, and medicine will be provided free in all circumstances.

BAGGAGE. Questions relating to Baggage should be referred to the Second Steward, who is the Ship's Baggage Master. Trunks, Chairs which Passengers may desire to leave in charge of the Company, should be properly labelled and handed to the Baggage Master on the Wharf at New York, and such articles will be stored entirely at owner's risk. It is necessary for passengers themselves to see all their Baggage is passed by the U.S. Customs Authorities on landing.

DECK CHAIRS can be hired at a charge of 4/- each for the voyage.

STEAMER RUGS can be hired at a charge of 4/- each for the voyage.

PASSENGERS are requested to ask for a Receipt on the Company's Form, for any additional Passage Money, Chair or Steamer Rug Hire, or Freight paid on board.

The Purser is prepared, for the convenience of passengers, to exchange a limited amount of English and American money, and he will allow at the rate of $4·80 to the £1 when giving American money for English currency, or £1 for $4·95 when giving English money for American money.

The following rates of exchange have also been adopted for American and French money—Eastbound, $1 = 5 francs; Westbound, Notes and Gold = 19 cents per franc, Silver = 18 cents per franc.

INFORMATION FOR PASSENGERS—CONTINUED

PASSENGERS' ADDRESSES may be left at the Enquiry Office in order that any letters sent to the care of the Company may be forwarded.

VALUABLES. The White Star Line has provided a safe in the office of the Purser, in which Passengers may deposit money, jewels, or ornaments, for safe keeping. The Company will not be liable to Passengers for the loss of money, jewels or ornaments by theft or otherwise, not so deposited.

TRAVELLERS' CHEQUES, payable in all parts of Europe, can be purchased at all the principal offices of the White Star Line. These Cheques are accepted on board White Star steamers in payment of accounts, but the Pursers do not carry funds to enable them to cash same.

AUTOMOBILE TOURS. Arrangements have been made whereby passengers by the White Star Line can hire automobiles to meet them on arrival of the steamer at Queenstown, Liverpool, Plymouth, Cherbourg or Southampton for tours in the British Isles or on the Continent. Orders may be sent from America or Canada through the White Star Line Offices, or direct from the steamer by the aid of the Marconi Wireless Telegraph.

DOGS. Passengers are notified that dogs cannot be landed in Great Britain unless a license has previously been procured from the Board of Agriculture, London. Forms of license must be obtained by direct application to the Department before the dog is taken on board.

This Steamer is fitted with Marconi's system of Wireless Telegraphy and also with Submarine Signalling Apparatus.

WIRELESS TELEGRAM RATES.

UNITED STATES.—The minimum Marconi rate via Seagate, Segaponack or South Wellfleet (Cape Cod), or through the medium of a passing steamer and one of these stations is 8s. 4d. for ten words. Each extra word 6d.; text only counted; address and signature free land charges additional; all charges must be prepaid.

The minimum rate via Siasconsett or Cape Race, or through the medium of a passing steamer and these stations is 12s. 6d. for ten words. Each extra word 9d.; text only counted; address and signature free; land charges additional; all charges must be prepaid.

The minimum Marconi rate via Sable Island, or through the medium of a passing steamer and this station is 16s. 8d. for ten words. Each extra word 1s.; text only counted; address and signature free; land charges additional; all charges must be prepaid.

CANADA.—The minimum Marconi rate via Quebec, Grosse Isle and Father Point, or through the medium of a passing steamer and one of these stations is 2s. 1d. for ten words; each additional word 1½d. The minimum Marconi rate via Clarke City, Fame Point, Heath Point, Whittle Rocks, Point Rich, Point Amour, Belle Isle, Cape Ray, or through the medium of a passing steamer and one of these stations is 4s. 2d. for ten words; each extra word 3d. The minimum rate through the Cape Race station, or through the medium of a passing steamer and this station is 12s. 6d. for ten words; each extra word 9d; text only counted; address and signature free; land line charges additional. All charges must be prepaid.

UNITED KINGDOM.—The rate via Crookhaven or other stations in the United Kingdom, or through the medium of a passing steamer, is 10d. per word; every word in address, text and signature counted; land charges additional; all charges must be prepaid.

SHIP TO SHIP The rate from ship to ship is 8d. per word; every word in address text and signature counted, and all charges must be prepaid.

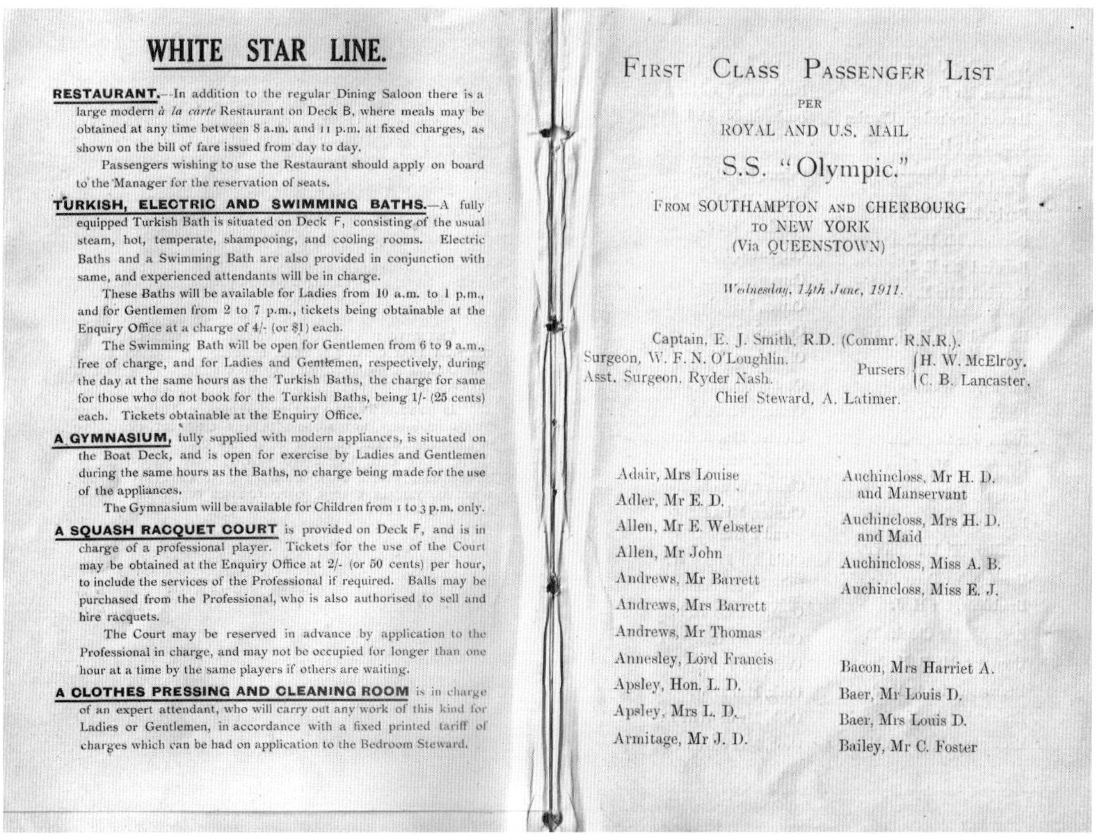

WHITE STAR LINE.

RESTAURANT.—In addition to the regular Dining Saloon there is a large modern *à la carte* Restaurant on Deck B, where meals may be obtained at any time between 8 a.m. and 11 p.m. at fixed charges, as shown on the bill of fare issued from day to day.

Passengers wishing to use the Restaurant should apply on board to the Manager for the reservation of seats.

TURKISH, ELECTRIC AND SWIMMING BATHS.—A fully equipped Turkish Bath is situated on Deck F, consisting of the usual steam, hot, temperate, shampooing, and cooling rooms. Electric Baths and a Swimming Bath are also provided in conjunction with same, and experienced attendants will be in charge.

These Baths will be available for Ladies from 10 a.m. to 1 p.m., and for Gentlemen from 2 to 7 p.m., tickets being obtainable at the Enquiry Office at a charge of 4/- (or $1) each.

The Swimming Bath will be open for Gentlemen from 6 to 9 a.m., free of charge, and for Ladies and Gentlemen, respectively, during the day at the same hours as the Turkish Baths, the charge for same for those who do not book for the Turkish Baths, being 1/- (25 cents) each. Tickets obtainable at the Enquiry Office.

A GYMNASIUM, fully supplied with modern appliances, is situated on the Boat Deck, and is open for exercise by Ladies and Gentlemen during the same hours as the Baths, no charge being made for the use of the appliances.

The Gymnasium will be available for Children from 1 to 3 p.m. only.

A SQUASH RACQUET COURT is provided on Deck F, and is in charge of a professional player. Tickets for the use of the Court may be obtained at the Enquiry Office at 2/- (or 50 cents) per hour, to include the services of the Professional if required. Balls may be purchased from the Professional, who is also authorised to sell and hire racquets.

The Court may be reserved in advance by application to the Professional in charge, and may not be occupied for longer than one hour at a time by the same players if others are waiting.

A CLOTHES PRESSING AND CLEANING ROOM is in charge of an expert attendant, who will carry out any work of this kind for Ladies or Gentlemen, in accordance with a fixed printed tariff of charges which can be had on application to the Bedroom Steward.

FIRST CLASS PASSENGER LIST

PER

ROYAL AND U.S. MAIL

S.S. "Olympic."

FROM SOUTHAMPTON AND CHERBOURG
TO NEW YORK
(Via QUEENSTOWN)

Wednesday, 14th June, 1911.

Captain, E. J. Smith, R.D. (Commr. R.N.R.).
Surgeon, W. F. N. O'Loughlin.
Asst. Surgeon, Ryder Nash.
Pursers { H. W. McElroy. / C. B. Lancaster.
Chief Steward, A. Latimer.

Adair, Mrs Louise
Adler, Mr E. D.
Allen, Mr E. Webster
Allen, Mr John
Andrews, Mr Barrett
Andrews, Mrs Barrett
Andrews, Mr Thomas
Annesley, Lord Francis
Apsley, Hon. L. D.
Apsley, Mrs L. D.
Armitage, Mr J. D.

Auchincloss, Mr H. D. and Manservant
Auchincloss, Mrs H. D. and Maid
Auchincloss, Miss A. B.
Auchincloss, Miss E. J.

Bacon, Mrs Harriet A.
Baer, Mr Louis D.
Baer, Mrs Louis D.
Bailey, Mr C. Foster

Top: The White Star Line provided all the 'Information for Passengers' that they would need for their crossing. (Günter Bäbler collection)

Bottom: *Olympic*'s first-class passenger list included many familiar names. Thomas Andrews appeared on the first page. The Ismays ('Ismay, Mr. J.B. and manservant' and 'Ismay, Mrs. J.B. and maid'). (Günter Bäbler collection)

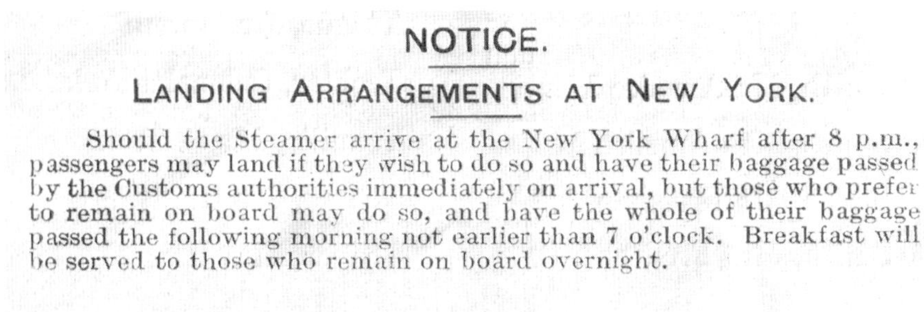

NOTICE.

LANDING ARRANGEMENTS AT NEW YORK.

Should the Steamer arrive at the New York Wharf after 8 p.m., passengers may land if they wish to do so and have their baggage passed by the Customs authorities immediately on arrival, but those who prefer to remain on board may do so, and have the whole of their baggage passed the following morning not earlier than 7 o'clock. Breakfast will be served to those who remain on board overnight.

Passengers were advised that there would be no logistical issues for them if the ship arrived at the dock in New York after 8 p.m. (Günter Bäbler collection)

Violet Jessop. 'We knew the excitement would be unprecedented and that our early morning arrival off quarantine would be more hectic than usual.' As expected, the call bells:

> ... rang from a very early hour. Strident and excited voices demanded food, for everybody aimed to be ready to greet their friends with an air of complete assurance as if to impress them that this successful maiden crossing was partly their personal achievement; we mere slaves obeyed instantly to assist them in the illusion.

When the crew mustered for the doctor's inspection at Quarantine, there was a 'nervous tension' in the air. Violet recalled passengers ringing their call bells, 'Passengers may make a hundred trips and know this inspection takes place regularly, in fact see the crew mustering, yet they will go straight to their rooms and ring their bells.' After a 'clean bill of health', *Olympic* 'proceeded majestically up the Hudson':

> [She was] escorted by all manner of official and unofficial small craft, up past the huge buildings on Battery Park, the wharves and ships, all of which appeared to shimmer with the moving, cheering mass of people gathered together. Not a window, however small, but had a little flag or handkerchief waving from it, as we slowly passed on to the accompaniment of shrill tootings of greetings from everything afloat that had a whistle to blow.

The *New York Times* reported that *Olympic* left Quarantine at 7.45 a.m. She 'steamed on slowly, while passing steam craft saluted, and all were answered, even down to the smallest motor boat, by Captain Smith,' according to the *New York Herald*. She closed in on the Chelsea Piers.

Cunard's *Lusitania* 'swung into the river at 9 o'clock, just after the *Olympic* had passed and gave the small army of photographers on the piers a chance for a good picture'. Unlike the liners docked along the piers, *Lusitania* 'did not salute'

White Star's 'SS *Olympic* entering New York Harbour on first trip'. (Library of Congress Prints & Photographs Division)

Excited crowds watch *Olympic*, accompanied by numerous tugboats, approaching the pier. The top coat of black paint on the bow near to the waterline has visibly worn away. This was commonly seen with new liners: older ships lost paint as well, but it was less visible because they had multiple coats. (Library of Congress Prints & Photographs Division)

Lusitania heads out to sea as *Olympic* comes abreast of the pier. (Library of Congress Prints & Photographs Division)

Olympic as she passed. 'It was thought that in the hurry of getting his ship under way and receiving congratulations on his coronation honour of Commander of the Bath, Captain Charles, skipper of the *Lusitania*, overlooked the *Olympic*.'

Frank Munsey recalled *Lusitania*'s arrival on her own maiden voyage almost four years earlier, 'A wonder-satisfied world paused to do homage to the monarch of the seas.' Now, 'few watched' *Lusitania* going, 'the deposed monarch passed out toward the ocean unnoticed and unsung, like a rooster put to ignominious flight, like a dog with tail between his legs in craven fear'.[12]

The *New York Herald* said 'the weather was perfect', with 'a bright sky and the spanking breeze which ruffled the waters of the North River adding to the picture'. Pilot Adler and Captain Smith were confronted with 'a strong ebb tide and stiff wind down the river, and it required more than an hour to bring the steamship around the end of the pier and into her berth'. During the operation, the pier extension came under pressure 'and the planking at the corner was crushed in by the great weight'.

Violet Jessop thought the tugs were 'haughty-looking' when they came to meet *Olympic*, 'thrusting their noses into the side of the great hulk, puffing frantically as they pushed her into her berth. Her bows came to a rest within a few inches of the dock wall and so the proud ship came successfully to port.'

Violet's description did not mention an unfortunate accident while *Olympic* was docking. Despite all the preparations, it did not go smoothly. William Marchant Smith, White Star's Marine Superintendent at New York, had phoned the New York Towboat Company the previous evening. He explained, '*Olympic* would probably be docking on a strong ebb tide in the morning, and we should want plenty of help.'

During fifteen years in New York, he had seen plenty of ships come and go. His duty when she arrived was 'to superintend the docking', so he was on the corner at the end of Pier 59 with a group of men under his command. Once *Olympic* 'made her landing on the corner [of the pier] we ran some lines from the bow of the ship on to the dock [...] from the starboard bow of the ship onto the pier'. There were three ropes and a gang of men 'attending to each rope to be carried up the dock as the ship went in'. Allen Luckhurst, who was William Marchant Smith's counterpart for the Red Star and American Lines, watched *Olympic* 'coming up to the piers at an angle of about thirty degrees, to the river; and at that time, at that particular time, she was about rounded on the end of her own pier'. She was then 'nearly motionless'.

That morning, Peter Cahill had received a request from the New York Towboat Company to assist *Olympic*. He sent the steam tug *O.L. Hallenbeck*, a medium-sized tug boat, 112ft long. Her commander, John D. Sjalund, was a licensed Master for the waters of New York and had been in command of tugs for fifteen years.

Sjalund made sure they were near the pier around 8.30 a.m. to wait for *Olympic*'s arrival. He noticed there was a north-westerly wind and estimated the wind speed at around '10 or 15 miles … something like that' and an ebb tide. The weather was clear as she came into view and slowly proceeded up towards the docks against the tide. Once *Olympic* lay 'perfectly still … lying along the end of the docks', with her bow pointing to starboard ready to move into the pier, *O.L. Hallenbeck* came 'up under her [aft] starboard quarter' and 'a little bit behind her after, stern smokestack'. *Olympic*'s decks were far too high for any tug boat to throw a heaving line aboard, so her crew threw a 6in hawser onto *O.L. Hallenbeck*'s fore deck. Two deckhands started working to secure it. Pilot Gus Sveberg was in the pilot house with Captain Sjalund. He recalled the helm was hard a'starboard (turning the tug to port) as [*O.L. Hallenbeck*]'s bow came against *Olympic* 'and shoved' to help get the liner into position. Captain Thomas Reynolds, aboard another tug, *Edmund Moran*, noticed that *Olympic*'s bow was 'maybe ten or fifteen feet' into Pier 59. Deckhand Yost Anderson was on top of *Edmund Moran*'s pilot house 'helping a young fellow [Hall] take pictures'.

Oceanic was docked nearby and her crew took a keen interest in the new *Olympic*. Chief Officer

Passengers in all three classes line *Olympic*'s upper decks. At the stern, one of the officers on the aft docking bridge is presumably First Officer William Murdoch. (Library of Congress Prints & Photographs Division)

Olympic's helm is hard a port as she prepares to turn to starboard, into the pier. (Mike Beatty collection)

Summers, First Officer Lightoller and Third Officer Shaw watched from the starboard side of the aft docking bridge. Lightoller thought *O.L. Hallenbeck* 'was in a useless position as far as the *Olympic* was concerned … her stern was towards the *Olympic*'. He assumed:

> She might have been after some barges or something in the dock with the object of towing them out, or going up the river, and it struck me that she had no business there at all to be moving around for barges at the time when a ship like the *Olympic* was docking.

Chief Officer Tennison was also watching from the Atlantic Transport liner *Minnehaha*, docked at the south side of Pier 54, with Chief Engineer McLean.

The hawser was not yet secured to *O.L. Hallenbeck* when *Olympic*'s starboard propeller began to move. She was following the usual practice for large ships described by William Marchant Smith, 'The port propeller would go ahead, while the stern [recte: starboard] one was going astern; that helps to threw the stern of the ship up.' Allen Luckurst thought *O.L. Hallenbeck*'s commander had 'handled his boat […] into a bad position; he was too far aft'. Her bow was close to *Olympic* and he 'thought the whole manoeuvre was wrong at the time; it was a very hazardous manoeuvre going right out into that ship in the way he did, and getting close to her wheels [propellers]'.

He was not the only one who saw impending danger. First Officer Murdoch was stationed on *Olympic*'s aft docking bridge. His duties as first officer included responsibility 'for the mooring or unmooring of the ship', making sure 'that the propeller notice [warning] boards are hung over the quarter' and seeing 'that nothing fouls the propellers when entering or leaving dock'. He saw the tug 'moving towards the [starboard] propeller' and ordered the engines stopped, an order which 'was promptly obeyed'.[13]

No matter how promptly the order was obeyed, stopping the engines took time. George Weber, an experienced New York pilot watching from *Edmund Moran*, saw 'a big whirlpool' as *Olympic*'s propellers began to move:

> I never see [*sic*] anything like it before – it made a big whirlpool and pulled the *Hallenbeck* right around in under the quarter of the ship and as she went in her flagpole broke right off and she stayed there for a while and then kind of went down in again, and then they must have stopped the steamer because I heard some hollering there and just as soon as they stopped she gave a swoop out and went out to the Anchor Line dock.

Yost Anderson saw that the stricken tug 'went around in a big circle' and 'went right in straight underneath' *Olympic*'s counter stern. She 'turned around two times after that.' To *Minnehaha*'s Chief Officer Tennison, it looked as if the tug 'was drifting down and when they she got abreast of the *Olympic*'s starboard forward [*sic*] propeller they moved their propeller and she got sucked in'. Lightoller saw the tug 'went in under the *Olympic*'s stern, quarter, that is, and I saw her flag staff carry away'.

Captain Sjalund saw the situation unfolding from *O.L. Hallenbeck*. *Olympic* 'started her propellers before the deckhand had time to get the heaving line fast [...] it seemed to me that they were all working'. *O.L. Hallenbeck*'s stern was being pulled towards the enormous liner. Sjalund ordered 'ahead full speed', but the tug simply did not have enough power to escape. *Olympic*'s starboard propeller broke the tug's propeller 'shaft, stern frame, rudder, and took [...] three plates out of her bottom'.

The tug's engineer, Edward Card, described how little time he had had:

> I just got her going full speed and I got another jingle and I put the by-pass over on her – that is, extra steam, and that threw her into the low pressure cylinder to give her more pressure [...] Then there was a grinding sound and the engine started to race, and I shut the engine off.

He realised 'it was time for me to get busy; the water started to come in the engine room through the plug holes'. The damage to *Olympic*'s starboard propeller was relatively slight, 'bending and injuring all three blades'. She continued docking, but *O.L. Hallenbeck*'s after hold was full of water. The tug 'drifted away and drifted ahead alongside of the French Line dock'.[14] *Olympic* was docked safely, although she had scraped a long length of her paintwork on the corner of the dock.

The difficulties in docking large liners at New York were addressed by Chester Carton in an article for *Technical World* magazine in 1912.[15] He felt that a 'lack of intelligence' had been 'displayed in arranging the docks at right angles to the current'. In Liverpool, Hamburg and Rotterdam, 'the docks are parallel with the flow of the tide, or nearly so. It took the vaunted ingenuity of the Yankee to figure out a scheme of docks athwart the current':

> There are two other wholly unnecessary drawbacks which makes the docking of a liner in New York the work of an expert. The first of these drawbacks is that the slips are only two hundred and fifty feet wide. At best this affords but scanty room for manoeuvring a vessel seven hundred feet or more long and eighty or ninety feet wide; but they rarely have even that much space. Nearly always there is another vessel at one side of the slip which with a flotilla of lighters and barges receiving or discharging cargo take up more than half the space. Added to this the Hudson [River] is so shallow that the big Hamburg-American liners, which dock at Hoboken, actually drag in the mud occasionally.

'The whistle blew for the gangplank at exactly seven minutes before ten o'clock,' wrote the *New York Herald*:

> ... and before this the Customs Department had its forces ready for the examination of the baggage of the passengers. There was some delay, caused because the steamship could not be warped far enough ahead to allow for more than one

Her helm and rudder have straightened up and *Olympic* is well into her pier, No. 59. (Robert L. Bracklow Collection/ Digital Culture of Metropolitan New York, New York Historical Society)

When she was docked, *Olympic*'s bow came to rest right at the end of the pier. Her draught is less than 29ft at the bow, because she has consumed so much fuel and supplies during the crossing. Note that crewmen are using the bow rails to air their clothing. (Mike Beatty/ Matt DeWinkeleer collection)

> gangway, but within a few minutes after landing the one hundred and fifty customs inspectors and thirty appraisers were at their work.

It took two and a half hours for the first and second-class passengers to disembark and leave the pier. By that time, the transfer of the third-class passengers to Ellis Island 'was well under way'. Now *Olympic* was docked, reporters were keen to seize their chance to check her out.

Pilot J.J. Adler 'left the bridge with a smile on his face' and told reporters:

> She steers like a catboat. I went onboard from [...] the pilot boat, at half past two this morning off the lightship, and we were anchored in Quarantine an hour later. She handled beautifully, and I only asked Captain Smith to slacken her down to about sixteen knots.

Smith himself was pleased with her performance, 'She has averaged more than twenty-two knots during a part of the voyage, and we started from Queenstown fully three hours late.' He said:

> *Olympic* can be depended on to be a Wednesday morning ship, just as she was designed for. To be sure there was no severe weather to try her out, for the passage was an ordinary June run. But we have power enough to take care of emergencies.

'Will she ever dock on Tuesday?' a reporter asked.

'No', Smith responded ('emphatically'), 'and there will be no attempt to bring her in on Tuesday. She was built for a Wednesday ship, and her run this first voyage has demonstrated that she will fulfil the expectations of the builders.' The *Trenton Evening Times* said Smith wore 'such a seraphic smile', 'If any person on earth was prouder than he, that person's identity hasn't been made known. His was the smile that couldn't come off.'

Smith's officers 'were enthusiastic over the voyage' and 'willing to show every courtesy to visitors, but preserved a sphinx-like silence on the subject of coal consumption'. Purser Hugh McElroy suggested that the daily tour of inspection, which he made at 10.30 a.m. with Captain Smith, Dr O'Loughlin and Chief Steward Andrew Latimer, 'was fully nine miles'. Captain Smith estimated the distance at 3 miles, but Dr O'Loughlin 'insisted that it seemed like nine miles'.

J. Bruce Ismay 'said he was perfectly satisfied'. He highlighted how popular the squash court had been during the crossing, 'It was that popular that we had to limit the time of each player to a half hour instead of an hour, as we at first intended.' The *New York Tribune* quoted Ismay saying:

> I think this ship is going to be hard to beat. She worked splendidly all the way over with a minimum of vibration. She was never put to her full speed, yet she is here on schedule time. We will never try to bring the *Olympic* up on Tuesday night, for she is a Wednesday boat. There are one or two minor details on her that will be corrected on the *Titanic*, her sistership.

Olympic had averaged more than her intended service speed. He said they would open her up on the return crossing:

> Leaving New York on the return voyage, the *Olympic* will steam about twenty-one knots the first day, and then gradually be increased until the maximum speed of the engines is attained, and then we shall have a good idea of what the ship can do.

He was 'pleased' to hear passengers' comments about the 'lack of vibration'. The *New York Times* reporter asked Ismay if he believed 'the combination of two reciprocating engines with a centre low-pressure turbine is the best method of propulsion for big liners?' Ismay responded:

> 'Well, we think so, and that is why we have ordered engines on the same principle for our big Australian steamship now building. In addition to the steadiness of the ship this method of having two reciprocating engines with one turbine in the centre is the more economical, as the turbine is fed by the exhaust steam from the reciprocating engines which would be otherwise wasted.'

> What is the rough cost of a round trip for the *Olympic*?
>
> 'About $175,000 [£35,000], including the wages of the crew and the cost of food for the passengers and the insurance.'
>
> Do you insure the *Olympic* in the company?
>
> 'We [IMM] cover a risk of $500,000 [*recte*: £500,000] and the rest is taken by the underwriters.'

Asked whether the 'limit in the size of ocean liners has been reached', Ismay argued:

> I do not see why ships should be limited in size in any way except in the facilities extended to them for docking. For instance, here in New York we had to put up a long expensive fight with the War Department to get a permit for the extension of a pier to accommodate the *Olympic* and then pay for the extension. It would be very dangerous to have a vessel moored to a pier in the North River with her stern projecting seventy feet out into the channel.

He was:

> ... very pleased at the manner in which the advent of the *Olympic* into the Atlantic service has been received by the travelling public [...] we have over seven hundred first and four hundred second cabin passengers booked for the eastward voyage next Wednesday, which will be the record number ever taken out of any port in the world on one ship as far as cabin passengers are concerned.[16]

Back in the United Kingdom, the *Southampton Times* said the maiden voyage:

> ... has apparently been a great success. The information from all sources seems to point to the fact that the vessel's performance on the 'mighty deep' has quite fulfilled the expectations of her owners and builders, and she has proved herself capable of a speed which has been extremely satisfactory for a vessel whose chief claim to public favour does not rest on her ability to do an extraordinarily quick passage.

The White Star Line released details of *Olympic*'s performance to the press and there were souvenir log cards for passengers with an 'Abstract of Log', showing each day's runs and the average speed for the crossing. They said she had covered 2,894 miles between the Daunt's Rock Light Vessel and the Ambrose Channel Light Vessel in five days, sixteen hours and forty-two minutes at an average speed of 21.17 knots. Unfortunately, there was a mistake in the calculation, which added 100 minutes to the total time taken. In reality, *Olympic* had taken only five days, fifteen hours and two minutes, and her average speed was 21.43 knots. (It appears the error came to the attention of someone in the White Star Line at some point, because the correct figures were used many years later in descriptions of *Olympic*'s career.)[17]

Some passengers reflected on the absence of vibration but reported that 'the thud of the wing propellers, however, could be plainly felt in the after end of the dining saloon and in the smoking room'. The *New York Times*' reporter described 'the impression of hugeness given by the external appearance of the *Olympic*', which was 'intensified on entering the interior of the ship where the amount of space on every deck cannot fail to impress the observer':

> The alleyways are so wide that there is no danger of that peculiar jostling that passengers inflect as they pass each other in rough weather on smaller ships. It would not be very difficult for two elephants to walk abreast down the main alleyways [...] When passengers were asked for an expression of opinion about the accommodations on the liner they referred to her as 'just a big hotel in which it was a hard task to imagine one's self afloat, she was so steady'.

Similar statements were made twelve years earlier when *Oceanic* came into New York for the first time. She was scheduled to leave that afternoon, with 885 passengers in all three classes. Anyone comparing the two ships saw an impressive increase in size: *Olympic*'s gross tonnage was over two and a half times her older running mate's.

There was plenty to do for *Olympic*'s crew while she was in port. Captain Smith had to complete an 'Affidavit of the Master or Commanding Officer, or First or Second Officer' to confirm that the ship's surgeon had examined all the third-class passengers:

> … and that from the report of said surgeon and from my own investigation, I believe that no one of said aliens is an idiot, or imbecile, or a feeble-minded person, or insane person, or a pauper, or is likely to become a public charge, or is afflicted with tuberculosis or with a loathsome or dangerous contagious disease, or is a person who has been convicted of, or who admits having committed a felony or other crime or misdemeanour involving moral turpitude, or is a polygamist or one admitting belief in the practice of polygamy, or an anarchist, or under promise or agreement, express of implied, to perform labour in the United States, or a prostitute, or a woman or girl coming to the United States for the purpose of prostitution, or for any other immoral purpose, and that also, according to my knowledge and belief, the information in said lists or manifests concerning each of said aliens named therein is correct and true in every respect.

The affidavit was a standard text, with Smith attaching his signature, the ship's name and last port of departure ('Queenstown'). He dated it 21 June 1911 and then one of New York's immigration officers certified it. Dr O'Loughlin completed an accompanying 'Affidavit of Surgeon', writing that he had 'thirty years' experience as a physician and surgeon, and that I am entitled to practice as such'. He confirmed 'that I have made a personal examination of each of the aliens' and that the information provided about them was correct.

Thomas Andrews wrote in a letter to his mother, 'We arrived early this morning after a very satisfactory voyage, quite the best I have ever made and a record in speed for the White Star Line.' He reported how pleased J. Bruce Ismay and passengers from all three classes were with the ship. 'I am looking forward to the equally successful homeward voyage.' Andrews wondered if his family enjoyed their time at the Coronation, knowing that his letter would not arrive until the event had taken place. He sent birthday wishes to his sister and signed off, 'Your loving son Tommie'.

J. Bruce Ismay lost no time. He cabled Lord Pirrie on the same day: '*Olympic* is a marvel, and has given unbounded satisfaction. Once again accept my warmest and most sincere congratulations. Will cable you full particulars speed, consumption later.' He then sent a message to Messrs Ismay, Imrie & Co., White Star's management company, at Liverpool:

> Arrived at Lightship 2.24, docked at 10 a.m. Everything worked most satisfactorily, passengers delighted. Passage 5 days 16 hours, 42 minutes, average speed 21.7 [*sic*]; coal consumption given 3,540 [tons], think it liberal estimate. Delayed by fog 1½ hours. Daily runs 428, 534, 542, 525, 548. Single ended boilers not lighted. Communicate this to Pirrie, Harland, Belfast.

He wrote a more detailed letter the next day. 'Everything on board the ship worked most satisfactorily, and the passengers were loud in their praises of the accommodation and table,' he began. 'The machinery worked excellently, and there was no hitch of any kind.' *Olympic*'s coal consumption was 'extremely low, averaging, as far as I can make out, about 620 tons a day instead of 720 as anticipated, and we arrived here, with approximately 1,300 tons of coal onboard the ship'. He arranged for plenty of coal to be taken on at New York 'to reduce the quantity required to be taken at Southampton'. The lower coal consumption represented an ongoing cost saving. (Cunard's *Mauretania* left Liverpool on 3 June on her fifty-first westbound crossing. She took four days, seventeen hours and one minute to cover the 2,889 miles between Daunt's Rock and Sandy Hook, averaging 25.56 knots. Her coal consumption was

1,090 tons every twenty-four hours. The cost of those extra 4 knots was substantial.)[18]

Olympic's performance in fuel consumption also compared very favourably with her builder's original expectations. Harland & Wolff had apparently anticipated a total coal consumption of between 825 and 850 tons in a twenty-four-hour period. They had advised the Board of Trade on 5 January 1911 that 'they estimate the consumption of fuel at about 850 tons per day'.[19] Then, by the time of the maiden voyage, as Ismay wrote, it was down to 720 tons.

***

Ismay echoed Andrews' comments. The first-class reception room was 'always crowded'. He confirmed an earlier order for fifty new chairs and ten new tables for the reception room. On the other hand, the corridor aft on A-Deck between the first-class lounge and the aft grand staircase 'was practically not used at all, and I think we must again consider putting staterooms in this space; but I propose to watch this matter carefully on the voyage home to see what use is made of the accommodation for the large number of passengers onboard'. He also thought the deck space, 'with the number of passengers on board going out, was certainly excessive, and I think, in another ship we might carry out the rooms on B-deck the same as those on C-deck'.

Arrangements were being made to fit the missing cigarette holders before the next round trip. A potato peeler was needed in the crew galley, as well as a Phillips water-tube steam oven for the bakehouse – 'experience shows that the oven at present in the ship is not capable of producing sufficient bread daily'. As Andrews had noted, some of the doors at the aft end of the first-class dining saloon were not needed. Ismay wrote, 'We have arranged to close the other doors, which will enable us to put in two additional tables.' He was more concerned with passengers' comfort while sleeping:

> The only trouble of any consequence on board the ship arose from the springs of the beds being too 'springy', this, in conjunction with the spring mattresses, accentuated the pulsation in the ship to such an extent as to seriously interfere with passengers sleeping, and we cabled, asking you to communicate with Pirrie that if he sees no objection, we would like to have lath bottoms fitted before the next voyage, and hope this can be arranged.

ALL-PURPOSES TWENTY-FOUR-HOUR COAL CONSUMPTION: SELECT PERFORMANCE DATA*

	Coal	Speed
Lusitania, voyage 1, westbound (September 1907)	863 tons	23.26 knots
Mauretania, voyage 40, westbound (September 1910)	1,142 tons	26.1 knots
Olympic, voyage 1, westbound (June 1911)	629 tons	21.43 knots
Aquitania, voyage 1, westbound (May 1914)	899 tons	24.2 knots
Vaterland, voyage 1, westbound (May 1914)	1,157 tons	23.2 knots

* 'The Cunard Steam Ship Company Limited: Atlantic Outward and Homeward Passages': Merseyside Maritime Museum, B/CUN/4/4/5 ('unbroken days' data used for the Cunard ships) and Braynard, Frank, *Leviathan: The World's Greatest Ship, Vol. 1* (South Street Seaport Museum, 1972), quoting a post-war newspaper report.

> The trouble in connection with the beds was entirely due to their being too comfortable.[20]

The American public were keen to take a look at the new ship. On Friday, 23 June, 4,000 people visited her, paying 50 cents each as an admission fee which was donated to various seamen's charities: 'No one was allowed aboard until 10 o'clock, from which hour until 5 o'clock they came in an endless chain, passing along every deck rapidly, and yet a complete view of the ship required more than one hour of fast walking.'

Even more looked over her on Saturday, so that the *New York Times* reported, '12,250 persons visited the *Olympic*, the world's largest steamship, on Friday and yesterday'. Passengers sailing on the eastbound crossing and their guests were able to apply for a total of 3,000 guest passes on Monday, 26 June, 'It was said that every one of them was used.'

The same day, J. Bruce Ismay and Philip Franklin entertained Mayor Gaynor, Robert Adamson, his secretary, Dock Commissioner Calvin Tompkins, State Engineer Benzel and other guests, including 'several members of the New Jersey Riparian Commission'. All of them had the opportunity to inspect *Olympic* after they had eaten lunch on board. Officials at White Star's New York office told the press that the 'gathering was entirely unofficial', but afterwards they said the meeting 'was arranged for the purpose of permitting those present to exchange their views regarding the extensions of the piers'. Ismay and Franklin wanted the temporary extensions made permanent.

White Star also took the opportunity to treat their extensive network of agents in the United States. Violet Jessop recalled:

> Men of every size and age appeared, prepared to enjoy wholeheartedly the lavish hospitality White Star was offering them. All that evening and well into the night, they were being conducted around (between drinks) to inspect various parts of the ship. As the hours wore on, both they and their steward escorts – by now more or less on a brotherly footing – began to show signs of wear.

By 3 a.m., she noted a 'lull in the convivial din'. Guests 'were found sprawling asleep in the most out of the way places', including 'inside baths and half under beds, where they seemingly had dropped in the tracks, giving the impression it was not an unfamiliar place for them'. Some 'wandered aimlessly' as they seemed lost, 'muttering for some guiding soul to rescue them'. Hours were spent simply trying to reunite guests and their coats. The next day, there were so many cigarette and cigar ends swept up that Jessop was surprised *Olympic* 'was not burnt to the water's edge'.

Their 'unbounded satisfaction' with *Olympic* was apparent from White Star's actions. They confirmed their intention to order a third sister ship. On 28 June, Harland & Wolff ordered the shipyard and engine works to proceed with all the necessary preparations for the new ship's construction and propelling machinery.

THE EASTBOUND CROSSING

It took a remarkable effort to get *Olympic* 'spic and span' for the return crossing. She had to take on plenty of coal and all manner of supplies for thousands of people for a week. On Wednesday, 28 June, passengers began to arrive and 'the volume of noise increased to a crescendo'. Violet Jessop described the 'strange medley of human feelings':

> From the prospective passengers, devastating leave-takings and unrestrained emotions, voices trying to rise above each other, screaming commonplace injunctions. That babble even blotted out thought, as perspiring masses of smartly dressed, over-scented humanity surged together: yet here and there, a really silent and pathetic farewell was taking place.
>
> Everybody was totally oblivious of the distracted stewards vainly attempting to move enormous pieces of baggage through the crush. Those in charge shouted orders and room bells rang impatiently for drinks, while the stewardesses' additional bugbear – flowers to arrange – arrived with the regularity of snowflakes. Boxes of every size were piled mountain high.

> The fact that they were in boxes saved them from utter destruction.
>
> It is puzzling to understand how otherwise considerate and thinking people fail to imagine what can possibly be done with the vast and expensive quantities of flowers they insist on sending departing friends. At times there might be as many as seventeen boxes for each person. In the confined spaces of a ship with, always, an inadequate supply of receptacles to arrange them in, it is and always has been a sailing day horror for stewardesses.

One dreaded fear was that a call bell would ring on a sailing day from a passenger requesting 'eight or ten vases' for flowers in their stateroom. That was the usual allotment for a stewardess' entire section. It took experience to realise the preferences of different nationalities: Americans considered it 'almost sacrilege' for the stems of 'American Beauty' roses to be cut down.

An issue for the return sailing was that a number of crew either deserted at New York or left the ship for other reasons (one of the firemen went into hospital, for example). The Swedish national L. Wilson, who had joined at Queenstown as an able seaman, deserted at New York.

The Engine Department had a particular issue, with a number of men deserting. There were also some deserters from the Victualling Department, which was a problem because the ship was so heavily booked. A large number of substitutes were taken on, which increased the number of crew in the Victualling Department by about twenty-five. Four substitutes were taken on for the à la carte restaurant: a storekeeper, a pantryman and two waiters. None of the existing restaurant staff had left the ship, but experience had shown that they needed to increase the number of restaurant staff from forty-five to forty-nine.

It appears that, as a result of crew leaving the ship and substitutes being engaged at New York, *Olympic* now had about sixty-four in the Sailing Department, 326 in the Engine Department and 487 in the Victualling Department, making a total of 877 crew.

The *New York Times* estimated that 'probably 10,000 persons' saw *Olympic*'s departure:

> Every pier head in the vicinity was crowded. The thousands cheered and waved handkerchiefs from the minute the biggest ship in the world began to move into the North River until she finally straightened out and headed down the bay. Many remained until she was lost to view.

There was an ebb tide when she was due to move out:

> *Olympic* was breasted off from her pier alongside the pier to the north to prevent her striking the new temporary pier extension work, as the tide carried her downstream. It took a dozen tugs to straighten her out, and, this done, she started off with her whistle answering the salutes which greeted her from shore and stream.

The *Southampton Times*:

> … computed that the vessel has forty or fifty millionaires on board, possessing an aggregate wealth of £60,000,000! A correspondent of the *Chronicle* says that the *Olympic* on this voyage is probably the 'richest' ship that ever sailed from New York, as apart from the millionaires, there are numerous wealthy families among the other saloon passengers.

Echoing the American newspapers, they said the send-off from New York was 'nearly as great […] as she swung out from her pier and sailed down the harbour on Wednesday as on her [maiden] arrival'.

She carried 736 first-, 491 second- and 1,052 third-class passengers for a grand total of 2,279. The eastbound season was picking up and *Olympic* was virtually full. (The previous day, Norddeutscher Lloyd's *Kronprinzessin Cecilie* had taken out 1,212 passengers, including 529 in first class – her best in first class that year. *Rotterdam* also sailed with 1,230 passengers. On 21 June, Cunard's *Lusitania* had departed with 1,516 passengers, whereas the older *Campania* sailed the same day as *Olympic* with 817.)

While she was proceeding out of New York, an aeroplane piloted by the British aviator Thomas Sopwith caught up with *Olympic*. The *London Evening Standard* carried a report that Sopwith 'dropped on board a pair of spectacles, carefully wrapped and addressed to Mr. W.A. Burpee, a prominent Philadelphia seed merchant'. It said that Burpee had sent them to Wanamaker's for repair and delivery to London, but they attempted delivery by aeroplane. Unfortunately, other reports stated that the package missed and went into the water!

Olympic passed the Ambrose Channel Light Vessel at 5.06 p.m. No attempt was made to test her speed at first: she logged runs of 389 and then 495 miles, averaging 21.2 and 21.3 knots. The fresh winds and moderate sea reported on her first day gave way to a 'rough sea' with south-easterly winds and clouds. Then, as Ismay planned, she was opened up gradually as all her boilers were brought into operation. Sea conditions moderated, too. By noon on Saturday, 1 July, she reported a daily run of 525 miles at an average speed of 22.6 knots. (On the eastbound crossing, the days were shorter compared to the westbound one, as the clocks were adjusted each day according to the ship's progress. That particular 'day' was twenty-three hours, sixteen minutes long as the clocks were altered by forty-four minutes.)

She continued to accelerate. On Sunday, 2 July, she posted a run of 526 miles. By noon the next day, she could report covering a further 532 miles at an average speed of 23 knots. She enjoyed a 'light westerly wind' and 'smooth sea'. Her final complete day at sea saw her slow slightly. Weather conditions remained clear and the sea smooth. This time, her performance was aided by the current, but it was an impressive result.

Her first-class passengers included the Booth family. George Gough Booth had purchased their tickets with the Christian Leidich Travel Bureau in Detroit, Michigan. Their car came along as well as their chauffeur, Parks. Fourteen-year-old Henry Scripps Booth kept a record of the crossing in his 'My Trip Abroad' journal, 'The *Olympic* is a fine ship. Florence was sick [the] first night and day. It was somewhat rough on June 29th [...] In the morning, I could hardly stand, because of the new sensation. I was somewhat sick, but ate lots and I felt fine.' He spotted Cunard's *Campania*, which came in view on 30 June (she had left New York over five hours ahead of *Olympic*, passing the Ambrose Channel Light Vessel at 11.47 a.m. on 28 June – although she had set speed records in her day, she was running at less than 20 knots and *Olympic* easily outpaced her).

George noted 'the ocean was as smooth as glass', with bright and sunny weather on 1 July, 'In the morning and afternoon [...] the second class people had races, and did all sorts of stunts'. There was an evening dance 'on the right hand deck.' The usual Sunday morning church service was held in the first-class dining saloon on the 2nd. He was in third place in the potato race on Monday morning. Then there was the Spar Pillow Fight, 'In Warren's case both fell off the bar at first but Warren was nocked [*sic*] off a second time. Warren did not win. James stayed on a number of times but was forced to give up.' Prizes were awarded in the reception room after dinner that evening.[21]

The eastbound crossing provided an opportunity to see how everything worked with a much larger number of passengers on board. The first-class passenger list far exceeded the number of seats in the first-class dining saloon, so White Star offered passengers a $25 (£5) discount provided they ate exclusively in the à la carte restaurant. White Star were being rather clever. The 'minimum rate per head' for three passengers having dinner in the restaurant worked out at over £0.60 each and so a passenger could spend the best part of £4 ($20) simply on their evening meals. Breakfast and lunch would easily take them over the value of the rebate they received. It also seems unlikely that they would be paying merely the 'minimum rate' all the time if they specifically chose to use the restaurant only. The average expenditure per passenger was a little under £5 ($24) and would have included passengers who used the dining saloon as well. By January 1912, the discount was given as £3 per adult for passengers whose ticket price was under £35, or £5 for passengers who paid over £35: 'This reduction in fare [...] can only be granted when passengers

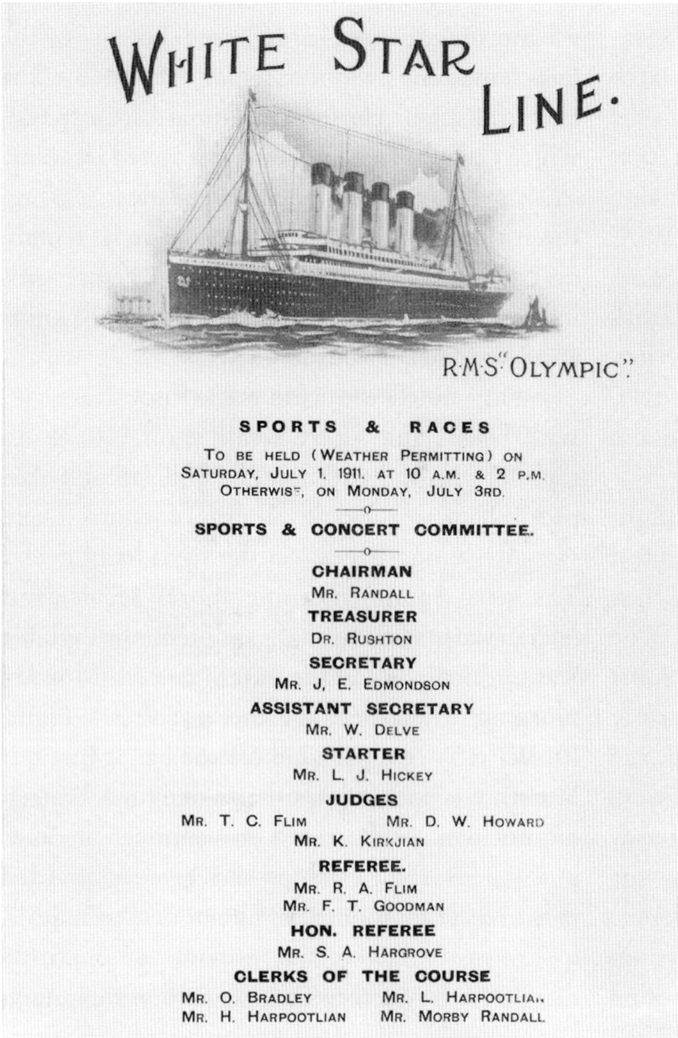

WHITE STAR LINE.

R·M·S "OLYMPIC."

SPORTS & RACES

TO BE HELD (WEATHER PERMITTING) ON
SATURDAY, JULY 1, 1911, AT 10 A.M. & 2 P.M.
OTHERWISE, ON MONDAY, JULY 3RD.

SPORTS & CONCERT COMMITTEE.

CHAIRMAN
MR. RANDALL

TREASURER
DR. RUSHTON

SECRETARY
MR. J. E. EDMONDSON

ASSISTANT SECRETARY
MR. W. DELVE

STARTER
MR. L. J. HICKEY

JUDGES
MR. T. C. FLIM MR. D. W. HOWARD
MR. K. KIRKJIAN

REFEREE.
MR. R. A. FLIM
MR. F. T. GOODMAN

HON. REFEREE
MR. S. A. HARGROVE

CLERKS OF THE COURSE
MR. O. BRADLEY MR. L. HARPOOTLIAN
MR. H. HARPOOTLIAN MR. MORBY RANDALL

The White Star Line advised passengers of the sports and races intended to take place on the eastbound crossing. The events included separate potato races for girls and boys; separate 50-yard dashes for girls and boys under 8 years old; a 100-yard dash for gentlemen; a wheelbarrow race for gentlemen; a skipping contest for ladies; a sack race for men; and a 'cigarette and tie race' for ladies and gentlemen. (Roy Derouin collection)

Passengers 'race' on the starboard Boat Deck amidships, heading aft. (Courtesy of Cranbrook Archives, Cranbrook Centre for Collections & Research, *Pleasures of Life, Vol.1*, p.11, Henry Scripps Booth and Carolyn Farr Booth Papers)

Looking aft from the forecastle deck on the starboard side, towards the bridge. The wide-open windows at the fore end of A-Deck are those which serviced the first-class corridors. It seems likely the stewards had opened them to help provide a breeze and cool the first-class accommodation in hot weather conditions. (Courtesy of Cranbrook Archives, Cranbrook Centre for Collections & Research, *Pleasures of Life, Vol.1*, p.11, Henry Scripps Booth and Carolyn Farr Booth Papers)

announce their intention to book without meals, and of making use of the restaurant, at the time of purchasing their ticket, and no rebate or reduction can be made under any other circumstances.') Over the westbound and eastbound crossings taken together, first-class passengers spent £1,730 in the restaurant.[22]

Olympic's swimming pool proved popular. On the eastbound crossing, over 200 men used it when they had to pay for a ticket and almost 300 when it was free (it was free from 6 a.m. to 9 a.m. or it was also included in the price of a ticket for the Turkish baths, so there may be more swimmers who used it as part of their Turkish bath ticket and who were not included in the figures). The swimming pool and Turkish baths were both more popular with male passengers. Almost four times as many men used the Turkish baths over the round trip and more than six times as many used the swimming pool, compared to the number of ladies who used them. The electric baths were not popular. Only twelve passengers used them on the westbound crossing and six passengers on the eastbound, but it worked out to an even gender split.

All of these extra facilities helped generate revenues. The revenue from the restaurant was equivalent to about 275 third-class passengers' average ticket fares. Then there was a trend to premiumisation. Passengers' tickets and accommodation was simply more expensive on *Olympic*. Revenues from the four 'parlour' suites were also equivalent to hundreds of third-class passengers' ticket fares. (Two years after *Olympic* entered service, the *New York Times* published an article: 'High-Priced Suites Pay on Steamships: The Past Season Saw an Exceptional Demand for Fine Accommodations.')[23]

One of *Olympic*'s first-class staterooms on C-Deck, with its own private bathroom and toilet, was more expensive for two passengers than the finest suite room on *Oceanic*. An ordinary first-class stateroom without private facilities cost almost 50 per cent more on *Olympic* compared to the closest equivalent accommodation on *Oceanic*. Second- and third-class tickets did not vary as much in price compared to the difference between the most expensive and the cheapest first-class accommodation. Nonetheless, their tickets were slightly more expensive on *Olympic* as well.[24]

Revenue was not the whole story. The bottom line was the operating profit earned, deducting the expenses from the revenue. It cost more to look after a first-class passenger compared to a third-class one, but White Star had targeted an increased number of first- and second-class passengers by building *Olympic*. They felt it was an attractive commercial strategy.

There was certainly comment in the press about her earning power. The *New York Times* reported estimates that the total revenues for the round trip might be 'from $325,000 [£65,000] to $350,000 [£70,000]', with expenses that 'would not amount to more than $175,000 [£35,000], including wages, coal, and food for crew and passengers'. The lower estimate for revenue would generate a profit of £30,000, whereas the higher estimate would represent a profit of £35,000. Whether the estimates were precisely correct or not, they were broadly credible for a well-booked round trip (Cunard's *Mauretania* generated an average operating profit of £19,435 on each round trip in 1911).[25]

Over on the west coast of the United States, the *Medford Mail Tribune* in Oregon claimed that food and drink worth $60,000 (£12,000) had been loaded for the eastbound crossing, including $5,100 (about £1,000) of poultry, $2,000 (about £400) of fish and $5,000 (about £1,000) of wine and spirits, 'and $2,000 [about £400] worth of cigars will be smoked'. Certainly, those estimates were very high compared to the known expenses of other ships.

***

During the crossing, J. Bruce Ismay received an interest payment of over £1,640 (before income tax was deducted) from his personal investment in the bonds that the White Star Line had issued to help finance *Olympic*. It was the interest due on the first half of the year up to 30 June. The company certainly had no problem paying its debts! He had grounds to be very pleased with her eastbound performance. *Olympic* arrived at the Eddystone Lighthouse, off Plymouth, at 4.36 p.m. on 4 July. She had completed 3,081 miles in five days, eighteen hours and thirty minutes at an average speed of 22.25 knots. ('The mail tender got the mail and other boats got passengers off to land. We then started across the channel,' noted Henry Scripps Booth in his journal.)

Clive Livingstone Du Val recalled a reunion of former Yale University students, held that evening. Not long after *Olympic* left New York, she was 'so steady and the weather so fair that it wasn't long before one or two of the older graduates onboard conceived the original idea of having an old fashioned reunion'. Their idea 'appealed to every Yale man on board' and they found out that there were thirty-seven graduates and undergraduates on board the ship. Two of them spoke to the chief steward and he allowed them to have the starboard veranda café 'for the evening before our arrival in Southampton'.

From 9 p.m., the tables were occupied by Yale men from the classes of 1872 to 1911. 'Many of the passengers gathered at the entrance to hear the songs and cheers and informal remarks. A number of our Harvard friends seemed especially envious because the same idea had not occurred to them.' By 11.30 p.m., it was time for 'lights out':

> The discipline of the ship could not be relaxed even for a Yale smoker; but before we adjourned it was suggested and unanimously carried that a wireless be sent to President Taft with the good wishes of the first Yale Ocean Smoker. Thirty-seven very happy Yale men then stood and sang 'Bright College Years' as the lights flickered and died out.[26]

Olympic made her next stop at Cherbourg as planned, returning to Southampton on 5 July. Her maiden voyage was a triumph!

10

THE HEIGHT OF THE SEASON

'The voyage was a complete success,' reported the *Illustrated London News* early in July 1911. *Olympic* was 'remarkably steady and seaworthy'. It commented on the impressive illuminated displays on Oceanic House, White Star's London offices, pointing out that they included a prominent model of the new ship. They anticipated a successful season as *Olympic* was prepared for her second round trip to New York.

Thirty-eight year old Henry Wilde took over from Joseph Evans, signing on as *Olympic*'s new chief officer on Thursday, 6 July. He transferred from *Cedric*. The same day, Superintendent Engineer Blake wrote from Southampton to the White Star Line's managers at Liverpool. He endorsed virtually all the recommendations Chief Engineer Bell had made during the maiden voyage.

One issue he highlighted was that some minor cracks had become apparent in the reciprocating-engine bedplates. They were 'nothing of a serious nature, and the American Line had the same experience with the *Philadelphia*'. The American Line had resolved the issue with strengthening plates, 'similar patches to those we are putting on [*Olympic*] now were fitted, and no trouble has since been experienced, nor have the cracks extended'. (Cuthbert Coulson Pounder, who worked for Harland & Wolff, recalled that 'the engines ran year after year' and 'often developed, for long periods, powers 20 per cent in excess of the design figures'. It was a remarkable performance.)[1]

Albert Ballin was in Southampton with a number of HAPAG officials. He was keen to see *Olympic* for himself. Philip Curry greeted them 'and under his guidance the magnificent vessel was thoroughly inspected' on the morning of the 7th. Ballin was 'very pleased with his visit, and he was lavish in his praise of the new steamer', according to the *Southampton Times*. There was plenty of interest in her second crossing: 'Between three to four hundred first class passengers have already booked their berths, and every post brings numerous additional applications.' White Star's managers regretted 'that owing to the amount of work that has to be done on board, they cannot allow visitors to inspect the vessel during her stay in port'.

All was ready. *Olympic* left on Wednesday, 12 July, calling at Cherbourg and then Queenstown the next day. She left the Daunt Rock Light Vessel at 2.02 p.m., contrasting with the delay on her maiden voyage. Her passengers included Lord and Lady Pirrie.

Edward Wilding was also on board to see her in action. He took measurements of vibration, noting that the optimum number of engine revolutions was about 76–77rpm, rather than 80–81rpm. As a result, the ship's engineers held the revolutions to 78rpm or less.

It did not stop her racing to New York. *Olympic* recorded a 560-mile run on her second day out, averaging 22.6 knots. Then she slowed gradually towards the end of the crossing, marked by her arrival at the Ambrose Channel Light Vessel at 10.08 p.m. on Tuesday, 18 July. She completed 2,891 miles in five days, thirteen hours and six minutes at an average speed of 21.72 knots, cutting about two hours off her maiden voyage time.

Passengers remained on board overnight, assured of a timely disembarkation the following day. 'Everything worked very nicely this morning,' Philip Franklin wrote to J. Bruce Ismay on 19 July. *Olympic* left Quarantine at 7.14 a.m., reached the dock at 8.16 a.m. and then the gangways were aboard by 8.50 a.m. All of the first- and second-class baggage had been landed by 10.20 a.m. and 'when I left the dock at 10.45 a.m. there were very few cabin passengers left there. The third class were all on the dock, ready to go aboard the steamer for Ellis Island.'[2]

Olympic's performance raised an issue for him. 'Although we were not so anxious to see the steamer arrive here this Tuesday afternoon,' Franklin wrote, 'she could have done it easily, and we think it would have been a good thing to have allowed her to have done it':

> I do hope, and strongly recommend, that next voyage you allow her to dock here on Tuesday evening, as it will materially assist in advertising the steamer, and help us very much in turning her round, and give everybody on shipboard a better chance, and I am sure it would please the passengers.

The next day, Franklin confirmed the coal consumption had averaged 650 tons daily. The weather was 'favourable' and 'passengers all satisfied':

> We have heard of nothing but very satisfactory reports from passengers and everybody else, so that apparently everything worked splendidly during this westbound voyage, and although the weather was very warm there were no complaints.
>
> [...]
>
> It is very apparent that the steamer could have docked here easily Tuesday evening, but you know our views regarding this, and we can only repeat that in the future we do hope you will allow the steamer to be docked here, weather permitting, Tuesday evening, as in addition to the excellent advertisement, the economy in working her, particularly on the short schedule, we believe the passengers will be better satisfied than to feel that they are on a steamer which is being deliberately held back.

J. Bruce Ismay confirmed to Franklin that his 'strong recommendation' would 'receive consideration' in a letter dated 27 July. He asked for clarification whether Franklin was recommending it for *every* westbound crossing. Ismay recognised the advantages, but had concerns:

> I at once admit that docking on Tuesday evening would help you in turning the ship round, and give those on board a better chance of getting the ship in good shape for the Saturday sailing, and further, that if we could make it a practice to do this, it would please the passengers, but as I have repeatedly stated, I feel very strongly that passengers would be far more satisfied to know, when they left here, that they would not land until Wednesday morning, rather than be in a state of uncertainty with regard to this for the whole of the trip. I do not think you can ever have experienced the miseries of a night landing in New York; had you done so, I think your views might be altered.

Four days later, Ismay wrote back to Franklin confirming he had received Franklin's second letter:

> If you keep writing me much more in regard to the *Olympic* docking in New York on Tuesday evening, I shall have to reply to you in the same manner as you did to Mr. Curry when he kept finding fault with the stamp of firemen you are supplying to the American Line steamers at New York.

Franklin had, apparently, been rather frustrated. Ismay concluded:

> As you are aware, I am not favourably disposed to trying to land on Tuesday afternoon, but if, after talking the matter over with Lord Pirrie,

> Captain Smith and Mr. Bell, the consensus of opinion is in favour of this being done, you may rest assured I will not allow my individual feeling to stand in the way.

On 4 August, Frederick Toppin responded on Philip Franklin's behalf while Franklin was travelling to San Francisco, 'We shall be very interested to hear the result of your deliberations with Lord Pirrie today on this subject.'

While discussions were going on ashore, *Olympic* was readied for her eastbound departure. She had another lengthy passenger list, carrying 1,846 passengers when she passed the Ambrose Channel Light Vessel just before 7 p.m. on 26 July. The weather was good with light to moderate winds.

Three days in, sailing through a 'calm to moderate' north-westerly wind and 'smooth sea', she completed a 540-mile run at an average speed of 23.24 knots. The North Atlantic rarely provided perfect sailing conditions for any length of time, nonetheless, a fresh to moderate gale and high seas encountered later during the crossing was not enough to stop her completing the passage in five days, seventeen hours and forty-six minutes at an average of 22.37 knots.

She arrived off Plymouth early on the evening of 1 August, dropping passengers and mails before continuing to Cherbourg and Southampton. 'People at Cowes, who rightly thought many of the big yachts there were floating palaces both as to magnificence of luxury and as to size, say they never realised how small they could look until the great *Olympic* passed that way the other day,' wrote the *Belfast Evening Telegraph*. 'She absolutely dwarfed the most imposing of the craft in the Solent; and for that matter she would dwarf almost any vessel with her huge size.' It said, 'Much interest was taken in New York in Lord and Lady Pirrie's enterprise in going out for a week and returning (so it was said) by the *Olympic*.'

It was Cowes Week, which made it particularly appropriate for *Olympic* to be in Southampton. On Sunday, 6 August, the King and Queen of Spain attended St Michael's Church, East Cowes. They were among a party of distinguished visitors who had lunch 'with the Duke and Duchess of Connaught on the Royal yacht *Alexandria*'. The *Cork Examiner* reported, 'The King and Queen of Spain and Princess Victoria afterwards went to Southampton in Princess Henry of Battenberg's yacht *Sheila* and inspected the *Olympic*, the largest liner in the world.' King Alfonso was 'deeply impressed'. They then returned to East Cowes and joined 'the Royal tea party there'. According to a report in the *Belfast Morning News* a week later, the Queen 'accepted from the White Star Line a porcelain cup and saucer, in a special case, from the restaurant of the *Olympic*, as a souvenir of the visit paid by Her Majesty and King Alfonso to the great liner'.

★★★

Following J. Bruce Ismay's discussion with Lord Pirrie, Captain Smith was instructed that 'it will be right for you to go full speed when on the short track, subject to you considering it prudent and in the interests of safe navigation to do so. This instruction applies to both the eastbound and westbound voyages when on the short track.'

Olympic left Southampton on 9 August for her third round trip to New York. Two days later, White Star's management sent a letter to their New York office confirming the authorisation. They had prepared a memo including the eastbound arrival times assuming *Olympic* averaged 22.5 knots on the short track, 'You will notice the excellent arrival times at Plymouth and Cherbourg.'

Captain Smith followed his instructions. *Olympic* left Southampton at noon, although she did not leave Cherbourg until 9.40 p.m. Perhaps, as the ship's crew got used to working the ship, the work of transferring passengers and mails at Queenstown speeded up, because *Olympic* left at 1.30 p.m. on the 10th and passed Daunt's Rock seventeen minutes later.

She averaged 21.7 knots on her first day out, accelerating gently until she covered 553 miles on her final day at sea. The 'Shipping and Mails' section of the *New York Times* listed her 'due

Wednesday' on 13 August. Two days later, *Olympic* was 'incoming from Southampton' and 'reported 755 miles east of Sunday Hook at 9.25 a.m. yesterday; due at her pier, if not delayed, tonight [Tuesday] or Wednesday morning'.

After her final day's run was posted, she had only 198 miles left to cover, arriving at the Ambrose Channel Light Vessel at 9.10 p.m. on 15 August. It all added up to a crossing time of five days, twelve hours and twenty-three minutes, averaging 21.83 knots. She did not dock on Tuesday evening. Her passengers had to wait to disembark until Wednesday morning.

She brought in 553 first, 485 second and 501 third-class passengers. Among the first-class passengers was J.P. Morgan himself, travelling in a suite on C-Deck. American newspaper reporters came aboard at Quarantine on the Tuesday evening and found him playing the card game Solitaire. Morgan's yacht *Corsair* came alongside but he remained on board *Olympic* overnight.

The *Sheffield Daily Telegraph* quoted a report from New York that two stowaways had escaped from a third-class cabin using a rope to climb out of the porthole, 'The stowaways apparently swam to Staten Island in the dark, a quarter of a mile away, where all trace of them was lost.' If true, it would have been a remarkable physical feat.

The decision to increase *Olympic*'s speed 'will insure her arriving here almost regularly on Tuesday afternoon, which will add materially to her attractiveness and popularity on this side,' wrote Frederick Toppin to J. Bruce Ismay on 18 August:

> It will also enable us to more satisfactorily handle her coaling her, as our experience is this voyage that if we can only get started coaling on Wednesday afternoon, we cannot get her finished in time on Friday to allow the ship's people reasonable opportunity for cleaning her down that day, and having her in thoroughly satisfactory shape for sailing about noon on Saturday.

The next day, Franklin 'was sorry the *Olympic* did not dock on Tuesday, but am pleased with her eastbound bookings'.

Olympic's next sailing reflected changes White Star implemented in August. Her first two eastbound departures from New York were both on Wednesday, but the company altered the schedule to Saturday. When *Adriatic* left New York on the 10th, it marked her switch to the secondary Liverpool to New York service. She would no longer sail from Southampton. The sailing day for the Liverpool ships changed from Saturday to Thursday 'to avoid competition with the Southampton service'. *Majestic* left New York on Wednesday, 2 August then *Oceanic* inaugurated the change when she departed on Saturday, ten days later. The *New York Times* noted:

> The Southampton service will be operated by the liners *Olympic*, *Oceanic* and *Majestic* until the *Titanic* comes out next March. They will depart from Southampton on Wednesday and have three days on this side to turn around instead of seven [...] By sailing from this port on Saturday the *Olympic* will avoid competition with the Cunarders *Mauretania* and *Lusitania*, and will run against the Hamburg-American steamships *Kaiserin Auguste Victoria* and *Amerika*, which she surpasses in size and speed.

In fact, *Olympic*'s first two eastbound departures from New York had clashed with Cunard's older *Campania* rather than *Lusitania* or *Mauretania*. None of the HAPAG ships the reporter named were due to sail on the same day as *Olympic*'s third eastbound departure.[3]

One of the passengers booked for *Olympic*'s Saturday, 19 August, departure from New York was Andre Jaeger-Schmidt. He had arrived in the city the previous evening. Some of his friends had arranged a supper in his honour, but the guests 'had to wait a while' because plenty of newspaper reporters wanted a story. The *New York Times* reported that:

> ... the young French journalist, who is attempting to get around the world in forty days ... will sail today at 1 o'clock on the *Olympic*, which will land him in Cherbourg almost a full day ahead of his schedule, allowing him twenty-four hours

> to reach Paris, a trip which can be made in about seven hours.

Jaeger-Schmidt himself said:

> I've had no experience to speak of, and no accidents ... Everything has gone smoothly enough with the exception of a storm in the Japan Sea, which delayed the boat fourteen hours, and once I missed a train which I should have connected with. That was arranged for, however, as I had a letter from the English [*sic*] Ambassador to Paris, which helped me everywhere. The English Consul arranged so that another train should take me. ... I carried as little baggage as possible with me, just one suit of clothes which I wear and the necessaries of the toilet in my bag. As for linen, I have bought that whenever I needed a change, and thrown away what I was wearing.

The customs officials at Vancouver had been helpful:

> I was then asked if I had any baggage, and I said only a small bag, and that I was on my way on a fast trip around the world, and then they let me through immediately.

He had originally intended to sail with the Allan Line from Quebec, but reached Vancouver ahead of schedule. His plans changed and he took *Olympic* 'as the next best boat [...] as I had plenty of time to do this, and a desire to go to Montreal, I thought I might as well go that way'. The *New York Times* continued:

> He should arrive in Cherbourg on August 25. The *Olympic* is now running on a course 125 miles shorter than the usual one. This ship has been habitually sailing from New York on Wednesday and landing passengers on Tuesday at Cherbourg. The present schedule of sailing on Saturday will make her reach Cherbourg on Friday, probably about 7 o'clock in the evening, unless the strike in England makes a delay in landing mail and baggage at Plymouth [...]
>
> He is due in his office Saturday evening, the 26th, before midnight, and there is no reason to believe that he cannot make it. If the *Olympic* lands in Cherbourg at 7 o'clock on Friday, the train for Paris will probably start two or three hours later, and he will arrive there early Saturday morning. He said last night that he would urge the Captain of the *Olympic* to make record time.[4]

PESKETT'S PLANS

Another passenger was Cunard's Leonard Peskett. On 14 June, the same day *Olympic* had left Southampton on her maiden voyage, he had written a memo about his recent visit to Clydebank. Progress on building Cunard's new *Aquitania* was under way, 'About 270 feet of flat plate keels (3 thicknesses) and about 180 feet of vertical keel and connecting bars are now in position on the blocks'. Technology was moving ahead quickly. He suggested that the double bottom should be riveted 'in such a manner as to make certain of the ballast tanks suitable for carrying oil fuel'. Cunard's new ship would be a coal burner, just like *Olympic*, but they foresaw she might be converted to burn oil fuel at some point during her life. His suggestion was approved, even though the builders thought the additional cost 'is not warranted by the likelihood of liquid fuel ever being used on this vessel'.

Cunard's plans became more ambitious over time. In January 1910, they envisaged a new steamship about 800ft long overall, displacing 40,000 tons and with a 'continuous sea speed' of 23 knots. Peskett noted that their plans would 'give an overall length of just over 800 feet for advertising purposes', even though he understood that *Olympic* was going to be about 50ft longer. By October 1910, he was aware of *Olympic*'s true proportions. *Aquitania*'s overall length ended up at about 901 ft.

He took a keen interest in *Olympic*. He suggested that Cunard reconsider their plans for *Aquitania*'s anchors and cables on the grounds that *Olympic*'s arrangement might be better. Even so, 'no cables or anchors could satisfactorily be

Cunard's Naval Architect Leonard Peskett took an interest in competitors and made many notes of his voyage on *Olympic* in August 1911. These included features to use in *Aquitania* and modifications to her design. *The Shipbuilder*, 1914. (Author's collection)

made to hold very large vessels when riding at anchor in a gale of wind, and that at all times when lying at anchor in bad weather the assistance of the engines and propellers will have to be depended upon'.

In August 1910, he had looked at the double porthole arrangement he knew had been used for *Olympic*'s first-class dining saloon. He thought a similar arrangement might be suitable for *Aquitania*. On 18 July, Peskett suggested extending *Aquitania*'s Boat Deck and A-Deck to 2ft 'beyond the line of B-deck', arguing:

> This will give additional width of promenade, and will also admit of the [life]boats being lowered with a less spread of davits. Even with the [...] increase of width on each side, the width of promenade will still be less than on the *Olympic*, as the latter has no tumble home and the two top decks project two feet on each side beyond the beam of the ship, whereas in the *Aquitania* there is a tumble home.

He argued the extra 4ft of width would add 60 tons to the ship's weight and 'slightly increase the cost, but in view of the advantages', it was worth it. Cunard's management approved the proposal.[5]

Peskett had the chance to examine all the accommodation in detail during the eastbound crossing. It was a valuable opportunity to inspect everything under true service conditions rather than when he had been aboard for only a few hours at Liverpool, weeks earlier.

When he boarded *Olympic*, he thought the coaling gear appeared to be 'very economical' and made a note that Cunard should fit a similar arrangement on their new ship. He had a good view of the ship's coaling outriggers when he boarded on B-Deck. Peskett noted *Olympic* had a good passenger list with 477 first, 205 second and 914 third class, for a total of 1,596 passengers (the North Atlantic Passenger Conference figures were slightly different, with 1,588 passengers).

A 'fairly large general cargo' and 7,000 tons of coal were loaded on board when she left the pier around 1 p.m. She passed the Ambrose Channel Lightship just over two hours later. Peskett kept an eye on her progress as she accelerated out of New York, increasing her day's run to 529 miles on the third day at an average speed of 22.74 knots and then slowing gradually towards the end of the voyage. *Olympic* was not being pushed. The reciprocating engines were never run at more than 78rpm.

He wasted no time in walking around the ship. Passengers used the raised roof over the first-class lounge for deck games but the Boat Deck 'was not used much as a promenade, and deck chairs were not allowed there'. A-Deck was popular for passengers to sit on deck and take a stroll but the enclosed promenade on B-Deck 'was deserted, and never at any time during the voyage was a first class passenger to be found there'. His observation echoed Ismay's suggestion that there was too much deck space.

Of the first-class public rooms, the reception room was 'perhaps the most popular feature of the ship':

> It serves as a meeting place before, and as a lounge after, meals. The band plays here after lunch, at afternoon tea, and after dinner. After lunch and at tea the room is full, but after dinner the place is crowded. All the surplus chairs and settees have been moved out of the en-suite rooms and placed here. Chairs are brought out of the [dining] saloon and every available space – main stairway included – is occupied by passengers. During the time the band plays in this room the passengers remain here, but as soon as it ceases (9.15) they disperse to their cabins and to the promenade and public rooms on A-deck. There seems to be no doubt that if the band played in the reception room until 11 o'clock the passengers would remain there, as only those who could not secure seats ever moved away until the music had ceased.

The band had a set routine:

> 10–11 a.m. second class companionway
> 11 a.m.–12 p.m. first class boat deck companionway
> 4–5 p.m. first class reception room
> 5–6 p.m. second class companionway
> 8–9.15 p.m. first class reception room
> 9.15–10.15 p.m. second class companionway

Peskett noted that *Olympic*'s reception room was going to be 'enlarged by taking in the space on each side of the main stairway'. He thought the general seating arrangement of the first-class dining saloon was good, writing 'very little fault can be found', but criticised the lack of an opening or well in the centre of the saloon, which would have aided ventilation, 'On this particular voyage the weather admitted of all the ports being open and the inner windows kept ajar – even under these conditions the place was very warm'. (The summer of 1911 was a particularly hot one and broke records in the United Kingdom.) The lighting was 'very effective':

> The shell [portholes] are larger than was supposed [...] The inner window is fitted with an inner and outer glass. To diffuse the light the glass next [to] the shell is made of Luxfer prismatic panes, placed in various positions. The inner part of the window is made up of leaded

> glass. When these windows are closed one cannot detect the outline of the ports, the light being well diffused over the whole surface of the inner glass. Inside the window and on each side of the frame there is a strip of Linolite fitted which proves very effective at night.

The service in the saloon was good, 'nothing better could be desired'. Peskett noted that the port side of the saloon was served from the port pantry and the starboard side from the starboard pantry. At lunch and dinner, soup was served from tureens on the sideboards. The saloon stewards' uniforms had light-blue facings, distinguishing them from other first-class stewards. He thought the standard of 'napery, cutlery and general service' was 'much higher in the Cunard than in the White Star – an opinion expressed several times by passengers on the *Olympic* who have evidently travelled by both lines'. The first- and second-class galleys impressed him, but the third-class galley and bakery were 'very hot and badly ventilated'. When he was on the fore end of the A-Deck promenade, on several occasions he caught the smell of 'cooking – and sometimes of burning fat' from the crew's galley under the forecastle.

The restaurant 'appears to be a success'. He was advised that the number of passengers taking meals there were 160 on the third westbound crossing (29 per cent of the first-class passengers) and 105 on the third eastbound crossing (22 per cent). The 'great demand' for seats had led to the number of tables being increased, 'The introduction of the restaurant appears to be creating a new class of passenger, who assumes an air of superiority and holds aloof from the ordinary saloon diner.' There was always 'that class of passenger who is always dissatisfied with the catering on every ship in which he may travel,' Peskett thought. 'The question, however, is whether, after having to pay a restaurant account for one voyage, he will not be more content to take, without demur, his food in the saloon.' The rebate from a first-class ticket, given to a passenger taking all their meals in the restaurant, 'would scarcely suffice for one day's patronage'.

White Star were already planning improvements for *Titanic*: the B-Deck landing of the aft grand staircase was going to be 'converted into a reception room or lounge for those who use the restaurant. Several cabins adjacent are to be dispensed with and [...] cabins are to be fitted as compensation in continuation of the officers' house on [the] boat deck'. Peskett argued the 'next step' in catering arrangements would be to look at 'running the main saloon as a restaurant. It would greatly facilitate matters if an experiment of this kind could be made before finally settling the catering arrangements for the *Aquitania*.' He was impressed by the maids' and valets' dining saloon on C-Deck – 'a [room] similar to this might be considered'. *Olympic*'s pantries were 'connected by a stairway from E to A-deck; there is also a lift in the galley. A similar stairway should be provided in *Aquitania*.'

The ventilation arrangements 'would be approved by most ventilating experts', he wrote. 'When one takes into consideration the enormous number of electric lights, the confined heat in the machinery spaces and the want of means of circulation, it is not surprising that a high temperature prevails in most of the cabins.' He noted the first-class suite rooms on C-Deck were 'ventilated by the ports', but the inside staterooms 'by trunks and louvres'. The fans 'supplying the trunks with air are in some cases so far away that the supply in some of the cabins is very feeble'.

Peskett thought:

> As an exhibition of electric light the ship is a great success [...] It is stated that there are 10,000 lights on the ship, and there is no reason to doubt the statement. In the saloon there are 404 bulbs in the ceiling, in addition to which there is the Linolite in the windows, 180 bulbs in the reception room, 330 bulbs in the first class smoke room – all other places being on the same lavish scale.

The two emergency dynamos were:

> ... supplied by steam from a pipe which is led along the beams in the working alleyway. This steam pipe has a connection to every section of

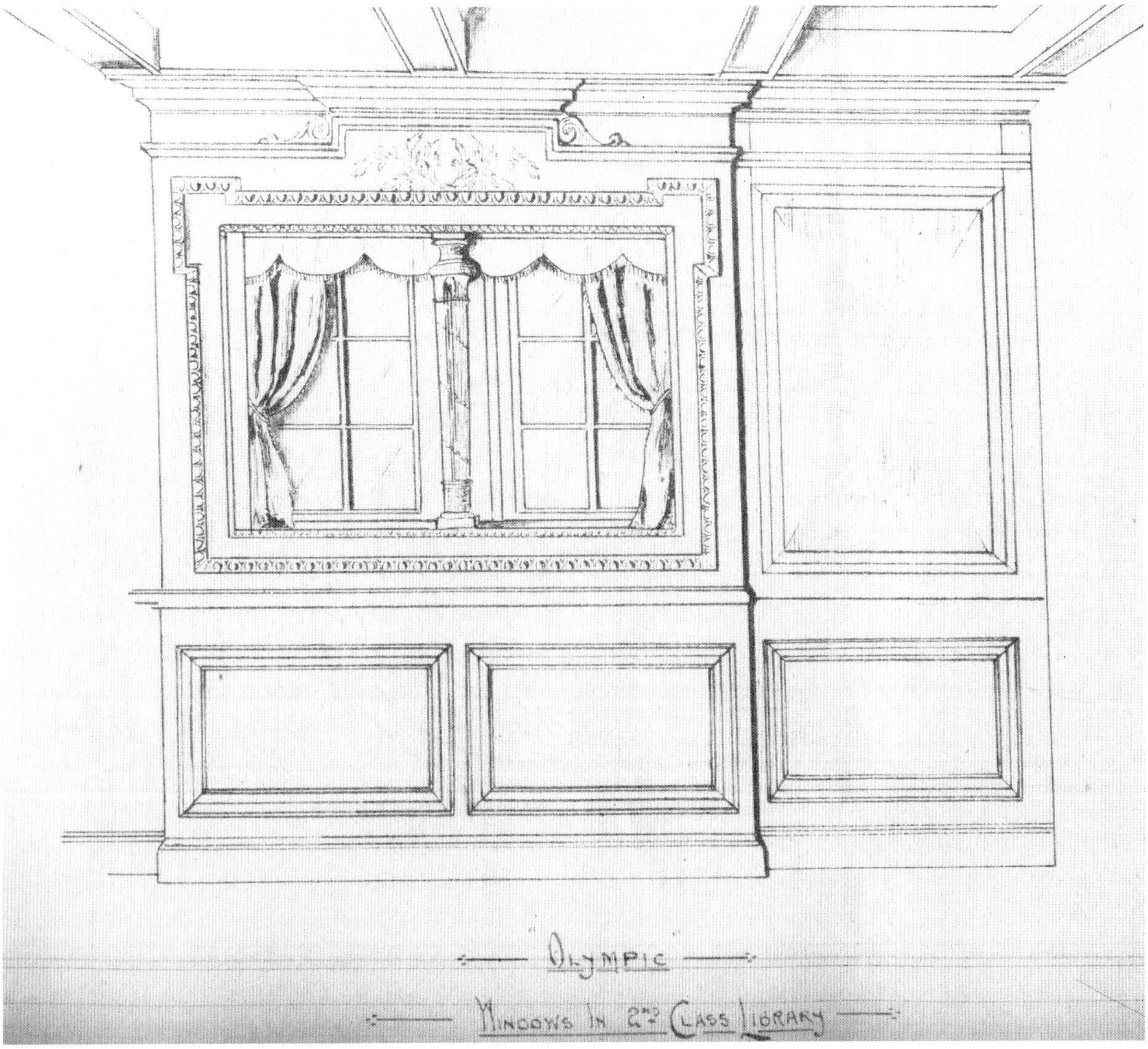

Peskett sketched several interior features, including the windows in *Olympic*'s second-class library. (University of Liverpool Library, Cunard Archives)

> boilers, so as to reduce to a minimum the chances of being without light. These dynamos supply the current for lighting the police light only.

He noted:

> The electric lights in the water closets [toilet cubicles] are put on automatically by closing the door. When the door is closing it presses on a button fitted in the architrave near the top hinge of the door, and puts on the light, the light going out immediately the door is opened.

It was 'better than a switch, as so many lights are left on when not required'.

The private bathrooms for the first-class suites were 'provided with fixed porcelain enamelled screens, and are fitted with sprays and showers', but the public baths in first class were 'fitted with hot and cold salt water valves only'. Hot and cold water was available to all first-class staterooms, except the interchangeable first- and second-class ones on E-Deck: 'All the best cabins have wash basins set in marble: the ordinary cabins have folding (clock) lavatories to which the hot and cold water service has been led, and overboard discharges fitted. The hot water service is clean and quick.'

The swimming pool was 'well patronised':

> When the ship was only slightly pitching there was considerable movement in the water, it being thrown up over the ends of the bath, consequently there is alternatively very little water at each end [...] At the shallow end the depth of water is 4 feet 5½ inches and 5 feet 1½ inches at the deep end. The water is being constantly sprayed in at the fore end of the bath and discharged under the steps at the after end. Marble steps are fitted at each end of the bath. Up to nearly the end of this voyage spring boards for diving were fitted at each end, but owing to there being three accidents to passengers through diving when the water was moving the boards were removed.

The swimming pool proved much more popular when it was available free of charge, with 223 first-class men using it compared to 109 when there was a fee. No first-class ladies paid to use it at all, although sixty-five used it when it was available free of charge. Whether the figures were for individual passengers or included some passengers who used it when it was free as well as when there was a charge, it was certainly popular.

By contrast, only eight ladies took a Turkish bath and five an electric bath during the third eastbound crossing. Fifty-seven men used the Turkish baths and six took an electric bath. Peskett theorised that one of the reasons the electric baths were 'not much used' was because 'they are too far away from the [Turkish bath] cooling room', which meant people were likely to 'get a chill in passing from one to the other'. He thought the cooling room itself was 'stuffy and needs more fans to exhaust the vitiated air'. *Olympic*'s first-class gymnasium 'was fairly well patronised and seemed at no time to be unoccupied'. The squash court was 'patronised and monopolised by about sixteen individuals who were continually playing during the hours the place was open'.

Although Mewes had praised the placement of the first-class lifts, which left the grand staircase free, Peskett felt that they were 'fitted in a bad position and too much cased in'. Passengers berthed amidships 'scarcely' saw or 'thought about' them. He thought they were 'very little used'. The second-class lift was 'in great demand for transportation from the saloon to the second [class] promenade' on the Boat Deck and 'more used than the first class' lifts.

He walked through the first-class accommodation and noticed that 'no storm rails are fitted in corridors of passenger accommodation, nor are there any handrails on the transverse framing of the first class stairways'. On deck, he liked the overhead battens fitted over the deck chairs which marked the numbers of the chairs. It saved passengers confusion and trouble in finding a particular chair when many of them were 'covered with rugs'.

He was not keen on the first-class suite rooms on B- and C-Decks, writing that they were 'overdone in detail and garish in decoration'. One criticism was a lack of storage space. While some suites had wardrobe rooms, those without had 'a lack of chests of drawers'. Oddly enough, he did not write in much detail about the passenger accommodation. First and second class were both discussed in one paragraph each, whereas third class warranted two paragraphs.

Second-class 'public rooms and corridor floors are covered with rubber cork linoleum tiles. The promenade and accommodation generally is not nearly so good as those proposed for the *Aquitania*.'

Olympic's large third-class enclosed space, forward on D-Deck was 'very popular', with 'tables and seats at the sides'. He noticed that the third-class dining saloon chairs were 'large, comfortable and spaced twenty-six inches apart centre to centre: this is rather more than is usually allowed'.

Single men travelling third class did not have the same standard of facilities. 'Wash basins are provided in the rooms intended for married people and single women, but no wash basins are provided in the rooms intended for men only. The beds are not supplied with sheets, only plain straw mattresses and blankets.'

The ship's hospital next to the second-class dining saloon pantry on D-Deck was 'tiled throughout with black and white encaustic tiles'. The wards were 'lofty and airy, plainly panelled, enamelled white'. He took an interest in the ship's dispensaries:

> Three dispensaries are fitted. The main one on the starboard side at the after end of first class accommodation on C-deck is laid with black and white encaustic tiles: lockers and drug racks are made of spelt iron painted white. There is a table with a plate-glass top, about 6 feet 6 inches by 3 feet, fitted in the centre of the room, and a plain mahogany seat, not upholstered. There is another dispensary and small waiting room fitted in the third class entrance on D-deck aft, and another under the forecastle forward for the use of the crew.

Following Ismay's recommendation after the maiden voyage, the spring bottoms beneath the Marshall mattresses had been removed and 'lath or wood bottoms fitted in their place'. The vibration was 'felt to a greater degree in the neighbourhood of the first class smoke room and restaurant, the centre of the ship, and the fore end of the deckhouses'.

Olympic was heavily loaded, which 'made it difficult to form an opinion of her capacity for rolling'. On Thursday, 24 August, as she neared the end of the voyage, Peskett noted she 'rolled from three degrees to five degrees in from eighteen to twenty seconds for each complete roll'. He thought that 'under ordinary conditions of lading', she would 'be tender towards the end of a voyage, and with a sea on the quarter would roll considerably, but easily'.

He thought the 'main engines work well and the boilers give satisfaction. The boiler power is ample, there being rather more than was required this voyage owing to the revolutions being limited to 78.' Anywhere above the starting platform in either the reciprocating or turbine-engine rooms was hot and 'in the stokeholds the space in front of the furnaces is rather confined: otherwise there is little cause for complaint'. He reported:

> The dummy funnel, from which so much in the way of ventilation was expected, has not proved a success. There are so many currents of air being driven in at the bottom of the funnel which are colder than that in the compartments below, that an air lock is formed at that particular place, and the heat on the top platforms of the reciprocating engine room and between the top of the turbine and bottom of the funnel is excessive. The theory of the cold air coming down the engine hatch and hot air going up the dummy funnel seems feasible, but in this case it does not work. This is probably due [...] to the colder air from other compartments being exhausted into the base of the funnel.

Peskett climbed 'to the top of the funnel, but could feel no great rush of air going out. The whole of the inside of the funnel was at a moderate temperature, and there was nothing to indicate the funnel was exhausting hot air from the machinery spaces.'

On the other hand, there was 'a good current of air' in the firemen's tunnel, deep in the bow. He noted that the firemen and greasers, berthed forward on D-, E-, F- and G-Decks, were 'allowed to promenade in the break between the forecastle and the bridge deck, but are not allowed on the forecastle head'. They were kept away from third-class female passengers, because only third-class men were berthed forward. The engineers had 'to climb through the engine casing to a space set apart between the first and second [class] promenade on the boat deck: evidently this was not appreciated, as only one engineer was observed using it during the voyage'. Peskett did not like the system of operating the watertight doors, which he saw in action twice, 'Once the door is released it cannot be stopped, and when it is down, it has to be raised by hand gear.'

Olympic reached the Eddystone Lighthouse at 4.17 p.m. and then dropped anchor at Plymouth at 5 p.m. on Friday, 25 August. All in all, she had completed the crossing in five days, twenty hours and twelve minutes. It was her slowest eastbound crossing to date.

The tenders *Smeaton*, *Cheshire* and *Sir Francis Drake* came to meet her. The mailbags had been placed on the fore well deck before the ship's arrival. They were dropped down two steep wooden chutes and offloaded onto *Cheshire*. Peskett thought the chutes were 'very steep, down which the bags were shot rapidly, getting rough usage in their contact with the deck on the tender'.

First and second-class passengers disembarked *Olympic* from the second-class entrance on E-Deck onto *Sir Francis Drake*, while third-class passengers disembarked from E-Deck forward for *Smeaton*. The more expensive passenger tickets had their privileges: *Sir Francis Drake* left at 6 p.m. and *Smeaton* thirty-five minutes later. Just after *Smeaton*'s departure, *Olympic* left for Cherbourg, where she arrived at 11.15 p.m. and anchored for the night.

By 7 a.m., *Nomadic* and *Traffic* were alongside. True to form, *Nomadic* took on first- and second-class passengers and left half an hour later; *Traffic* left with third-class passengers at 7.50 a.m.

Olympic left for the short run to Southampton. (While she was in the English Channel, a search of the ship was 'made for dutiable articles' but 'nothing was found concealed by any member of the ship's crew'. Captain Smith, Chief Officer Wilde, Chief Engineer Fleming and Purser Hugh McElroy made a report and signed the entry in the ship's log.) She docked at 1.20 p.m. The first London-bound train departed at 2.40 p.m. with first- and second-class passengers, getting them into London's Waterloo Station at 4.15 p.m.

Peskett and the majority of the other passengers were probably unaware of a sad event which occurred during the crossing. Sixty-nine-year-old passenger William Williams passed away from 'cirrhosis of the liver and heart disease' only a day out of New York. He was 'buried at sea, with the rites of the Church of England' the following evening. His death was apparently the first to occur on board *Olympic* and it had to be reported to the authorities at Southampton, where an official certified that 'the entries in the official log book are correct' on 31 August.

By the time the passengers disembarked on the afternoon of Saturday, 26 August, there were only three complete days left before her fourth westbound departure on the 30th. The task of readying *Olympic* for her next sailing was a huge logistical challenge at the best of times, but the height of the westbound season was approaching. The number of first-class passengers she carried increased, this time to 718. She carried 1,931 passengers in total, which was double the number *Oceanic* had taken out the previous week. The first-class passenger list on its own represented 'the largest number in that class ever taken across the Atlantic from Europe in one steamer', according to the *Henley & South Oxfordshire Standard*.

She was slightly delayed leaving Queenstown, passing the Daunt's Rock Light Vessel at 3.11 p.m. on 31 August. Sailing conditions were satisfactory, except for a strong south-westerly gale and high sea on her third day out. It calmed to a 'smooth sea' by the end of the crossing but not before she encountered some showers. Her fourth westbound crossing marked *Olympic*'s first crossing on the 'short track' or Northern Track, which shaved off about 110 miles from the westbound passage compared to the southern route. The agreement between major shipping lines on North Atlantic routes specified that the shorter route for westbound crossings would be in force from 15 August to 14 January inclusive. Eastbound crossings were altered similarly, with a shorter route in force from 24 August to 14 January inclusive.

J. Bruce Ismay wrote to Frederick Toppin, 'We have withdrawn all restrictions with regard to speed this trip, so all well, I expect she will put in an appearance in good time on Tuesday afternoon.' He wrote to Franklin with an update on 5 September:

> Your sorrow that the *Olympic* did not dock on Tuesday night last voyage will, I hope, be mitigated by her docking on Tuesday this voyage, as we have just received a cable that at 9 o'clock last night [Monday, 4 September] she was 271 miles east of Nantucket, which we calculate would make her due at Ambrose Channel at 6 o'clock tonight, and I presume she will get up to the dock at about 10 o'clock, which will make an extremely comfortable (!!!) landing for passengers, and I am sure they will much prefer this to dawdling away time and landing on Wednesday morning, to say nothing of their having had the pleasant uncertainty, from the time they left

> here, as to whether they would land on Tuesday evening or not.

Olympic reached the Ambrose Channel Light Vessel at 5.40 p.m. on the 5th. She had covered 2,781 miles in only five days, seven hours and twenty-nine minutes. Her arrival early that evening allowed plenty of time to get her docked later that night. Franklin was certainly satisfied:

> We were very glad to have this steamer docked Tuesday evening, and although there was more tide than expected, requiring more time for docking, the baggage was all gotten out promptly considering the very large quantity, and the cabin passengers were well away before 11.30 [p.m.] and everybody seemed to be well satisfied and there were no complaints.

The *Daily Telegraph*'s correspondent reported, 'Her human cargo was enthusiastic in its admiration of everything on board.' One passenger who 'attracted a good deal of attention from the Customs officials was Miss Emily Grigsby […] beneficiary under the will of the late Charles T. Yerkes'. She had 'forty pieces of luggage, besides a leather covered tin box in which was jewellery valued at £160,000', but was able to satisfy them that 'the collection had already been in this country, and, however grasping the Customs officers may be, no gems are dutiable twice'.

Olympic carried 1,215 passengers for her fourth eastbound crossing. There was a natural lull in eastbound traffic heading into autumn. Alexander Carlisle was on the first-class passenger list. Second class was particularly quiet and her passengers were mostly third class. She passed the Ambrose Channel Light Vessel at 1.45 p.m. on 9 September. Smooth and slight seas gave way to rough seas and 'fresh southerly and strong easterly winds', which slowed her down as she neared the end of the crossing. Her first eastbound crossing on the shorter Northern Track saw her cut her time to five days, fourteen hours and thirty-two minutes. She passed Eddystone Light House at 9.17 a.m. on the 15th. (The memo the White Star Line had prepared early the previous month had assumed *Olympic* leaving New York at noon on the 9th. Then she would arrive at Plymouth around 5 a.m.; Cherbourg at 12 p.m.; and Southampton at 6 p.m. on the 15th. She was running over four hours behind what they had assumed.)

★★★

Olympic had already completed four successful round trips to New York and carried 12,779 passengers. All the signs were that she was popular with the travelling public. Her propelling machinery was performing well. She completed 23,679 miles over her first four round trips at an average speed of almost 22 knots. White Star looked forward to *Titanic*'s completion, confident that she would replicate *Olympic*'s success. By spring 1912, they would have two popular new ships operating the Southampton express service with the venerable *Oceanic*.

What they did not know was that the first of many unfortunate events was just around the corner.

11

'SPINNING RIGHT ROUND': *HAWKE* STRIKES

Olympic's fifth westbound crossing was set for 20 September 1911. White Star had arranged for her to sail about an hour earlier than her normal time, taking advantage of the high tide. There were apparently some signs of unrest among the crew, with reports of union men refusing to sail because of two non-union men who had signed on. It was reportedly resolved when the non-union men left.

Company officials were 'congratulating themselves on the fact that another record had been created by' *Olympic*, 'in that she had onboard the largest number of first class passengers which had ever left an English port in one ship. Seven hundred and thirty-two was the official number, so that the vessel broke her own record of a month ago' (it was said there would be over 2,100 passengers in total).

The huge number of first-class passengers, in particular, was a challenge for Captain Smith. One of his hobbies was reading. *The People* reported that 'he always keeps a good library at sea'. He had to be 'careful, too, to keep himself as well posted on American as on English politics, for, as he says: "To converse intelligently with my passengers, I must know the political game".' More passengers represented a challenge, not just in finding varied topics of conversation but keeping track of who had sailed with him before.

Nineteen-year-old Zetie Duncan 'eagerly climbed up' the gangway with her sisters, Rose and Katherine, her half-sister, Vera Parker, her stepfather, Philip Walter, and her mother, Matilda. They were sailing as second-class passengers. She made her way on deck and watched as the remaining passengers came aboard. *Kaiser Wilhelm II* was docked nearby. Zetie remarked to another lady that she had seen pictures of the German liner and thought it might be nice to travel on her. 'Not for me,' the lady replied. 'I only travel on ships commanded by Captain Smith. I do not feel safe with anyone else.'[1]

Once all the usual preparations were completed, *Olympic* was ready. Her officers made sure that they were at their stations. Captain Smith was on the bridge with Pilot Bowyer. Chief Officer Wilde was stationed at the bow, on the forecastle head; First Officer Murdoch was stationed aft, on the poop deck; Second Officer Hume was in the crow's nest with two lookouts; Fourth Officer Alexander was standing by the telegraphs 'in the midships portion of the bridge' to signal the pilot's orders; Fifth Officer Tulloch's station after leaving dock was in the conning tower (or 'compass platform'), on the raised roof of the first-class lounge amidships; and Sixth Officer Holehouse was on the bridge, maintaining the rough copy of the deck log and 'taking the whole of the notes from the time of going on stations'. The *Southampton Times* reported that 'a deputation from the Harbour Board were present to witness the departure'. They were examining whether the authorities needed to improve the marking of the swinging ground off the docks.

Olympic left the dock at 11.10 a.m. Any observant spectators might have seen that she was drawing 33ft 6in at the bow and 34ft at the stern. The weather was fine and clear with a moderate west-south-westerly breeze. The second high tide had been at 10.44 a.m., and she was set to have an ebb tide of 'about a knot' with her as she went down Southampton Water. Bowyer would guide her down into the Solent. By 12.27 p.m., as she came down Southampton Water, her engines were ordered 'half ahead'.

★★★

Quartermaster Albert Haines was at the ship's wheel on the bridge. It fell to him to enact all the helm orders given by Pilot Bowyer.

Two minutes later, Calshot Castle was off *Olympic*'s starboard side. At 12.30 p.m., the reciprocating engines were slowed to 'slow ahead' and the turbine was cut out as she passed the Blackjack Buoy to make a sharp turn around the Calshot Spit into the Thorn Channel. It was the start of a complex series of manoeuvres where she had to do a reverse 'S'-shaped turn to take her into the East Solent:

12.34 p.m.	Calshot Spit abeam.
12.35 p.m.	Steady on South 65° West (magnetic). 'Full speed ahead' and turbine started.
12.37 p.m.	North Thorn buoy abeam.
12.40 p.m.	Thorn Knoll buoy abeam. Helm starboard-easy. Port engine 'slow ahead' and turbine stopped.

OLYMPIC ENGINE ORDERS

In 1911, the agreed procedure was that the ship's engineers would work the reciprocating engines up to a particular number of revolutions when an order was telegraphed down to the engine room.

Telegraph Order	Revolutions	Speed
'Full ahead'	>75rpm	>21.8 knots
'Full ahead' (coastal waters)	68rpm	20 knots
'Half ahead'	50rpm	15.5 knots (turbine engaged)
'Half ahead'	50rpm	13.4 knots
'Slow ahead'	30rpm	8.7 knots

The low-pressure turbine driving the centre propeller was not engaged until the reciprocating engines were running at 'half ahead'. The engineers did this routinely when 'half ahead' was ordered. The turbine's revolutions were proportional to the reciprocating engines: the ratio to a single revolution of the reciprocating engines was about 2.2 revolutions for the turbine. Senior Second Engineer John Thearle noted that the turbine would be making 175 to 180rpm if the reciprocating engines were running at 80rpm. It is known that *Olympic*'s reciprocating engines had attained 83rpm during her early months of service and Thearle stated that the turbine revolutions were about 190rpm when the ship was at her maximum speed.*

* Further detail and calculations comparing revolutions with speed can be found in Sam Halpern's online article, 'Speed and Revolutions: The Development of a Slip Table for the SS *Titanic*', revised in October 2015.

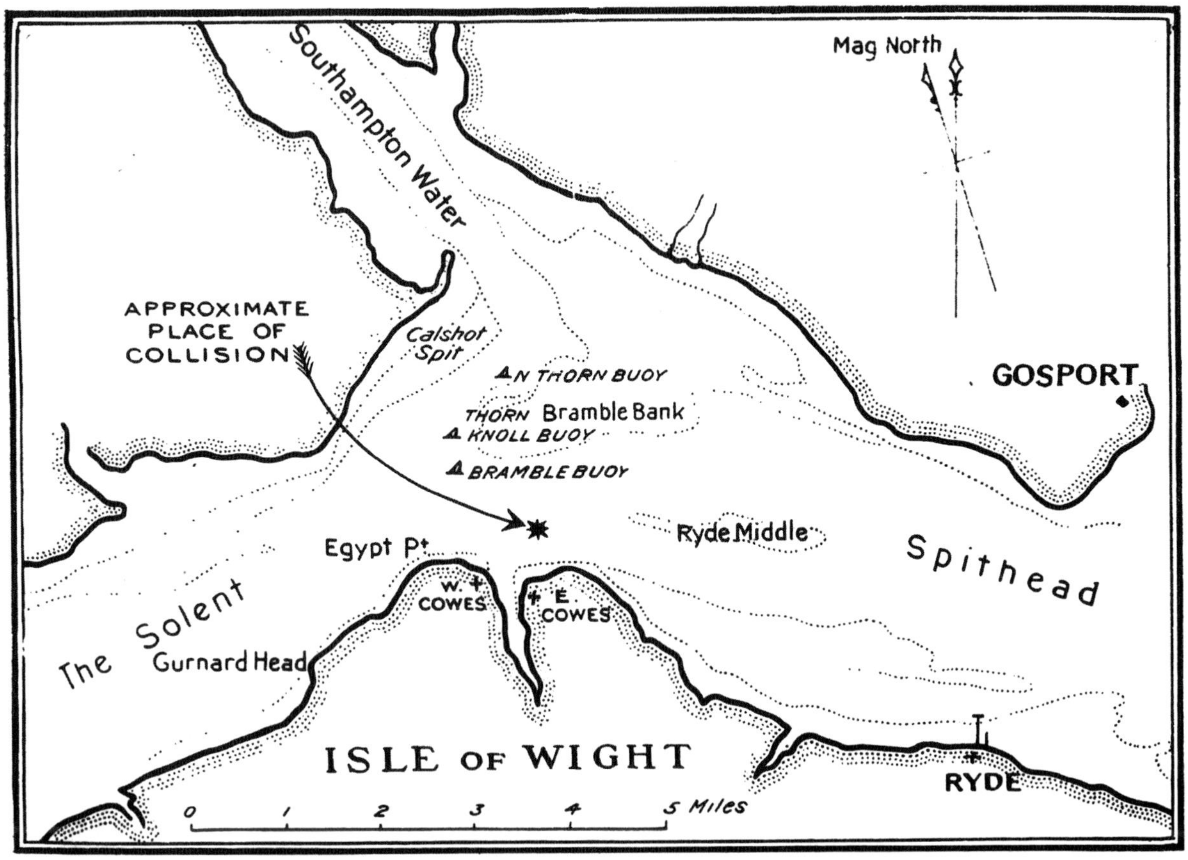

MAP OF THE SCENE OF THE COLLISION.

Olympic headed down Southampton Water, turned to the south-west to go round the Bramble Bank, then steadied on a course of South 59° East to take her South of Ryde Middle as she proceeded towards Spithead. She had Calshot Spit off her starboard beam at 12.34 p.m., passed North Thorn Buoy off her port side at 12.37 p.m., Thorn Knoll at 12.40 p.m. and the [West] Bramble Buoy at 12.42 p.m. Meanwhile, *Hawke* had been heading in a north-easterly direction up the Solent, then towards Spithead using the same channel as *Olympic*. All points are approximate. *Journal of Commerce*, 1911. (Author's collection)

12.42 p.m.	West Bramble buoy abeam. Helm hard a'starboard. Port engine 'stop', then 'half astern' and 'full astern'.
12.43 p.m.	Steady on South 59° East (magnetic). Port engine 'full ahead'.
12.44 p.m.	Turbine started.[2]

Fifth Officer Tulloch, on the compass platform above the first-class lounge, played a key role in conning the ship so that she was steadied on the correct course:

> There is a bell in the conning compass, and we have to press a button there; when we want to bring the ship's head round to port we ring two bells, and when we want to bring the ship's hard round to starboard one bell, and when the ship is on her right course we ring three bells.

Olympic steadied on the correct course, South 59° East, and he rang three bells. On the bridge, Sixth Officer Holehouse heard the three bells. He noted the course 'on the bridge compass and passed that to the quartermaster'. In the engine room, the engineers got to work. Senior Second Engineer Thearle described 'gradually opening up' the stop valve, which regulated the supply of steam, 'and giving her more steam to increase the speed'. The available steam pressure was 'within ten pounds' of the full 215lb per square inch working pressure of the boilers and the volume of steam being fed to the engines steadily increased. (When *Olympic* was in coastal waters, she worked to a 'reduced speed' under 'easy steam' of 20 knots at 'full ahead'. She only accelerated up to about 22.5 knots after leaving the Nab Light Vessel for the open sea.)

Olympic was accelerating into the channel towards Spithead, heading out to the Nab Light

Vessel, which was located off the eastern coast of the Isle of Wight. As usual, Pilot Bowyer would disembark and then she would cross the English Channel over to Cherbourg. Unfortunately, a series of events were unfolding which were not usual at all.

Captain Smith first noticed the naval cruiser *Hawke* as *Olympic* approached the Thorn Knoll Buoy. Bowyer had noticed her around the same time and given 'orders to blow two blasts [on the ship's whistle], a starboard helm signal, to let the cruiser know that I was on the starboard helm and that I was bound for the eastern passage'. Sixth Officer Holehouse carried the order out. 'After we were on our course and the engines rang full speed ahead,' recalled Fourth Officer Alexander, 'I noticed the *Hawke* was 2 [22.5 degrees] to 3 points [33.75 degrees] on our starboard quarter.'

Hawke's commander, William Frederick Blunt, had plenty of experience, commanding various Royal Navy vessels since 1896. That morning, she had been on power trials. The first trial, with her two reciprocating engines averaging 92rpm, was 'satisfactory'. Her speed was 17 knots rather than the 18.4 knots she 'should' have attained 'under favourable conditions' (Blunt attributed the discrepancy to the fact that she had been 'out of [dry] dock for one year'). He was on the bridge the entire day, except for meal breaks. At 12.25 p.m., he went down below for a quick lunch, 'I was keeping my eye on the wardroom clock, which was directly in front of me; and I took the time when I finished [...] the time was then 12.37 p.m. exactly [...], [I] folded up my napkin, got up and went on the bridge.' The navigator, Lieutenant Aylen, was there.

Lieutenant Bashford, who had arrived a few minutes earlier, was getting ready to take on his duties as 'officer of the afternoon watch'. Bashford drew his attention to *Olympic* 'coming out of Southampton Water; I turned my head over my shoulder and saw her'. He judged that she was 4 points (45 degrees) off *Hawke*'s starboard bow. She was between the Thorn Knoll Buoy and West Bramble Buoy (about 12.41 p.m.). He used his binoculars to look. 'She was actually on the turn when I looked at her, my first impression is that I saw rather more of the port bow [...] but, roughly speaking she was dead on to us.' He even saw 'two puffs of steam from her foremost whistle [...] distinctly two blasts, to show that she was going to take the eastern channel'. That meant that both ships were going in broadly the same direction.

'She is going to take the eastern channel, and she will not have much room to turn,' Bashford remarked as he put his binoculars down. *Hawke* altered course to come round Egypt Point, but he changed the course to South 74° East rather than South 85° East 'because of the presence of the *Olympic*'. The two ships' courses were intersecting at an angle of about 15 degrees. He remarked that '*Olympic* was coming unpleasantly close' and ordered the helm put 5 degrees to port (a turn to starboard) until he gave a further command. *Hawke* turned slightly to starboard but 'commenced to pay off back to port, not in the ordinary way of a swing, but what I should call a swerve', towards *Olympic*! 'What are you doing?' he called out to the helmsman. 'Port, Port, hard a'port!'

Blunt realised 'with the increasing velocity of this swerve, that no helm the ship could carry was ever going to take her to starboard'. He ordered the port engine stopped and 'full astern' on the starboard engine. Petty Officer Ernest Hunt put the helm hard over but it jammed at 15 degrees and he called out, 'Helm jammed!'

Blunt ordered 'full astern' on both engines, personally pressing the lever on the port-side engine telegraph down. Meanwhile, Lieutenant Bashford and Leading Seaman Henry Yeates had joined Hunt at the wheel and tried to force the helm over. Blunt 'even went over to the starboard side' to double check that the starboard engine telegraph was set to 'full astern'. He could tell 'by the vibration' that the engines were running astern. *Hawke*'s speed had been reduced from about 15.25 knots to 'about 8 knots', but it was too late.

Captain Smith had seen *Hawke* 'running practically on a parallel course'. 'She continued to overhaul until she got with her stem almost abreast of our bridge,' he recalled. 'I could see the

An artist's impression of the collision. The scale is not quite correct but *Olympic* was certainly much larger than *Hawke*. (Jerry N.J. Vondeling collection)

stem through the aperture on the bridge shelter.' The two vessels seemed to 'run about even [in] speed [for] some little time … then we gathered speed, drew ahead a little, or she dropped astern'.

Bowyer had been on the port side of the bridge and walked back towards the middle (adjacent to the ship's wheel) when Smith drew his attention to *Hawke* on the starboard side. Bowyer saw her 'swinging very fast, as though on a hard a'starboard helm'.

Smith had a better view from the starboard side of the bridge. He saw *Hawke* turn 'very quickly indeed' towards *Olympic*. It 'seemed inconceivable, a manoeuvre I could not understand'.

'If she is going to strike let me know in time to put our helm hard a'port,' Bowyer requested. A few seconds passed and he asked again, 'Is she going to strike us or not, sir?'

'Yes Bowyer, she is going to strike us in our stern.' (Smith recalled saying, 'He is starboarding [turning to port] and he is going to hit us.')

Bowyer ordered the helm hard a'port (a turn to starboard), hoping to swing *Olympic*'s stern away from *Hawke*. Quartermaster Albert Haines put the helm hard over. Haines was not watching the compass, but he judged that *Olympic* 'just began to feel the helm' ('it may have been a point, perhaps half a point') when he felt the collision 'a few seconds' later.

Chief Engineer Fleming was standing on the top platform in the reciprocating engine room when he felt the collision 'slightly'. By the time he had walked down to the middle platform, he 'noticed the starboard engine 'pulled up … just for a moment, and then cleared itself'. Bowyer ordered the engines stopped 'almost immediately' after the collision. Fleming saw the 'stop' order for both engines when he reached the bottom platform.

Bowyer estimated that the force of the collision made *Olympic*'s bow 'sheer to starboard' by about three points (33.75 degrees).

Zetie Duncan and her family had sat down at a table in 'the lovely' second-class dining saloon. She happened to glance toward one of the large portholes and 'was startled to see the prow of another ship […] a few seconds passed as we glided along. Then came a terrific crash.'

Frank Beken used exactly the same term. He saw the collision from a motor boat. 'We could see right into the *Olympic*. The *Hawke* presented a terrifying spectacle, listing over to such a dangerous angle that I thought she was going to capsize.'

First-class passenger Waldorf Astor, MP, saw *Hawke* immediately before the collision. 'She just butted into us, and caught us absolutely at right angles,' he recalled later to a Press Association special correspondent. 'The cruiser was going along parallel, and was trying to get in apparently behind us, but did not clear.'

Another passenger interviewed by the Press Association, Luscombe Carroll, saw *Hawke* 'steaming altogether too close'. He told his friend that a collision was inevitable:

> I leaned over the side and waited for the impact [...] When it came there was not much of a shock. We heard a crunching noise. I then saw the cruiser back off and her bow plating fell away like crumbling wood. Then she heeled over. We were going at half speed, 10 or 12 knots at the time, and the cruiser was going slightly faster and gradually gaining [...] She tried to make a half circle to clear our stern ... it was impossible.

Meanwhile, fellow passenger Edgar Davis praised the 'splendid' behaviour of everyone. 'There was no sign of panic among the passengers of any class.'

Commander Blunt 'purposely kept [*Hawke*'s] engines working astern till the ship had sternway'. He wanted to make sure that there was as little pressure on the watertight bulkheads as possible. *Hawke* 'cut across the *Olympic*'s wake, as she backed'. Her stern pointed in a north-north-easterly direction as she finally came to rest. Meanwhile, *Olympic*'s own log recorded:

12.46 p.m.	Struck on starboard quarter by HMS *Hawke*.
12.47 p.m.	Stop; Warning bell rung and watertight doors closed.
12.48 p.m.	Stop engines.
12.50 p.m.	'Full astern'.
12.51 p.m.	'Stop'.

Olympic began to settle at the stern, a sure sign that she was taking on significant water. The large watertight compartment between bulkheads P and O flooded rapidly. One crewman, standing by the watertight door in O bulkhead at the aftermost section of the propeller-shaft tunnel, saw the rush of water coming through the ship's side after *Hawke* disengaged from *Olympic*, 'The automatic release [to close the doors] from the bridge had not yet been worked, and he took the hand lever, standing on the fore side of the door, and released that door.' The door 'fell and closed properly, but during the time it took to do so sufficient water had come through the door to bring about three feet of water into the next forward section of the tunnel – some three or four hundred tons of water had come through'.[3]

Frank Munsey, founder of *Munsey's Magazine*, had been watching events from first class:

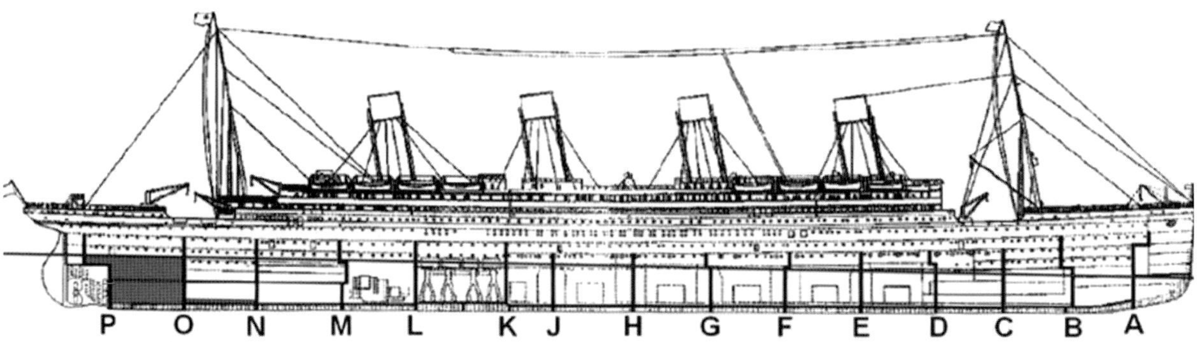

The collision resulted in serious flooding. *Olympic*'s aftermost hold was completely flooded (between watertight bulkheads P and O) and several hundred tons of water entered the adjacent compartment (between watertight bulkheads O and N) before the watertight door closed. Diagram by Sam Halpern. (Author's collection)

> It was at about 12.30, while I was in the Ritz [*sic*] restaurant, that someone told me a big cruiser was running alongside the *Olympic*. I left the table and went to a window. There I saw the *Hawke*, running parallel with us. I watched her for a minute or two. We were then just past Cowes.
>
> I went back to finish luncheon, and in less than two minutes felt a tremendous wrenching of the ship, which sounded as if it had struck a rock. I did not know what the trouble was. I went out on deck and found the cruiser had swung around and rammed the *Olympic*. Our ship then pointed toward the shore and cleared the cruiser, which had her nose badly smashed.
>
> There was no confusion or panic aboard. Everyone seemed to realise that it was best to take things calmly. Of course there was plenty of excitement, the novelty of the situation insuring this. We just remained where we were with anchor dropped until 4 o'clock.
>
> As the afternoon wore on the *Olympic* began to list badly to starboard, the aft list being perhaps some twelve feet before we left her, although the watertight compartments worked admirably. Still, a good deal of water got in. Quite a number of cabins were smashed aft [...] although, fortunately, I believe, no one was hurt.
>
> It was at 4 o'clock this afternoon when we had the first definite news of the extent of the injuries. The captain addressed the passengers, saying everything was all right and there was no cause for alarm, but if any of us wanted to go ashore tonight, he had wired for a tender to take us, but he thought the wisest thing to do was to remain aboard tonight, and he would guarantee to land us [...] at Southampton tomorrow morning.[4]

Over the afternoon, Zetie Duncan watched from one of the second-class decks, 'Professional divers came from Southampton to check out the gaping hole in the stern [...] Dripping steerage baggage was being hoisted from the hold.'

J. Bruce Ismay addressed a telegram to the naval authorities:

The damage to *Olympic*'s hull above the waterline was clearly visible. It was only when divers inspected her after the collision that the full extent of the damage beneath the waterline became apparent. (© TopFoto)

Hawke also looked very much worse for wear after sustaining significant damage in the collision. (Author's collection)

Olympic settled at the stern but remained afloat and stable. She was never in any danger of sinking. (Ioannis Georgiou collection)

> Liverpool
> Secretary Admiralty Whitehall LDN [London] SW [South West] =
> Referring serious collision between *Olympic* and *Blake* [*sic*] will greatly appreciate your instructing Portsmouth afford *Olympic* any assistance required = Ismay +

Hawke's log recorded that her collision mat was put over the bow at 12.50 p.m. Forty minutes later, she proceeded slowly towards Portsmouth. The *Belfast Telegraph* reported that two dockyard tugs were being sent to assist, but they were not needed.

The *Cork Examiner* described the scene aboard *Olympic* that afternoon. Captain Smith and Purser McElroy 'ascertained that owing to business engagements across the water nearly a hundred passengers were anxious to get ashore to Southampton so as to catch the earliest boat for New York'. They arranged a 'special dinner' at 6 p.m. 'for those leaving the ship'. Passengers at Queenstown who were expecting to board *Olympic* the next day reacted with dismay. A number of people had booked their tickets 'months ago – in fact, it was impossible to get a berth on her months ago, so great was the run on her fine accommodation'.

A telegram to Lloyd's reported that *Olympic*'s draught was 39ft aft and 31ft forward, 'Considerable water in after hold. Expect will proceed to Southampton next tide.' The *Belfast News-Letter* reported that 'between sixty and seventy' of the ship's passengers reached Southampton that evening and boarded a train for London.

Olympic's officers had time to reflect on what had happened. Pilot Bowyer remembered Fourth Officer Alexander commenting that 'he thought we were going a little faster [than *Hawke*] at the last moment', when they were 'talking the matter over probably that same night [...] It may have been that same evening we anchored in Cowes Road, when we were in the officers' room smoking, the conversation came up.'

Perhaps Captain Smith pondered the events prior to the collision and what he had thought was *Hawke*'s 'inconceivable' manoeuvre. Meanwhile, passengers settled down for a night aboard. Zetie Duncan recalled, 'All through the night the *Olympic*'s crew were dismantling the rooms and stacking furniture in the corridors.'

'A STRICT AND CAREFUL ENQUIRY'

'I consider the SS *Olympic* to blame in not observing the Rule of the Road by not easing her engines or waiting till there was sufficient sea room to pass *Hawke*,' Lieutenant Aylen concluded his written statement on the day of the collision. It was one of several witness statements from those on board *Hawke* that were recorded for the Admiralty, while the details were still fresh in their memories.

The following day, Admiral Moore – the Commander-in-Chief at Portsmouth – directed Captain Henry Grant and Captain Edward Booty to investigate the circumstances of the collision, 'You are to assemble onboard the *Hawke* at 10 a.m. on Friday, the 22nd September 1911, and hold a strict and careful enquiry into the circumstances of the case, being guided in your proceedings by Chapter XVIII of the King's Regulations.'

Commander Blunt's report, as well as the other statements from *Hawke*'s officers and crew, were forwarded with Moore's request for an investigation. A single day of hearings was held on 22 September. The witnesses were all asked questions based on their statements. Only 187 questions were asked, some of which merely requested that witnesses confirm their statement as being accurate 'to the best of your knowledge and belief'.

Blunt himself answered sixty questions. 'Have you any remarks to make as to how the Rule of the Road was observed or not observed by the *Hawke* and *Olympic*?'

'Yes', he replied:

> When I first saw the *Olympic* I was under the impression that she was going down through the Needles Channel and it was not till she sounded two blasts that I realised that she was going to the eastward, as she was then on starboard helm

and having her starboard side to me, on a closing course I considered that under Article 19 of the 'Rules for the Prevention of Collisions at Sea' it was not my place to give way, and later when the turn of the *Olympic* was completed and she was in close proximity on my port hand I could hardly realise that she intended to cross my bow and take the channel to the south side of the Ryde Middle, when she had all the rest of the channel open to her. With reference to the *Olympic*'s turn round the Bramble Buoy she was followed some time after by the German liner *Konig Wilhelm* [*sic*] and I noticed that while we were lying at rest after the collision her own turn brought her a full half mile to the north of us although we were then lying further to the north than at the time of the collision.

[...]

In my opinion, the collision was caused by the *Olympic* misjudging her turn round the Bramble Bank and thereby coming so close to the *Hawke* that as she forged ahead past her the suction due to her great mass drew the *Hawke* in on her. The *Hawke* could give no more room owing to the proximity of the Prince Consort Shoal.

Able Seaman Alfred Crookham described *Olympic* and *Hawke* 'steaming together for about five minutes'. The enormous liner 'gained on us and gradually drew ahead. I then noticed our ship's head pay off to port and next moment she struck the *Olympic*.'

'The actual position of the collision was not ascertainable owing to the rapid oscillation of the compass card, caused by the impact and the attraction of the moving iron of the ship,' testified Lieutenant Aylen. He made the remarkable observation that 'my notebook was knocked overboard by the collision'. Consequently, he was unable to provide the exact 'fix' he had taken for *Hawke*'s position before the collision. He confirmed his written statement when he said that 'directly *Olympic* came up abreast the compass started oscillating'. Just before the collision, *Hawke*'s compass was 'spinning right round'. At that time, *Hawke* was being drawn into *Olympic*'s side, 'The ship did not stop slewing but continued more rapidly as time went on till we struck the *Olympic*. It was apparent it was actuated by some outside force. The swerve was very quick and no helm could have brought her round so quickly.'[5]

Captain Grant lost no time in forwarding the court's conclusions to Admiral Moore the very next day, which read, 'We are of the opinion that from the evidence heard that the *Olympic* alone is to blame.' Blunt appeared:

> ... to have done all in his power to give the *Olympic* as much room as possible, and was as close as he could be to his starboard side of the channel. We consider that he could not have done otherwise than he did and that he is in no way to blame.

Grant and Booty summarised *Olympic*'s responsibility:

> The *Olympic* came out of Southampton Water, and was throughout on the port hand of HMS *Hawke*. The *Olympic* should therefore have kept out of the way of the *Hawke*.
>
> The *Olympic* on ranging up abreast of the *Hawke* passed too close to her.
>
> The *Olympic*, though perhaps not the overtaking ship according to the definition laid down, had excess of speed over the *Hawke* and could have reduced speed to keep astern of her in the narrow channel, seeing that she was obliged by the Rule of the Road to keep out of the way. Instead of which she attempted to pass the *Hawke*.

'A FEARFUL EXPERIENCE HAPPILY SURVIVED BY ALL'

After spending the night anchored off Cowes, *Olympic*'s passengers were keen to return to Southampton. Press reports stated 'there were twenty millionaires' among them, 'whose aggregate wealth is said to be five hundred million dollars'. When the *New York Times* reporter arrived off Cowes just before 8 a.m. on

21 September, they saw the wounded liner getting ready to return to port:

> Her decks were alive with passengers, fully resigned to the inevitable. Half a dozen tugs were in attendance, two at the bow and two on either side.
>
> […]
>
> The journey was necessarily slow, for, though the great liner drew but little more aft than forward, the margin of water beneath her keel was very little and the navigation of so huge a vessel was a matter of some difficulty.
>
> The crippled liner presented a very striking spectacle as she crept slowly up the estuary. By 10.30 she was at the entrance to the dock, and fifteen minutes later was safely tied up. She was slightly down at the stern, showing that a good deal of water must have got into the aft holds.

The White Star Line arranged for divers to 'ascertain the extent of the damage beneath the waterline'. 'It will be all of three months before she is in commission again,' said Harold Sanderson. 'One of the divers, who spent three hours underwater examining the injuries, told me the damage was every bit as severe below the waterline as above.'[6] *Olympic* would need to be repaired temporarily before returning to Belfast and going into the only dry dock that was large enough to accommodate her.

Journalists spoke with Alexander Carlisle, even though he had been retired for more than a year and had no direct knowledge of the situation. 'If any of her plates are broken in,' he told *The People*, 'it will be at least a £25,000 job.'

On the dock, the Duke and Duchess of Manchester had been waiting. The previous day, they had missed seeing off Eugene Zimmerman, the Duchess's father. Now, they saw him return, 'We knew everyone was safe.' Zimmerman was lucky to arrange accommodation on the American Line's *St Louis*, leaving on Saturday, 23 September.

Another passenger, John J. Sinclair, who had been returning home to New York with his daughter and son-in-law, commented that there had been 'absolutely no confusion and no nonsense after the collision'. He thought Captain Smith was 'the greatest captain on the ocean':

> I saw the affair, and it seemed to me that the man in charge of the cruiser must have gone crazy. She was running parallel with us when all of a sudden she turned at right angles and came straight for us. Please say for me that the officers and men of the *Olympic*, from Captain Smith down, behaved splendidly. In fact, there was so little excitement that I met a well-known man an hour after the accident who told me he had slept all through it, and did not know anything had happened until he came on deck and found us anchored.[7]

Zetie Duncan recalled a 'gloomy breakfast' and the slow journey back to Southampton. They disembarked into:

> … a large dock shed. A special train had been ordered to take us back to London. We waited and waited …
>
> There was much grumbling about having given up hotel rooms, and not getting home for special events. Our family was fortunate to have Aunt Emma's home to return to.
>
> Hours went by and still no train. Everyone was tired and hungry […] Bengal tigers were drawn through the shed, pawing at the bars of their cage on wheels. Suddenly my father became very angry. He shouted to an officer: 'What kind of steamship company are you running, letting all these people go hungry while you have tons of food on that ship?' The officer came out of his daze … Within a few minutes waiters came running down the gangplank with trays fastened around their shoulders. The trays were loaded with cakes, biscuits and fruit. They continued coming until everyone had eaten all they desired. The Bengal tigers, benefited by the 'awakening to food', stopped grumbling. My father had become the hero of the hour.[8]

The train came 'at last' to take them to London. In the end, they sailed on White Star's *Celtic*,

Spectators on the dockside could see the full extent of the damage above the waterline close up. (© Mary Evans Picture Library)

Looking into the third-class passenger accommodation through the ship's damaged side, the corridor framing and part of a stateroom are visible. (Author's collection)

which had brought them over from the United States. Alternative travel arrangements had to be made for the passengers. White Star reportedly booked up every empty berth on all the liners departing during the next ten days. The third-class passengers were going to be accommodated temporarily in Southampton, then transferred to other ships, including *Majestic*.

Accommodating all of *Olympic*'s passengers was no mean feat when it was peak westbound season. *Adriatic*'s first- and second-class accommodation was essentially full when she arrived in New York on 22 September. *Mauretania* brought in 505 first and 451 second-class passengers on the 28th.

Olympic's mail was landed and 'placed in a special train for London, to be despatched by other routes', reported the *Irish Times*. Her cargo had to be shipped by other liners. The Post Office announced that the Canadian mail would be shipped by *Empress of Britain* from Liverpool and United States mail would be shipped by *Mauretania*.

***

All of the logistical problems to sort out were frustrating enough, but there were a number of positive comments in the press. *Olympic* had been seriously damaged and taken on a significant amount of water, but was never in any danger of sinking. The *Cork Examiner* admired *Olympic*'s safety:

> *Olympic* not only withstood it, but returned from the scene of the accident unassisted. It speaks well, too, for the system of watertight compartment doors which she has, that they splendidly stood the test they were subjected to. Altogether, the fact that there was no loss of life is highly satisfactory, and goes to prove two things – first, that the *Olympic* is unsinkable, which for the travelling public is an assurance of incalculable worth; and, secondly, that the handling of her could not be excelled, which is only what might have been expected from a master mariner of the status of Captain E.J. Smith RNR, who needs no eulogy, as his record is second to none.

The *Dublin Daily Express* reflected that the adoption of Marconi wireless telegraphy had 'materially diminished the dangers of the sea', citing the case of *Republic* summoning help almost three years earlier, 'Watertight compartments, as was shown in the recent case of the *Olympic*, are another great safeguard.' The *Belfast News-Letter* said *Olympic*'s 'stiff frame stood the shock well, and the watertight doors held the compartments hermetically [*sic*] sealed'. The *Western Echo* said that the 'moveable [*sic*] bulkheads [...] ensured the safety of the *Olympic*, and the confidence of the passengers in the vessel, and those who were controlling it prevented anything like a panic'.

Newspaper speculation began about the cost to the White Star Line. A late edition of *The Scotsman* noted that *Olympic*'s insurance excess was £150,000, meaning that her owners would have to cover repair costs up to that amount themselves:

> The sinking of the *Republic* in January 1909 [...] is often cited as an illustration that watertight compartments are not a complete safeguard against loss, but the financial effect of that disaster was a fleabite as compared with what would have happened had the *Olympic* come seriously to grief in the Atlantic.[9]

By 23 September, the divers' survey revealed:

> On starboard quarter at break of poop [deck], a large triangular hole about 12 feet in length at upper part fore and aft, and extending from just above waterline to D deck, a height of about 14 feet. The shell plating is crushed and bent inwards for a distance of about 8 feet, also deck plating, beams and frames in vicinity of same broken and started; cabin bulkheads and fittings in wake damaged. The shell plating from the large hole is scored slightly for a distance of about 60 feet aft. It is at present anticipated that eight bottom plates will require renewal and three removed faired and replaced, but this may require modification on opening out. The stuffing box bulkhead, and gland to starboard shaft is strained, and water finding its way through

fittings into the dynamo room, but temporary measures have been taken to stop leak as much as possible.

Divers employed by the White Star company report that there is a large pear shaped hole in shell plating below water, and immediately under the large hole previously referred to, and this was apparently caused by point of ram of HMS *Hawke*. Also the [propeller shaft] boss plating around starboard shaft badly strained for a distance of about 12 to 14 feet; and blades of starboard propeller broken.

Temporary repairs included wooden patches over the hole above the waterline. *Olympic*'s log reported that she left Southampton on 3 October, but press reports said she left on the morning of the 4th. She was relatively lightly loaded, drawing 29ft 9in at the bow and 30ft 11in at the stern. She anchored in Belfast Lough on the 6th. The *Irish Times* reported:

The White Star liner *Olympic* was berthed at the deep water fitting out wharf in Belfast Harbour on Saturday morning [7 October]. The liner left her anchorage in the Lough between eight and nine o'clock. The five Mersey tugs and Queen's Island *Hercules* were engaged in the task of towing the liner along; three were positioned ahead of her, one was to the starboard, and the other two were to the aft. The Liverpool salvage steamer *Ranger* was also in attendance in case her assistance was necessary [...] as it turned out, her services were not required. Having arrived opposite her station, the *Olympic* was moored at the wharf in the short space of fifteen minutes, the entire operations of her transfer having been completed at 10.11 a.m. The crowds who thronged the new dock are commanded an excellent view of the berthing scene, and, with the tugs steaming to and fro, the spectacle was one of extremely interesting animation. Lord Pirrie was present during the berthing operations.

She was moved into the dry dock on the morning of Wednesday, 11 October. The *Belfast News-Letter* noted 'a trifling matter' of one of the ship's hull plates scraping the dry dock entrance as she came in. By midday, everything was complete and the workers were getting ready to pump the dock dry. On 12 and 13 October, Robert Steele inspected the damage in detail. Naval architects Harry Roscoe and John Denholm Young attended on behalf of White Star as well as a government representative (Roscoe was a consulting naval architect from the Liverpool firm, Roscoe & Little).

Harland & Wolff had a lot of work to do, repairing the obvious external structural damage and remedying the damaged interior fittings, including:

Accommodation on Starboard Side.

D deck. Woodwork and fittings in rooms nos. 85 and 89 broken and damaged, remainder of rooms from frame 93 to 133 damaged.

E deck. Woodwork and fittings in rooms nos. 145 and 151 damaged and destroyed, remainder of rooms from frames 103 to 139 damaged.

F deck. Woodwork and fittings in rooms and passages from frame 93 to 139 damaged.

G deck. Wood bulkhead of room no. 207 broken [and] remainder of rooms from frame 89 to 139 damaged.

The refrigerated cargo space from frame 93 to 111 between G and tunnel decks has been badly damaged through the effect of the silicate cotton becoming saturated with water.

Wood sparring from frame 111 to 118 between G and tunnel decks badly broken and damaged [...] Litosilo on D, E, F and G decks for the extent of deck and shell damage has been destroyed.

Plumbers needed to replace pipes and sink discharges. Deck plating was damaged from D-Deck down to the propeller-shaft tunnel deck. It was not merely a question of replacing the outer shell plating. Beams supporting the interior decks were damaged; beam knees (beneath the end of a deck beam where it was connected to the side plating) were injured; hull frames were twisted; and rivets were started.

The full programme of repair work was described by one of Harland & Wolff's managing directors, George Cuming, in a conversation with an Engineer Commander, Director of Dockyards (on behalf of the Admiralty) on 14 October.[10] There was bad news and good news: 'There are no marks on the propeller blades of the centre and port shafts to show that these have been touched by anything at the time of collision.' Unfortunately, the starboard propeller blades were damaged: 'The three blades have been removed; they are damaged towards the tips. They are probably bent as well although this is not obvious. Mr. Cummings' [*sic*] proposal is to scrap these three blades, appropriate three spare and replace the spare blades used. The blades are […] manganese bronze.'

The propeller boss itself, which the propeller blades were fixed to, was 'apparently undamaged' (it was made of solid cast steel), but either of *Titanic*'s propeller bosses were available as a replacement in any case. They needed to examine thoroughly the starboard engine, 'as that engine was supposed to be taking full steam when the engines brought up'. Meanwhile, the aftermost end of the propeller shafting needed to be removed. A section of it was bent. The process was complicated:

> The tail shaft can be withdrawn into the dock and so removed to the shop, the three pieces forward of this necessitate that certain plates should be removed from the ship's side so as to pass them out into the dock and so send into the shop.
>
> Where the shafting passes through bulkheads, the plating has had to be cut in order to uncouple and pass the shafting to be removed through the orifice being cut in the ship's side.
>
> It is not expected that these four lengths will be in the shop for another seven or eight days, and so the renewal necessary as regards them is unknown. As a precautionary measure a forging has been ordered for one length of shafting. The shafting is hollow and Messrs. Harland & Wolff do not consider that if any length is bent it can be made serviceable by straightening.
>
> The *Titanic*'s shafting is available if necessary but if used would entail considerable delay in that ship's completion, as the engines are now being put into her.

Harland & Wolff were 'taking the opportunity of altering the pitch of the port propeller blades; the cost of this will be kept separate'. The replacement starboard propeller blades were undoubtedly fitted accordingly so that both port and starboard propellers had the same pitch (increased from 33ft to 34ft 6in).

They completed the mammoth task of repairing all the damage within weeks. The *Belfast News-Letter* reported *Olympic* 'was removed from the new graving dock' to the fitting out wharf on 3 November. The newspaper had already announced on 20 October that *Olympic* would 'definitely' be ready to sail from Southampton on 29 November. She left Belfast for Southampton on 20 November. The paper noted that *Titanic*, which had originally been moved to the Alexandra Wharf early in October to make way for *Olympic*'s arrival and dry docking, would be moved back to her fitting-out berth on 21 November.

AN ACT OF DESPAIR

The successful repairs and her departure from Belfast masked an eventful day aboard *Olympic* on 18 November. Foreman Labourer James Stewart was on Queen's Road around 4 p.m. when he met Edward John Wilson. Wilson explained to Stewart that 'he had been paid off [dismissed] and that his son had been paid off on the night before'.[11] He asked Stewart if Joseph Sharpe was working on *Olympic* and Stewart said Sharpe would be working until 'about 5 or 5.30 [p.m.] according to the time he would get the job finished'. Wilson showed Stewart a revolver, 'saying he was going to blow Sharpe's brains out'. Stewart 'said to him not to be making a fool of himself'. They were standing 'about a quarter of a mile' from *Olympic*, but Wilson headed towards the ship. Stewart followed.

Boilermaker William Warmsley was working in the main engine room with Joseph Sharpe.

He recalled that it was about 4.10 p.m. when Wilson came into the engine room 'and went round to where Sharpe was standing'. Warmsley had known Wilson 'for a considerable time'. He knew that Wilson's wife was 'lying ill and her life despaired of'. Wilson was not normally 'a man who was inclined to quarrel with his fellow workmen', but his demeanour was different now.

The situation escalated quickly.

'Come into the stokehold, I want to speak to you,' Wilson asked Sharpe.

'Say what you have to say here,' Sharpe replied.

'If you don't come into the stokehold you'll be sorry for it,' Wilson responded.

Warmsley saw that:

> [Wilson] then immediately drew a revolver from one of his pockets. Sharpe who was about a foot away from him grabbed him above the wrist and pushed him back to the bulkhead[.] The two of them wrestled for a few seconds, and as a result they both fell on the floor, Sharpe fell on top of [Wilson] who had the revolver still in his right hand [and Wilson] rolled over to his right side and fired the revolver. I [...] rushed forward and caught hold of [Wilson's] right arm and he fired a second shot, a number of men who had been working on board came to my assistance and [...] I was able to get the revolver from [Wilson].

By the time Stewart got down into the engine room, he was just in time to see Wilson and Sharpe 'struggling on the floor with a crowd round them'. Sharpe 'was crying out' when the first shot went into his left thigh. 'While I was lying on the ground and those people around us, he fired another shot striking me on the right leg below the knee,' he recalled. 'After that [...] a third shot was fired. I was then lying on my side, the bullet struck the skin of the right leg on the outside opposite the second wound.' He was taken to the Royal Victoria Hospital.

Belfast Harbour Police Constable Alexander Shannon went down to *Olympic* around 5.15 p.m. He saw a 'crowd of people' disembarking via the gangway, then arrested and cautioned Wilson.

'I am very sorry for what I have done,' Wilson explained. 'Is the injured man dead? Constable, blow my brains out with that revolver.'

Wilson was brought before the magistrates at Belfast Custody Court on Monday, 20 November, charged that he did 'feloniously, wilfully, and of his malice aforethought shoot at one Joseph Sharpe with a revolver loaded with gunpowder and bullets with intent to kill and murder'. He was remanded in custody, but bailed early in December after a further hearing. Meanwhile, Sharpe was recovering in hospital.

Wilson, Stewart, Sharpe, Warmsley and others all submitted depositions. Sharpe said he had 'known the prisoner ten or twelve years. On that day he had some liquor taken, and I had a little dispute with him over a little overtime. He was not dismissed as a result of that dispute, but he was dismissed on that Saturday at about 12 o'clock.' He said, 'I have no ill feeling against him. I was aware previous to the Saturday that his wife and youngest child were seriously ill. He had told me that they were on the point of death.' Joseph Turner said Wilson was 'a quiet decent hard working man and as far as I knew all parties had been good friends'. Wilson himself gave a statement saying he was 'very sorry' and 'I must not have been in my right senses'.

The *Northern Whig* reported on the case coming to court on Thursday, 21 March 1912. Wilson pleaded guilty and expressed remorse. His defence admitted 'the case was a very sad one, and might have been much more serious'. He 'did not really know what he was doing at the time'. Wilson had worked for thirteen or fourteen years at Harland & Wolff. He was a 'father of eighteen children, eleven of whom were alive'. His wife was 'seriously ill suffering from cancer, and for several weeks prior to this occurrence he had been sitting up night after night with her'. After he had been 'told that his services would no longer be required' earlier that fateful Saturday, he faced 'the prospect of complete starvation'. His son had also been laid off.

On the way home, he 'took some drink', even though he was normally teetotal. He told his wife what had happened, then 'in a moment of madness he went out and got more drink. He bought a revolver.' He thought Sharpe 'was to blame for getting rid of him and his son's services'. He had not fired deliberately. The revolver went off in the struggle.

Reverend G.R. Wedgewood had known Wilson for five years, 'He had never known anything against him [...] He was the sole support of a large family, the youngest child being scarcely two years old.'

Lord Justice Cherry said he was 'satisfied that he committed the act in a sudden fit of passion and under very trying circumstances. He was sure he did not intend any serious crime.' He thought 'he was justified in pardoning the accused under the Probation of Offenders Act'.

A CHANGE OF PLAN

White Star had already released their sailing list covering the winter. It showed *Titanic*'s maiden voyage on 20 March 1912. By 11 October 1911, the company had realised that she would not be ready in time. It was put back to the date of what would have been her second voyage – 10 April. Meanwhile, they had a temporary gap in their express service on the Southampton to New York route.

Two more of *Olympic*'s round trips, scheduled to leave Southampton on 11 October and 1 November, had to be cancelled. There were worries that she might not make her intended departure on 29 November either. The *Southampton Times* reported, 'Every effort will be put forward to complete the repairs at the earliest moment, but the opinion has been freely expressed that the vessel will not be back again before December.' It said that *Olympic*'s scheduled sailing on 11 October would be taken by the American Line's *St. Paul* (she arrived in New York on the 19th, carrying 1,002 passengers). None of White Star's express liners sailed from New York between *Oceanic*'s departure on 4 November and her departure on the 25th, but travellers did have the option of sailing by *Cedric* or *Baltic* on their Liverpool service instead.

They pressed ahead with their plans for a sister ship to join *Olympic* and *Titanic*. The shipyard and engine works had already been instructed to proceed with all the necessary preparations on 28 June and the letter of agreement was signed on 23 October. Her keel was laid in No. 2 Slip (where *Olympic* had been built) on 30 November: 'Yard Number 433'.

'IS SHE NOT A WRECKED SHIP?'

Another headache for White Star was that they faced legal action from *Olympic*'s crew. On 29 September 1911, Fireman Douglas Fraser and Seaman William Weller brought 'claims equivalent to a month's wages' on behalf of the different sections of the crew. Their total claim amounted to £4,758 under the Merchant Shipping Act of 1894. Following the collision, the White Star Line had offered three days' pay, but their offer was rejected on the basis that 'they were entitled to the damages they suffered by being suddenly thrown out of employment'.

When their case went to court in Southampton, Mr S.H. Emmanuel prosecuted and Mr Charles Lamport appeared on behalf of White Star's defence. The *Daily Telegraph* called it 'a test case of considerable importance'. Legally, it amounted to whether '*Olympic* was or was not a wreck'. The crewmen's case amounted to arguing that *Olympic* was not a wreck and therefore they were 'entitled to damages for the termination of their engagement. The real definition of a wreck was that the vessel must be cast up upon land and between [a] high and low water mark.'

White Star argued that there was no legal definition of what a wreck was under the Merchant Shipping Act. They opted for a 'plain, simple, English definition', arguing the case did not fall under Section 262 of the Act but rather Section 158. The chairman and eight magistrates were 'equally divided' as to whether *Olympic* was legally a wreck or not. The case was referred to the Probate, Divorce and Admiralty Division.

Early in March 1912, the case came before Mr Justice Bargrave Deane. He asked if a ship caught fire and 'burned out, is she not a wrecked ship?'

'In this case I am not asking your Lordship to confirm the meaning of the word "wreck",' Emmanuel explained. 'I am only going to show that what happened did not make the *Olympic* a wreck.'

Emmanuel tried to argue that *Olympic* was not a wrecked ship under the legal meaning, because she had returned to Southampton 'after the collision under her own steam'. The judge responded, 'The word "wreck" was used so generally that its proper meaning was rather obscured. He always had an idea that a wreck meant something damaged.'

Mr Laing appeared for White Star. He pointed out that the crewmen had 'admitted at the trial of the collision action that they would not have proceeded on the *Olympic* in her damaged condition'. *Olympic* 'was a passenger ship and only got her certificate back on 29 November 1911, after she was repaired'. Laing argued the voyage had been 'brought to an end by a peril of the sea, and that made the vessel a "wreck" within the meaning of the statute'. White Star was not liable to pay damages and even if the court decided differently, it was for the court to decide the exact amount.

Philip Curry, White Star's Southampton manager, testified that he had gone out to the stricken liner on a tugboat as soon as he had learned about the collision. That afternoon, the Board of Trade had contacted him and 'required the passenger certificate to be surrendered'. Curry asked for it in writing, then returned it on the advice of the company's solicitor. He thought it had been returned 'a day or two' before *Olympic* sailed on her first commercial voyage after being repaired.

By 1 April 1912, the court was ready to pass its verdict. Mr Justice Bargrave Deane said that *Olympic* had been a 'wreck' under the law, which meant that her crew were not entitled to wages or compensation 'beyond the time at which the "wreck or loss" takes place'. White Star proved their case and 'a stay of execution pending an appeal was granted'. Legally, the company had already exceeded their obligations. They had offered three days' wages to the crew, which covered the period after the collision.[12]

'THE MOST CRUEL CASE'

On 21 September 1911, the White Star Line filed their 'statement of claim', delivered on 26 October, that the collision was 'solely caused by the negligent navigation of the *Hawke*', arguing that 'a good lookout was not kept' by the cruiser; she neglected to keep out of the way; the cruiser's helm had been 'improperly starboarded or her head was improperly allowed to come to port'; her engines 'were not eased or stopped or reversed in due time or at all'; the cruiser 'failed and neglected to indicate her manoeuvres by the appropriate whistle signals'; she 'improperly and at an improper time attempted to pass to the southward of the *Olympic*'.

The government issued their own, alleging the opposite: that the collision was 'caused solely by the negligent navigation of the *Olympic*'; *Olympic* had not kept out of the way of *Hawke*; she had come 'too close to the *Hawke* and going at a speed which in the circumstances was excessive and improper'; 'entering the Solent main channel at an improper time or in an improper manner'; 'taking too wide a sweep round the West Bramble Buoy'; 'not starboarding sufficiently or in due time'; 'proceeding at an excessive speed'; and not reducing her speed or signalling properly her manoeuvres. The '*Olympic* with her starboard quarter struck the port side of the *Hawke*'s stem a heavy blow doing considerable damage'.[13]

Each side's legal team considered which witnesses could help their case and whether some witnesses could be helpful at all. The government side thought that Lieutenant Aylen's account proved their argument as to *Olympic*'s whereabouts when she was first seen. A large-scale chart and models would be prepared to help present the case:

> The witnesses ought, if possible, to be in agreement about their estimates of distance between the two vessels [...] the witnesses [...] should be given the opportunities of making themselves familiar with the handling of models to avoid the mistakes which a witness frequently makes when asked in the witness box to show by the models the respective positions of two ships at a given moment.

They felt their witnesses appeared 'in complete accord as to the orders given and the rapid sequence of events before the collision'. Commander Blunt would 'have to explain why he did not sound a port helm signal when he ported round Egypt Point, and later on when he ported near the East Council [*recte*: Conical] Buoy [also referred to as the Prince Consort Buoy]'. Expert witnesses 'who are likely to be able to explain the nature and effect of the forces which attract two vessels travelling at high speed in about the same direction' needed to be contacted quickly. Until they had been spoken to, it was difficult 'to say who should be called without knowing what theories experts will advance, and whether their theories will agree, or will afford an explanation of what in fact happened'.[14]

The White Star Line's team undoubtedly went through the same process before the case began. Their core argument was that both ships were sailing in parallel and *Hawke* had been attempting to overtake *Olympic*. Therefore, it was *Hawke*'s duty to keep out of the way, because Article 24 of the Regulations for Preventing Collisions at Sea commanded: 'Every vessel, overtaking any other, shall keep out of the way of the overtaken vessel.'

Key to their argument was demonstrating that *Hawke*'s speed was greater than *Olympic*'s. The government's argument was that the two ships had been on converging courses, crossing each other. Therefore, *Olympic* should have given way to *Hawke*, under Article 19: 'The vessel which has the other on her own starboard side shall keep out of the way of the other.'

***

The hearings began on Thursday, 16 November 1911, in the High Court of Justice's Probate, Divorce & Admiralty Division. The presiding judge ('The President'), Sir Samuel Evans, was supported by Captains Thomson and Crawford.[15] The plaintiff's (White Star's) legal team was led by Mr F. Laing, Mr D. Stephens and Mr H.C.S. Dumas; the defendant's (government's) was led by Attorney General Sir Rufus Isaacs, Mr Butler-Aspinall, Mr A.D. Bateson and Mr C. Robertson-Dunlop. Sir Rufus was effectively the United Kingdom government's chief legal advisor.[16]

Captain Smith was sworn in as the first witness. He confirmed that he had been sailing 'in and out' of Southampton for four years. *Olympic* had left Southampton under the charge of a compulsory pilot, George Bowyer. Asked whether 'the combined power of your engines is equal to 59,000 horsepower', Smith responded, 'Yes, something like that.'

He was asked, 'Does the *Olympic* steer well?' and answered, 'Very well.' Then he confirmed that he remained on the bridge with the pilot as she steadied on her course in the minutes immediately prior to the collision. Laing continued:

116. Tell us what happened with regard to the *Hawke* from your own observation?
After we got on our course and she was on the quarter, I turned round and looked forward for a short time, to see the ship steadied – to see she was going right. It was a short time after that I saw she was on our beam.

117. Which beam?
On the starboard.

118. All that time she had drawn up on your beam. From your vision was she going faster than you or slower than you, or the same speed?
She was going faster than we were.

119. When you saw her drawing up on your beam, as you describe it, how far away from you was she? I mean, having regard to the line of the ship, how far laterally was she from you?
From 1 to 2 cables [about 600 to 1,200ft].

120. When you saw her drawing up on your beam, had you any anxiety about her at that time?
No, not at all.

121. Will you tell us in your own way what happened next?
She continued to overhaul until she got with her stem almost abreast of our bridge. I could see the stem through the aperture on the bridge shelter.

122. Up to that time did he appear to be on what sort of course in relation to yours?
Running practically on a parallel course.

123. You saw her stem through the aperture, as you have described. What followed?
For an appreciable time we seemed to run about even speed some little time. Then we gathered speed, drew ahead a little, or she dropped astern.

124. [*The President*] Did you say 'and she dropped astern' or 'or she dropped astern'?
She seemed to drop astern, or, in other words, we gained on her in speed, and immediately after her bow came to port, as if she had starboarded, she turned very quickly, and stuck us on the quarter – apparently to me, a right-angle blow almost.

125. [*Mr Laing*] When she commenced to turn towards you, as you describe, whereabouts was her stern – you described she dropped off again?

I should have thought between our second and third funnel.

126. [*The President*] How many funnels have you got?
Four funnels.

The witness described to the learned Judge with a model.

127. It would be exactly in the middle of the ship?
About that.

128. [*Mr Laing*] Had she still maintained the lateral distance or had it been altered, as far as you could judge?
As far as I could judge, it remained about the same.

129. I think you told us she seemed to turn quickly?
Very quickly indeed.

130. Did you say anything to the pilot when you saw that happen?
Yes; when she commenced to fall in, I called out to the pilot that he was starboarding [turning to port] and that he was going to hit us.

Oddly enough, when Attorney General Sir Rufus Isaacs began cross-examination, Smith said that he had been on his fourth round trip on *Olympic* rather than his fifth. There was a lot of discussion about speed, complicated by the issue of how much time elapsed between the ship steadying on course and the collision:

353. Did it occur to you at all there was any danger of her [*Hawke*] being sucked into your quarter [aft starboard side of *Olympic*]?
No.

354. Not at any time?
There was nothing in the conditions up to the moment she started to fall into us that led me to think we were in any danger at all.

355. It is not quite the question I am putting to you. Did that possibility of force of suction drawing her into your quarter ever occur to you?
No.

356. It did not at all?
No.

357. I suppose it is something which you would take into account if a vessel is sufficiently close to you?
Yes, I should take that into account.

358. Supposing she had been within 100 yards of you, there was a possibility of this vessel operating, was not there, in your view?
I know practically nothing about it really; I should think probably there would be.

Smith confirmed *Adriatic* was a 'very' large vessel when she entered service, but 'not so big' as *Olympic*: 'I think it is 25[,000 gross tons] as against 43[,000 gross tons] or something of that sort – the *Olympic* is one third larger, anyhow [*sic*]'.

Laing asked, 'In these large vessels, in your experience, have you ever yourself felt, or seen, other vessels affected by suction?' and Smith responded, 'No, that is not my experience.'

438. Do you come up the [River] Mersey with these vessels?
I was in the Mersey once with the *Adriatic* and once with the *Olympic*.

439. I think before the *Olympic* at all events, you had been working the Mersey for a number of years?
Yes.

440. Big ships?
Yes, the *Baltic*, the next largest ship to the *Olympic* [*sic*].

441. The Attorney-General asked you if it ever occurred to you to starboard your helm and give more room to the *Hawke* as you were passing along – did that ever occur to you?
No, I saw nothing in the conditions to lead me to anticipate trouble.

There was a lot of discussion about the ship's clocks. *Olympic*'s master clock displayed seconds but other clocks only displayed minutes, including the wheelhouse clock. The minute hand 'jumped' every sixty seconds so that a time of '12.40 p.m.' might be out by seconds or it might be almost a minute out. Normally, it made little difference, but whether *Olympic* had been accelerating for three or four minutes after completing her turn was potentially quite significant in working out her speed.

Pilot George Bowyer was called the same day. He confirmed he had been a Trinity House pilot for thirty years. Sir Rufus asked:

622. What I do want to put to you is this: were you gathering speed right up to the collision?
Yes.

623. Increasing your speed all the time?
Yes.

624. And at any time before the collision were you gaining on the *Hawke*?
No, she was gaining on us.

625. All the time?
Up till the last moment.

Laing asked about 'vessels being drawn together':

695. From the time you first saw the *Hawke*, until the time when Captain Smith drew your attention to her coming in did she give you any anxiety at all?
Not the slightest.

696. Do you know anything about suction, have you ever experienced it?
No.

697. Perhaps I ought to have used a technical term, I think it is called interaction between vessels ... Have you ever experienced the vessels being drawn together?
Yes, I know, I never had one sucked in yet.

698. And you have had a good many vessels under your charge?
Yes, we often pass another in a 600-feet channel.

The next day, Sixth Officer Harold Holehouse and Fourth Officer David Alexander testified. First Officer William Murdoch followed them. He recalled, 'Two short blasts were sounded on

Captain John Pritchard, a master mariner with eighteen years of service with Cunard, had been retired for two years. He testified about his time commanding *Mauretania*. While navigating in narrow waters, such as the River Mersey, he confirmed the practice was to go 'slow' ('about 13 or 14' knots) or 'slow-slow' ('about 8 or 9 knots'). 'Half speed' was 'about 17, I suppose'.

> 2767. Have you, in your experience, ever come across suction?
> Never.
>
> 2768. Or interaction between vessels, or whatever it is called?
> No, I have never seen any trouble, in all my career.

He said *Mauretania* steered 'very well' and confirmed one occasion when he had passed HAPAG's *Amerika* at a distance of about 150ft. His ship had been doing 'about 12 [knots] till we got up to her, and I gave her a little more [speed] to get past her'.

Other comments were less helpful to White Star's case. He was asked whether he would 'go past at 18 or 20 knots [...] in a narrow place?' and his response was, 'No, I do not think that would be very safe.' His concern was 'a ship not answering her helm in such narrow waters, and not much water under a ship, only about 7 or 8 feet'. 'When you are passing another ship it is certainly taking the water from her,' Pritchard explained, 'and she might not steer sometimes.'

Richard Lewis, an experienced first-class Liverpool pilot, recounted passing the Cunard ships in the Mersey. They had passed 50 or 60ft away. He never noticed 'their steering affected'.

Then Harry Roscoe, a consulting naval architect for the firm Roscoe & Little, based in Liverpool, took the stand. He confirmed he had 'over 25 years of surveying large steamers' and 'very extensive experience in damage cases'. He testified he had surveyed *Hawke* on 22 September and 31 October 1911 (at Portsmouth) and *Olympic* on 22 and 23 September 1911 (at Southampton) and 'afterwards at Belfast in dry dock'. Roscoe was asked about 'the angle of the first impact when first they came together' and testified he was 'fairly sure that it was never less than 5½ points [about 62 degrees]; it may have been something more than that.'

Thomas Andrews was asked to provide *Olympic*'s body plan and displacement scale, but had to explain, 'I have not been subpoenaed for the displacement scale'. The body plan was available. Andrews asked, 'As the matter of handing them over is a matter of great importance to the builder, as I am sure your Lordship will understand, may I ask that no copies be taken of them, or models made therefrom?'

The court needed information about *Olympic*'s draught and displacement. Andrews confirmed he knew the information 'very well' and he was sworn in to give it in testimony. Her displacement was 50,500 tons at a draught of 33ft 6in and 51,340 tons at a draught of 34ft. He confirmed the ship's 'moulded breadth' was 92ft and the 'extreme breadth' 92ft 6in.

'What is the difference?' Sir Samuel asked.

'The one, My Lord, is the overall of the plating on the ship's side. The other is from heel to heel of frame,' Andrews responded.

Andrews was only asked a few questions. 'I daresay you could have got [this information] in Belfast,' Sir Samuel suggested. 'However, this gentleman has had a pleasant trip.'

Roscoe's testimony concluded the next day. Then John Denholm Young, another naval architect practising at Liverpool, confirmed he had participated in the surveys of both ships. He had a slight disagreement with his colleague, estimating the angle of first impact at 79 degrees.

The first witness for the defendants was Commander Blunt, who confirmed he had commanded various naval ships 'from destroyers upwards, since the year 1896'.

> 3316. Which vessel was going the faster at the time the blow was struck?
> In my opinion, undoubtedly the *Olympic*.
>
> 3317. Was she going a little faster, or much faster, or what?

> No, in my estimation her speed was anything from 19 to 20 knots.

Blunt was very clear that *Hawke*'s helm had not been starboarded (a turn to port), 'It was never starboarded by my order, and neither was it ever starboarded to my knowledge.' Under cross-examination, he argued *Olympic* should have given more room:

> 3509. [*Mr Laing*] Now tell us, what do you say the *Olympic* did wrong?
> I say that the *Olympic* could and should, as the ship having me on her starboard hand, under the crossing rule, have given me more room, especially seeing that she had a mile to a mile and a half of deep water to port of her while I was hugging the buoys as close as possible.
>
> 3510. Now, is that a complaint under the crossing rule [i.e. Rule 19]?
> Yes.

Lieutenant Reginald Aylen testified about *Hawke* swerving to port before the collision, 'It was an unnatural swerve; it showed me, with my experience of ships turning on the helm, it was some outside influence turning her.' He put her speed at 'just over 15 knots through the water'. His note of the angle of impact was 'six points' (67.5 degrees), which was between Roscoe and Young's estimates.

Aylen's testimony ran into the fifth day, followed by Lieutenant Geoffrey Bashford and Lieutenant McGregor Robertson. Quartermaster Ernest William Hunt confirmed he had carried out the orders given, including immediately before the collision:

> 4956. When you got the order, 'hard a'port' what did you do with your helm?
> Put it over, put my wheel to starboard.
>
> 4957. From left to right?
> From left to right.

The helm was over 15 degrees before it 'brought up' and he did not get it any further, even when his colleagues tried to help. Leading Seamen Harry Yeates and Alfred Crookham concluded the day's witnesses.

Hawke's engineers accounted for much of the sixth day's testimony, confirming details such as the number of revolutions that her engines had been making. Commander Robert Wilberforce Myburgh, captain of HMS *Hood*, testified about the collision between HMS *Wizard* and the German liner *Kronprinz Wilhelm* in October 1902, in Southampton Water:

> I was going up the starboard side of the channel and she [*Kronprinz Wilhelm*] was overhauling me [*Wizard*], and when her bow was abreast of my stern […] When our bridges were abeam I saw we had closed in a lot […] to about 50 yards, and I gave the order to port the helm. The helm was ported 15 degrees, and I saw there was no movement of the ship to starboard, so I immediately gave the order 'hard a'port' and put my starboard engine to full speed astern, but we closed very rapidly, swinging in towards her at the same time […] and hit her about 60 feet before her stern.

Wizard acted against her helm and Myburgh attributed it to 'the rush of water to take the place of a vacuum caused by the *Kronprinz Wilhelm* going at high speed through a narrow and shallow channel'.

Robert Steele, practising 'as a consulting naval architect and engineer for over twenty years', agreed with Harry Roscoe about the first angle of impact. He thought the damage to *Hawke* was consistent with *Olympic* going faster at the time of the collision. Then George Baker, the superintendent of the William Froude Tank at the National Physical Laboratory at Teddington, provided details of the model experiments in the test tank.

The seventh day kept the court busy. On Friday, 24 November, David Watson Taylor, a naval constructor in the United States Navy, testified. He spoke about a ship advancing through the water and that 'water shall flow sternwards in order to fill up the space which she leaves'. It was proportional to the area of the vessel and her speed, indicating *Olympic* 'would be throwing aft [...] 5,400,000 cubic feet of water per minute'.

> It develops what would be called in-currents in the water, and in deep waters where there is plenty of room for it to take place the velocity is comparatively small, and in shallow water the velocity is necessarily increased by the area, and in narrow and shallow water the velocity is still further increases, and the result is, if you have two ships side by side parallel, there must be a greater velocity and flow of the sternway between the two ships than there is outside.

This 'greater velocity' led to the tendency of the ship on the starboard side to be drawn in. Professor John Harvard Biles, a Professor of Naval Architecture and Marine Engineering at Glasgow University, testified that he generally agreed with Taylor's evidence. Biles had 'seen ships in the Clyde subject to the action of suction occasionally go ashore on account of it'. He testified that 'it takes four minutes to get from 12 to 18 knots', which agreed with Captain Smith's estimate. Then Professor John Joseph Welch, at the Armstrong College at the University of Durham, testified to confirm that he agreed with Taylor as well. He also endorsed Biles' calculations as to speed.

After a weekend break, on Monday, 27 November, the court resumed. There were no more witnesses. It was time for each side's closing arguments. Mr Laing began for White Star:

> If it should be found that the *Hawke* was the overtaking vessel, there is no need to discuss the question of suction, or to discuss anything else. It was the *Hawke*'s duty, if she was an overtaking vessel, to keep clear; '*ex concessis*' she did not.

Laing argued that *Olympic* had:

> ... records of every time, of every order, and of the passing of every mark, and of the setting of every course from the time they left Southampton to the time of this collision, and your Lordship has seen the record taken by the officer who was noting these things, and we know that the times were taken by this electric clock, which therefore shows the same time in the engine room as it does on deck, and the records are therefore of the most accurate, I suppose *the* most accurate description that could possibly be devised. On the *Olympic* there were seven officers with master's certificates – most of them had extra master's certificates, and a pilot of the Trinity House, who is perhaps the most experienced pilot with large vessels for these waters, and my submission is, as a general observation, that nothing could very well exceed the way, the skill, the caution, and the care, with which the different things were being observed onboard the *Olympic*, and therefore the evidence of the *Olympic*'s witnesses is, I submit, likely to be correct.
>
> On the other hand, with regard to the *Hawke*, it is a most unfortunate thing that the one document which would have told us everything in this case that we want to know, has been unfortunately lost. The log book of the *Hawke* which we were enabled through the courtesy of the Admiralty to see, has one entry, and one entry only, of any real importance in this case, and that entry is the entry relating to the place where the collision happened, and, strangely enough, that entry is said to be wrong.

He pointed out that Blunt had only come on duty about five minutes before the collision, and highlighted some confusion as to who had been the officer of the watch:

> Lieutenant Bashford thought he was not, because he was not in pilotage waters. Lieutenant Aylen said he thought he was automatically officer of the watch, but he was going to lunch, and only stayed on deck because he was interested in watching the *Olympic*; and, beyond these three gentlemen [including Blunt], there was no other officer on the deck of the *Hawke* at all the critical moments.

Laing cited Colonel Sexton White's evidence, stating that he was:

> ... a gentleman of position, an independent witness, a person showing the sights of the Solent to his daughter, and a scientific man who was perfectly able to appreciate the distances and bearings, and who has had a vast experience in running trial ships and various other ships, and therefore able to judge distances, and so forth.

White had described *Hawke* as being on a parallel course compared to *Olympic*, laterally 'at least 300 yards' away from the White Star liner, and the 'overtaking vessel up till the time she came approximately abeam'.

Laing had 'a few words' to say about whether the *Hawke* had starboarded immediately before the collision:

> The eyewitnesses from my ship [*Olympic*] and the shore, all they can say is she [*Hawke*] looked as if she had starboarded, and beyond that I cannot go, but that there was something peculiar or unusual going on with the helm by the time when this vessel commenced this sheer, or whatever it was, is beyond doubt, because the helmsman was relieved and another helmsman took his place, and we had afterwards the jambing of the helm. Something was going on, at all events, quite unusual with the helm, and your Lordship will remember the shout of Captain Blunt, which he described as being in these words: 'What are you doing? Port, hard aport,' but Captain Blunt's recollection does not agree about this shout with Lieutenant Aylen's, and those who were on the bridge – their recollection of the shout was: 'What are you doing with your helm?' That is quite another matter to 'What are you doing? Port, hard aport.' This vessel was a twin screw, and when this sheer was observed by Commander Blunt, undoubtedly the order was given, whether carried out effectively or not, is another thing. The first order that was given was: 'Stop the port engine.' I submit, stopping the port engine, if it was done, was not a seamanlike thing to do, that the port engine should have been kept going full speed ahead, and that the starboard engine should have been reversed if the proper manoeuvre had been carried out. Stopping the port engine would, I submit, be helping the cant towards the *Olympic*. The port engine was stopped, as we know by the evidence of the man who stopped it, and it was standing at stopped when the order to go astern both was given.

Laing concluded:

> We say the whole question turns upon a question of fact, which was the overtaking vessel. If we establish the *Hawke* was the overtaking vessel, there is an end of all further discussion, because no inevitable accident is pleaded. It was her duty to keep out of the way, and she failed.

Mr Butler-Aspinall agreed that:

> [the] collision ought never to have happened. We have here two fine vessels – the *Olympic*, perhaps the last word in American liners – both commanded by experienced and capable navigators, and navigating in well-known waters, and in broad daylight they are found in collision, and what one asks oneself is, what probably was the cause of this collision.

His characterisation of *Olympic* as an 'American liner' would not have gone down well with many in the United Kingdom.

Pilot Bowyer was 'an experienced pilot, but the man had no great experience of her [*Olympic*]; he had taken her out on four previous occasions, I suppose he had her in his charge on each occasion some few hours':

> Two ships were entering this strait about the same time – we [*Hawke*] were there a little the first, and being there a little the first, although we never got into such a position as to be overtaking ships within the rule, viz., two points abaft the beam, she came in just about the same time, and having come in about the same time, she was just on the point – probably succeeded

> – in straightening herself, and was on the course which, in fact, was intersecting us, and, if so, of course her duty would be [...] to keep out of our way, and our duty would be, until a very late period, to keep our course and speed.

Hawke's speed had been steady – 'about 15 knots' and was 'always a crossing vessel'. *Olympic* was going faster. He dismissed the suggestion that *Hawke*'s helm had been starboarded prior to the collision. Even if *Hawke* had been the overtaking vessel, 'He had no right to increase his [*Olympic*'s] speed. What does he do? He puts his turbine into operation and goes on accelerating his speed. He introduces a new factor into force in his engine room, and introduces a new factor into his speed. That is wrong.'

Laing got up to make counter-arguments. *Olympic* had simply been resuming her previous speed, because she had slowed to manoeuvre, 'It is not a breach of the rule to regain your natural speed after you come round a corner, and get into the straight, especially if you are in the position of the overtaken ship, as my case is.'

***

Sir Samuel Evans handed down his verdict on 19 December. The collision 'ought not to have happened, but it did happen, and I have, to the best of my ability, to decide upon the evidence how and why it happened, and who are, or is, to be blamed for it, according to the law of Admiralty', Evans remarked:

> The collision took place on a fine day, not long after midday, in an open channel. The situation was free, and was not complicated by the presence, or navigation of any other vessels. One of the colliding vessels was the largest and finest product of the shipbuilding enterprise and skill of the first maritime nation of the world; the other was one of the protected cruisers of her navy. The contemplation of the calamity, and of the damage which resulted from it, cannot but produce a feeling of regret, and even a sense of pain.[17]

Evans described *Olympic* as 'a veritable "Leviathan that maketh the deep to boil like a pot"'. The evidence from *Olympic* placed her further north than the accounts from *Hawke*, putting *Olympic* further south in the channel and leading Evans to conclude, 'The explanation [...] is, I think, that the pilot with this huge vessel made too large and sweeping a turn round the Brambles.' Worse still, 'what tide there was – half a knot – would also tend to set her to the south'. He also thought that she had been going faster than Laing had argued. *Hawke* was the slower vessel immediately prior to the collision:

> The *Hawke* was *not* [original emphasis] an overtaking vessel within the meaning of the rule.
>
> If either vessel was, it was the *Olympic*, which overtook the *Hawke*.
>
> I think that in approaching the place where the vessels collided, they were crossing vessels, and, as the *Olympic* had the *Hawke* on her starboard side, she ought to have got out of the way.

Both ships' witnesses had been broadly consistent up until *Olympic* completed her turn round the Brambles and proceeded into the eastern channel. *Olympic*'s witnesses played up *Hawke*'s speed, arguing she was intending to overtake them but dropped back and turned into the stern. *Hawke*'s witnesses described *Olympic* coming out of the turn behind them, accelerating and passing them too close. Evans did not accept that *Hawke*'s helm had mistakenly been starboarded prior to the collision. Her officers and crew had testified to the contrary. Captain Smith himself had called it 'an inconceivable manoeuvre':

> I am of the opinion that in the exceptional conditions which prevailed, the forces set up in the water are sufficient to account for the *Hawke* being carried towards the *Olympic* in a swerve that was beyond her control.
>
> The *Olympic* had ample room and water in the channel to the northward. She came much too close to the cruiser on the south side of the channel. She did not take proper steps to keep out of the way. She might have averted the collision

> right up to the last if she had put her helm hard a starboard. Even when the pilot saw the *Hawke* come towards his vessel, he delayed action; and even when he took it, he ordered the helm hard aport, which was a very doubtful manoeuvre. In the agony of the situation, certain orders were given on the *Hawke*. They did not avail; but they were not wrong. In carrying them out the helm was jammed, but this was an accident for which no one is responsible, and which did not affect the collision.

Evans established:

1. That the *Olympic* has failed to establish her contention that the *Hawke* was an overtaking vessel bound to keep out of the way;
2. That the vessels were crossing vessels, and that the *Olympic* having the *Hawke* on her starboard side, should have kept out of the way of the *Hawke*;
3. That the *Hawke* was not to blame for what she did, or omitted to do;
4. That the cause of the collision was the faulty navigation of the *Olympic* by her pilot in going dangerously near the *Hawke*, and the non observance of the rule which required her to keep out of the way.

Although *Olympic* was to blame, Evans noted that she was 'in the charge of a duly licensed pilot in a compulsory pilotage district; that all the orders preceding the collision were given by the pilot; and that all his orders were properly and promptly obeyed'. Evans mused, 'it may be very desirable that the law as to the defence of compulsory pilotage should be altered [...] But it is for the legislature to make the law. My function is to declare and administer it as it now stands.' Therefore, the White Star Line's defence on the grounds that their vessel was under compulsory pilotage succeeded.

***

The defence of compulsory pilotage resulted in a big public debate. On 29 December, Sir Edward Seymour, Admiral of the Fleet and now on the retired list, wrote to *The Times*. He highlighted that 'if such a [merchant] vessel is under the charge of a Trinity pilot, in compulsory pilotage waters, her owners cannot be made to pay a penny for damage to the other injured ship'. *Hawke* was owned by the British government, which had ample financial resources to pay for repairs, but 'were she the property of a small company, or owner, however poor, the scheme of law holds good'. Seymour argued, 'This may be law, but I cannot call it justice':

> My second point is this: in a man of war, though she may be navigated by a pilot, her captain and navigating officers are still held responsible to a considerable degree, which I consider to be perfectly correct.
>
> In the merchant ship her master (or captain) who in a vessel like the *Olympic* is doubtless a first rate seaman, and would perhaps have avoided the accident had he been alone in charge, is a nonentity, and is no more responsible than if he were a mere passenger.

His letter prompted W.R. Kennedy, a retired admiral, to respond on 1 January 1912:

> The captain might as well be drinking champagne with his passengers in the saloon [...] instead of being on the bridge, looking after his ship. For, supposing the pilot was drunk, or lost his reason temporarily, or was running the ship into danger, or risking a collision, is the captain to do nothing, because the pilot was in charge? If he did, I should fancy his directors would soon dispense with his services.

Kennedy's letter then prompted another response from James J. Page, a retired pilot. He recounted giving evidence as an expert witness for an Admiralty case in the High Court in Calcutta, 'The Judge subsequently asked me if there were no conceivable occasions on which a captain would be justified in interfering with a compulsory pilot, and I replied I could think of two only – should the pilot be obviously drunk or insane.' Page quoted an unnamed 'acknowledged authority':

> The primary rule is that the pilot supersedes the master in all matters connected with the command (conduct) and navigation of the ship. His authority is supreme, his orders must be implicitly obeyed, and any negligence in carrying them out, or interference with him in his duties, will make the owners liable in case of collision.

Another letter submitted by the Blue Funnel Line (Alfred Holt & Co.), argued:

> Surely nothing can be more unjust than to compel the owners of a merchant vessel to entrust her to a navigator whose character and professional abilities the owners are not permitted to test in any way, and then ask them to pay for his misdeeds.

The state of the law had been under discussion for years, including by the Departmental Committee on Pilotage, which reported in March 1911. On 15 August 1911, the Pilotage Bill was introduced in the British Parliament to abolish 'the immunity of a ship carrying a compulsory pilot'. On 26 February 1912, Conservative MP Almeric Paget

> … asked the Prime Minister whether he proposes to introduce legislation to amend the law relating to collisions at sea, having regard to the findings of the Admiralty Court in the recent *Hawke–Olympic* case, in which it appeared that when a vessel under charge of a Trinity pilot in compulsory pilotage waters causes a collision her owners are under no legal liability to pay compensation for the damage done, even though the vessel under such pilotage may be solely to blame for the collision?

Liberal MP John Robertson, in his capacity as Parliamentary Secretary to the Board of Trade, directed him to the proposals made several months earlier. The defence of compulsory pilotage was abolished in the United Kingdom within Section 15 of the Pilotage Act, which received Royal Assent on 7 March 1913.

'THE STORM OF INDIGNATION'

The White Star Line's defence of compulsory pilotage had succeeded, but they tried to clear their ship entirely. On 13 March 1912, their chance came. In His Majesty's Court of Appeal, 'On appeal from the High Court of Justice (Admiralty Division)', Lord Justices Vaughan Williams, Farwell and Kennedy sat to listen to the arguments. As the appellants, *Olympic*'s owners were represented by Mr Laing and Mr Stephens; the respondent, Commander William F. Blunt 'of HM Cruiser *Hawke*', was represented by Attorney General Sir Rufus Isaacs, Messrs Butler Aspinall, Bateson and Robertson Dunlop.

If Laing could convince the court that the appeal should be heard, then they would be able to introduce new evidence.[18] Laing recalled that following the verdict, White Star swept the channel for any wreckage that would help to pinpoint the exact place where the collision occurred. The water was 80ft deep. Several pieces were found, including one dubbed 'the large piece', which was 'undoubtedly part of the forefoot or the lower part of the forefoot of the *Hawke*, and it weighs seven-and-a-half tons'. The damage from *Olympic*'s starboard propeller was also visible on it.

The Lord Justices allowed the appeal. Several days of hearings, held days and weeks apart, followed. Joseph Gibney, a salvage contractor based in Liverpool, was called by Laing as the first witness on 15 April 1912. He confirmed he had been asked in early January 1912 to advise whether 'there was any probability' of finding the wreckage. Gibney and diver Peter McKenna hired a tug and began their search on Friday, 9 February, finding it two days later.

Edward Wilding testified on 17 April. Harland & Wolff had sold *Olympic*'s damaged plating to a scrapyard, but 'nearly all of it' was recovered for examination. Wilding believed that the main wreckage had fallen from *Hawke* 'at the conclusion of the third cut, and just as the next blade [of *Olympic*'s starboard propeller] was beginning the fourth cut on the body of the *Hawke*'. He thought the starboard engine had been making about 64 revolutions at the time of the collision:

> … there was no question that the plating, when the vessel arrived in Belfast, was driven hard in, and the frames doubled up inside by pressure of the fore foot on the boss plating, and that the boss plating and framing had been driven down on a big loose coupling which was beneath them, and that the [*Olympic*'s starboard] engine, in its effort to go round, or to continue going round, when the pressure came on it, had torn and done very considerable damage to the framing inside the structure of the *Olympic*; and it is quite in my mind conceivable – although, of course, it is not certain – that that was sufficient to bring up the engine momentarily. Then, as the pressure of the *Hawke*'s fore foot was lifted off by her movement over the big propeller casting, that the engine was sufficiently free to be enabled to go on again.

Frederick John Blake, the White Star Line's Superintendent Engineer at Southampton since 1907, described the salvage operations and the position where the 'big piece, the fore foot' was found on Sunday, 11 February. The position had been taken on board the tug, and then the party went on board HMS *Dryad*, whose officers were in agreement.

During the fifth day's hearings, on 5 July, Colonel Hobart, a lieutenant colonel in the Territorial Force and a holder of the Distinguished Service Order (DSO), testified that he had been watching the new *Titanic* 'as she started from her berth in the new wet dock' on 10 April. His testimony about the near-collision between the *New York* and *Titanic* was potentially very damaging to White Star's case, so Laing objected immediately.

> Mr. Bateson: We are proposing to prove that these large ships do affect other vessels when they are passing them and to give an instance which is well known as the *Titanic*.
>
> Mr. Laing: I can only protest, but it does seem monstrous that we should go into this question of suction.

Ultimately, the appeal failed. On 5 April 1913, Lord Justices Vaughan Williams, Kennedy and Parker dismissed it. Kennedy noted, 'It is important to bear in mind that the burden of proving that the *Hawke* was an overtaking vessel lay upon the Plaintiffs [*Olympic*]. If they had made this out, I think that the *Hawke* must have been held to blame.'

White Star kept trying. By 9 December 1913, it had been discovered that the position of a number of the navigation buoys was marked incorrectly on Admiralty charts. *Hawke* was therefore nearer to the middle of the channel than previously assumed. Any hopes that this could help overturn the verdict were dashed on 9 November 1914. The House of Lords, as the United Kingdom's highest court, dismissed the White Star Line's final appeal.

TRIUMPH AND DISASTER

It does not pay to continue this race for the largest ship.

Albert Ballin, HAPAG General Director, 7 December 1911

It seems scarcely necessary to say much about the *Titanic*. The privileged few who have had the pleasure of visiting the ship since her arrival [...] have been at a loss to express their admiration. One person said that the *Olympic* was all that could be desired, and the *Titanic* was – well, something even beyond that!

Southampton Times, 6 April 1912

During the coming season, there will be a scent of battle all the way from New York to the shores of this country – a contest of sea giants, in which the *Titanic* will doubtless take high honours.

The Cork Examiner, 12 April 1912

I thought her unsinkable, and I based my opinion on the best expert advice. I do not understand it.

Philip Franklin, IMM Vice President, 15 April 1912

In the *Titanic* we produced a ship which we believe had only her sister vessel, the *Olympic*, for a peer, and even this latter vessel was, in some important respects, surpassed by her newer sister. We knew we had put two ships in service which surpassed all others in size and magnificence. But more than this, we believed that we had made these ships practically unsinkable, and absolutely safe for ocean travel.

Harold Sanderson, White Star's General Manager, 22 April 1912

Before travelling on this steamer [*Olympic*] I had no idea what a splendid, comfortable and marvellous steamer she is but all the time I cannot help thinking of your goodself – the man that had the nerve, and ability to order and plan her, and then of what has happened, and it all seems too cruel.

Philip Franklin to J. Bruce Ismay, 8 August 1912

12

'THE STORMIEST TRIP'

Captain Smith needed to be on board *Olympic* by 6 a.m. on 19 November 1911, in preparation for her departure from Belfast. Firemen were told to be 'on board' for 4 a.m. the following day, then the seamen had to be ready four hours later. Fourth Officer David Alexander returned to *Olympic* after a short period serving on *Zealandic*, while First Officer William Murdoch had not been to sea since leaving *Olympic* some weeks earlier. (Arthur Holme joined as chief officer, from *Majestic*. His appointment was a temporary measure until Henry Wilde could sign on at Southampton.)[1]

She was logged leaving Belfast at 4.31 p.m. on 20 November, carrying 'no passengers'. Smith reported her safe arrival at Southampton two days later, where preparations began for her scheduled departure on the 29th. The sailing date represented what was originally intended to be her eighth westbound crossing. She had missed three round trips but even so, Harland & Wolff had done a sterling job. It was a little over two months since she had started out on her fateful voyage; Harold Sanderson's original fear was that it would take three months of repairs to rectify the collision damage.

Once again, it was not to be smooth sailing. Southampton saw a prevailing fog that was 'denser than has been seen […] for years'. *Olympic* did not leave until around 7.30 a.m. the next day.

Perhaps Captain Smith was keen to make up for some of the lost time. She made a 'remarkably fast voyage from Southampton', steaming at 23.16 knots 'once past The Needles' and arriving at Cherbourg at 12.15 a.m. There were no fewer than 425 passengers waiting for her, who 'had to pass the night in Cherbourg on board the tender *Nomadic* and in the compartments of the special train', according to the *New York Herald*. Passengers for the Norddeutscher Lloyd liner *Kronprinz Wilhelm* were also held up, but the German Line 'got ahead of the White Star representative by engaging accommodations for the *Kronprinz Wilhelm*'s passengers on their arrival from Paris'. Cherbourg simply did not have sufficient hotel accommodation to cope with more than 500 unexpected passengers. The *New York Herald*'s reporter was 'able to learn that the officers and crew are anxiously awaiting the delivery of the judgement in the suit between the British Admiralty and the White Star Line'. It all made for less than a happy start to the crossing.

Olympic left at 2.35 p.m. While she was between Cherbourg and Queenstown, Harland & Wolff took the opportunity to record trial data, capturing how she was performing with the altered port and starboard propeller blades. The port engine was developing 18,700 indicated horsepower at 78 revolutions and the starboard engine 18,900 indicated horsepower at 79.3 revolutions; the low-pressure turbine recorded 18,500 shaft horsepower at 169 revolutions. All the indications were that the change was beneficial.[2]

Her passenger list of 1,090 was the lowest westbound total since she had entered service. It was the low season, but her first-class bookings were healthy with 551 passengers. The *New York Times*

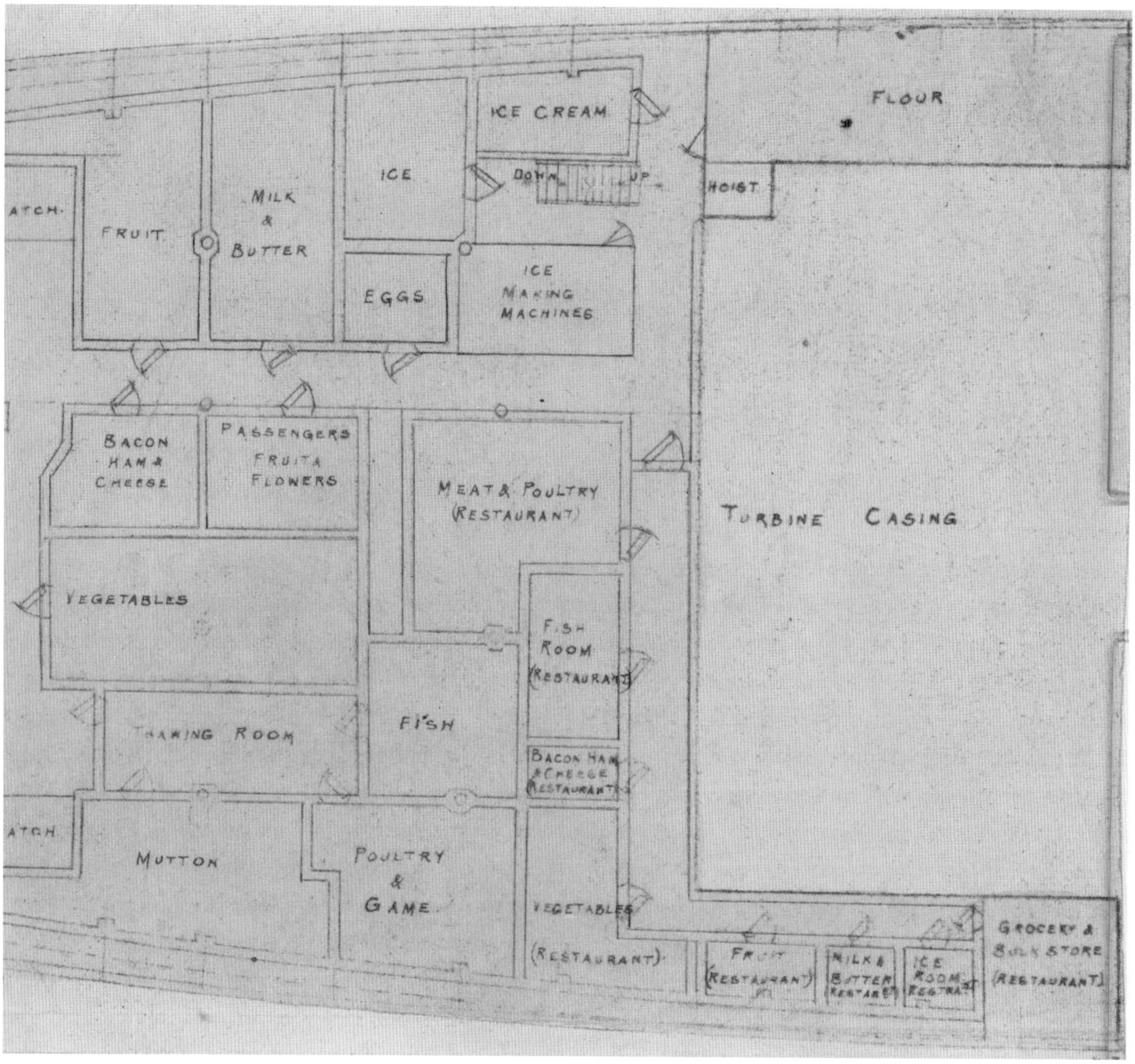

During the repairs following the *Hawke* collision, Harland & Wolff modified the refrigerated storage areas on G-Deck, aft of the turbine engine room. The many changes included the original 'mutton' and 'fish' rooms being converted into a single 'vegetables' room and the original vegetable room became 'meat & poultry (restaurant)' (see page 304). (HOYFM.2010.73.161, National Museum NI, Harland & Wolff Collection, Ulster Folk & Transport Museum)

reported an unusual number of pet dogs travelling in first class, 'They included all breeds, from a King Charles spaniel to a Mandalay Ju-Ju, a species of brown fluffy dog with smart whiskers.'

Olympic soon ran into 'very rough weather'. On Saturday and Sunday, 'Her forward decks were swept by great waves rolling over her bows … ventilators were smashed and one sea poured down the [crew] galley funnel [on the forecastle deck] and washed out the cooks as they were preparing breakfast, and another wave broke three of the steerage ports.' Her officers estimated 'some of the combers were as high as the crow's nest'.

When she arrived at New York on 7 December, she was 'welcomed by shrill blasts from many tugboats and the hoarse salutes of the ferryboat foghorns'. Unfortunately, fog delayed her at Quarantine and 'made her two hours late in reaching her pier'. Then, when she did finally dock, the customs service was slow at the pier, 'There were only twelve appraisers instead of thirty, and about one hundred inspectors instead of 150 to handle a total of [approximately] 750 cabin [first and second class] passengers.' There were still passengers waiting over two and a half hours after she docked, and some ladies 'were so tired with the long wait that they had to get substitutes to take their places' in the queue. Passengers 'had much the same kind of luck as the ship and grumbled loudly at the arrangements for their accommodation at Cherbourg' to newspaper reporters, then complained about the delays at New York. (One lady 'said she managed to get a table in the dining room of an inn in the outskirts of Cherbourg, but had to wait until 3 o'clock in the morning till the dining room was empty'.)

The press reported that 'two groups of professional gamblers' had been on board, 'one party consisting of three men, and the other of four'. Neither group was aware of the other. After approaching other first-class passengers without success, they sat down together on Sunday 'to play poker'.

One man lost $1,000 and 'then trouble started which necessitated the interference of the smoking room steward'. Mr Grahame-White, an English aviator, was 'one of the victims'. They had been playing cards in his first-class suite 'and one of his friends, a young Englishman, had given a check for $840 in addition to the ready money he had lost'. Their bridge game turned into poker, but he knew the stakes were high, 'it was quite all right'.

Less than two hours after the first- and second-class passengers had disembarked, a notice came from White Star's New York office that 1,000 members of the American Engineers' Society would visit *Olympic* at 2 p.m. 'What Purser McElroy, known familiarly as "the Eastern Despot", and Chief Steward Latimer said when they had this news was improper [impolite], and what Chief Engineer Fleming, who had got all his staff busy below overhauling gear, said was unthinkable,' according to the *New York Times*. When the engineers' party asked Chief Engineer Fleming about seeing the engine room, he was determined not to allow it. The newspaper outlined the story:

> 'You see', said the Chief Engineer, opening a blue book on his desk, 'here it is in Schedule C, Section 439, Article 57: "Barbers on the White Star liners at sea are not permitted to cut hair or trim the whiskers of passengers, officers, or crew before the sun has passed the meridian, and the commander has made it so by having eight bells struck". No, that's not it. Let me look in Schedule B, under "Engineers". This looks like it. Section 210, Article 39: "Passengers on White Star steamships taking coal from the cook to throw at sharks must be charged for same at current coaling rates". No, it's not that—'
>
> The Chief Engineer in despair turned over several more sections until he came to one setting forth in bold-faced type that no persons were to be admitted to the engine rooms on White Star Line steamships without special permit from the office, in addition to that admitting visitors to view the vessel.
>
> 'There you are', said Mr. Fleming pointing out the regulation. 'You must wait until the *Olympic* is in port another trip and come down, all of you, with the special permits. Don't wait until the ship has been in dock for three or four hours, but be waiting for the gangway to be run out'.[3]

Baltic arrived on 9 December. Her officers 'reported the worst weather she has encountered since her entry into the Atlantic service'. There were 'heavy gales and mountainous seas the whole voyage, with waves at times rising thirty feet above the bridge, which is sixty-four feet above the water line'. The sea had 'smashed the after port rail on the bridge'. She carried 2,000 canaries from Lancashire, Yorkshire and Norfolk 'on her main deck, aft'. They 'did not seem to be at all affected by the stormy weather and sang blithely each day, to the joy of the steerage passengers'.

Meanwhile, *Olympic* was preparing to sail at noon. Her late arrival meant that one of the three days 'to get her cleaned, stored and coaled for sailing time' had been lost. She had 1,918 passengers booked ('a record number for this season'), which only added to the amount of work to do. Fourteen gangs of coalmen worked 'day and night to put 6,000 tons of coal into the ship to get it all finished early' and enable the sailors to 'wash decks and clean ship before the passengers came aboard'. Luckily for them, *Olympic*'s average coal consumption of 650 tons every twenty-four hours was 'considered very economical for her size and speed'.

She got away in good time. Smooth seas on her first two days at sea with 'gentle' winds gave way to rough seas and 'moderate to strong winds'.

Among her first-class passengers was Willem Frederik Piek Junior, who wrote extensive notes about *Olympic* for the Holland America Line (Mrs Piek was also on the passenger list).[4] The Dutch company had been discussing plans for an 'improved *Rotterdam*' with Harland & Wolff. They were making a significant effort to compete with the British and Germans for first- and second-class passenger traffic. They saw potential in attracting passengers from the United States.[5]

A month earlier, on 7 November, Lord Pirrie had written to Mr J.V. Wierdsma, the Holland America Line's president. Pirrie reported that 'although the preparation of the designs for such large passenger steamers "now-a-days" entail so much work, I was pleased to find good progress had been made' with 'the alternative designs for your proposed new vessel'. He explained that it was 'quite impossible, with the arrangement of boilers now decided upon', to have the necessary cargo capacity and a speed of 17 knots 'on a vessel of less than about 740' x 86' x 47'', which equated to 'just about the 30,000 tons gross desired'. Commercial reality dictated that a larger ship was necessary, 'and I think you are quite correct, in view of other competitive tonnage building, to consider a vessel of this size'. He suggested that 'with some modifications which I thought well to suggest from experience gained during my recent visit to America, I think this works out extremely well'.

Harland & Wolff were rushing to complete the plans ('together with the alternative design of the smaller steamer'). Pirrie was keen for Thomas Andrews to discuss it with Holland America Line representatives in Rotterdam:

> I am particularly anxious that in such large vessels, the designs should be as fully considered as possible, prior to work being actually commenced, so as to avoid alterations as far as possible [...] I have also instructed Andrews to take with him as many detail ¼" scale plans of the various staterooms, etc., for discussion, as they can have ready, which will give you a much better impression of the size of the rooms.

Andrews had then met with them in Rotterdam on Monday, 4 December, presenting coloured concept plans 'of alternative designs B and C' and detailed plans of the passenger accommodation. They decided that 'the vessel [is] to be fitted with three funnels, the third to be a dummy funnel and to be fitted over the engine room casings'. Recommendations included adopting features from *Olympic*:

> Two steering engines, in place of one, to be fitted as in the SS *Rotterdam*, but with improved gearing, as in the SS *Olympic*.
>
> Harland & Wolff's pattern of watertight doors, electrically operated, as in SS *Olympic*, to be fitted fore and aft on the tank top [...]
>
> Vertical sliding windows as in the SS *Olympic* to be fitted on the port and starboard side of the upper promenade deck, as shown on rigging plan, also across fore end.
>
> The fore and aft stateroom passages on the lower promenade deck to be decorated if possible in a similar manner to the SS *Olympic*.

They discussed 'fitting a swimming bath', but decided against it 'as it cannot be worked in without dispensing with cargo capacity'. The 'type of [lifeboat] davits to be fitted' would be 'considered later' and the plan for the gymnasium on the boat deck was approved.

They held another meeting in London four days later, when Pirrie joined them. The ship's depth was increased slightly and the watertight bulkheads extended one deck higher, ensuring she would remain 'afloat with any two compartments flooded'. Later that month, Wierdsma informed Pirrie that the Holland America Line's board had approved the decision to proceed ('at the price named of £767,000'). They continued to work on details: 'We are now [...] studying the cabin arrangements [...] and comparing which system of cabins will show the largest amount of passage-money without losing sight of proper, but not exaggerated comfort.'

Piek's experience and observations on *Olympic* undoubtedly helped them in many ways. When he embarked at New York, 'everything went very smoothly and pleasantly'. The steward helped with hand luggage and the elevators came in useful to convey the suitcases to the right deck. He noticed that the ship's crew used a protective wooden device to make sure the elevator (lift) car's panelling was not damaged. The linoleum tiles in the corridors were not a bright colour and he thought that was good because 'when leaving or [in] wet weather the dirt is not so eye-catching either. The floor of the dining saloon has the same advantage and does not stand out as beautiful, but does not bother and does not at all make the impression of a lesser kind of flooring'. He thought that 'the vestibules and larger rooms with linoleum tiles look bleak because of the drab colour'. Piek made a note of a 'black curved line' on the front of the bottom step of the stairs. It 'had the very good effect of being careful when going down the stairs', but he doubted that was the reason it was there.[6]

He thought it was a mistake to allow smoking in the first-class reception room: couples 'and also gentlemen smoking alone (even pipes) from lunch onwards'. The smoke made its way throughout the ship and he thought it was objectionable for ladies who had to pass through the reception room when they arrived or departed the dining saloon at meal times.

The service in the dining saloon 'was good' but there was only one steward serving their table of four. He had 'a major complaint' that there was 'a lack of serving tables', so adjacent tables in the saloon were 'robbed for spoons ... etc'. The seats in the saloon were 'loose but can be fixed to a special cross between the legs, which allows the distance of the table to vary, with a bolt in the floor. The chairs sit very easily but are not nicely shaped and seem to me heavier than necessary.'

In contrast to the service in the dining saloon, he had seen no deck or lounge stewards when leaving New York, 'There are only two deck stewards and one of those sitting in his office while the other was probably somewhere else than where I was walking. It was only after departure that they came into view.' Piek liked the deck chairs. They were 'heavy but substantial and not fitted with the metal frame for the [passenger's] name. This is pleasant because with us [Holland America Line ships] the frame hinders sitting right in the middle when the passenger leans [back] and people complain of this.' The arrangement of *Olympic*'s deckchairs was better and 'someone who wants their seat marked gets a label on it'.

He did not like the decoration of the first-class public rooms, criticising the 'colours and combinations of them': 'bad taste in an attempt at [being] beautiful that failed'. He wrote that it was hard to believe that the ship was so new, 'The furniture looks worn out and the panelling shows cracks and tears in many places. There are also signs of poor hanging, e.g. in the reading room a cornice sagged and in the dining saloon a piece of ceiling.'

The à la carte restaurant was 'best decorated' and the 'chairs are nicely upholstered', but he thought the wood panelling 'could have been quieter and with less gold'. He found that the food was 'good but one does not get the impression of an ostentatious kitchen'. (The *carte du jour* for dinner on 15 December included eleven different fish options, including turbot, halibut and sole.)

'During the day one misses the cheerfulness of rooms like the palm court and library on the *Rotterdam*,' he wrote. 'I attribute this mainly to the promenade deck which also makes our social halls gloomy.' On the other hand, 'the large windows one has everywhere are a big advantage'. The 'windows are cheerful and well fitted. On deck, the bay windows do little harm and make the rooms friendly.' He liked the fact that the windows of the first-class public rooms on the promenade deck extended through the Boat Deck, but they gave 'little extra light during the day'. Then there were the double portholes in the first-class dining saloon, which made the sides of the saloon 'cheerful and light [...] the effect on the coloured window is as one mass of light penetrating it'. The veranda cafés 'are really part smoking room and the so-called ladies' side no more pleasant than the others'. (He disliked the 'poor separation' between first- and second-class passengers, because the first-class promenade aft on A-Deck had a second-class promenade on the decks both above and below.)

The maids and valets had their own dining room on C-Deck, 'but otherwise no room where they can stay so one finds them hanging around in the halls all the time. Were this dining room a bit better located or decorated they would surely be easy to pull together there.'

Piek noticed the 'great attraction of the swimming pool as a morning pool. Coming out you can take a freshwater shower. The Turkish bath was busier than I thought.' The squash court 'was used', but 'is not general entertainment'. The gymnasium 'is a playground for old and young and it occurred to me that few used it regularly but many tried it once or spent some time there watching'.

Piek found the beds 'very pleasant':

> Originally, the beds had a wire spring on which this mattress rested, but this sprung so excessively that on the first trip in New York, all the wire springs were supported with slats. Returning they got all the wire springs out and replaced them with a fixed weave of galvanised iron bands [...] It is very sturdy, flattens the bed [...] moreover, one does not have the grinding and squeaking that the wire spring gives. If the mattress is strong, it is certainly a point that deserves very close attention because it will also prevent the older beds from getting a dimple in the middle as is common with the wire spring.

The cotton bedding 'had the advantage of not feeling damp and cold'. The electric heaters in the bathrooms were pleasant and 'the chamber pots in the cabins have a much better model than ours and cannot topple over'. On the other hand, the dressers 'in the cabins are nowhere near as good as ours on the Rotterdam where drawers give a great advantage. The hanging cupboards in the cabins are better than ours because one can hang clothes out there for a long time, which is mostly impossible on ours.'

After *Olympic* reached Plymouth and then Cherbourg, Piek was able to visit 'many' of the first-class staterooms. They were 'all quiet [...] and seem to avoid all items that rattle. Hooks and clippings are all silent.' The same was true of the bathrooms and toilets 'where, moreover, they have the added advantage of being less in the way than our catch hooks on the toilets'. If the fittings were quiet and did not rattle, that was not the full story, 'The cabins are noisy, this applies mainly probably to the regular cabins and one can also hear the water running away from the upstairs neighbours ad nauseam.' He noticed:

> The bathtubs are enamelled iron and this is presumably lighter or has other advantages [but] I cannot admire it and [it] does not look like it will please in the long run. The bath I used already showed a rust streak. The nickel work for plumbing etc. looks heftier and nicer than I saw anywhere on a ship.

Another issue, particularly pertinent for a cold winter North Atlantic crossing, was hot water: 'Our cabin (C.24) had running water fitted in an old model 2-person sink [...] The hot water is hot in name only'. They had to fill and dispose of four or five bowls of water before it began to warm up, 'Actual *hot* [original emphasis] water we never tapped and if all sinks suffer from the same objection it must lead to a big waste of water.' It was better in the bathrooms, 'perhaps' because they were used more frequently.

He admired the electrical lighting, which was 'neatly fitted throughout the ship. Simple porcelain sockets with frosted glass wire lamps look very neat for corridors, toilets, bathrooms, etc.' The light switches in the staterooms were 'neatly placed'. Unlike Leonard Peskett, Piek did not like the lights turning off in the toilets – 'not pleasant and, in my opinion, not conducive to cleanliness'. He criticised 'a certain frugality with light'.

Another contrast was their views on ventilation. Several months earlier, Peskett had been aboard in the warm summer weather and criticised the relative lack of ventilation, whereas on Piek's winter voyage he thought, 'The artificial ventilation seems excessive to me', 'Sunday and Monday were uncomfortably cold on board'. On Tuesday, apparently after some adjustments, 'it was better and the grille in one cabin suddenly gave hot air', before getting cold again. Another issue was that the ventilation system blew dust throughout the ship as well as air: near the lubricated grates, 'a lot of dirt had accumulated in the oil or grease in a short time'. He disliked the constant 'hissing', and 'on days of cold ventilation [...] the constant cold wind around feet and ankles was more than unpleasant'. 'There was also a very noisy ventilation in the reading room that was much like a constant storm.' The dining saloon 'was quiet' with 'no ventilation noises', as was the smoke room and restaurant. 'The lounge was less quiet but here it was more the usual sound of steam heating.'[7]

This reflected the logistical challenge of ventilating and heating a huge ship like *Olympic* and trying to make sure she was relatively cool in summer and warm in winter. An article published in *Modern Sanitation* rated *Olympic* highly. Herbert M. Lome praised the sixty-four electrically operated fans on the boat deck and the absence of cowls as seen on other liners, claiming that the ship's interior was:

> ... sweet and fresh under all conditions of outside temperature or local circumstances. In the case of the staterooms, the air enters by means of a 'hit or miss' ventilator that is placed in some convenient location. If the passenger finds that his room is becoming too airy, he may shut off the supply altogether. If the weather is frigid and the current too cool in consequence, he touches a little lever connected with the ventilation and a portion of the incoming air is made to pass over steam-heated coils that are connected with the supply trunk. And if in spite of all this he continues to shiver he may turn on the electric heater.

Lome spoke to one of the officers, who said *Olympic* 'is the only ship in existence in which the occupant of a stateroom can make the temperature and the incidental ventilation suit his bodily idiosyncrasies [...] In the past it has been pretty hard to suit everybody [...], but I think that this has been accomplished.'[8]

***

By the time *Olympic* reached Eddystone Light at 7.42 a.m. on 15 December 1911, she had cut her time to five days, twelve hours and sixteen minutes, setting a record for her fastest eastbound crossing so far. Her average speed was 22.49 knots. Piek thought the 'call in Plymouth went very quickly with three tenders', one for receiving the mails, one for the luggage from the ship's hold and one for the passengers and their hand luggage.

Her next departure from Southampton on Wednesday, 20 December made it inevitable that *Olympic*'s passengers and crew would celebrate Christmas at sea. Three days later, the *Bournemouth Daily Echo* commented, 'Spending Christmas day at sea onboard the largest vessel in the world' was an alluring prospect. The reporter imagined the Christmas Day activities that a first-class passenger might enjoy:

> Rise, say, at eight o'clock, and enjoy a swim in the incomparable bath [...] before partaking of breakfast, during which greetings are exchanged with relations and friends. During the early morning listen for a while to the ship's orchestra: then for the needful exercise walk a mile or two round the glass enclosed promenade deck should the weather be cold, or the open deck above in the event of mild breezes. The more active will perhaps visit the gymnasium and enjoy 'punching the ball' or indulge in rowing or any of the numerous aids to physical fitness. Luncheon follows, and you can have it either in the superb dining saloon or in the Louis XVI restaurant, where the famous chef will rise to lofty heights on Christmas day and win encomiums from the most critical of gourmets. A Turkish bath will appeal to

> some during the afternoons [...] whilst others will prefer a cigar and a quiet chat in the smoke room, and again a couple will doubtless be found in the squash racquet court [...] Many will later forgather in the beautiful reception room, in anticipation of the great event of the day – Dinner, with a large D – when absent friends will be remembered with pity but toasted with enthusiasm. A quiet game of bridge in either the lounge or smoke room will bring the evening to a close. In the second and third class quarters special arrangements will also be made for the enjoyment of passengers.

The *New York Times* noted that the passengers included 'a number of prominent persons', including Wilhelm Backhaus, the well-known pianist, and Georgette Leblanc (Madam Maurice Maeterlinck), 'en route to Boston, where she will sing in the forthcoming production of *Pelléus et Melisande*'.

Olympic arrived in New York safely, but it was another quick turnaround and she left again on 30 December. Passengers would also celebrate the New Year at sea! The *New York Times* reported that J.P. Morgan boarded fifteen minutes before departure. His son, J.P. Morgan Junior, and son-in-law were among those seeing him off. Morgan had booked the starboard side 'parlour suite' on C-Deck. There, he told reporters that he had originally intended to sail on *Adriatic* at the start of December. He had no definite plans about when he was going to return to the United States.

She reached Plymouth on the morning of 5 January 1912, where 253 passengers disembarked and 2,319 bags of mail were unloaded. She was quickly cleared for Cherbourg (320 passengers and 764 bags of mail) and Southampton (291 passengers and ninety-six bags of mail, plus specie valued at $550,000). The press noted J.P. Morgan's name on the passenger list, a number of British aristocrats and officials, including the British Trade Commissioner of Canada and the secretaries of the American Legations in London and Rome.

Morgan's 'special train met with a slight accident' after he disembarked at Cherbourg. The *New York Times* recorded that it had been 'passing through the rail-laid streets of Cherbourg [when] it collided with a tramway car, but fortunately no one was injured'. He reached Paris safely not long after midnight.

The *Journal of Commerce* noted that 'the palatial White Star steamer *Olympic*' had made an 'excellent' eastbound crossing 'during which her speed averaged over 22 knots'. She 'gave further evidence of her popularity with ocean travellers by bringing the splendid total – for a winter voyage – of 440 first class'. 'Many distinguished people were aboard, the diplomatic corps in particular being well represented.' It reported that she had made six round trips to New York and back, carrying a 'splendid average' of more than 1,000 first-class passengers on each round voyage.

All was soon ready for *Olympic*'s seventh westbound crossing, departing Southampton on 10 January 1912. The *New York Times* reported she was set to sail 'with a big list of passengers'. There were almost 400 booked in first class but second and third were relatively quiet. She carried 894 passengers in total. J. Bruce Ismay was on board for the first time since the maiden voyage.

Olympic did not reach Cherbourg until 8.21 p.m. The run to Queenstown was better and she left on time. The first two days out saw 'rough seas' with a 'fresh south-westerly to strong north-westerly wind' and snow on 13 January. It grew steadily worse as the North Atlantic hurled all its fury with high seas and gale force winds. *Olympic* completed 485 miles on her first day at sea, falling to 471 and then 389. On Sunday afternoon, *Olympic* 'dipped her nose' under a 'monster' wave, 'which curled on the forward deck and tore off the four ton steel cover of hatch No. 1, hurling it ten feet in the air and dropping it twenty feet or more further aft'. One newspaper reported:

> The thunder of the hatch cover and the assault of the towering sea brought the diners to their feet and startled the whole ship's company. The cataract [waterfall] that roared through the open hatch got down into the stokehold and engine room [*sic*] and flooded the steerage quarters. Ports were smashed and ventilators carried away. Nobody was hurt ... but everybody for a moment was wrought up.

Olympic 'was slowed down to mere steerageway and kept so until the weather moderated'. By the time the day's run was reported at noon on Monday, she had covered a mere 301 miles. Her average speed was 12.3 knots. Captain Smith 'said that the seas were the worst ever'. Some of his officers estimated that the great wave 'was at least sixty feet tall'.

According to a report in the *New York Times*, one of J. Bruce Ismay's first-class stateroom portholes was 'smashed by the big sea'. 'He found that, in spite of her size, the ship rolled and pitched a good deal, but that her promenade decks were dry and that the majority of the passengers were able to go into the saloon for their meals.' (*Olympic* carried portable deadlights on E-Deck, with one for every 50ft of her length, but none were provided for D-Deck or above. They were available to help secure damaged portholes if needed. If it was true that one of Ismay's portholes had been damaged, the crew would probably have undertaken a temporary repair. The ship's carpenter had some hinged wooden plugs and a number of 'spare glasses for each type of sidescuttle [porthole] throughout the ship'. On occasion, the ship's crew had 'to renew a glass [porthole] under conditions of considerable difficulty'. The porthole glass was typically 1¼in thick, illustrating the sort of force necessary to fracture or smash it.)

In a sensationalised account published several months later in the *Seattle Star*, Fred L. Boalt wrote, 'Steerage passengers fell on their knees and prayed, believing the end had come.' He recalled being on the Boat Deck, 'where the four giant stacks towered above us, and the rigging of the wireless sang in the wind. There were long avenues of lifeboats up there, and we amused ourselves speculating on with what success they could be lowered':

> When you leaned over the rail of that deck and looked down, it made you dizzy. Her sides were like the sides of a steel-plated cliff, punctured far below with tiny holes, which were portholes.
>
> When she mounted a wave, it seemed she would never stop until she scraped the sky. When she sank in a trough, it seemed she would never stop until she buried her nose in the ocean floor. The greatest strains seemed to come when, after wallowing in the depths, she struggled to rise. Then the grinding and groaning in her vitals were terrible and indescribable.[9]

Conditions improved somewhat. The *New York Times* carried a report on 15 January: 'SS *Olympic*, Southampton to New York, 1,476 miles E. at 6 p.m.; due 17th, noon. White Star Line.'

Sea conditions presented a 'moderate wind and sea' on 16 January and then moderate winds but a 'smooth sea' on the last day of the crossing. *Olympic* finally reached the Ambrose Channel Light Vessel by mid-afternoon on 17 January. It took six days, six hours and sixteen minutes to complete the crossing. Her average speed of 18.62 knots was destined to be one of the slowest crossings of her entire career. She did not leave Quarantine until nearly 4 p.m.

Unfortunately, docking successfully presented another challenge. She 'was jammed by the flood tide against another pier on the opposite side of the slip, and the tugboats pushing against her huge sides made about as much impression as a policeman might make with shoulder against St. Patrick's Cathedral'. Hundreds of people were waiting 'on the pier in the cold' and 'it was only by sagging away, ahead and astern, and then using warps and winches, that the ship was docked' at exactly 8 p.m. The delay of two and a half hours prompted much comment from steamship men who were concerned about 'the inefficiency of New York as a port for the accommodation of the new, colossal steamships now being added to the Atlantic fleet'. One suggestion reported in the *New York Times* was that New York needed 'a common landing stage in the lower section of the harbour near the Battery where four big liners could go alongside to land their passengers and mails without difficulty at all stages of the tides'. It would remove the delay at Quarantine 'to take off the mails in the Post Office tenders, and also it would enable the dispatch of mails for points beyond New York'.

Cedric left Queenstown the day after *Olympic* and also 'experienced strong gales and high seas the whole way across'. She was a slower ship in any case; nonetheless, she was thirty hours late reaching New York on Sunday, 21 January. Captain H. Smith told reporters that she had 'shipped a sea on the starboard beam that carried away the iron ladder on the promenade deck aft, and a second sea smashed No. 3 lifeboat, and also badly battered Nos. 5 and 7, hanging on the davits'. The seas were 'fully sixty feet high' and *Cedric* only completed 167 miles between noon on Monday, 15 January and noon the following day. On the morning of Wednesday, 17 January, a 'sea came over' the promenade deck on the starboard side and 'smashed a window of one of the staterooms with such force that a large piece of the glass was hurled across the cabin and embedded in the panel of the door'. It was only luck that nobody was injured, because the stateroom's occupant was out at the time. Two ladders leading to the boat deck were 'washed away by the same sea and were driven along the promenade deck at such a velocity that they burst through the iron gates dividing the first and second class decks and bent a strong iron ladder right aft'.

That week, *Nieuw Amsterdam*, *Bretagne*, *Laurentic* and *Columbia* all arrived late, 'owing to the stormy weather on the Atlantic and with their decks' rigging and boats covered all over with ice and snow'. The *New York Times* said *Nieuw Amsterdam* 'had fully a foot of frozen snow on her decks and men had to cut away the ice from her hatches with axes before the baggage could be got up on deck'. Each ship's commander and crew 'agreed that it was the worst weather they had ever encountered, as there was no let up. It was all bad and the vessels were tossed up and down by the high seas as if they had been corks.' (Almost exactly two years earlier, *Lusitania* had been 'battered' by an 'inferno of wind and wave' on the North Atlantic. Her experience on 10 January 1910 was dramatic. A wave over 80ft high smashed the bridge and wheelhouse, wrenching off the ship's wheel, then rolled aft along the boat deck. The foremost two lifeboats on the starboard side 'were lifted from their chocks' and smashed as the sea bent the davits, the officers' quarters were flooded and documentation from the chart room was swept out to sea.)[10]

These experiences illustrated the harsh conditions the liners faced. It was a happy coincidence that the White Star Line had already planned that *Olympic* would have a week in New York on 'each trip until March, instead of turning round in three days, as she has been doing'. The extra time was all the more welcome. Earlier that month, the *New York Times* reported the change would 'give the officers and crew a chance to rest and overhaul the ship thoroughly'. They noted that Norddeutscher Lloyd 'takes off all its fast ships in January and February except the *Kronprinz Wilhelm*', which was scheduled to arrive in New York early in February and early in March. (There was a gap of more than a month between their express sailings from New York: *Kronprinzessin Cecilie* sailed on 4 January then *Kronprinz Wilhelm* on 13 February.)

Olympic left her New York pier at noon on 24 January, passing the Ambrose Channel Light Vessel shortly before 2 p.m. Her passenger list included 563 first class. The weather on her first days at sea included 'moderate to fresh' northwesterly winds and 'snow squalls'. 'At the outset of the voyage the comfort of the passengers was affected by the prevalence of snow squalls, but later, despite strong winds and high seas, she made such capital progress that better time could not have been kept even by a railway train', one reporter wrote, 'At midnight on Sunday Captain E.J. Smith [...] calculated that his vessel was 756 miles west of the Eddystone [Light].' Smith estimated *Olympic* would arrive at 12 p.m. on Tuesday, 'So well did the liner keep her time, that when she dropped anchor in Cawsand Bay it was just two minutes after midday.' She completed the crossing in five days, sixteen hours and thirty-six minutes, averaging 22.56 knots – her fastest eastbound crossing to date.

Four tenders were 'quickly [...] moored alongside' to receive over 400 passengers and the bags of mail. J. Bruce Ismay 'took a very keen interest in the arrangements'. The transfers went very well 'and before half past one the *Olympic* was steaming full speed across the Channel for

Cherbourg. There another three hundred passengers landed, but by the time they reached the French port those who had disembarked at Plymouth had practically arrived in London.'

The Great Western Railway Company ran 'three special trains' from Plymouth's Millbay Docks and the last was away at 3.30 p.m. Thirty minutes later, Norddeutscher Lloyd's *George Washington* arrived. She had left New York the day before *Olympic* and her passengers 'anticipated that they would have reached Plymouth' first, 'but a strong north-west gale dashed their hopes […] *Olympic* swept by'.

The *Western Daily Mercury* headlined an article '*Olympic* in a Race Across the Atlantic', noting *Olympic* had run 539 miles in one day compared to her German competitor's 447 miles. In truth, there was nothing in it. *George Washington* averaged 4 knots less than *Olympic*. They were not racing each other and there was no reason for *Olympic* to slow down and delay her arrival. Perhaps J. Bruce Ismay took the opportunity to do some personal bookkeeping towards the end of the eastbound crossing. His expenses in January 1912 included $100 in 'cash advanced at New York' (£20 10*s* 8*d*); expenditure of $170.30 at New York's Ritz Carlton restaurant (£34 19*s* 4*d*); and a bill of more than 13*s* paid to 'Purser McElroy SS *Olympic*, clothes pressed'.

Olympic arrived at Cherbourg at 8.30 p.m., returning to Southampton on 31 January.

***

Storms certainly caused disruption in January 1912, but there was plenty to celebrate. The *New York Times* reported a first-class passenger 'record' for White Star in 1911, claiming they carried 'the largest numbers taken by any British or Continental steamship company' between European ports and the United States.[11] Whether the report was correct or not, White Star certainly carried 32,407 first-class passengers on its services from Southampton and Liverpool to New York and from Liverpool to Boston; Cunard carried 31,662 on its Liverpool to New York and Boston services; Norddeutscher Lloyd carried 28,783 on its 'regular' and 'express' services between Bremen and New York; and HAPAG carried 19,696 on its Hamburg to New York and Philadelphia services.

White Star rose to the top in 1911, from third place in 1910. Its total first-class passenger numbers rose 16 per cent while their rivals' fell. They owed their prime position in large part to *Olympic*'s popularity. She carried 40 per cent of the first-class passengers on White Star's Southampton express service. Her performance was all the more impressive because she only entered service in June and then she missed the height of the westbound season due to the *Hawke* collision.[12] In contrast to its strength with first-class passengers, White Star's second- and third-class passenger carrying lagged rival Cunard. They carried 66,203 third-class passengers against Cunard's 87,303. The German Lines competed aggressively with the support of their government, capturing a vast share of third-class traffic from the Continent.

Olympic was successful and compared well to her nearest rivals. Looking at the arrivals and departures from New York in 1911, her sailings were most commonly scheduled closest to Norddeutscher Lloyd's *Kronprinzessin Cecilie*. *Olympic*'s first-class passenger lists were higher than all her competitors' closest sailings on eleven out of twelve occasions and second class on six out of twelve occasions. The same was only true for third class on a minority of occasions because the German lines performed so strongly.[13]

***

J. Bruce Ismay was considering his future. Early in January 1912, he proposed to Harold Sanderson that he would retire as IMM's president on 31 December. Sanderson would then take over. Following discussion with others within IMM, Ismay then suggested postponing his retirement until 30 June 1913, but Sanderson was not happy. After a great deal of discussion back and forth, they agreed Ismay would retire on 30 June 1913, but continue as a member of IMM's British Committee. Sanderson would have six months of leave from IMM, working on other directorships, before becoming president from 1 July 1913. They agreed that the arrangement would be kept private until 31 December. Charles Steele, secretary of Morgan & Company in New York, wrote on 2 March:

> I quite agree there is no reason whatever why it should be made public at the present time [...] I think it much better that nothing more should be said on the subject until the time arrives to act. I trust everything will run along smoothly now. There are lots of things that may happen before 30 June 1913.[14]

Ismay had plenty to keep him busy meanwhile. Throughout the second half of 1911 and early months of 1912, he was dealing with business concerns related to huge ships like *Olympic*. On 1 August 1911, Ismay had written to Cunard's chairman Alfred Booth on an 'absolutely private and confidential matter'. He explained that 'for some time past, I have been endeavouring to arrive at some scheme with the object of eliminating undue competition among the steamship companies most largely interested in the North Atlantic trade' (Cunard, HAPAG, Norddeutscher Lloyd and White Star). Ismay's idea was that they could set up a steamship company with each of the four competitors having an equal interest. It would:

> ... go a very long way towards ensuring peace in the North Atlantic trade, and would, at the same time, reduce the heavy depreciation which we have at present to face owing to each company having to build new tonnage to enable it to compete with the tonnage being put into the trade by its competition.

He asked for Booth's opinion before taking it up with others, 'There may be absolutely nothing in the scheme, but the more I think of it, the more I feel it would be the means of preserving harmony in the trade.'

Booth responded by wondering if Ismay's proposal was enough, because it only eliminated competition 'by tying up the money'. He thought they might need to agree on abstaining from building new tonnage for the New York trade 'for some definite period'.

By early December 1911, Ismay and HAPAG's Albert Ballin were in further discussion by telegram. Ballin had seen press reports that White Star had ordered 'another steamer of the "*Olympic*" type'. On 7 December, Ballin suggested it would be:

> ... [a] wise policy if we agree not to exceed certain dimensions. Propose 900 feet by 100 feet. We are discussing contract for a third boat of this type, and shall probably have to order a fourth ship like that within [the] next six months to secure a weekly service from April 1915. Think it does not pay to continue this race for the largest ship.

His proposal was out of tune with HAPAG's own shipbuilding plans, because they were already building ships even larger than *Olympic* to run from Hamburg to New York.

The next day, Ismay sent a response to Ballin:

> Have seen no newspaper report which gives correct information concerning dimensions [of] our new steamer. Would welcome understanding amongst leading companies which would restrict competition for owning largest ship, and will gladly discuss question with you and others [at the] first suitable opportunity. Feel strongly [that the] only means [of] preventing unnecessary construction of huge vessels would be some arrangement of joint ownership by [the] principal companies engaged in North Atlantic trade, as suggested to you in London. We are considering building three additional steamers, in order [to] run semi-weekly service between Southampton and New York.

Ismay wrote to Booth and asked, 'Would it be practical to form two groups – you and ourselves on the one hand, and the Hamburg Packet [HAPAG] and Norddeutscher Lloyd on the other, and come to some arrangement?' He suggested agreeing that 'any new tonnage of the exceptional type' would be 'jointly owned'. Booth was 'in entire sympathy' with the views Ismay expressed to Ballin, '... no doubt the details of any concrete scheme would be difficult to work out'. He understood Ismay was proposing 'some definite agreement as to the number of steamers of the largest size which the trade will warrant, and, [...] that there should be joint interest in some form or other in these steamers'.

Albert Ballin wrote to Ismay from Hamburg on 16 December, confirming 'with great pleasure' that he was ready to join discussions. He proposed meeting 'in the second half of January in Paris'. Unfortunately, Ismay was already due to sail on *Olympic* and would not return before the end of the month. Ballin was not available later in February, so he suggested early that month.

On 10 February 1912, Ismay responded, 'I cannot arrange to come to Paris between the 12th and the 20th [March ...] as my daughter is going to be married on the 21st, and, naturally, there will be a lot of matters calling for my attention.' He explained, 'We shall have to defer our discussion until some time in May, as it is my intention, all being well, to make the first voyage on the *Titanic* leaving Southampton on the 10th April, being due back about the 27th.'[15]

***

Insurance was another area where the shipping lines might benefit from cooperation. White Star arranged *Titanic*'s insurance cover. On 9 January 1912, broker Willis Faber & Co. approached Lloyd's of London. All of the marine underwriters and the majority of London marine insurance companies subscribed, underwriting between £200 and £75,000 each, so that the risk was well spread. Twelve insurance companies and more than fifty Lloyd's syndicates pooled the risk.

Two days later, all was complete. *Titanic* was insured for £1 million, which compared to £6.75 million for the 'total insured marine losses' for the whole of 1912 (the same cover applied to *Olympic*). *Titanic*'s policy ran from 30 March 1912 to 30 March 1913. The excess was £150,000.[16] The insurance policy represented approximately two-thirds of her cost, then the remaining balance was carried by IMM's own insurance fund, which pooled the risks of the shipping lines in the combine. The combine 'inaugurated the system of insuring its own ships to a large extent', with the constituent shipping lines paying insurance premiums. Any losses were charged to the fund. Philip Franklin did 'not believe there is any company crossing the Atlantic that carries such a large proportion of its own insurance as the subsidiary companies of the International Mercantile Marine'.[17]

Ballin wrote to Ismay on 14 February, 'Do you not think that the companies which own these large ships of the "*Olympic*" and "*Mauretania*" type might agree upon an insurance arrangement calculated to reduce the expenses for insurance by some mutual arrangement?'[18]

Ismay said that Ballin's idea 'is one that I have had in mind for a long time ... it would be very advantageous'. Booth agreed, citing worries with the insurance of *Lusitania* and *Mauretania*:

> In order to avoid paying exorbitant premiums, we carry a larger total loss risk than I at all care about, and I should certainly much prefer to see this risk spread over eight or nine ships rather than concentrated on two or three. In any case the subject is well worth further consideration.

Towards the end of March, White Star and Cunard were looking for insurance experts to advise them. It was a striking fact that White Star's 'steamships, piers, wharves, buildings and furniture' were valued at £8,136,685 at the end of 1911. *Olympic* alone represented around 19 per cent of the company's assets and with *Titanic*'s completion, an even more significant proportion of White Star's capital assets would be tied up in just two ships.[19] However low the risk of a total loss was, it was worth addressing.

***

Olympic continued the 'North Atlantic Ferry'. She left Southampton on 7 February for her eighth round trip to New York. Despite moderate gales and rough seas, she reported a run of 563 miles on her fourth day out at an average speed of 22.7 knots, which may have been the fastest westbound day's run since she had entered service. She reached the Ambrose Channel Light Vessel safely at 9.37 p.m. on 13 February with her 1,015 passengers. It had taken her only five days, eleven hours and forty-eight minutes to complete the crossing, averaging 21.94 knots.

While he was at New York, Captain Smith took the opportunity to write a letter replying to his nephew, Frank Hancock. He apologised for the delay in answering, 'When I thought of your letter it was somewhere where it was not convenient to write, and then it would slip from memory, so there you are, it was not want of appreciation, I assure you.' The aftermath of the *Hawke* collision rumbled on, 'You have no doubt heard we are appealing the case. I have not much hope as it is hard to upset a verdict in England; however it will let them see we are not going to take it lying down.' Smith lamented the previous crossing: 'We had a poor trip as far as weather was concerned.' He explained, 'I leave this ship after another [round] voyage and bring out *Titanic* on April 10th from Southampton,' then signed off, 'Your affectionate Uncle, Edw. J. Smith'.[20]

When *Olympic* left New York on Wednesday, 21 February, she had 1,037 passengers on board, including 546 in first class. She carried '2,900 bags of mails for Plymouth, and 900 for Cherbourg', according to a report in the *Hampshire Independent*'s 'Shipping News' column, three days later. All seemed set for a successful voyage.

It soon became clear *Olympic* was 'doing exceptionally well'. For 'two days and four hours', she 'steamed at a speed higher than she had previously accomplished', logging a day's run of 546 miles. Depending on the precise clock adjustment made that day, it was an average speed of around 23.5 knots. Passengers 'anticipated the giant liner making her record trip'.

By 4.26 p.m. on 24 February, *Olympic* had reached latitude 44° 20' North, longitude 38° 36' West. She was steaming at 'about 23 knots' when there was an unfortunate shudder, which brought her good performance to an end. She had lost a propeller blade and the port engine 'raced' for a short period. Captain Smith wrote in the ship's log that 'a blade from the port propeller was cast off'. *Olympic* proceeded at reduced speed 'of about twenty knots' for the remainder of the crossing.

Passengers near the stern 'felt the shock, but others who were forward knew nothing of the mishap until the next day's run was recorded as being considerably lower than those that had gone before'. There was no 'excitement or panic'. One passenger said, 'Except for a slightly reduced speed [*Olympic*] travelled as steadily and well as she ever has done.'

The loss of a propeller blade was not the only unfortunate event. Two days later, third-class passenger James Kneetone was lost overboard. Kneetone 'was deported as mentally deranged' from the United States and 'confined in hospital under watch' aboard ship. Watchman Ronald Lamont left him 'alone for about three minutes and during that time he escaped to the deck'. Although a thorough search was made of the ship, he was not on board and the following day, Smith reported, 'Nothing has been seen of Kneetone since.' Rumours circulated ashore that *Olympic* had been in a collision, prompting the White Star Line to send a telegram from Liverpool which was printed in the *Belfast News-Letter*:

> *Belfast News-Letter*, Belfast. *Olympic* not in collision. Merely threw off blade of port propeller. Will call inwards as usual at Plymouth, Cherbourg and Southampton, afterwards proceeding to Belfast, where due on Friday to fit blade. Sails again for New York from Southampton on Thursday 7th of March – Signed, Ismay.

Olympic reached Plymouth on the morning of Wednesday, 28 February. The *Western Daily Mercury* carried a report that she had made the crossing in six days, three hours and forty-five minutes at an average speed of 20.9 knots.

She proceeded to Cherbourg and Southampton, dropping off her remaining passengers while preparations were under way to receive her at Belfast for repairs. No time was lost and she left

Southampton for Belfast, where she arrived on 1 March.[21] (The same day, a question was asked in Parliament whether 'the sunken wreck which has recently caused an injury to the SS *Olympic* is still a possible cause of danger' and whether other ships could be warned. The Board of Trade replied that they were 'informed by Messrs. Ismay, Imrie and Company that the report that the SS *Olympic* struck a sunken wreck is devoid of any foundation, though [...] she lost a blade of her propeller'.)

Her running mate, *Oceanic*, was also in the news. She returned to Southampton 'shortly after ten o'clock' on Saturday, 24 February, having experienced 'during four or five days of the passage from New York gales of exceptional violence'. On one evening, 'the seas were so high that the wireless equipment attached to the foremast was carried away'. It did not stop her leaving Southampton as scheduled on 28 February. She called at Cherbourg as usual that evening and then left Queenstown late afternoon on the 29th. Not long after leaving, *Oceanic* also lost a blade from her own port propeller. Captain Haddock advised her speed had been reduced to around 17 knots instead of her more usual 19.5 knots. One of White Star's frustrated officials quipped that 'the bed of the ocean must be covered with propeller blades in the tracks of the steamships east and westbound'.[22]

Olympic arrived in Belfast 'about half an hour too late to enable her to take advantage of the morning tide,' according to the *Belfast Weekly News*. An initial proposal to take her into dry dock that evening was 'not considered advisable', so she stayed in Belfast Lough until early on Saturday, 2 March. She 'steamed slowly up the Lough', accompanied by the tugs *Herald*, *Hercules*, *Herculaneum* and *Hornby*, reaching the dock entrance a little after 9 a.m. Many spectators had gathered to watch her enter the dock and the Harbour Police were 'on the scene to prevent any interference, through over curiosity, with the duties of the workers'. Harland & Wolff officials, including Thomas Andrews, superintended proceedings as *Olympic* entered the dry dock. 'So successfully was this accomplished that there was not even the slightest scraping of the sides of the huge vessel against the walls.'

Around 10 a.m., *Olympic*'s 'vast towering mass' was in position as workers 'set about arranging the caisson' and making all the necessary preparations to secure her in position. Once the dock had been drained, the shipbuilder set to work replacing the missing port propeller blade and all was ready to enable *Olympic* to leave the dry dock on 4 March.

It looked like all was set for *Olympic* to return to Southampton so she could sail as planned. Unfortunately, things did not go as intended. She was 'taken in tow by five tugs, and proceeded slowly down the channel stern first; but after she had reached the second buoy it was decided to bring her back'. One of Harland & Wolff's officials informed the press that 'the weather became boisterous, and as the tugs were not quite able to handle her satisfactorily in the channel owing to the strong wind, it was deemed advisable, in order to obviate any further trouble, to bring the liner back'. The problem was that the only place *Olympic* could be accommodated was back in the dry dock, because *Titanic* was occupying the only berth that was suitable.

She was taken back temporarily. Rumours spread in Belfast that she 'had got out of the channel and touched the sandbank', but Harland & Wolff 'emphatically denied' it. The *Belfast Telegraph* did report that 'the graving dock was pumped dry to enable an examination to be made of the propellers', which had not 'sustained a scratch'. It said officials 'made a close examination of the vessel, and, we understand, were thoroughly satisfied with the survey [...] The massive rudder is as perfect as the day it was first shipped.'

White Star had hoped that *Olympic* could sail on Thursday, 7 March, a day later than her original schedule, but it was forced to change its plans. The sailing was simply cancelled and there was no equivalent White Star sailing from Liverpool either (Cunard's *Lusitania* did sail from Liverpool for New York with a healthy passenger list; she arrived on 15 March).

Olympic finally left Belfast three days later than intended, lightly loaded. She arrived at Southampton on 8 March. Three days later, an 'official announcement' was made at Southampton. The *Cork Examiner* reported that the American liner *St Paul*'s scheduled 23 March sailing for New York was being cancelled and *Oceanic* would not sail as scheduled on the 27th. Then the *Liverpool Daily Post* reported further cancellations of White Star's Liverpool sailings, including *Majestic*'s 10 April departure for Boston. It advised that *Oceanic* passengers could travel instead via '*Baltic*, leaving Liverpool on the 28th [...] or the *Olympic*, from Southampton, on the 3rd April'.

The key issue for the shipping lines was the availability of coal supplies. A national strike began in Great Britain late in February. It was set to last more than a month. Luckily for *Olympic*'s passengers, she left Southampton as planned on 13 March. Her 1,586 passengers included 1,137 in third class, which was the highest number she had carried westbound. First class was quiet by comparison.

She was 'held up for a couple of hours by a heavy fog off Calshot' and then made her usual calls at Cherbourg and Queenstown. Once again, she experienced bad weather and for her first three days at sea she battled high head seas, gales and 'strong' winds. They kept her progress each day to less than 500 miles. Towards the end of the crossing, the weather cleared up somewhat, but it cut her average speed down to 20.84 knots. She did not arrive at the Ambrose Channel Light Vessel until 6.03 a.m. on 20 March. The same day, a report by the Associated Press from London said that 'considerable numbers of miners' had already restarted work in 'some Scottish coal fields'. Some men in Yorkshire collieries were 'anxious to resume today'.[23]

While *Olympic* was at sea, the *Lyttelton Times* published an account attributed to 'Mr J.H. Harris, one of Messrs. Anthony Hordern and Sons' managers', who had recently returned to Sydney, Australia, 'after a trip round the world'. His travels had included a voyage on *Olympic*:

> He was simply astounded at its spaciousness and luxuriousness. 'With all those passengers onboard', he said, 'you could look along the decks at times and think they were almost deserted. The expanse of deck was phenomenal, and the fittings of the vessel almost beyond description. In a two berth cabin such as I occupied hot and cold water were laid on, there were several electric fans, a couple of couches, chairs, a radiator, and every convenience one could desire'.

The rush to get *Olympic* ready for the return voyage presented the American workers with the added challenge of loading additional coal supplies. *Olympic* had arrived in New York with a surplus of 'not more' than 500 tons. They had arranged for her to take on coal wherever it could be stored, including the reserve bunker, stokehold plates, Boiler Room 3 and even the third-class dining saloon ('9,210 tons stowed'). Her third-class passenger list was very low and the full dining saloon capacity was not needed.

When she left New York on 23 March, *Olympic* carried 663 passengers – her lowest to date. It reflected the general uncertainty.

White Star issued a 'special letter' to their New York office, copied to both Captain Smith and Captain Haddock, near the end of the month, noting that 'the trouble now existing may continue in a more or less acute form for some time'. Normal coal supply conditions would not return 'for an indefinite period' and they resolved to prioritise *Olympic* and *Titanic*, 'cancelling the *Oceanic* as may be necessary'. There was 'sufficient coal at Southampton for the next sailing of each ship but this was not a safe position', so they instructed that speed should be reduced to 20 knots to save fuel, 'thus ensuring future sailings'.[24]

Although *Olympic* encountered 'strong' west-south-westerly winds on 26 and 27 March, otherwise 'the weather was fine'. It showed in her average speed for the eastbound crossing, which was reported as 22.7 knots in the press. The *Western Daily Mercury* stated that 134 passengers disembarked and $700,000 worth of silver was offloaded at Plymouth. The majority of her passengers disembarked at Cherbourg and Southampton, including the Secretary of the German Embassy.

White Star reassured the travelling public 'that arrangements have been made to ensure the scheduled sailings from Southampton of the mammoth liners *Olympic* and *Titanic* throughout the season'. One newspaper reported on 30 March, 'The usual speed of these vessels is close on 23 knots, but in view of the difficulty now experienced in securing coal supplies [...] these steamers will, while the trouble continues, be run at a speed of about twenty knots.' The *Southampton Times* noted that White Star 'seem to be very confident' in *Olympic* and *Titanic* sailing on schedule, but *Oceanic* might stay 'on the idle list' throughout April. They reported further cancellations of American Line sailings.

While *Oceanic* was laid up at Southampton, White Star took advantage of Haddock's availability to assign him to command *Titanic* while she neared completion at Harland & Wolff. When he returned to Southampton, Captain Smith had to rush to Belfast to take over. The *Southampton Times* said it 'has been definitely decided that Captain Smith, who has commanded the *Olympic* since she was first placed in commission, will be transferred to the new steamer *Titanic*'. It 'understood that Captain Haddock will be promoted in succession to Captain Smith', from *Oceanic* to *Olympic*, 'He, too, is an old and trusted servant of the White Star.'

Oceanic lost a number of her crew. Fifth Officer James Moody wrote to his family to say he would have preferred to stay with her, rather than transfer to the new *Titanic*. He was 'glad they are keeping me on the Atlantic run as I think I deserve a summer here after weathering such a rotten winter'. He continued, 'I don't know whether I explained to you that the *Titanic*, *Olympic* and the *Oceanic* are going to be the three White Star ships on the S'ton–New York route for at least all next summer, unless of course any unforeseen diversion turns up.'[25]

Late in March, the Belfast Steamship Company's *Patriotic* left on her sea trials. Lunch was served on board afterwards. Thomas Andrews spoke for Harland & Wolff. He said he was 'very sorry that Lord Pirrie could not be present, as otherwise the duty with which he was entrusted would have been in much abler hands'. Harland & Wolff's prosperity 'was never greater'. They were expanding elsewhere, but he was sure their headquarters would always be in Belfast. The total wage bill had just set another record, reflecting a growing workforce. 'A good deal of that had been involved in the completion of the two first class passenger ships – the *Patriotic* and the *Titanic* [laughter]. It was a heavy task for the firm to complete the two ships in one week.' Andrews looked forward to *Titanic*'s own sea trials, scheduled for 1 April.[26]

Hopes were high. On 12 April, the *Cork Examiner* referred to 'the fight during the coming season'. 'There will be a scent of battle all the way from New York to the shores of this country – a contest of sea giants, in which the *Titanic* will doubtless take high honours.'

13

'A CONTEST OF SEA GIANTS'

Titanic's hull presented an impressive sight, riding high at the fitting-out wharf after her launch, but there were occasional difficulties. On 16 June 1911, Harland & Wolff wrote to Harbour Master Captain McIntyre:

> No doubt it has been reported to you that two cross channel steamers passed [...] this morning at a high rate of speed, so much so that the *Titanic* ranged heavily on her moorings, and one of the 5 inch long fore and aft springs parted, and is this is most dangerous to men working on the wharf and on the scows at this vessel, we shall be glad if you will please take the necessary steps to prevent a recurrence.

They had another complaint on 21 July, when a similar incident happened and caused a wire hawser to fall, 'Some of the men narrowly escaped being killed and knocked into the water.' The Belfast Harbour Commissioners issued a circular to the steamship companies and commanders the next day. It led to 'good results', but one of *Titanic*'s mooring lines parted on another occasion in mid-December. Captain McIntyre attributed it to a 'very heavy gust of wind', which coincided with a steamer passing at a normal speed.[1]

Harland & Wolff's workmen had the gargantuan task of turning her empty hull into a floating city, but the shipbuilder was simultaneously making significant alterations to *Titanic*'s design. They were under consideration as soon as *Olympic* completed her maiden voyage, based on J. Bruce Ismay's and Thomas Andrews' observations. At the fore end of the Boat Deck, the officers' quarters deckhouse was enlarged, enabling six additional first-class staterooms, and forward on B-Deck, seven first-class staterooms were added on either side at the expense of the enclosed promenade, which was little used on *Olympic*.

The port and starboard 'parlour suites' on B-Deck were enlarged with an enclosed private promenade deck (the first seen on an ocean liner). The first-class staterooms amidships were all moved out to the ship's side, so that they enjoyed a remarkable sea view, and there were more wardrobe rooms than on *Olympic*. Twelve new single-berth staterooms were added inboard, ideal for accommodating maids and valets.

The staterooms near the aft grand staircase were rearranged and a reception room introduced around the staircase. The restaurant was expanded to the port side of the ship at the expense of the second-class promenade, reflecting its popularity on *Olympic*; the pantry and galley were enlarged in consequence, and separate facilities were added, including storage for meat and poultry, fish and a chef's larder. On the starboard side, the enclosed promenade was initially intended as a 'restaurant promenade'. It was still shown on first-class accommodation plans as late as January 1912, but became the Café Parisien by March. Another late addition just above were new first-class staterooms with bathroom facilities, one on each side of the ship off the aft grand staircase on A-Deck.

Titanic outfitting in September 1911. The modifications to B-Deck are already under way. The original windows for the enclosed promenade deck are being cut away fore and aft. Those abreast the grand staircase remained, because that area became an entrance hall; those abreast the port-side 'parlour suite' also remained, because the area became the suite's private promenade deck. (Günter Bäbler collection)

Many details were incorporated following recommendations Thomas Andrews had made for *Olympic*, such as fitting first-class staterooms instead of a large C-Deck cloakroom and increasing the size of the first-class reception room on D-Deck by reducing the entrance halls on either side. One of the electric baths was removed (reflecting their lack of popularity on *Olympic*) and the layout of the Turkish baths rearranged. The cooling room layout was altered so passengers no longer had to walk through it to reach the shampooing rooms.

Despite the loss of some of the second-class promenade space on B-Deck compared to *Olympic*, second-class passengers had little to complain about. There was plenty of deck space available.

Titanic's plans reflected a significant improvement for some of her third-class passengers. When *Olympic* entered service, she had two sections of open-berth third-class accommodation forward on G-Deck: the 124-berth Section 'D' and 40-berth Section 'F', as shown on third-class accommodation plans issued in April 1911. Further aft, there were eighty-six third-class berths in Section 'S', right at the stern, which were portable and could be removed so that the area could also be used for cargo. On *Titanic*, Sections 'D' and 'S' became permanent third-class staterooms and Section 'F' became portable third-class staterooms, built to the 'White Star Pattern'.

Other changes helped the crew working the ship. The layout of the refrigerated storage area on G-Deck, abaft the turbine engine room, was altered in line with changes made to *Olympic*, including moving the vegetable storage room.

All these changes entailed plenty of preparation. The work needed was no mean feat, even for one of the world's premier shipbuilders, but perhaps it helps explain Pirrie's comments to the Holland America Line early in November 1911. He sought 'to avoid alterations as far as possible' (see page 286) in such large ships.

★★★

The heavy machinery, including the boilers and engine parts, was lifted by crane then lowered into the hull. By the end of July 1911 work was continuing on the deckhouses and superstructure, and on 20 August, *Titanic*'s foremast was fitted. Nine days later, Chief Engineer Joseph Bell took up his position.[2]

In September, work began in earnest reconfiguring B-Deck, changing the layout of the rectangular windows so that they suited the new first-class accommodation rather than the previously planned enclosed promenade space.[3] (On the 30th, H.P. Mutters en Zoon wrote to Harland & Wolff concerning 'the new order for the alterations of the staterooms on B-deck'.)[4]

'The electric turbine [*sic*] and auxiliary and other machinery, and all the boilers are onboard,' reported the *Manchester Courier* on 19 October, 'the funnel uptakes complete; and one of the funnels is erected in position, as also are the masts [...] on the principal passenger decks, accommodation is very far advanced, and the work is proceeding speedily'. The *Belfast News-Letter* confirmed 'that the main staircase is in position; also the electric elevators'.

J. Bruce Ismay had said to the *New York Times* on 30 June 1909 that both ships would be in service in 1911. When *Olympic* was launched, the White Star Line had announced to the American press that she would be in service from summer 1911 and then *Titanic* from the autumn. However, their sailing list, which was published 11 January 1911, did not show *Titanic* sailing that year at all. Although *Titanic*'s construction was faster than her older sister's up until the completion of the double bottom, by the time of launch her pace of construction was running about four months behind. The gap grew further during outfitting, so that it took about seven months longer to complete *Titanic*.[5] Following *Olympic*'s return to Belfast to repair the damage sustained in the *Hawke* collision, White Star announced that *Titanic*'s maiden voyage would be 10 April 1912 (the date originally given as her second voyage).[6]

The first-class rates published in January showed that they had taken the opportunity to increase *Titanic*'s earning power. One of the two B-Deck 'parlour suites' on *Olympic* cost about $2,300 for up to two passengers in summer season, with an extra charge of $50 for each additional passenger. On *Titanic*, the new private promenade deck added to the same suite and the use of a servant's room allowed the company to charge $4,350 for 'one or two passengers and servant' at the same time of year. On *Olympic*, stateroom B54, with a private bath, toilet and wardrobe room, could be booked for two passengers at $1,050 in the summer season. On *Titanic*, stateroom B76 was situated in a similar location on the port side of B-Deck amidships, with a private bath, toilet and wardrobe room, and could be booked for two passengers at a slightly higher rate of $1,100. These staterooms could also be let as suites. B57 and B59, 'comprising two bedrooms, with private bath and toilet (each room being fitted with a wardrobe room) and a servant's room', were available to 'one or two passengers and servant' for $2,000 and '$50 extra for each additional person'.[7] For the ordinary first-class staterooms forward of the grand staircase on B-Deck, the rates were essentially the same but there was an increased number of them on *Titanic*. The net result was that *Titanic* had the potential to earn significantly more first-class revenue than her older sister.

While White Star focused on the passenger accommodation, Harland & Wolff always had a keen eye out for technical improvements. Edward Harland had explained in May 1873, 'After the first few voyages of a ship we get returns to guide us in the structure of future vessels.' They learned lessons from previous large ships when designing *Olympic* and her sister. Now, they had an opportunity to review *Olympic* and apply refinements to *Titanic*.

Her record was pleasing. In May 1915, Edward Wilding discussed *Olympic* from the point of view of her general strength, 'We have had less repairs to the *Olympic* than to any large ship we have ever built, due to external causes, of course.' The Board of Trade's principal ship surveyor later referred to *Olympic* and said she 'has, I think, proved to be a

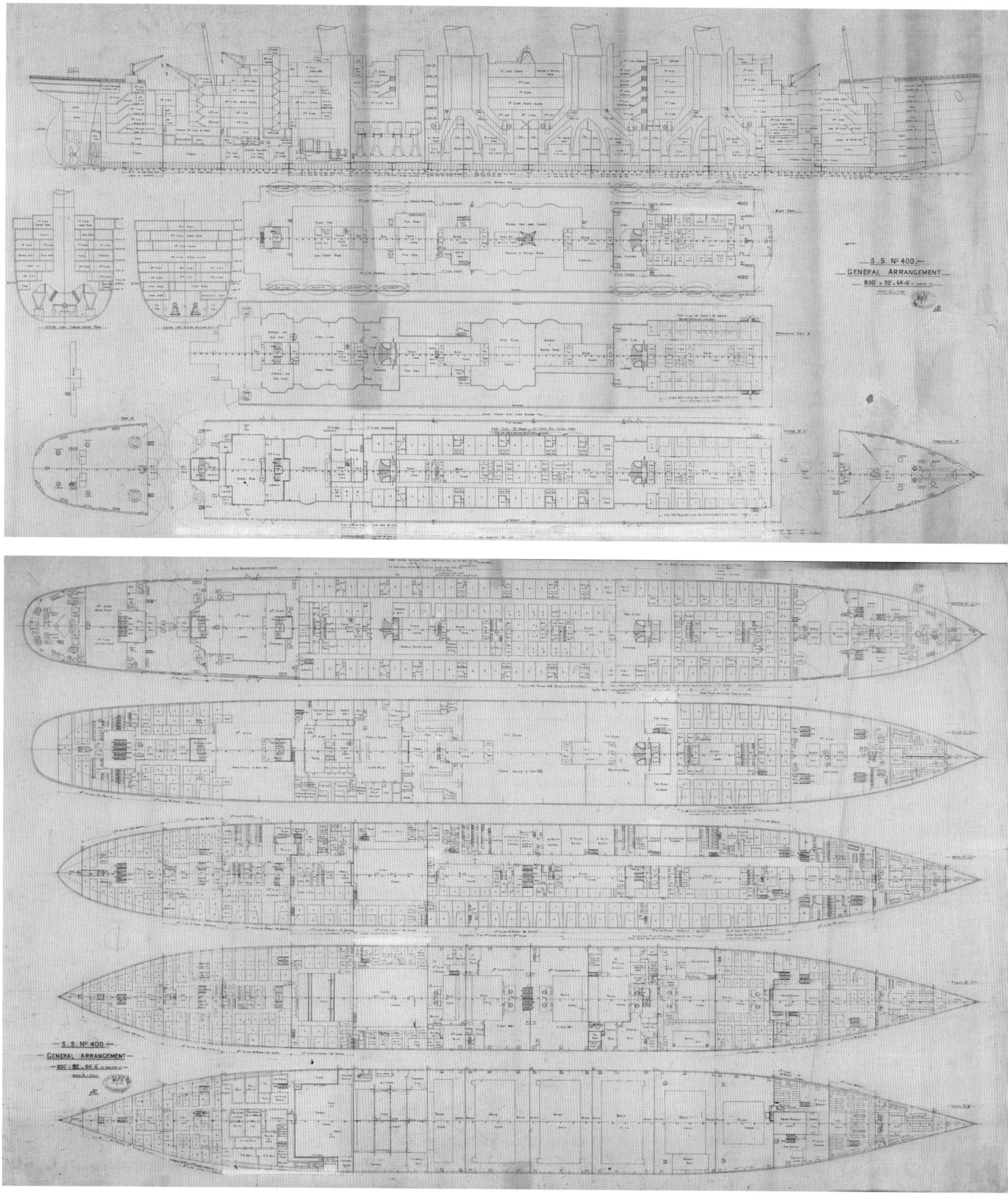

Olympic 'SS No. 400 General Arrangement' plans bearing the Harland & Wolff stamp and dated 10 October 1911. The originals were scaled to 1/16in to 1ft. They show her before the modifications made in October and November, including the removal of the cloakroom on the port side of the grand staircase on C-Deck or the enlargement of the first-class reception room on D-Deck. (HOYFM.2010.73.161 & 162 © National Museum NI, Harland & Wolff Collection, Ulster Folk & Transport Museum)

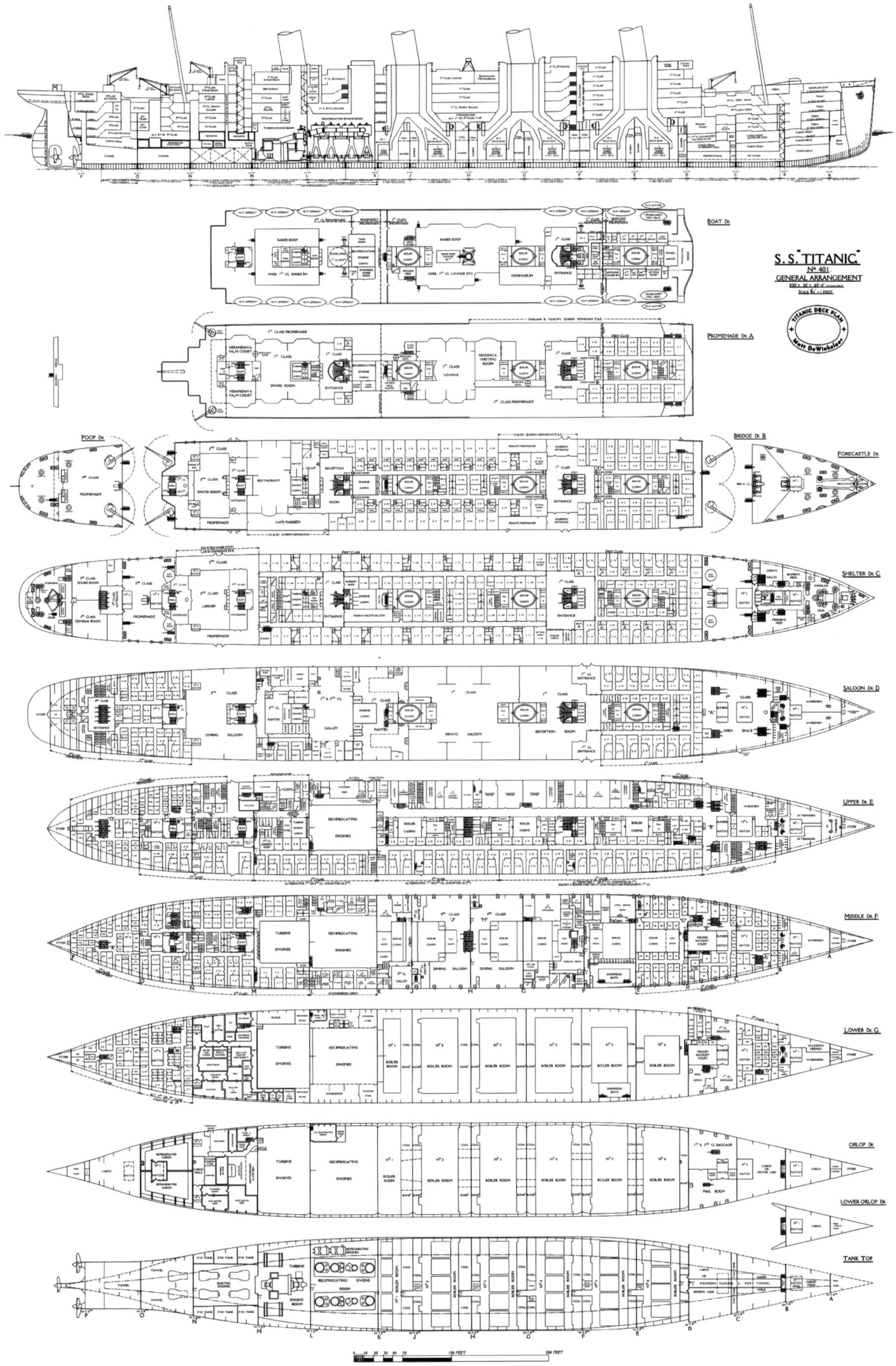
S.S. "TITANIC"
Nº 401
GENERAL ARRANGEMENT
TITANIC DECK PLAN
Matt DeWinkeleer
BOAT Dk
PROMENADE Dk A
POOP Dk
BRIDGE Dk B
FORECASTLE Dk
SHELTER Dk C
SALOON Dk D
UPPER Dk E
MIDDLE Dk F
LOWER Dk G
ORLOP Dk
LOWER ORLOP Dk
TANK TOP
100 FEET
200 FEET

successful ship in the matter of strength'. Cunard took an interest, too. Leonard Peskett compared *Olympic*'s structural design to *Aquitania*'s. The Cunard ship's design was then modified to bring her closer in line with her rival.[8]

Thomas Andrews had a reputation as a perfectionist. He made all manner of small modifications to *Titanic*. On 5 January 1912, he approved changes to the beams supporting A-Deck, near to the aft mast, which may have been intended to reduce vibration. On 14 February, an amendment was made to the ship's plans – a new 'weather wall' with a door and window was shown at the fore end of the A-Deck promenade on each side. Harland & Wolff were preparing to install a screen with windows extending for about 200ft on each side of the ship, providing an enclosed promenade area. Early in March, scaffolding was being prepared so that work could get under way; by the end of the month, the screens were completed.

All the improvements to the first-class accommodation on B-Deck had left *Titanic* with no enclosed promenade space for first class, even though second class still had their own enclosed promenade aft on C-Deck. In all likelihood, the new enclosure on A-Deck was intended to remedy that.[9] (It is unlikely the modification was very costly. On 26 June 1913, Harland & Wolff costed a modification to *Statendam*, which they were building for the Holland America Line. It consisted of 'lengthening [the] boat deck and extending [the] steel screen at fore end of promenade deck right across vessel' for £300.)[10]

The heavy storm that *Olympic* experienced in January 1912 demonstrated the North Atlantic's fury at its strongest. She came through it well, but it did lead to Harland & Wolff making changes to the fore hatch and its cover for *Titanic*. Francis Carruthers also noted some improvements to *Titanic* on 13 February, noting that they were 'fitting a strap [steel plating] 1 inch thick over the [riveted] landings at the upper turn of the bilge' on the port and starboard side, 'in way of no. 6 boiler room and extending three frame spaces forward of the watertight bulkhead at the forward end', and the same further aft, 'in way of the turbine room and extending two frame spaces into the reciprocating [engine] room' (making the riveted landing there a quadruple-riveted one rather than a treble-riveted one). Following the storm, a small number of *Olympic*'s rivets had needed repair or replacement. Harland & Wolff were taking action to stop it becoming a recurring maintenance nuisance during the ship's overhauls and to prevent it happening on *Titanic*. (When *Olympic* was dry-docked early in March 1912, Francis Carruthers took the opportunity to do a hull inspection. He 'carefully inspected the vessel inside in the neighbourhood of these slack rivets but found no further signs of stress'.)[11]

Then there were the ship's propellers. On 17 January, Harland & Wolff requested that the Belfast Harbour Commissioners provide divers 'with the necessary gear, to be at the new deep water wharf at 9 a.m. tomorrow (Thursday), to complete the removal and recovery of the launchway packing pieces' from *Titanic*'s bottom. The Harbour Commissioners confirmed they could provide the necessary apparatus but that Harland & Wolff would need to arrange qualified divers.

On the morning of Saturday, 3 February, *Titanic* entered dry dock for exactly two weeks to have her propellers fitted and painting completed. Harland & Wolff's records show that *Titanic*'s propellers differed from *Olympic*'s. During the *Hawke* collision repairs, the pitch of *Olympic*'s port and starboard propeller blades had been increased from 33ft to 34ft 6in. Her performance must have been pleasing because the pitch was increased even further on *Titanic*, to 35ft. The pitch of *Titanic*'s centre propeller was the same as its counterpart on *Olympic*, but its diameter was 17ft and the number of blades was three instead of

Previous Page: *Titanic* General Arrangement Plans show many differences with her sister *Olympic*, including lesser known details such as the additional first class staterooms adjacent to the aft Grand Staircase on A-deck and changes to the Turkish and Electric Bath complex. (These comprehensive plans also include the Orlop Deck, Lower Orlop Deck and Tank Top. which are not included in the *Olympic* General Arrangement Plans.) (© Matt DeWinkeleer 2025)

four (see page 7 of the colour section).[12] J. Bruce Ismay recalled that Harland & Wolff advised the White Star Line that *Titanic*'s speed would be 'a little bit better' than *Olympic*'s.[13]

***

Everything was coming together. On 6 February 1912, Harland & Wolff's managing directors held their monthly meeting and confirmed they were working to a date of 1 April for the ship's trials. They noted that the workmen employed on *Titanic* were 'idling' and it 'has become very marked', so they arranged to meet again in three days to discuss what to do. Another important question came up – 'improving the staff dining room menu'. John Kempster and Thomas Andrews were left 'to deal with' it!

Titanic's four Engelhardt collapsible boats were despatched to Belfast individually on 6, 9, 13 and 15 February, from Greenock. Her loadline certificate was signed off by Walter Howell, one of the assistant secretaries to the Board of Trade and Chief of the Marine Department. It was valid for four years, until 23 February 1916.[14]

Titanic left the dry dock on 17 February 1912 and returned to the deep water wharf, but there were logistical complications when *Olympic* had to return to Belfast. *Olympic* was dry-docked on 2 March. When she came to leave the dry dock on the 4th, *Olympic* had to proceed stern first because *Titanic* was in the way and preventing her being swung around. *Olympic* returned to the dry dock due to unfavourable weather conditions, leaving for the second time on 6 March. *Titanic* was moved into dry dock on the same day, taking *Olympic*'s place. The dock was pumped dry merely to inspect her bottom. *Olympic* then left Belfast on the 7th. *Titanic* was ready to leave the dry dock on the 8th but was

Titanic (left) getting ready to enter dry dock. *Olympic* (right) has just exited the dry dock to make room for her sister. The date is 6 March 1912. (Davison Photo, 'Kempster Album')

delayed for a day due to unfavourable weather conditions. She then berthed at the deep water wharf.[15]

Despite these challenges, on 11 March, Harland & Wolff noted that 'as far as could be seen at present No. 401 [*Titanic*] would be finished and ready to leave on the 1st April'.[16] Nine days later, they asked the Belfast Harbour Commissioners for 'permission for three open fires at the new deep water wharf, well clear of the wooden portion, for heating Bitumastic Composition', for use on board *Titanic*. They assured them that a man would always be there when the fires were lit, to keep an eye on things. Permission was granted.[17]

***

At the end of March, *Titanic*'s crew began to sign on. The majority were merely delivery crew, signing on for the trip to Southampton.

Early in November 1910, J. Bruce Ismay had been quoted in the *New York Times*. According to the newspaper, he anticipated *Oceanic*'s commander, Captain Haddock, would command *Titanic*. Despite the arrangements they had made for Captain Smith to take *Titanic* and Captain Haddock to take over *Olympic*, Haddock had the honour of signing on as *Titanic*'s first captain. He signed the particulars of engagement on 25 March, serving for barely a week until Captain Smith could take over. William Murdoch transferred from *Olympic* and was promoted to chief officer; First Officer Charles Lightoller transferred from *Oceanic*, as did his colleague,

TITANIC'S GROSS TONNAGE

Titanic's gross tonnage (a measure of the total enclosed space) was calculated as 46,328.57 tons when her details were added to the British Registry on 25 March 1912. By comparison, *Olympic*'s gross tonnage was 45,323.82 tons when she was completed in 1911. *Titanic*'s gross tonnage was therefore 1,004.75 tons greater than her older sister's. It placed her as the largest ship in the world.

It is a popular myth that the enclosure of part of the A-Deck promenade was largely responsible for increasing *Titanic*'s gross tonnage. In fact, her registration specifically stated that the 'open space on promenade deck, abreast windows port side – 198 feet long' was 'not included in the cubical contents forming the ship's register tonnage' (the same applied for the starboard side).

The real reason for *Titanic*'s increased gross tonnage was the expansion of the first-class accommodation on B-Deck, which accounted for the majority of the increase. This passenger accommodation counted as enclosed space, whereas the enclosed B-Deck promenade on *Olympic* did not (a fact specifically noted on *Olympic*'s registration). Other changes, such as an enlarged officers' quarters deckhouse, contributed as well.

In March 1913, *Olympic*'s own gross tonnage increased to 46,358.70 tons, following a major refit which included expanding the restaurant and adding a Café Parisien on the starboard side of B-Deck. These modifications ensured that the remainder of the enclosed promenade space on this deck was now counted as enclosed space in the gross tonnage calculation. She emerged from the refit with a higher gross tonnage than *Titanic*, even though her A-Deck promenade was never enclosed. (If *Titanic*'s enclosed promenade space on A-Deck had been included in the gross tonnage calculation, it would have added 720.51 tons, increasing her gross tonnage to 47,049.08 tons.)

OLYMPIC AND *TITANIC'S* DISPLACEMENT (WEIGHT)

Contrary to popular belief, *Olympic* and *Titanic* weighed the same overall, because their displacement (weight) at their designed load draught was the same: 52,310 tons. However, the composition of that weight differed, because of the changes made to *Titanic*.

The load displacement was made up of two basic elements: lightweight tonnage and deadweight tonnage. The lightweight was the weight of the ship herself without cargo, fuel, water, ballast, passengers or crew. The deadweight was the weight the ship could carry, loaded to her designed draught, including the weight of cargo, fuel, water, provisions, passengers and crew.

Titanic's lightweight was about 480 tons greater than *Olympic*'s, but their total weight (displacement) was the same because *Titanic*'s deadweight was correspondingly lower.

	Olympic	*Titanic*
Deadweight	14,030 tons	13,550 tons
	+	+
Lightweight	38,280 tons	38,760 tons
	=	=
Displacement	52,310 tons	52,310 tons

It is interesting to compare them to *Adriatic*. She had a deeper draught, a lightweight of 22,370 tons and a deadweight of 18,420 tons. Her design allowed for a remarkable cargo capacity even though she was significantly smaller.

David Blair, serving as second officer. Others followed.[18] The ship's crew all had to be on board for 4 a.m. on Monday, 1 April, for trials and the voyage to Southampton.

'While the vessel was fitting out at her berth, the boats were swung out at various times,' wrote Francis Carruthers, and on 25 March, 'a final inspection was made when all the 16 boats under davits were swung out and lowered and hoisted again into position, all the gear proving satisfactory'. Despite the last-minute preparations, a number of invited guests toured the new liner on the 30th. 'The visitors [...] were deeply impressed by the immense size of the *Titanic*,' wrote the *Belfast News-Letter* on 1 April, 'as well as by the excellence of the accommodation.' The newspaper said she was due to leave Belfast that day for Southampton, but the *Liverpool Daily Post* reported her departure was delayed 'due to stress of weather.' High winds made it a dangerous to manoeuvre the ship and so *Titanic*'s sea trials were postponed for a day. Harland & Wolff had already recorded the ship's delivery date as 1 April. They crossed it out and amended to 2 April.[19]

14

'THE DANGERS OF THE SEAS'

'As the hawsers cast off the *Titanic* – the largest vessel in the world – floated proudly on the water, a monument to the enterprise of her owners and the ingenuity and skill of the eminent firm who built her,' according to the *Belfast Weekly News*. Crowds of spectators, 'who had assembled on both sides of the river raised hearty cheers as she was towed into the channel. The mammoth vessel presented an impressive and picturesque spectacle, looking perfect from keel to truck.' She 'left the deep water wharf shortly before ten o'clock' on 2 April 1912.

'The trials consisted of running from slow up to full speed ahead, manoeuvring with the turbine cut out, going astern and swinging the vessel at full speed ahead with the helm hard over to test the steering gear,' wrote Francis Carruthers afterwards. 'After sea trials, the vessel was brought back to Belfast Lough.'[1] The centre anchor had been 'lifted over the bows and the wire cable brought out through the stem and shackled on'. The anchor crane, steam capstans and engines for working the drum all worked 'satisfactorily'. Then, on completion of her sea trials, 'the port and starboard anchors were dropped in Belfast Lough and lifted again'. Compasses were adjusted. Everything was 'satisfactory'.[2]

Carruthers busied himself completing all the paperwork. All the watertight 'doors and frames for the doors were substantial and well made; when completed tests were made of the closing arrangements and the fit after closing of all the doors; all proved satisfactory'. The steam-pumping arrangements were 'more complete than usual'. (When *Olympic* entered service, she carried her four Engelhardt collapsible boats on the Boat Deck. *Titanic*'s configuration changed. Collapsibles B and A were placed atop the officers' quarters on the port and starboard side, abreast the first funnel. Carruthers noted, 'The intention was to lower the boats to the boat deck and then move them over to the nearest set of davits.')

Thomas Andrews was 'busy' the entire day, inspecting and superintending all manner of work finishing the final details aboard ship. 'We got away this morning in fine style and have had a very satisfactory trial,' he wrote to his wife. 'We are getting more ship shape every hour but there is still a great deal to be done.'[3] He had eight colleagues with him, forming a 'guarantee group': William Parr, assistant manager of the Electrical Department; Roderick Chisholm, the ship's draughtsman, who had led the design of the lifeboats; Anthony Frost, outside foreman; Robert Knight, leading hand engineer; William Campbell, joiner apprentice; Alfred Cunningham, fitter apprentice; Frank Parkes, plumber apprentice; and Ennis Watson, electrician apprentice.

TURNING CIRCLES AND STOPPING DISTANCE

During her sea trials, *Titanic* 'made a very small [turning] circle, I observed, and it was noticed,' recalled Francis Carruthers. Data for similar passenger liners is comparatively limited but supports his observation. The turning circle diameters are expressed as a multiple of the ship's length between perpendiculars.

Ship	Speed	Turning Circle
Lusitania	24 knots	4
Titanic	c. 20 knots	4–4.5
Queen Mary	c. 28 knots	4.5–5

(Based on more limited secondary source information published in the *New York Times* in 1913, it appears *Imperator*'s turning circle was about 5.2 times her length between perpendiculars.)*

Titanic also performed creditably when her engines were reversed 'full speed astern', bringing her to a stop in only 3.5 times her length between perpendiculars.

Ship	Speed	Stopping Distance
Titanic	c. 20 knots	3.5
Lusitania	22.8 knots	6

* *Lusitania* data taken from Bell, Thomas, 'Speed Trials and Service Performance of the Cunard Turbine Steamer *Lusitania*', read at the spring meetings of the Forty-Ninth Session of the Institution of Naval Architects, 9 April 1908, p. 8. For the turning circle, 'The resulting circular path was approximately 1,000 yards, or four lengths of the ship, in diameter.' For the stopping trial, her engines were running at 166rpm when 'full astern' was ordered and she stopped in about three-quarters of a nautical mile (around 4,560ft). She came to a stop in three minutes, fifty-five seconds.

Titanic data taken from Edward Wilding, 'Running straight with the helm put hard over she turned in a complete circle; and the distance across the circle from the original course was between 4 and 4½ lengths of the vessel [3,400 to 3,825ft], in diameter; that is, the tactical diameter.' For the stopping trial, he testified before Lord Mersey (Question 25,295) that *Titanic* stopped in three minutes, fifteen seconds and 'the distance run was just over 3,000 feet'. He said then that the starting speed was about 18 knots, but three years later, he gave the speed as 20 knots before the Limitation of Liability Hearings. *Titanic*'s engines were not reversed as quickly as *Olympic*'s had been the year before: *Olympic* was doing about 22 knots and 'stopped in just about the same distance; rather less, although the speed was higher', 'Deposition of Edward Wilding': The National Archives and Records Administration – Northeast Region (New York).

Queen Mary data is taken from a letter from John Brown & Co. Ltd, 30 October 1945, noting, 'In the case of Queen Mary, a half circle trial at 21 knots was carried out and observations made indicate that diameter of turning circle was about 4 to 4½ lengths. It is deducted that diameter of circle at full speed would be 4½ to 5 ship's lengths': Merseyside Maritime Museum, B/CUN/3/6/10/1. *Imperator* data has to be treated with some caution because it was more limited and from a secondary source. However, the information in the *New York Times* report indicated she had a *c.* 4,450ft diameter turning circle at *c.* 23 knots, which equates to about 5.2 times her length.

Titanic heads out for her sea trials on the morning of 2 April 1912. (Davison Photo, 'Kempster Album')

Titanic's 'parlour suite' on the starboard side was one of two suites to have its own private promenade deck – a new feature for ocean liners. (© TopFoto)

The 'parlour', decorated in Adam style, was spacious and comfortable. *L'Illustration*, 1912. (Author's collection)

The Café Parisien was a new feature for *Titanic*, occupying space that was originally envisaged as an enclosed restaurant promenade area. *L'Illustration*, 1912. (Author's collection)

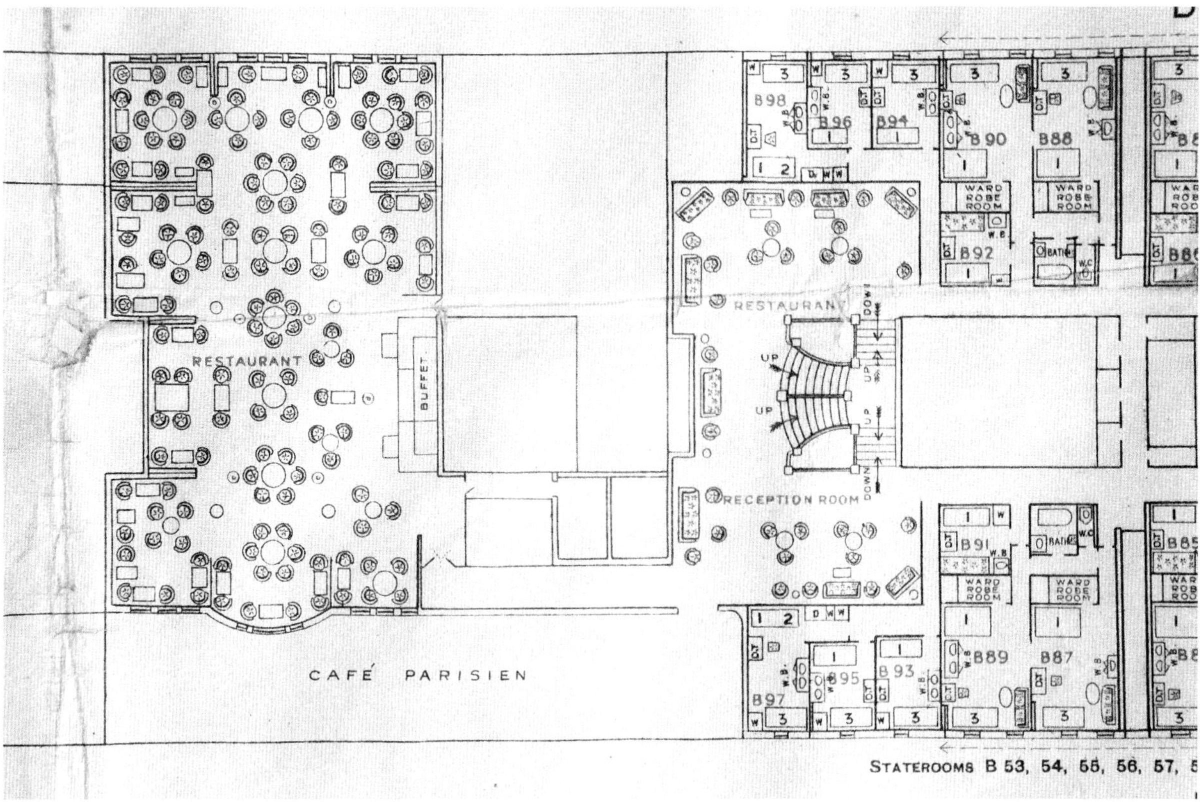

Titanic's expanded à la carte restaurant and reception room were a marked improvement on *Olympic*'s. (George Behe collection)

Harold Sanderson went on board at Belfast. That evening, *Titanic* sailed for Southampton. (There is a popular myth that a solitary first-class passenger, Wyckoff Van der Hoef, was on board. In reality, he booked his ticket for the maiden voyage in Belfast *after Titanic*'s departure.[4] No tickets were sold by the White Star Line because it was simply a delivery voyage.)

On 3 April, Harland & Wolff did further tests of the ship's electrical installation, checked the sanitary arrangements 'and various other matters of that sort'. *Titanic* was lightly loaded to a draught of about 30ft as she sped south in 'fine and bright' weather: 'for a big ship, it was fine weather; if you had been out in a cockle sheel you would have called it rough'. Her engines were running well. 'We obtained a speed of about 23¼ knots for several hours,' Edward Wilding recalled.[5]

Lloyd's List reported *Titanic* off the Lizard, on the southernmost tip of the Cornish coast, at 10.30 a.m. on 3 April, but it seems probable that this should have been 1.30 p.m.[6] The port authorities logged *Titanic*'s arrival at her Southampton berth as 1.15 a.m. on the 4th.[7]

Later that day, *Titanic* was fully dressed with all her flags hoisted (rather than Good Friday, 5 April, as popularly believed).[8] When the crew particulars of engagement relating to the delivery trip were received by the authorities in Southampton, they noted Captain Smith had not signed them. He was 'informed that he has not signed agreement and he sent a message that if he could not call before he sailed he would do so on the return of the ship'.[9]

'From an early hour' on 4 April, Thomas Andrews spent time working aboard. He wrote home, 'I think the ship will clean up all right before sailing on Wednesday,' noting that Lord Pirrie's doctors had refused to allow him to make the maiden voyage. Until sailing day, 'he was always busy, taking the owners round ship, interviewing engineers, officials, agents, managers, sub-contractors'. His secretary, Thompson Hamilton, observed that Andrews was 'never for a moment idle'. He noted problems with the hot press in the restaurant galley and looked at

Titanic docked at Southampton days before sailing. She was docked stern first, in a contrast to *Olympic*'s maiden voyage, which meant her departure would be faster and easier. Workmen are still touching up the funnel paintwork. (Restoration by Steve Hall; Mark Chirnside/J. Kent Layton/Daniel Klistorner collection)

reducing the number of screws in the stateroom hat hooks. Then there was the colour of the pebble-dashing on the private promenade decks ('too dark') and the notes he made to stain the wicker furniture green on one side of the ship. On 9 April 1912, he wrote, '*Titanic* is now about complete and will I think do the old firm credit tomorrow when we sail.'

Before sailing, Chief Officer Henry Wilde replaced William Murdoch, who moved down to first officer; First Officer Charles Lightoller moved down to second officer; and Second Officer David Blair left the ship.

All manner of small details were still being attended to. First-class saloon steward Frederick Dent Ray took home a memento, acquired either during the final days in Belfast or at Southampton. One of the workmen was cutting the carpet for a first-class stateroom on C-Deck, but he had to cut 6in off. He told Ray he would 'throw it away', but Ray asked to take it home 'to show my wife what nice carpets they are putting in the ship'.[10]

On 6 April 1912, the *Southampton Times* noted that a 'third *Olympic*' had been ordered and the White Star Line's intended 'weekly service of mammoth ships' would be 'challenged at every step by the Germans'. Their rivals were still under construction, so the 'present score is two to nil'. Its reporter called *Titanic* 'a wonderful ship':

> In a port where the magnificence of the appointments of the *Olympic* are so well known, it seems scarcely necessary to say much about the *Titanic*. The privileged few who have had the pleasure of visiting the ship since her arrival [...] have been at a loss to express their admiration. One person said that the *Olympic* was all that could be desired, and the *Titanic* was – well, something even beyond that!

The day before sailing, General Post Office officials visited *Titanic* to inspect her Post Office, in the same way they had examined *Olympic* before her maiden voyage, 'The arrangement of the sorting room and mail room is similar to those provided in the *Olympic*.' Harland & Wolff had appropriated part of the mail room for stowing parcels sent by the 'Semi Official Service'. There were usually about ten (five from Southampton and five from Liverpool embarked at Queenstown), advised on a Way Bill to the care of the Purser. The postal service suggested that if an enclosure was necessary for these mails, it should be in the baggage room adjacent to the mail room. The door in the bulkhead between the mail room and baggage room was fitted with ordinary locks, but 'for greater security they should be replaced by bars and padlocks similar to those on the doors from the trunk hatchway'.

Two United States Sea Sorters and a representative from the British service were on board. They showed the officials some features of the sleeping accommodation and dining arrangements, 'which they regarded as objectionable', 'some [...] were merely frivolous [...], but their complaint regarding the noisy condition of their sleeping cabins should perhaps be brought to the company's notice'. The cabins were:

> ... situated among a block of third class cabins, and it is stated the occupants of these [...] who are mostly low class continentals, keep up a noisy conversation sometimes throughout the silent hours and even indulge at times in singing and instrumental music. There is, too, a door on the other side of the inboard bulkhead of the larger cabin through which these steerage passengers are constantly passing and the sorting clerks complain that this door bangs at night and keeps them awake. These remarks were prompted by their experiences in the *Olympic* where the accommodation is similar.
>
> I think that in justice to the men these conditions should be represented to the company as if their work during the day is to be performed efficiently it is essential that they should enjoy decent sleep at night.
>
> They complained about their dining room and desired to be established in the saloon with the passengers, but I consider that the mess room provided is suitable, convenient and in accordance with the understanding between the United States Post Office and the company, and no change is

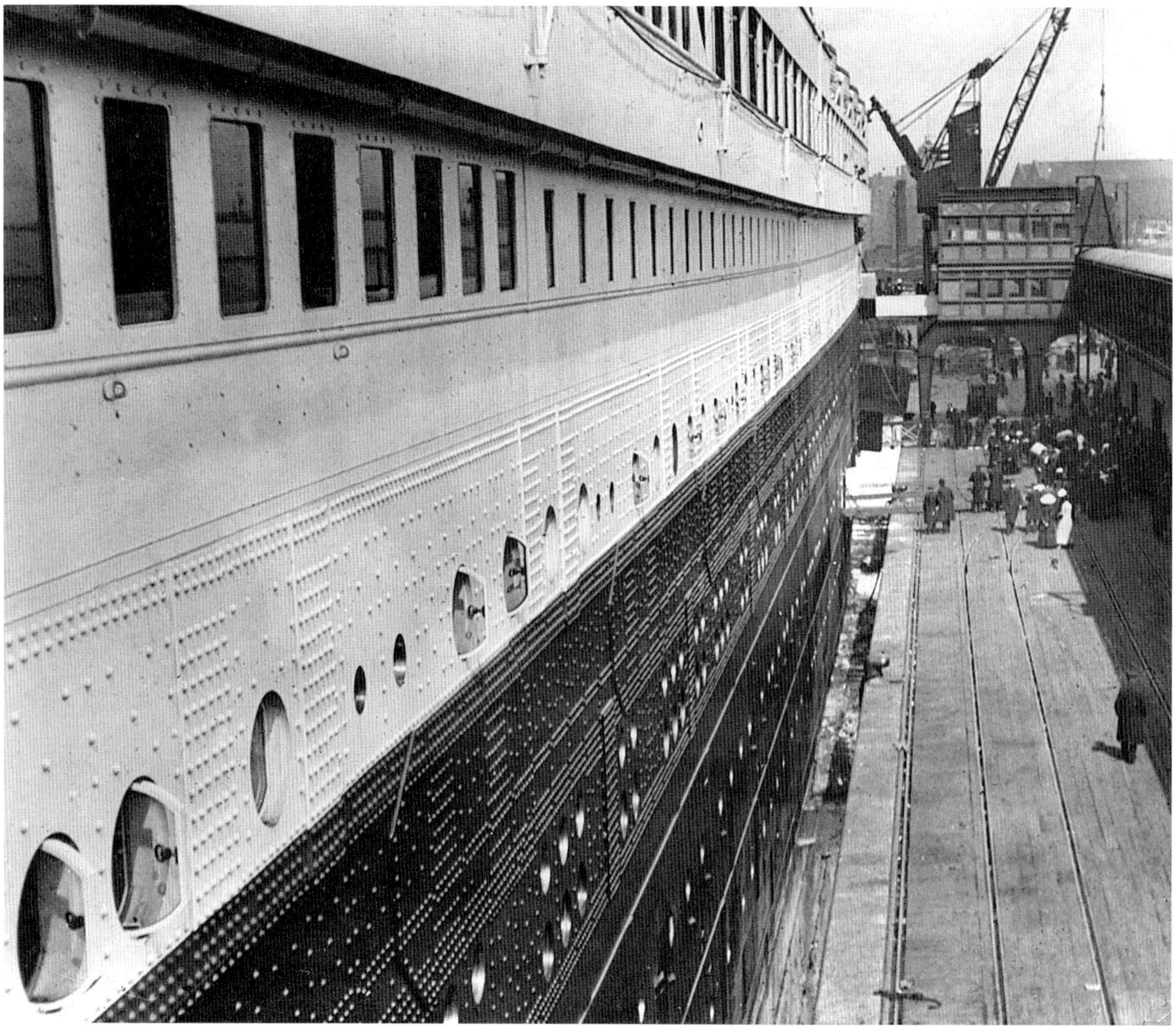

Titanic towers over the dockside at Southampton. This photo was taken from the first-class gangway leading to B-Deck, by first-class passenger Francis Browne. Second-class passengers can be seen boarding on C-Deck, aft. Browne disembarked at Queenstown on 11 April 1912. (© TopFoto)

> required. It might be suggested, however, that the doorway leading from the valets and ladies maids' dining room should be fitted with a door in order to prevent the present practice of using the mail sorters' mess room as a thoroughfare.[11]

Altogether, the accommodation was 'far in excess both as regards size and fittings of any other British packet carrying a Sea Post Office on the Atlantic'. Nonetheless, a letter was drafted to the White Star Line, requesting that some changes be made for both ships, dated 12 April.

All manner of supplies were being taken aboard while *Titanic* was in port, from coal to food. A local poulterer, E. Folliot Ltd, kept a record of what they supplied to *Titanic* on 9 April, including 400 plover eggs and twenty-four pigeons to the à la carte restaurant. The total order for the restaurant came to over £53 and the order for the dining saloons stood at over £619.[12]

***

Sailing day dawned: Wednesday, 10 April. Andrews came aboard at 6 a.m. and spent the morning inspecting the ship. J. Bruce Ismay, his wife, Florence, and children, Tom, George and Evelyn, had stayed overnight at Southampton's South Western Hotel. (Ismay's monthly office account book contained an entry for the hotel bill amounting to £25 14*s* 5*d*.) They came aboard around 9.30 a.m. The children explored the ship before leaving with their mother for a motoring holiday in Wales.

Passengers were arriving. Henry Julian found his first-class stateroom 'more like a bedroom than a ship's cabin. I have just been all over the ship ... It is all most luxurious.'[13] Another first-class passenger was Jhr. Johan George Reuchlin, a director of the Holland America Line. He was travelling across the Atlantic on business, as a guest of the White Star Line. Ismay had personally made sure he was assigned a suitable stateroom. (They had a treasure trove of information from Willem Piek's voyage on *Olympic* four months earlier. Undoubtedly, Reuchlin was taking notes on how *Titanic* had been improved.)[14]

As usual, Captain Smith had to complete a short report to the White Star Line before departure:

> 10 April 1912
>
> I herewith report this ship loaded and ready for sea. The engines and boilers are in good order for the voyage, and all charts and sailing directions up-to-date.
>
> Your obedient servant, Edward J. Smith.[15]

She had 5,892 tons of coal on board, 'sufficient to take the ship to her next coaling port', according to Maurice Harvey Clarke, Assistant Emigration Officer to the Board of Trade at Southampton. Even on the most conservative assumptions, that meant that she would have about two days' spare coal consumption left over on arrival in New York.[16] She had plenty of coal. (During the course of twelve westbound crossings from May 1911 to April 1912, *Mauretania* arrived in New York with an average of 663 tons left over, which represented less than a day's steaming at her normal service speed).[17]

The *Daily Express*'s correspondent 'looked down on the world' from the Boat Deck, then went down to the quay and 'looked up at the projecting heads of the passengers. It was like standing by the wall of St. Paul's Cathedral and craning your neck to get a glimpse of the Apostle on the roof.' He watched *Titanic*'s departure and a narrowly averted accident:

> It was just noon when the vast steel wall in front of us began to move. For the first yard a caterpillar might have raced the *Titanic*. It was difficult to imagine such a tremendous object moving so slowly. I walked along to the end of the deep water dock, and saw her come by, at a slow pace, within a stone's throw of the quay. Her propellers churned the green sea up to liquid grey mud.
>
> She had to go round a bend to the left [...] in order to clear the end of the long quay, which juts out slantwise into Southampton Water. It was while trying to round this bit of a bend that the *Titanic* pulled the [...] New York from her berth.
>
> [...] Hundreds of townsfolk had already gathered at the quay end [...] When I came out on the quay end an astonishing spectacle held the gaze of the crowd, for between the *Titanic* and the quay – a distance of two or three hundred yards – the *New York* was drifting stern first towards the outgoing new liner.
>
> The *Oceanic* lay moored along the quayside. A few minutes before the *New York* was moored close beside her. Now the *New York* was adrift and sweeping towards the *Titanic*. What was said to have happened seemed a fantastic absurdity until I saw the frayed end of a steel wire hawser, about as thick as a man's wrist, lying on the quay.
>
> 'It snapped like the crack of a gun', a man told me who saw it break. Broken hemp cables hung down the *New York*'s side.
>
> The crowd was breathless with excitement [...] As soon as the *New York* broke loose the *Titanic* reverse her engines, and in a brief space of time stopped dead and began to back.
>
> Then the tugs Neptune and Vulcan raced at the *New York*, caught her with ropes [...] and tried to lug her back to her place [...]
>
> But no one in uniform was hurried. A master of port navigation, with a megaphone, stood solidly on the quay, issuing orders across the water as calmly as if he were having his tea. He had the *New York* pulled back, across the *Oceanic*'s bows, and round the bend to the quay, and there moored securely, and then he let the *Titanic* come on again towards the open water

As *Titanic* moves past, *New York* breaks loose from her moorings. *Oceanic* is visible to the extreme left. A collision is narrowly averted. *L'Illustration*, 1912. (Author's collection)

> [...]
>
> It was a relief to everyone when the *Titanic* at last passed the bend, and glided slowly away to sea, with the Royal Mail liner Tagus following her, like a maid of honour holding the train of a queen.

At Cherbourg, Andrews thought the two tenders 'looked well'. 'We expect to arrive at Queenstown about 10.30 a.m. tomorrow,' he wrote home. 'The weather is fine and everything shaping for a good voyage. I have a seat at the doctor's table.'[18]

The *Belfast Evening Telegraph* published a passenger's account describing the run to Cherbourg and Queenstown. From the time she cleared the eastern end of the Isle of Wight and headed down the English Channel towards Cherbourg, *Titanic* gave the impression of 'wonderful steadiness. Were it not for the brisk breeze blowing along the decks, one would have scarcely imagined that every hour found us some 20 knots further upon our course.' On Thursday, 11 April, they had a 'morning plunge' in the swimming pool, 'where the ceaseless ripple of the tepid sea water was almost the only indication that somewhere in the distance 72,000 horses [*sic*] in the guise of steam engines fretted and strained under the skilful guidance of the engineers'. Half an hour in the gymnasium 'helped to send one's blood coursing freely and created a big appetite for the morning meal'.

William Henry Harrison, J. Bruce Ismay's secretary, completed a personal letter to his 'own dear Nancy', so that it could be posted from Queenstown. He recalled his 'most rotten journey' between Manchester and Derby on the way down to London, 'The train was absolutely packed.' (He thought 'the fool who sat by me was drunk. He rode nearly all the way with his head out of the window.') To make matters worse, they were 'an hour late' into London's St Pancras Station, 'Those who travel much by rail these days are learning not to expect too much.' Ismay had been keeping him busy with work: 'I didn't get to bed till after midnight ... the amusing part of it is that people think it is a holiday' He signed off, 'There dear, I have told you all that has happened to me. Nothing interesting as Nelly says, no it only remains for me to say goodbye with best love and lots of kisses for you all and to remain your own faithful devoted Will.'[19]

'She dropped anchor at 11.55 a.m. about a mile outside Roche's Point,' wrote the *Cork Examiner*, 'and as one saw her steaming slowly into view – a majestic monster, floating, it seemed irresistibly into the harbour, a strange sense of might and power pervaded the scene.' The paper described her interiors and first-class suite rooms as 'equal to bedrooms in the best modern hotels ashore'. It imagined a 'typical day' for a first-class passenger: 'He may first have a turn in the splendidly equipped gymnasium; next a game in the squash racquet court.' *Titanic* 'weighed anchor at 1.25 [p.m.] and proceeded on her first western journey'. She promised a comfortable, safe crossing to New York, 'The dangers of the seas are practically non existent on these latest magnificent vessels.'

***

The first report of ice 'unusually far south' came the same day *Titanic* left Queenstown. In several previous years, the Southern Route was moved further south 'as soon as it was discovered that the ice was coming to the south of its normal limits' (it was changed as early as 14 April in 1905).[20] Captain Smith had seen ice during at least fifteen of his thirty-one years of White Star service, including two large icebergs on the eastbound leg of *Adriatic*'s maiden voyage almost five years earlier. It was not necessarily a great surprise to see ice that far south early in the season. In 1912, there appear to have been more icebergs than normal and they were also further south, but it was not unprecedented.[21]

On *Titanic*'s first full day at sea, the *Cork Examiner* reported on the improving situation for coal supplies. 'Most of the North Wales collieries are at full work' and men were returning 'to the pits in large numbers'. The Midland Railway Company announced that 'a number of trains, which were temporarily withdrawn in consequence of the miners' strike will be restored' from 15 April. It boded well for the availability of coal during her first transatlantic season.

Passengers settled in to what was shaping up to be a pleasant voyage. Second-class passenger Kate Buss was happy, 'The second class accommodations surpassed that of many first class hotels [...] I loved the beautiful and extremely wide staircase and seldom used the elevator.' The first days at sea were 'marvellous: perfect weather, calm seas, wonderful food and a lovely crowd of people. I did not miss a meal or any of the between-meal "lunches", I had a nice double cabin to myself. I remember it had an extra sofa and a roomy wardrobe closet.'[22]

'Each day, as the voyage went on, everybody's admiration of the ship increased,' recalled Second Officer Charles Lightoller, 'for the way she behaved, for the total absence of vibration, for her steadiness even with the ever increasing speed, as she warmed up to her work.' *Titanic* covered 484 miles on her first day's run, 519 miles on her second and 546 miles on her third. She had left Queenstown over two hours ahead of *Olympic* on her maiden voyage. Her average speed up to the posting of her third day's run was very slightly ahead of her sister. By the late evening of Sunday, 14 April, she was going through the water at about 22.5 knots.

Fireman George Kemish thought 'she was a good job in the stokeholds', 'not what we were accustomed to in old ships – slogging our guts out and nearly roasted with the heat'. He recalled, 'We just had to keep the furnaces full and not keep working the fires.'[23] She looked set to outperform her older sister and arrive at the Ambrose Channel Lightship on Tuesday evening.

That Sunday, *Mauretania* left Queenstown. Her passengers included Alfred Booth, Cunard's chairman. He intended to have an overnight stopover in New York and return on *Titanic*'s eastbound crossing to size 'up the new monster of the White Star Line fleet'. His plans meant both White Star's and Cunard's chairmen would be sailing home together on 20 April 1912. Booth was following in the footsteps of Leonard Peskett.[24]

★★★

The realisation that something was wrong came differently to people. Some were awakened when *Titanic*'s engines stopped late that Sunday evening. Word spread that she had struck ice.

Stewardess Annie Robinson saw Captain Smith and Thomas Andrews on E-Deck, returning from the vicinity of the mail room. Andrews told Smith, 'Well, three have gone already, captain.' It was apparently a reference that they knew at least three watertight compartments were flooding. Robinson looked down a staircase. The seawater had risen to within six steps of E-Deck.

Smith went up to the bridge while Andrews remained below. Later, first-class passenger Dorothy Gibson saw Andrews rushing up the grand staircase, 'three or four steps at a bound'. He now knew that the ship could not remain afloat and was rushing to inform the captain.[25]

When the time came to load and lower the lifeboats, many felt safer on the ship. Third Officer Herbert John Pitman, who was ordered to take charge of Lifeboat 5, 'did not like the idea of going away at all, because I thought I was better off on the ship'. Murdoch wished him 'good luck'. Pitman thought that several watertight compartments were flooded and that the ship would settle lower in the water, but remain afloat. It struck him from Murdoch's manner that the first officer did not expect to see him again.

Two of *Titanic*'s lifeboats approach *Carpathia*. Collapsible D is on the left and Lifeboat 14 on the right. (Library of Congress, Prints & Photographs Division)

First-class passenger Emily Ryerson felt the engines stop:

> I got up and looked out, down the corridor, there was no sound, I opened the window in our room, and there was still no sound, and I put on an extra wrapper because it was bitterly cold, much colder than it had been. When we went to bed it was bitterly cold, I couldn't get to sleep I was so cold. I looked out a second time and saw some people going on deck, and wondered why that was, and a man passenger ran up and said 'Put on your life-belt and come up on the boat deck.' I said, 'Where did you get those orders?' and he said, 'From the Captain.'[26]

Her lifeboat, No. 4, was finally lowered shortly before 2 a.m. She said:

> The water was washing in the portholes, and later I think some of the square windows seemed to be open, and you could see in the cabin and see the water washing in and the gold furniture and decorations, and I remember noticing you could look far in, it was brilliantly lighted.[27]

She 'lost hope [...] this was about the first thing that made me realize the ship was gone'.

Able Seaman Edward Buley watched *Titanic* settle by the bow:

> First of all you could see her propellers and everything. Her rudder was clear out of the water. You could hear the rush of the machinery, and she parted in two, and the afterpart settled down again, and we thought the afterpart would float altogether [...] She uprighted herself for about five minutes, and then tipped over and disappeared.[28]

★★★

The commander and officers of Cunard's *Carpathia*, who had responded immediately to *Titanic*'s distress calls, heard the news from Fourth Officer Joseph Boxhall. Once the occupants of Lifeboat 2 had all been taken on board, *Carpathia*'s Second Officer James Bisset accompanied Boxhall to the bridge. Captain Arthur Rostron 'burst out, excitedly', asking, 'Where is the *Titanic*?'

'Gone! She sank at 2.20 a.m.'

There was a moment of 'stunned silence'. Boxhall's voice 'broke with emotion' as he explained that more than 1,000 people might have perished.

Daybreak revealed a number of icebergs nearby. *Carpathia*'s own passengers lined the decks early that morning, 'joined by increasing numbers of rescued people, anxiously watching each boat arrive. The rescue operations proceeded in a deathly silence.'[29]

★★★

Philip Franklin, IMM's vice president, was woken up in the early hours by the telephone ringing at his home in New York. It was a reporter who had heard that *Titanic* was sinking and sending out distress calls. He tried to get more information from officials at the dock, who told him several reporters had been in touch with them. Then he called the Associated Press. He asked if they 'could not [...] give out such an alarming report until they could see whether it could be confirmed', but the news had already been released.

Franklin made several attempts to get the news confirmed and requested that *Olympic*'s captain provide an update. To avoid alarming Captain Haddock, he merely asked if he could ascertain *Titanic*'s position. He reassured reporters throughout the day, telling Mr Byrne and Mr Smallwood of Dow, Jones & Co.:

> We cannot state too strongly our belief that the ship is unsinkable and the passengers perfectly safe. The ship is reported to have gone down

> several feet by the head. This may be due from water filling forward compartments, and [the] ship may go down many feet and still keep afloat for an indefinite period.

'During the entire day we considered the ship unsinkable,' Franklin recalled, 'and it never entered our minds that there had been anything like a serious loss of life.'

Early on the evening of Monday, 15 April, he received the report from Captain Haddock that *Titanic* had foundered with heavy loss of life, 'It was such a terrible shock that it took us a few minutes to get ourselves together.' Holding the telegram in his hands, Franklin read it out to reporters. Later, he said, 'I thought her unsinkable, and I based my opinion on the best expert advice. I do not understand it.' He wept as he spoke.[30]

15

'A MOURNFUL VOYAGE'

Captain Herbert James Haddock returned to Southampton so that he could take command of *Olympic* for her 3 April westbound departure. He was eleven years younger than his predecessor but an experienced White Star commander. His most recent five-year stint was as *Oceanic*'s captain. Dr Beaumont described him as 'a great navigator and sailor'. He was a 'gentleman, yet he had his "moods" when he could be peevish and inscrutable'. Haddock remained 'calm and dignified, never raising his voice beyond his usual soft tones'. Beaumont was proud to join him on his new ship, serving as chief surgeon.

Olympic left Southampton around noon. Her *process-verbal* (wireless log) recorded '*Titanic* off Lizard' on the southern Cornish coast at 1.35 p.m. That afternoon and into the evening, she was in contact with *Titanic* and exchanged several messages.[1] She called at Cherbourg as usual, leaving at 6.49 p.m. for Queenstown. She left Queenstown for New York at 3.30 p.m. on 4 April. The westbound passenger list was her highest to date: 323 first, 592 second and 1,128 third class, for a grand total of 2,043.

One of the first-class passengers was B.J. 'Bud' Hauser, who was a professional gambler and confidence trickster. He travelled under the name 'Barton J. Harvey', spending time in the smoke room. The 'reckless' way he played cards attracted comments from other passengers, as did his alcohol consumption (including absinthe, whisky and champagne). It was reported that, during private card games in his stateroom, Hauser and accomplices managed to fleece thousands of dollars from other passengers. Hauser's lifestyle caught up with him on the morning of 10 April, as *Olympic* approached the Ambrose Channel Light Vessel. He passed away aged 37, due to 'paralysis of the heart'. The press reported rumours 'that there was something strange about the death', but Dr Beaumont was satisfied that there was nothing untoward when he recorded it in the ship's log (the entry recorded the death of 'Barton J. Harvey or Isaac M. Hauser').[2]

The *New York Times* reported *Olympic* steaming into port late in the afternoon of 10 April: 'Her sister ship the *Titanic* started on her maiden voyage from Southampton [...] Therefore the *Olympic* has made her last trip to New York as the biggest ship afloat.' Lord Hyde and Lord Somers 'were among the most interesting passengers [...] they are leaving England to start life anew in Canada as fruit farmers'. They blamed their departure on Chancellor David Lloyd George's taxation policies.

James Bisset, serving as second officer of Cunard's *Carpathia*, went on board *Olympic* for the first time 'for a short visit of curiosity, and that was the first time that I ever trod the deck and bridge of a super-liner of over 40,000 tons – a "leviathan liner", the first of her kind'. Years later, he recalled, 'Everything was colossal, awe-inspiring, and, as I thought, unwieldy. I was old fashioned enough to think that she must be "too big to handle" – but that was only my innocence and perhaps envy.' One of the ship's

Passengers line *Olympic*'s decks as she comes into her New York pier on 10 April 1912. The spectators are on an upper level of the pier, high above the waterline. (Library of Congress, Prints & Photographs Division)

young officers showed him 'her splendours' and 'explained that she was unsinkable. She had fifteen watertight bulkheads.'[3]

Everything was made ready for her next departure. *Olympic* 'got off' at 3 p.m. on 13 April, 'with a list of passengers indicating that the summer rush to Europe is beginning'. The *New York Times* described 'a large crowd' at the pier to watch her leave. As she headed eastbound, the wireless picked up messages from liners including *George Washington* and *President Lincoln*. It was late on the evening of 14 April that Ernest James Moore, the senior of the two Marconi wireless operators on board, picked up a message from 'MGY' (*Titanic*). He updated the wireless log, 'Hear MGY signalling to some ship and saying something about striking an iceberg, not sure if it is MGY who has struck an iceberg.'

The atmospherics were bad and other wireless stations were active, making it difficult to determine what was being said by which ship. Ten minutes later, he heard the distress call 'CQD'. Another ten minutes passed and *Titanic* responded, giving 'his position 41° 46' North 50° 14' West and says we have struck an iceberg'. Moore reported immediately to the bridge, but *Olympic* was about 500 miles away from *Titanic*'s position. He woke up Alec Bagot, his junior colleague. It took some moments to convince him that the situation was serious but he got dressed within minutes, putting his uniform on over his pyjamas.[4]

Bagot arrived to find Moore hard at work. *Titanic* had advised that the weather was calm but passengers were being evacuated in the lifeboats. Moore handed him an urgent message to take to the bridge. Bagot 'sped along the port passage way leading from the wireless room [...] through the chart house to the bridge'. Captain Haddock, First Officer Robert Hume, Second Officer Albert Fry and two junior officers were there. As Bagot handed over the message, he asked whether *Titanic* was sinking.

'Sinking? No, of course not. How could she? But they say she's struck an iceberg and may need our help. Wait while I give this to the commander.'

Haddock 'tore open' the envelope containing the message and asked Bagot to come to the chart room with him, while he dictated a reply to tell Captain Smith that *Olympic*'s boilers were all being lit as fast as possible. (It was unfortunate that, if Moore's recollection was correct, *Olympic* was running at a slower speed than normal. It would take time to get all the boilers contributing their maximum steam to the engines.)

Haddock ordered them to keep him informed.

'Yes, sir,' Bagot responded.

He turned to go, but Haddock added, 'There is no need, of course, to say no mention of this business may be made outside.'

Bagot privately thought, 'What does he take me for?'

'All went smoothly till the Sunday night,' Dr Beaumont recalled:

> Around midnight, [Purser Claude] Lancaster and I were in his room, having a pipe together, when the door opened, and in walked Bob Fleming, the chief engineer. 'Boys', said he, 'there is awful trouble. The *Titanic* has been struck by an iceberg, is badly damaged, and we are going up north to her assistance'. He then left us and went below to light up more boilers and drive the ship to her full capacity.

Sleep 'was out of the question' that night for any of the officers on board.

Bagot had been reassured by the report from *Titanic* that the weather was 'clear and calm', but her wireless signals came to an end. It was a worrying sign. Both ships had the most powerful Marconi wireless apparatus afloat. They kept calling *Titanic*, but there was no reply. *Hellig Olav* confirmed she had also heard nothing from *Titanic*. 'It was an uncanny, unforgettable night,' Bagot wrote. *Olympic* was 'trembling with energy':

> The sea was calm, yet the mighty vessel gathered rhythm, with a noticeable up and down movement, as though leaping a series of invisible hurdles. Outside our cabin along the boat deck, sailors were removing tarpaulins, swinging davits, testing falls. Every now and then an officer, or A.B. [able-bodied seaman], would come to the wireless room seeking news. Always the same reply. 'We've lost touch.'

They received Philip Franklin's request for *Titanic*'s position.

By breakfast, Beaumont recalled, 'everyone knew about it and worked up to a pitch of intense excitement':

> [I] spent most of the day with my staff, preparing some 200 emergency beds in the large public third class rooms in the after end of the ship.
>
> [...]
>
> [The] firemen and trimmers down below were straining every nerve to drive the ship, and it was reported that during twelve hours, i.e. from 5 a.m. to 5 p.m., the *Olympic* covered 300 knots, an average speed of 25 knots per hour. Even then she remained steady, but vibrated more than usual.

The *Belfast Evening Telegraph*'s report that 'she tore along at a rate of 24 knots' was probably closer to the mark, 'The stokers and engineers were working like Trojans to get the last fraction of speed out of the ship's engines.' (Harold Sanderson said in 1915 that the best average speed *Olympic* had achieved up to that time was 24.2 knots over a twenty-four-hour period in the Atlantic.)[5]

Haddock sent a message that morning: 'ISMAY, New York: Since midnight, when her position was 41° 46' North 50° 14' West, have been unable to communicate. We are now 310 miles from her, 9 a.m., under full power. Will inform you at once if hear anything. Commander.'

Parisian indicated that *Carpathia* had 'picked up 20 boats with passengers' and *Baltic* was returning to *Titanic*, 'Don't know if latter sunk.' An ice report was received from *Mesaba*. Then, the

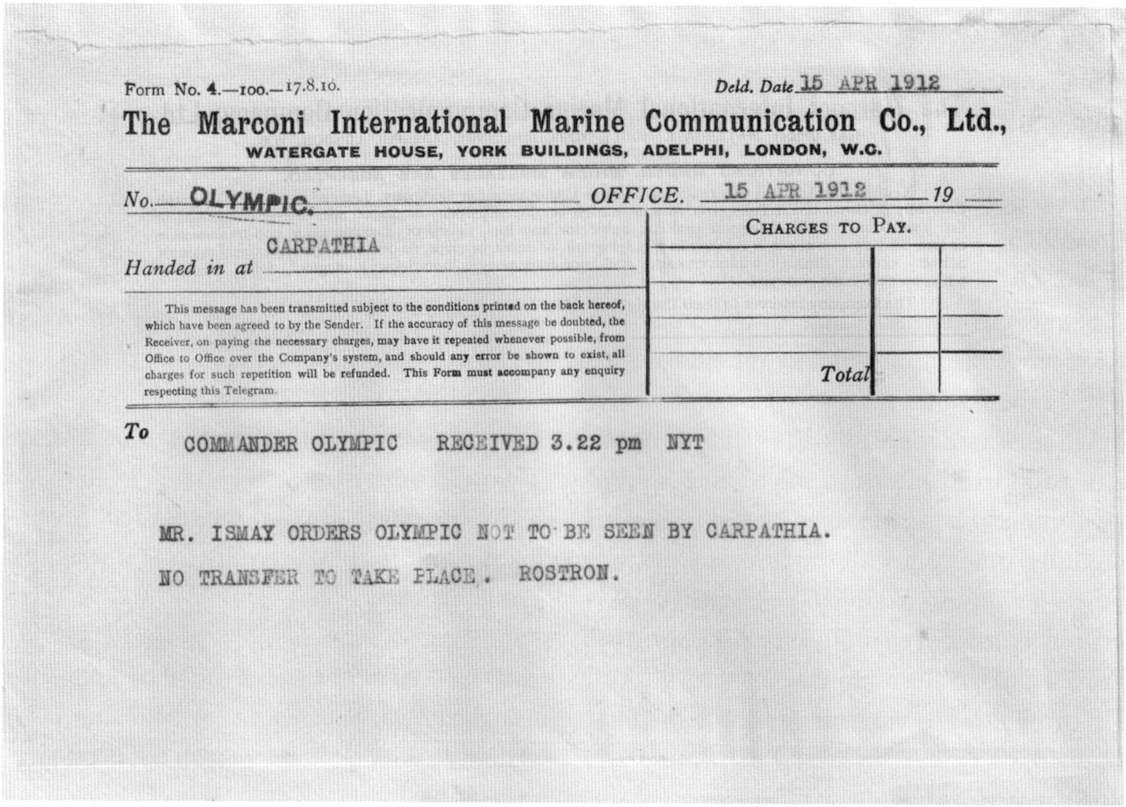

Form No. 4.—100.—17.8.10. *Deld. Date* 15 APR 1912

The Marconi International Marine Communication Co., Ltd.,
WATERGATE HOUSE, YORK BUILDINGS, ADELPHI, LONDON, W.C.

No. OLYMPIC. *OFFICE.* 15 APR 1912 *19*

CHARGES TO PAY.

Handed in at CARPATHIA

This message has been transmitted subject to the conditions printed on the back hereof, which have been agreed to by the Sender. If the accuracy of this message be doubted, the Receiver, on paying the necessary charges, may have it repeated whenever possible, from Office to Office over the Company's system, and should any error be shown to exist, all charges for such repetition will be refunded. This Form must accompany any enquiry respecting this Telegram.

Total

To COMMANDER OLYMPIC RECEIVED 3.22 pm NYT

MR. ISMAY ORDERS OLYMPIC NOT TO BE SEEN BY CARPATHIA.

NO TRANSFER TO TAKE PLACE. ROSTRON.

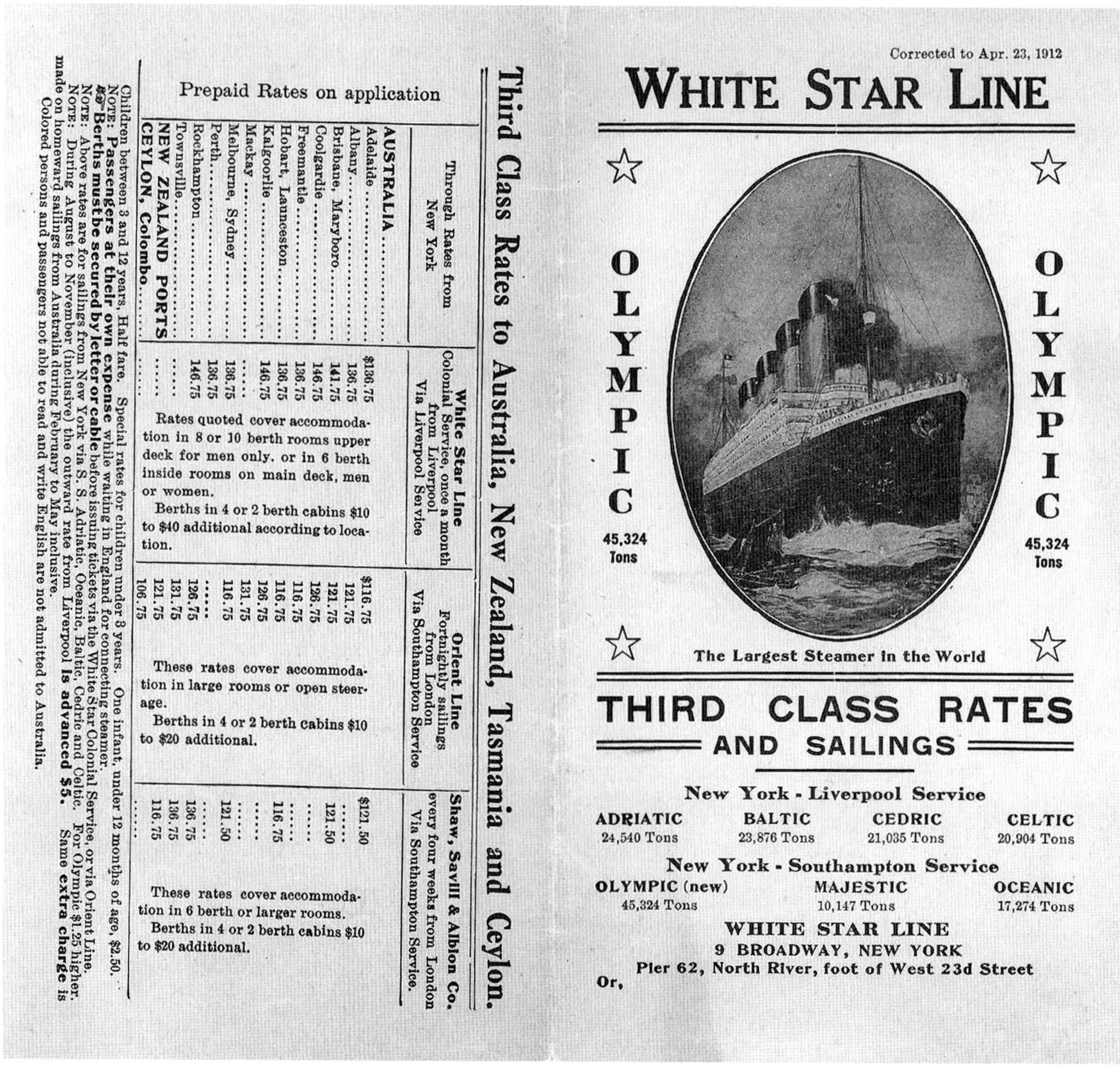

Third Class Rates to Australia, New Zealand, Tasmania and Ceylon.

Prepaid Rates on application

Through Rates from New York	White Star Line Colonial Service, once a month from Liverpool Via Liverpool Service	Orient Line Fortnightly sailings from London Via Southampton Service	Shaw, Savill & Albion Co. every four weeks from London Via Southampton Service.
AUSTRALIA			
Adelaide	$136.75	$116.75	$121.50
Albany...........................	136.75	121.75	......
Brisbane, Maryboro...........	141.75	121.75	121.50
Coolgardie	146.75	126.75	......
Freemantle	136.75	116.75	......
Hobart, Launceston...........	136.75	116.75	116.75
Kalgoorlie	146.75	126.75	......
Mackay	......	131.75	......
Melbourne, Sydney............	136.75	116.75	121.50
Perth.............................	136.75	......	......
Rockhampton	146.75	126.75	136.75
Townsville.......................	......	131.75	136.75
NEW ZEALAND PORTS	......	121.75	116.75
CEYLON, Colombo.......	......	106.75	......
	Rates quoted cover accommodation in 8 or 10 berth rooms upper deck for men only. or in 6 berth inside rooms on main deck, men or women. Berths in 4 or 2 berth cabins $10 to $40 additional according to location.	These rates cover accommodation in large rooms or open steerage. Berths in 4 or 2 berth cabins $10 to $20 additional.	These rates cover accommodation in 6 berth or larger rooms. Berths in 4 or 2 berth cabins $10 to $20 additional.

Children between 3 and 12 years, Half fare. Special rates for children under 3 years. One infant, under 12 months of age, $2.50.
NOTE: **Passengers at their own expense** while waiting in England for connecting steamer.
☞ **Berths must be secured by letter or cable** before issuing tickets via the White Star Colonial Service, or via Orient Line.
NOTE: Above rates are for sailings from New York via S. S. Adriatic, Oceanic, Baltic, Cedric and Celtic. For Olympic $1.25 higher.
NOTE: During August to November (inclusive) the outward rate from Liverpool **is advanced $5.** Same **extra charge** is made on homeward sailings from Australia during February to May inclusive.
Colored persons and passengers not able to read and write English are not admitted to Australia.

Corrected to Apr. 23, 1912

WHITE STAR LINE

OLYMPIC 45,324 Tons

The Largest Steamer in the World

THIRD CLASS RATES AND SAILINGS

New York - Liverpool Service

ADRIATIC 24,540 Tons | BALTIC 23,876 Tons | CEDRIC 21,035 Tons | CELTIC 20,904 Tons

New York - Southampton Service

OLYMPIC (new) 45,324 Tons | MAJESTIC 10,147 Tons | OCEANIC 17,274 Tons

WHITE STAR LINE
9 BROADWAY, NEW YORK
Or, Pier 62, North River, foot of West 23d Street

wireless station at Cape Race confirmed they had no further news of *Titanic*. Haddock estimated that they would reach *Titanic*'s distress position around midnight. *Olympic* and *Parisian* exchanged messages concerning where ice had been sighted because Haddock needed to plot the safest course to reach the distress position.

Early that afternoon, *Carpathia* confirmed the awful reality of what had happened. Captain Rostron advised that J. Bruce Ismay was 'under opiate'. There was a consensus that it would be unwise for *Olympic* to meet *Carpathia* or for *Titanic*'s survivors to see her nearly identical sistership. *Carpathia*'s wireless operator, Thomas Cottam, began to transmit the names of first- and second-class survivors. ('Please excuse sending, am half asleep.')

Bagot recalled the initial shock, 'One part of my brain kept saying "It can't be – it can't be", whilst another voice, even more compelling, repeated: "It's true. It's true. It's true".' Haddock ordered an able seaman stationed just inside the wireless room door to support their work. (Three weeks or so earlier, Captain Smith had 'casually' made an unusual, unannounced visit to the wireless room to bid them goodbye and inform them he was transferring to *Titanic*. Standing in the same place, Smith had been playing with a toothpick when he commented that *Olympic* was a 'happy ship. I'll be sorry to leave her.' He had offered to recommend that both men transfer with him to *Titanic*, but they had 'spent months getting our wireless equipment into good order'. They were not keen on starting again. The toss of a coin settled the dilemma and they remained on *Olympic*. Bagot pondered what would have happened if they had transferred to *Titanic* and whether they would both still be alive.)

On Tuesday, 16 April, Charles Moore and his travel companions on *Olympic* had breakfast in their staterooms. When they went to read the list of survivors received from *Carpathia*, he saw Frank Millet's and Archie Butt's names were not included: 'My steward is in grief; his son was a steward on the *Titanic* and has gone down. The ship is in gloom; everybody has lost friends, and some of them near relations.'

Beaumont also described the state of shock: 'Not one of us could realise really what had happened. There was nothing to do but steam to Southampton.'

For the remainder of the voyage, *Olympic*'s wireless log recorded plenty of transmissions. Alec Bagot recalled:

> We may have slept, Ernie Moore and I, during the next three [*sic*] days to Southampton, but if we did, I have no recollection of it. Eight hours on watch; eight hours off; four hours on; four off was our ordinary day. During our long watch 'below', we normally attended to the motors, wrote up our abstract of messages, did odd jobs and got our 'long' sleep. On this memorable voyage, we seemed never to stop sending or receiving Marconigrams. In addition, we received orders from London to prepare a complete copy of our wireless log. I made an extract for Moore and myself. Next voyage, in New York, a journalist offered $2,000 for it. 'Heck, son, ain't yer gotta price?' he asked, when I explained, patiently, 'we weren't allowed to sell'.

The *Belfast Evening Telegraph* reported that *Olympic*'s musicians did not play for the remainder of the voyage. Events on board were cancelled as a mark of respect. On Friday, 19 April, the White Star Line posted a notice to the ship's

Opposite:
Top: *Carpathia*'s Captain Arthur Rostron had to consider his options, eventually choosing to return to New York with the *Titanic* survivors. J. Bruce Ismay turned down emphatically the idea that they be transferred to *Olympic*. (Mike Beatty collection)

Bottom: White Star Line third-class rates 'corrected to Apr. 23 1912'. *Titanic* has been hastily removed. (Roy Derouin collection)

Olympic at Plymouth. Workers are organising sacks of mail in the foreground. Further aft, the second-class gangway on E-Deck is open for passengers to disembark. In 1912, the average number of mail bags she carried westbound was 2,885, whereas the eastbound number was only 1,577. (Royal Mail Group, courtesy of the Postal Museum)

passengers expressing the company's 'high appreciation of the fortitude with which they received the painful news of the disaster to the *Titanic*, and of their kind and empathetic attitude to the company both on board and ashore in this time of trouble'. It commended the 'magnificent total of £1,347 6 shillings towards the Relief Fund' for 'the relatives and friends of members of the crew who perished'.[6]

Meanwhile, *Mauretania* arrived safely in New York. Her captain, William Turner, went as much as 65 miles further south to avoid the ice. It added 26 miles to the total distance but she actually completed the crossing more quickly than her previous one, because her average speed was about half a knot faster.[7] According to press reports, *Titanic* had been well booked for her eastbound New York departure scheduled for 20 April, including 600 passengers in first class. Those passengers had to make alternative arrangements. (They had a number of choices, including Holland America, Cunard and White Star liners. *Rotterdam* left New York on 23 April, carrying 1,031 passengers; *Mauretania* departed the next day with 1,458 passengers; and *Celtic* followed a day later with 731 passengers.)

20 April 1912: the ensign at *Olympic*'s stern flies at half-staff (half-mast) as a mark of respect for the *Titanic* disaster. (Günter Bäbler collection)

Olympic's logbook recorded her departure from Plymouth at 8.01 p.m. [recte: a.m.] on 20 April and her departure from Cherbourg at 2.17 p.m. the same day. Captain Haddock wrote a 'Confidential Report' to the Marconi International Marine Communication Company Ltd, in London, on 'Mr. E.J. Moore & A. Bagot'. He rated their attention to duties, general conduct, personal appearance and sobriety as 'very good'. 'Can scarcely speak too highly of these officers' conduct at a very trying and anxious time. Mr. Moore, especially, worthy of promotion. Thoughtful and reliable. H.J. Haddock, Commander.

According to the Southampton Harbour Master's logbook, *Olympic* docked at 2.30 a.m. on 21 April. However, there was more sad news to come. Greaser A. Galpin had been taken ill during the voyage. He was transferred to the ship's hospital, but 'gradually got worse'. Dr Beaumont thought he was 'suffering from paralysis of the legs due to disease of the spinal cord'. Once *Olympic* docked, Galpin was transferred by ambulance to the Union Infirmary in Southampton, but he died there the following evening. It was not until 15 May that his death was notified officially to the Board of Trade. 'Owing to the stress of work and worry on board connected with the *Titanic* disaster the entry [...] in the log book was inadvertently overlooked,' Dr Beaumont wrote.[8]

16

THE 'SCALLYWAGS' OF PORTSMOUTH

'Captain Haddock looks extremely ill, worried and fagged [exhausted],' reported the *Daily Mirror*. 'He was so fatigued with his extra work during the last twenty-four hours and so upset by his distress at the terrible news of the *Titanic* [...] that he slept all through this morning.' He had little time for more rest. *Olympic* was scheduled to sail from Southampton in three days, on Wednesday, 24 April 1912. Aside from all the usual preparations needed to get her turned around, J. Bruce Ismay had issued an instruction that no White Star liner or other ship sailing for an IMM company should leave port without enough lifeboats for everyone on board.

Early that Sunday, 21 April, Captain Maurice Harvey Clarke, a Board of Trade Assistant Emigration Officer at Southampton, learned that 'a large additional number of collapsible boats' were going to be placed aboard *Olympic*. He arranged to visit the ship and 'satisfy myself that the boat deck would not be too hampered' or obstructed by the additional boats. On his way, he met Philip Curry, White Star's manager at Southampton, who was concerned that they were short of time. Marine Superintendent Captain Benjamin Steele was overseeing the placing of an additional forty collapsible boats. Harland & Wolff confirmed that the boats were 'all from His Majesty's Transports'. Clarke was in a strange position because he 'felt I had no official authority at the time to increase the seating boat accommodation of the *Olympic* or to accept these boats'. Her existing lifeboats already exceeded what the law required.

Orders came down from the White Star Line's head office at Liverpool that 'not more than 24 collapsible boats were to be placed on board'. This would mean that there were plenty of lifeboats to accommodate the total number of passengers and crew on the next voyage. It would also free up space on the Boat Deck to work and load the lifeboats if needed. Unfortunately, the instruction meant that 'quite a large number' of the additional collapsible boats already taken on board would need to be taken off the ship on Monday, 22 April. It started the rumour mill in Southampton 'that the Board of Trade were condemning several of the boats', which was not true.[1]

Clarke was on board *Olympic* that evening to see the new equipment and 'wire falls that were to be supplied to enable the collapsible boats to be launched'. Thomas Lewis, president of the recently established British Seafarers' Union, 'had demanded more deck hands to handle the extra boats'. (Captain Steele assured him that *Olympic* would take on four extra able-bodied seamen to do so.) The next day, Clarke was kept busy on Cunard's *Ultonia*, 'to carry out the full emigration survey and in view of the public unrest I gave special care to the boats and other life saving appliances'.

On sailing day, Clarke boarded *Olympic* at 7 a.m. He instructed Chief Officer Robert Hume that he required one of the new Berthon collapsible boats to be swung out, together with the wooden lifeboat abreast of it. They were chosen at random. 'The first boat was swung out and lowered into the water in three and a half minutes, and the collapsible boat was swung out opened out and lowered into the water in nine minutes.' Clarke was 'quite satisfied with both officers and deck crew'. Their colleagues were standing by watching.

Clarke also explained in detail 'the working of collapsible boats'. He was satisfied with everything he had seen. At 11.50 a.m., he was about to hand in the certificate of clearance to permit *Olympic* to go to sea. There were 946 passengers on board and hundreds more expected to join at Cherbourg and Queenstown, taking the total number to around 1,400.

It was at that moment, minutes before *Olympic* was due to sail, that Philip Curry had to request Clarke's permission for *Olympic* to proceed instead 'to a safe anchorage', because 'all the firemen had gone ashore'. The firemen were concerned about the seaworthiness of the additional collapsible boats. Curry asked Clarke if he would come with him to the quayside and discuss the matter with a deputation of the firemen to try to reassure them.

Curry explained that it was difficult to obtain ordinary wooden lifeboats at such short notice, but the collapsibles were up to Admiralty standards and had been passed by the Board of Trade. Clarke backed him up. He assured the firemen that he was happy to remain on board while *Olympic* proceeded to an anchorage off the Isle of Wight. Once she was there, they could do a working demonstration of the boats. The firemen 'continued to insist on wooden lifeboats'.

Dr Beaumont took a rather uncharitable view of the firemen's motives. He described them as:

> A weird lot of human beings, ignorant and stubborn, impulsive, unreasoning and easily swayed by any agitator, who has only to voice a supposed grievance for the men to follow him like a flock of sheep, often enough not knowing what they are striking for.

He thought the recent disaster 'had been too much for their nerves'.[2]

***

Olympic left the dock and proceeded down Southampton Water to her intended anchorage

Captain Maurice Harvey Clarke checks one of *Olympic*'s lifeboats as part of his inspection before issuing her clearance to go to sea. *Illustrated London News*, 1912. (Author's collection)

Olympic's passengers line her decks. They are increasingly anxious about when she will sail. (Günter Bäbler collection)

Olympic leaves the dock at Southampton to proceed to a safe anchorage. (© Chronicle/ Alamy Stock Photo)

early that afternoon. First-class passenger J.H. Turner described the 'great crowd' seeing them off: they 'followed right to the end of the wharf, still waving handkerchiefs'. The gong sounded for lunch and the passengers thought they 'were fairly on our way', but it became clear to them later that afternoon that there was a problem.[3]

Clarke confirmed with his superiors that he could clear *Olympic* to proceed to Plymouth, if needed, but not Cherbourg. The White Star Line sought help from the Board of Trade. Philip Curry even requested whether the government could provide stokers from the Royal Navy to replace the firemen, but a telegram was sent to him at Southampton around 4 p.m. confirming that this would not be possible. The *Edinburgh Evening News* quoted a Central News telegram from Cherbourg reporting that 'several hundred passengers were waiting here' for *Olympic*. They were set to wait overnight, hoping she would arrive the next morning.

Early the following morning, Thursday, 25 April, Clarke was on the Boat Deck. Hands were turned out to lower some of the collapsible boats. Two were lowered into the water, but 'passengers were beginning to move about' and they decided to 'do nothing on the boat deck which might arouse suspicion'. Union delegates came aboard and asked to see some collapsible boats tested and rowed. They chose six, one of which ('No. 38') Clarke had already seen lowered into the sea and rowed around earlier.

Two hours later, the delegates pointed out to Clarke that one of the collapsibles was leaking. There was 'a small hole in the inner skin through which the water very slowly trickled'. 'We found a tiny hole in the outer skin which looked as if it had been done by a nail or something of the sort,' Clarke later wrote. He put two men in the boat and they bailed out the water in less than three minutes.

The delegates, Philip Curry and Captain Haddock went to the captain's quarters to discuss the situation. Clarke was there merely as an observer. The delegates were happy, with the exception of the 'slight leak' and another boat which they thought was 'unseaworthy'. They said they had no other complaints. Curry suggested that a replacement collapsible should be put on board. Lewis said that if the boat was replaced, he would ask the firemen to return to *Olympic*.

All of this activity had taken up much of the day. By 5.30 p.m., Clarke was able to report to the Board of Trade that replacement firemen were expected to embark by 8 p.m.

Dr Farquharson, the Board of Trade Medical Officer, and Captain Clarke inspected about 100 firemen who had been brought alongside by a tug. They were also approved by Chief Engineer Fleming. It was starting to look positive that *Olympic* would be able to sail, but around midnight they suffered a setback. The ship's seamen disembarked onto the tug, because they objected to the replacement firemen. Captain Haddock ordered them to return to duty, 'but they unanimously refused to return to the ship'.

It was now Friday, 26 April. Haddock signalled the cruiser HMS *Cochrane*, at Portsmouth: 'Crew deserting ship, request your assistance, Haddock, master.'

Around 12.50 a.m., a tug from Portsmouth came alongside with a further 168 firemen, who the White Star Line had recruited from further afield (some from Sheffield). Dr Farquharson passed them as satisfactory firemen and trimmers. The problem was that the White Star Line now needed to replace the seamen.

Shortly after 1 a.m., *Cochrane*'s commander, Captain W.E. Goodenough, arrived. His pinnace was made fast to the tug alongside *Olympic*, then he crossed the tug and boarded the ship. Captain Haddock apologised for bothering him and the three men consulted in the captain's quarters. Goodenough spoke to the men and asked 'what their grievance was'.

Although there was some comment about the new collapsible boats, Goodenough later wrote that 'the main question in reality was evidently the quality of the men who were being shipped as firemen from Portsmouth'. He noted, 'The men were perfectly respectful both to me and their own captain.' There was 'no question of any act of violence' and he did not think it

Olympic's passengers kept themselves entertained, consuming more food and drink supplies than a typical day when she was crossing the Atlantic. (Roy Derouin collection)

'necessary to place a guard on board'. Captain Haddock was also 'unwilling that I should use any force'. Goodenough thought their concerns about the qualifications of the replacement firemen were 'genuine', but 'it was quite evident to me that the seamen, though anxious of themselves to return to work, were more afraid of offending their union by working with non-union men than of offending the law by breaking it'.[4]

There was nothing further he could do. He decided to return to *Cochrane*. When he was leaving, some of the men 'told me that the new firemen were the "riff raff of Portsmouth, discharged second class stokers from the service, etc."', but 'that, to me, did not appear to bear on the case'. Goodenough told the men that, regardless of the law, he thought they did not have the right 'to leave a ship in a roadstead, full of passengers, in the middle of the night'.

Around 10 a.m., police officers from Portsmouth arrived to take the men ashore in a tug. Captain Haddock detailed Fifth Officer John Edward Withers to accompany the men until they were landed ashore under police custody. An hour later, a boat came alongside with

The scenes on *Olympic*'s Boat Deck offered an eerie illustration of how *Titanic*'s Boat Deck might have appeared on the night of the recent disaster. (Günter Bäbler collection)

Passengers begin to appear on deck and take an interest in the lifeboat-lowering operations. *Illustrated London News*, 1912. (Author's collection)

A 'slight rent' in one of the collapsibles resulted in further protest from the crew. *Illustrated London News*, 1912. (Author's collection)

Right: Captain Herbert James Haddock spent many anxious hours trying to resolve the situation. (Ioannis Georgiou collection)

Far right: Long after she should have left Southampton, *Olympic* remained stranded off Ryde. (Roy Derouin collection)

replacement seamen from Portsmouth. Clarke inspected them:

> Most of them held naval rating discharges [...] They were duly mustered on the navigating bridge, where in company with the captain, officers and boatswain, I put them through a satisfactory examination as to their fitness to be put upon the [ship's] articles [crew] as efficient deck hands. A few [...] did not satisfy me and were rejected.

He had requested permission from the Board of Trade to clear *Olympic* to proceed to Cherbourg and Queenstown, subject to the seamen being replaced with competent replacements. There were going to be 914 crew, consisting of seventy in the Sailing Department, 360 in the Engine Department and 484 in the Victualling Department. It was higher than normal, reflecting that they were taking on additional seamen to man the lifeboats and more firemen and trimmers so that the ship's speed could be increased to try and make up for some of the lost time.

Another concern came to light. The *Daily Mirror* carried a report about 'the question of replenishing stores'. The passengers on board were 'consuming rather more per head [...] while the ship is at anchor than they would be if she were on the Atlantic'.

That Friday afternoon, the White Star Line came to an inevitable decision. *Olympic* was over two days behind schedule. They cancelled the voyage. The announcement came from Liverpool shortly before another wireless message was received confirming that the remaining replacement seamen were about to leave Portsmouth so they could join *Olympic*. Clarke reported, 'Efforts were then made by wireless to cancel Liverpool decision, but this was not possible.' A notice was posted throughout all passenger areas: 'It is with the greatest regret that the White Star Line announce to passengers of *Olympic* that they are compelled to abandon voyage notwithstanding every effort made to obtain a crew.'

Captain Haddock decided to return to Southampton. A 'loyal quartermaster' was posted at the ship's wheel and *Olympic* returned to Southampton around 7.30 p.m. One of the seamen told the press that 'none of us has any ill feelings against' Captain Haddock, but that the strikers 'did it for our own safety, the safety of the passengers and everybody concerned'.

There was plenty of discussion in the city. The *Southampton Times* quoted an 'expert acquaintance', who speculated that the unrest 'might not have originated with the White Star Line or the *Olympic*'. 'The fact is that there has been a great lack of confidence among seafaring men for some time in the Mercantile Marine Department of the Board of Trade,' he explained:

> Grievances against the department have been expressed in various ways and have not received an effective hearing. Now, in consequence of the most deplorable maritime disaster [...] the attention of the public has been secured, and if members of the crew of the *Olympic* have recognised that and decided that now is the time to shout [...] then I, for one, shall not blame them.

The *Belfast Evening Telegraph* reported that 'the first batch' of first-class passengers reached Southampton after 7 p.m. aboard the paddle steamer *Duchess of Kent*. The majority stayed on *Olympic* and 'followed later', when she docked. 'They all spoke in glowing terms of their treatment,' the reporter explained, 'The company [...] had behaved magnificently all through. In fact, their trip had been more like a two days' pleasure cruise. They placidly accepted the situation.' Even so, there was a 'state of suspense' because they did not know if or when they would sail.

An American passenger spoke to the *Daily Mirror*'s reporter, claiming, 'If I do not make Minneapolis by Sunday week I shall lose £10,000.' One passenger spent much of the time walking the decks 'in a state of furious indignation'. Others made use of the gymnasium. There was also a honeymoon couple who said they would have to cancel the party they had planned on their return to the United States.

J.H. Turner and his companions arranged to sail on *Philadelphia*, but were able to remain on board *Olympic* meanwhile, 'We were sorry to leave the *Olympic* as we had made many friends and she is a much finer vessel [...] The various rooms are more like the rooms in a large hotel.' He wrote that the rooms were 'panelled and beautifully carved, no two rooms being alike'. When they did sail the following week, they found *Philadelphia* 'convenient and comfortable', but she was significantly slower.

Olympic's mail had to be reassigned to other ships. The *New York Times* reported early in May that *Lusitania* had arrived in port with 5,210 sacks of mail (more than the Christmas peak). It included her own mail cargo as well as *Olympic*'s. She also carried a number of additional lifeboats.

***

Fifty-three men appeared in court at Portsmouth that Friday afternoon, before *Olympic* had even returned to Southampton. *The Times* reported that 'the defendants included 35 able seamen, five quartermasters, five lookouts, two lamp trimmers, four greasers, an engine room storekeeper, and another'. They were charged under Section 225, Subsection B of the Merchant Shipping Act (1894): 'That if any seaman was guilty of wilful disobedience to lawful command he should be liable to be imprisoned for a period not exceeding four weeks.' All pleaded not guilty. Mr Hiscock was prosecuting the case on behalf of the White Star Line; Dr C.A. Emanuel represented most of the defendants on behalf of the British Seafarers' Union; and Mr G.H. King appeared for the others.

Hiscock opened to summarise his case. Fifth Officer John Edward Withers gave evidence, describing events on board *Olympic*. Under hostile questioning from Emanuel and King, Withers frequently used phrases such as 'I cannot say' or 'I don't know'.

'Don't you know there was a consultation at Southampton and there was another long conversation last night?'

'I was in bed,' Withers replied.

'Oh! It didn't interest you very much, then.'

'Well, my watch finished.'

King found Withers' answers frustrating. 'Don't you wish to answer questions?'

'I am doing my very best.'

'Well, were there not 284 men left at Southampton?'

'You cannot judge numbers,' Withers replied.

'Don't say I cannot judge numbers,' King admonished him. 'I am asking you.'

Hiscock stepped in. 'It's quite obvious what he means. When he says "you" he means "I cannot, or anyone else cannot, judge numbers to one or two".'

King put it to Withers that the replacement men were 'a scratch crew'.

'Well,' Withers smiled back. 'I don't know.'

The defendants were granted bail until Tuesday, 30 April.

Captain Haddock testified at length. It was put to him that he was a popular commander. 'I'm very pleased to hear it,' he responded.

The *Southampton Times* reported an exchange where he was questioned whether he had 'advised' the men to return to work. 'Certainly not,' Haddock replied. 'I used the word "order".'

'If men between 40 and 50 years of age came on board and said it was their first ship, you would not care about them?'

'The engineers are perfectly competent to look after their staff,' replied Haddock.

'And if their staff was incompetent?'

'They would let me know.'

'A scuffle with the police', *Illustrated London News*, 1912. (Author's collection)

Olympic's arrested seamen under escort by policemen to Portsmouth Town Hall. (Günter Bäbler collection)

Some of the seamen sitting before the magistrates in court at Portsmouth. *Illustrated London News*, 1912. (Author's collection)

Haddock explained that the shortage of crew had consequences. When *Olympic* moved from the dock to go to her anchorage, some of the ship's officers had needed to let go of the ropes at the stern; some of the stewards had helped in the boiler rooms.

Captain Clarke testified that *Olympic*'s existing lifeboat complement already exceeded the regulations. They were in 'satisfactory condition', even before the extra 24 Berthon boats were put on board. After further testimony, hearings resumed on Saturday, 4 May.

Junior Second Engineer Charles McKimm testified that he and another officer from the American liner *St Louis* had selected the substitute firemen and greasers. They had examined them as they normally did to make sure that they were competent for the job.

Mr Norman Raeburn, who had taken Mr Hiscock's place, asked, 'Would you call "firing a ship" skilled labour?'

'No, sir,' McKimm replied. 'Not skilled labour, but hard work.' His response prompted laughter in court.

Under questioning from Dr Emanuel, McKimm said he had selected 168 out of 200 men sent from Sheffield. Many of them did not have 'books', but they were 'mostly [coal] pit hands, used to hard work'. Some had been at sea on *Adriatic*. Then Mr King stepped in and quoted a newspaper report that the 'men were ill dressed, looked as if they had not seen soap for years, and were the class of men who spent their time at street corners and in the lowest public houses'. McKimm replied that it was not a fair description.

Magistrate Sir Thomas Bramsdon, chairing proceedings, asked whether the replacement men would have been taken on 'under ordinary circumstances'. McKimm conceded that 'I suppose we should not have taken them if there were others knocking about', but he thought that they would have served satisfactorily.

Superintendent Engineer Frederick John Blake testified that he had asked the men if they had gone to sea before and every man had said, 'Yes'.

'Were you content to take as firemen criminals and men of that description?' asked King.

'I am afraid we have to take all kinds,' Blake responded. There were no convictions recorded in any of the discharge books.

Purser Claude Lancaster testified that 101 of the 162 Portsmouth men had been to sea previously, but 'a great many stated that it would be their first voyage'. Then William Joyce, one of the engineers on *St Louis*, said that the men were good firemen. He would have taken them on *St Louis*.

There was a testy exchange between Philip Curry and Dr Emanuel. 'Did you threaten to have the men locked up if they did not go on board at once?'

'Emphatically no. It would have been rather a difficult proposition to get them locked up,' Curry argued.

'I can quite understand that you found it difficult to do it legally,' Emanuel remarked.

Captain Benjamin Steel, who was not only the White Star Line's marine superintendent at Southampton but also a commander in the Royal Naval Reserve, provided details of the Berthon boats that were taken on board. They had come from various naval transport ships. Some were about a year old. Ten from HMT *Soudan* were 10 years old.

Mr Raeburn protested to the magistrates that the White Star Line was not on trial for either over or undermanning *Olympic*. The defendants were on trial for disobeying Captain Haddock's lawful commands. Sir Thomas Bramsdon replied that the full circumstances were relevant.

Dr Emanuel tried to raise an objection that the men had been aboard a tug when they disobeyed Captain Haddock, so they had not been serving on *Olympic* at all. He was overruled.

Quartermaster Martell testified that the men had taken a ballot and voted to go ashore. He was 'thoroughly disgusted' with some of the replacement men, who were 'all the "scallywags" of Portsmouth'. He asserted it would take twelve hours to launch the twenty-four Berthon boats.

Martell described Haddock's command to return to the ship as 'more like giving advice than giving an order'. He said, 'I should not have been afraid if there had been no boats at all; but I wouldn't sail with incompetent men.' Quartermaster J.W. Cox, Assistant Storekeeper J. Middleton and others gave evidence in corroboration.

To close the defence case, Dr Emanuel continued to argue that the order had not been given on *Olympic* at all. He said that Section 458 of the Merchant Shipping Act obligated shipping lines to make sure their ships were seaworthy. His argument was that the White Star Line were in breach because 'the placing of an incompetent crew on board amounted almost to rendering her as unseaworthy as if she had a hole in her hull'.

Mr King said the men had been loyal. They were 'anxious and willing to serve the company but they refused to endanger their lives with the scallywags of Portsmouth and elsewhere, very many of whom were known to the bench and the police'.

The magistrates spent two hours considering their verdict. The *Telegraph*'s report summarised the outcome. They agreed that the White Star Line had proved their case, but decided that 'it would be inexpedient to send the men to prison or to impose a fine'. It 'could not be called a victory for the men, because they had been found guilty of wilful disobedience [...] The bench, however, hoped the differences would be settled, and that the men would return to work.'

The *Aberdeen Daily Journal*'s coverage emphasised that the men had acted following a 'terrible catastrophe, for which no precedent was to be found'. The magistrates were 'greatly pleased by the way that the men had behaved in court, and by their good relations with the captain'. They thought the circumstances were 'totally different from everything which had occurred before, or they hoped would occur in the future'. It was a pragmatic verdict in the circumstances.

(On 1 May, several days before the verdict, the Hammersmith Club had written to Board of Trade President Sydney Buxton, advising him that their political committee had passed a resolution congratulating the seamen 'on the stand they have taken and we enter our protest against any punishment being meted out to them, as we consider that in the light of recent events the men only done [*sic*] what was natural in the interests of the travelling public and themselves'. It was a rather pointless protest, because the court judgement was not something that the Board of Trade could influence.)

Captain Clarke's efforts did not go unrecognised. He had remained 'on duty for nearly 60 hours continuously, largely without sleep'. On 29 June, the Board of Trade confirmed that the Treasury had 'sanctioned the grant of a gratuity of £5' to recognise his 'exceptionally arduous and important work'.

***

The White Star Line were confronted with another problem dating from *Olympic*'s maiden arrival in New York. Late in April 1912, *O.L. Hallenbeck*'s owner, Peter Cahill, decided to take the company to court for damages to the tug. He argued the 'collision and damage were not produced or caused by any negligence upon the part of' his tug, but 'wholly produced by the negligence' of the White Star liner. He said *Olympic* was negligent 'in starting the starboard propeller in reverse motion' and 'failing to give the necessary orders to stop said propeller'. The repairs to the tug and loss of revenue while it was out of use amounted to 'about the sum of $10,000 [£2,000]'.

White Star responded with a counter-suit early in June, filed in the Southern District Court of New York by Burlingham, Montgomery & Beecher and signed by Philip Franklin, acting as agent. They argued negligence insofar as the tug 'came close to the steamship's side and disregarding the warning signs, negligently got into a position so near the starboard propeller, which was then in motion, that she was caught by the swirl and currents generated'. She 'was not in charge of a competent pilot' and 'did nothing to avoid a collision'.

Claim and counter-claim followed throughout 1912. White Star had spent 'the sum of £1,500 [$7,500], as near as can be now estimated' on repairs to *Olympic*'s starboard propeller, 'no part of which has been paid though payment [...] has been duly demanded'. They highlighted the 'large warning signs with letters at least six inches high, reading: "Notice, This Vessel Has Triple Screws: Keep Clear"'. There was a debate about whether the tug had even technically been engaged in assisting *Olympic*.

It was the start of yet another legal entanglement for the White Star Line which dragged on into 1913 and 1914, when witnesses were trying to recall specific details several years after the event. Cahill's arguments were dismissed by District Judge Hazel on 4 November 1914, 'I think he [the tug's commander] should have realised that it was quite probable that the steamer would render assistance to the tugs with her propellers, and should have governed himself accordingly, either by keeping the tug under control or taking a position farther from the stern.'

'It is customary in this port for masters of tugs to select their positions in rendering assistance to steamers, and in doing so they are required to have regard for their own safety,' Hazel noted. 'In my opinion the *Hallenbeck* sustained injury from the swirl of water caused by the motion of the starboard propeller of the steamer *Olympic*, because she was too close to the stern of the steamer and did not manoeuvre properly to arrest her sheer to port.' There was insufficient evidence to place blame on *Olympic* for starting her starboard propeller. An earlier court case 'held that a tugboat assumes the risk of being sucked in by the propeller of a steamer if she goes too near it'.

Cahill's claim was rejected, with no costs toward either party, but he appealed against it on 17 November 1914. That final appeal was dismissed by Circuit Judge Lacombe on 12 May 1915, this time with total costs of $94.85, 'with interest thereon until paid'.[5]

17

'THE SHADOW OF A GREAT CALAMITY'

'We meet here today under the shadow of a great calamity,' Harold Sanderson told a special meeting of the White Star Line's managers on 22 April 1912. He reflected on the company's progress since he had joined seventeen years earlier:

> We have seen the ships grow in size and magnificence, step by step, from the *Teutonic* and *Majestic*, until we reached what we thought was ... an epoch in the success of the company's career by the production of the *Olympic* and *Titanic*.
>
> [...]
>
> In that great ship [*Titanic*] we invested not only an amount of capital which is equal to the cost of seven ships which would have been classed as great only a few years ago, but we devoted to her production all the knowledge and experience that we, as shipowners, and our friends, Harland & Wolff, the greatest shipbuilders in the world, had acquired during a generation of experience [...]
>
> In the *Titanic* we produced a ship which we believe had only her sister vessel, the *Olympic*, for a peer, and even this latter vessel was, in some important respects, surpassed by her newer sister. We knew we had put two ships in service which surpassed all others in size and magnificence. But more than this, we believed that we had made these ships practically unsinkable, and absolutely safe for ocean travel.

Titanic's crew represented 'everything in the way of skill', 'a finer set of men never manned a ship, nor could be found in the whole of the mercantile marine of the country, and no higher testimony than this can be paid to the worth of the crew'. Her loss was a 'stupendous calamity', confirming 'that in spite of all we can do, the Atlantic is, and will always remain [...] our master, and a very hard and cruel one'. Their deep grief for 'the terrible loss of life' was coupled with 'respectful admiration for those splendid men of the White Star Line who have lived up to the best and highest tradition of the British service, by dying so nobly at their posts in order that passengers might be saved'.

He appealed to his colleagues to 'resume work with determination to do all in your power to help the White Star management to live down the effects of this great calamity' and to maintain the company's 'proud position which it has held for so long in the shipping world'. The managers recorded their 'heartfelt sympathy' for J. Bruce Ismay's 'terrible experience' and, as a mark of respect, adjourned the meeting until the next day without conducting any business.[1]

The same day they were meeting, the Honorary Secretary of the Southampton Post Office wrote to the Postmaster, General Post Office, London. As well as discussing a suitable memorial for their lost *Titanic* colleagues, the staff had been alarmed at:

> … the conditions under which our respected colleagues met their death on the SS *Titanic* and as a protest and safeguard respectfully request that the Postmaster General be asked by telegraph not to allow any sea post staff to sail on the SS *Olympic* until the sleeping quarters on that ship are altered.[2]

He responded:

> The sleeping cabins, two in number, are situated on the middle deck [F] and provide accommodation for four clerks, two in each cabin. They are conveniently near the Post Office and are self-contained being completely separated from the passenger accommodation of the ship. Both the cabins are fitted with port holes and are light and airy. The ship is ventilated throughout […] On the score of the safety of their position there is absolutely no cause for apprehension, and, even if the most expensive suites were allocated to the Post Office staff, no greater immunity against loss of life in case of shipwreck would be secured to them.

However, comments about the noise from third-class passengers had been made before *Titanic*'s departure and some changes were already being arranged on *Olympic*. On 29 April, the White Star Line acknowledged that they had 'given instructions for the necessary alterations' and on 13 May explained:

> Under the circumstances we will be pleased for the present to arrange for two staterooms to be placed at their disposal when there is the accommodation vacant, and will certainly arrange this for the next voyage leaving Southampton on Wednesday. We will look into the question as to whether it is possible to make some permanent change. [*Olympic* sailed with two E-Deck staterooms allocated to the Sea Post Clerks: No. 5 for the British and No. 6 for the Americans.][3]

All the shipping lines agreed to move the course for eastbound steamers to 'the extreme southerly track' from 16 April; westbound steamers departing on 25 April or later were also to follow a more southerly course.[4]

Olympic remained at Southampton until her next scheduled sailing and on Saturday, 4 May, Captain Haddock attended the Wreck Commissioner's Court, which was the British investigation into the *Titanic* disaster. Two days later, Wreck Commissioner Lord Mersey and his expert assessors went aboard *Olympic*. According to a report in the *Daily News*, Mersey inspected the lifeboats, engine room and stokeholds and the third-class accommodation. Captain Haddock and other White Star officials were 'busily occupied for some time in answering Lord Mersey's questions'. After lunch, they watched several lifeboats launched.

Only a few weeks earlier, J. Bruce Ismay had been expecting to return to Southampton aboard *Titanic* on 27 April. He eventually returned to the United Kingdom on board *Adriatic*, arriving in Liverpool on Saturday, 11 May (his wife Florence had joined him at Queenstown). He lost no time in writing to the Lord Mayor of Liverpool the same day.

His father had founded the Liverpool Seamen's Pension Fund in 1887 to provide pensions 'for deserving seamen of whatever rank sailing from the port of Liverpool, who are past work' (later, the Margaret Ismay Widows Fund was established to provide for their widows too). The Mercantile Marine Service Association administered the fund as trustees and by 1912 it had grown to about £69,000 (£52,000 of which came from contributions either by the Ismay family or the White Star Line). Now, he proposed a fund providing 'for widows of those whose lives are lost while they are engaged upon active duty […] upon the mercantile vessels of this country'. The 'terrible disaster to the *Titanic*' highlighted the 'necessity of such a fund' and he proposed to contribute £10,000 with a further £1,000 from his wife. It would 'continue for all time'.

News of Ismay's letter apparently reached the press and the Earl of Derby, Lord Mayor of Liverpool, sent a telegram to Ismay asking if he could publish it, 'saying I have gratefully accepted your offer?' Ismay responded, 'Please act in whatever manner you think best, leave myself entirely in your hands.'[5]

Oceanic took the next express sailing from Southampton to New York, where she arrived safely on 16 May. During the westbound crossing, her crew sighted *Titanic*'s abandoned Collapsible A adrift. They recovered the decomposed remains of three victims, who were buried at sea. (There were sensationalist press reports that some of the victims had starved to death. In reality, they were already deceased before they could be taken on board *Carpathia* and so the boat had been abandoned at sea. White Star had to assure the public that it was not true.)

White Star had written to the Southampton Harbour Board asking whether they still intended to charge them £87 7*s* for *Olympic*'s sailing, which had been 'abandoned under the special circumstances'. On 13 May, the Finance Committee voted on it and agreed to only charge them half.[6] It was a small piece of good news for them as *Olympic* prepared to sail from Southampton two days later.

This time, *Olympic* sailed as planned but, as she went past Berth 38, a coaster 'came very close' to 'breaking adrift'. Marine Superintendent Benjamin Steel was concerned enough to raise it with Philip Curry. Any ship at Berth 38 was in 'a dangerous position' when one of White Star's large express ships was passing, 'as in the case of the *Titanic–New York* incident'. The channel was narrow and each ship:

Philip Franklin (left) and J. Bruce Ismay (right), photographed while the United States Senate Inquiry into the *Titanic* disaster was under way. (Library of Congress, Prints & Photographs Division)

The proceedings before the British Wreck Commissioner, Lord Mersey, took place in the Scottish Drill Hall in London. *Illustrated London News*, 1912. (Author's collection)

> … has to make a considerable turn to port to take the bend of the Channel, and in executing this manoeuvre has at times to go back hard [astern] on the port engine, the backwash striking the wall with great force and rebounding on to the quay side of the vessel lying at this berth.[7]

One of the passengers wrote an account for *The American Marine Engineer*.[8] When *Olympic* arrived off Queenstown, she was 'put through a series of unexpected manoeuvres', including:

> [swerving] west, and from the west to the south, and so round till we had described a complete circle. Thus, for an hour, travelling now at full speed, now at half speed, the ship circled this way and that in the water. It described perfect circles; it drew the figure '8' and the letter 'S'; she turned upon her trail and met again the line of white water which showed her course […]
>
> We all observed the marvellous rapidity with which this prodigious ship turned from side to side, answering to its steering gear like a yacht answering to the tiller.
>
> It was a manoeuvres through which this ship had never before been put. The captain himself, issuing his instructions from the bridge, was playing with the monster as with a toy which had developed a new trick.

It was probably on this occasion that they tested the time *Olympic* took to turn 2 points (22.5 degrees) to port. Lord Mersey had already heard testimony estimating that *Titanic* had turned 2 points before the collision with the iceberg. Edward Wilding later testified to that, based on a test conducted with *Olympic*.

Even after taking on passengers at Queenstown, she only carried eighty-eight first, ninety-seven second and 342 third-class passengers. The total of 527 passengers was the lowest since she had entered service. Whether it was the result of the recent disaster or whether prospective passengers had been concerned that she would not sail as planned was a finely balanced question, 'It was not so much that the public feared to travel in an "unlucky" ship; they were merely afraid that she would not sail.' The first-class dining saloon 'had no more than a sprinkling of diners. The band played in an almost empty reception room.' The gymnasium and other public rooms were largely deserted, as were the Boat Deck and the enclosed promenade on B-Deck.

The ship's crew seemed to be making a special effort to reassure passengers, including oiling the lifeboat davits, 'Wherever you looked your eye encountered lifeboats, circular lifebelts and jacket lifebelts […] Aft on the third class deck there was a phalanx of collapsible boats.'

Then, after Sunday morning service, they held a boat drill and a fire drill. He also noticed how far south *Olympic*'s course had taken her. Porpoises 'leapt from the water in the morning', but the 'monotony of the sea was only varied by sunshine and cloud'. He was able to visit the engine room, to see the boilers 'where the firemen are at work, and deep down in the bows where firemen and seamen sleep':

> It would be hard to find a more striking contrast than that between the upper and the lower decks, between the Sicilian luxury of the one and the toil and squalor of the other […]
>
> I believe that there is no ship in the world where the firemen are better provided in respect of space, food and washing utensils, but, but it at the best, they work in a grim underworld; they eat and sleep in the part of the ship most exposed in the event of collision; they live in sweat while the first cabin passengers are listening to the band or taking Turkish baths.
>
> That is what they felt – as many of them told me – when they came out into the open at Southampton [on 24 April] and demanded that safety should be provided for them as well as for passengers.

Olympic arrived at the Ambrose Channel Light Vessel at 11.06 a.m. on 22 May. The southerly course she had taken extended the westbound crossing to 3,093 miles (it was 200 miles more than normal and added over nine hours to the crossing time, taking it to six days, one hour and forty-four minutes).

There were three days to get the ship turned around before her eastbound departure on Saturday, 25 May. All was nearly ready when Senator William Alden Smith, who was leading the United States Senate Inquiry into the *Titanic* disaster, paid a visit to *Olympic* at short notice. The senator even went down below to the boiler rooms to take testimony from Frederick Barrett, who had been serving as a leading firemen on *Titanic*. He recalled that 'a large volume of water came through' about 'two feet above the floor plates' on the starboard side of *Titanic*'s Boiler Room 6.

Olympic left New York with 887 passengers. The eastbound crossing itself was relatively uneventful but the distance of 3,247 miles reflected the new southerly route. She reached the Eddystone Lighthouse at 9.05 p.m. on 31 May. *Lloyd's Weekly News* carried an article headlined: 'Eight Liners Race Across Atlantic: *Olympic* Outpaces Rivals with Speed of 22½ Knots'. It said she had landed 400 passengers at Plymouth as well as '£50,000 in specie, and a specially addressed mail of about 400 bags'.

If one second-hand account, published many decades later, is to be believed, there was an unfortunate navigational error towards the end of the crossing which left *Olympic* much further north than she should have been. The error came to light when the lookouts spotted broken water ahead, where the sea was breaking at the foot of the cliffs near Land's End. The officer of the watch ordered the engines reversed, so that the danger could be avoided and the correct course resumed.[9]

Following the call at Cherbourg, she returned to Southampton and docked on Saturday, 1 June. Captain Haddock complained that 'a small steam yacht was anchored in the "swinging ground" off the dock'. The tugs were well positioned 'and the docking would have been easily carried out, but for the yacht, with whom we had to close to, about, a distance of thirty feet. This is a very narrow margin [...] and, under the circumstances, totally unnecessary.'

***

White Star's express service was getting back into a normal routine. *Majestic* was called upon to replace *Titanic*, but she was past her prime. Her first westbound arrival in New York on 30 May brought in only fifteen first, eighty-one second and 261 third-class passengers. *Oceanic* followed her on 5 June and then *Olympic* arrived in New York seven days later. Once again, she had a three-day turnaround, but she left New York with 1,296 passengers. This time, first-class bookings were very healthy with 666 passengers.

Olympic got back to Southampton safely and left for another round trip to New York on 26 June. While she was making the call at Queenstown, Captain Haddock was concerned that she had touched bottom. After much investigation, including discussion between Captain Haddock, Captain Bartlett and the pilot, Bartlett concluded that she had not, but he suggested that she had been anchored 'too far' eastward of the deeper channel 'and must have been very close to the bottom as she swung [...] Captain Haddock avoided touching by moving ahead as required.'[10] She arrived safely in New York on 3 July.

When she left New York on the 6th, *Olympic* carried 1,425 passengers. It was an indication that she was gradually regaining her popularity. Mike Kinney recorded the crossing in a series of letters to 'My Dearest Maude', which were published in *The Gimlet*.[11] 'This enormous seagoing hotel was backed out from the pier by numerous busy little tugs,' he wrote:

> We were dragged and pulled out into the channel of the Hudson [River].
>
> [...]
>
> We at last got straightened out and started down the channel. Then we stuck in the mud. My, how busy the little tugs did get! More tugs came. They grabbed hold of the floating Marlborough-Blenheim everywhere. They snorted and pulled and puffed. They filled the air with smoke. Passengers escaped – coughing – below. [A seaman later told him that she had stopped because repairs were necessary to 'her steering machinery' and that the tugs were 'keeping her steady'.]

CALLING AT QUEENSTOWN

Captain Haddock's experience on 27 June 1912 highlighted an increasing concern for the White Star Line. For some time, they had been considering the westbound call at Queenstown and whether it made sense from a business point of view. Harold Sanderson calculated that *Olympic* lost an average of seven hours, fifteen minutes on each westbound crossing by calling at Queenstown, compared to if she had headed for New York direct from Cherbourg. The passenger traffic from Queenstown was largely third-class passengers leaving for the United States.

White Star had a good secondary service from Liverpool, provided by the 'Big Four', and Sanderson thought that 'the slower service [...] is the favourite service for the third class passenger'. He pointed out that the ticket costs 'are slightly lower; they are lower than the *Olympic*'.

Cunard's chairman thought:

> The passengers who take up the fastest steamers are only those who are rather in a hurry. They are either the very rich or the very busy, or people who have to cross very often from one definite point to another definite point, from London or Paris to Berlin or New York.

The 'easy going travellers' preferred to go by 'the rather slower steamers'. They were not necessarily emigrants and might sail in first or second class.

Captain J.T.W. Charles' impression was:

> I do not think ships of the class of the *Lusitania* and *Mauretania* are favoured as much by tourists as the slower ships. You find tourists who say that they do not want to cross the Atlantic too quickly, particularly in the summer, when they say they like the sea voyage.

Sanderson cited complaints from passengers 'who did not like [...] being carted about after they got on board ship – they wanted to go to America'.

By March 1912, White Star's management decided they wanted to stop the call. The problem was that they were bound by a contract to carry mail from Queenstown. The postal authorities asked them to reconsider.

'Before we had time to do anything more,' Sanderson recalled, 'our unfortunate calamity happened and we lost the *Olympic*'s sister ship'. They continued on as before, but Captain Haddock became increasingly concerned. On one occasion when Sanderson was on board *Olympic* at Queenstown, Captain Haddock requested he come to the bridge. 'I do not feel at all comfortable about bringing this ship into Queenstown Harbour,' Haddock explained. He showed Sanderson the chart and explained the area to manoeuvre *Olympic* inside the harbour was too small. Sanderson reassured him that the call at Queenstown was going to end in the near future and cautioned him to 'run no risk that he could possibly be blamed for' in the meantime.*

---

* These comments are taken from the testimony of Alfred Booth, 16 June 1914; Harold Sanderson, 18 June 1914; Captain J.T.W. Charles, 21 July 1914, pp. 22, 26, 28 and 34 in particular.

Photographed in May 1912 at New York, Lifeboat 16 can be seen on the aft port side of *Olympic*'s Boat Deck. One of the additional collapsible boats is on the deck inboard, numbered '16A'. (Author's collection)

Once *Olympic* was on her way, he found 'a great crowd' waiting for their mail at the enquiry office on C-Deck:

> Then down to our cabins on the lowest and cheapest deck, E. This time we decided to try the very cheapest first class cabins. On the chart these cabins were shown without any outside ventilation. Imagine our surprise to find outside portholes and nice, large, roomy cabins. Right here, my dearest Maude, let me tell you when you go on the *Olympic*, get the cheapest cabin. The high priced cabins on the upper promenade decks have windows that open on the decks, and as people are promenading the deck at all hours of the day and night you have no privacy or quiet. This pointer alone is worth a subscription to *The Gimlet* for a year.
>
> The stewards were putting our steamer trunks under our berths. While there were [almost] 700 first class passengers everything was working as if oiled. We checked our trunks from our homes right to the cabin of the *Olympic*. Each trunk was tagged with our names, the ship, date of sailing and number of stateroom – and they arrived there via St. Louis Transfer Co., and ship's stewards, just as soon as we did. This means modern system, and such system is not equalled anywhere else in the world. Europe can learn much from our American checking system ...
>
> Then we had to secure our seats in the dining room. There was a long line. Some were terribly particular about where they sat and with whom. You know they might have gotten out of their set for a whole week. Wouldn't it be terrible, Maude? We just took what they gave us and later with a party all our own we had a table outside of the main dining room where we could sit, chat and smoke just as long as we pleased.

He was up early on Sunday, 7 July:

> At seven this morning I donned my bathrobe and descended to the salt plunge bath. The room was all of tile. There were little dressing rooms along one side. Opposite were the portholes looking out upon the blue ocean. The pool itself was about 30 by 20 feet and four to six feet deep. The salt water came in at one end direct from the ocean. It was changed rapidly, so it was always fresh.
>
> First we stepped under the shower of cold fresh water then plunged into the pool. It was just cold enough to give one a shock. The salt water was very buoyant and you almost felt like having your bacon and eggs served as you floated. This swimming pool on a ship has caused a good many smiles but it is a great institution and was very well patronized by both ladies and gentlemen, of course at different hours.
>
> Another fresh water shower, a hard rub down with rough towels and then the run to our cabins. Don't you think this a fine start for the day's strenuous work?
>
> On the top deck is a small gymnasium with a director in charge. Here too, there are hours for ladies and gentlemen. The various apparatus run by electric power. The plan is to give one involuntary exercise. People who cross the Atlantic haven't will power enough to exercise themselves, so it must be done without any exertion of will on their part.

'I rather liked the looks of the second class passengers better than the first,' he wrote. 'They certainly were not so beastly fat and seemed a far more friendly, natural and interesting lot.'

By Monday, 8 July, *Olympic* was well out into the Atlantic and running at around 23 knots.

> It is wonderful how time slips away on shipboard. A large percentage of our fellow passengers doze the time away in their steamer chairs. What a shame! A lot of these old mummies should stay at home and let young people come over and enjoy the sights. They – the old mummies – could sleep just as well on the back porch. But the old have the cash while the young have youth and their hopes.
>
> The ocean to-day is as blue as the wash water on Mondays. We have leaned over the rail and watched schools of dolphins or porpoises. We have seen a few flying-fish, but no whales or sharks. We are running northward and the air for the first time is fresh and cooler.
>
> Monday the deck games started. There is 'shuffle board', 'bull board', and the 'rings'. Shuffle board is the popular game [...]
>
> It is growing cooler and the sea is somewhat rougher. It does seem strange to ride up and down in an elevator in a rolling ship – still, with so many decks the elevators are a great convenience, especially for elderly people ...
>
> You wanted to know how we liked such a large ship! I much prefer a smaller vessel. This ship is like the Marlborough-Blenheim Hotel at Atlantic put out to sea with nothing changed. There are cafes, palm rooms, card rooms, parlors, manicures, pedicures, barbers, masseurs, etc. etc. The hallways are endless. Except when you are out on deck you do not realize you are on a ship. There are so many passengers and the ship is so large that you do not see the same people for days. Then you are constantly meeting new people you have never seen before. Naturally the passengers have settled down into little sets or coteries and there is little general mixing. In this respect it is very different from a small vessel, where all the passengers soon become well acquainted.
>
> I have not seen the captain yet. He has not been down to meals in the cabin. The officers have nothing to do with the passengers and are seldom seen. Passengers are left to amuse themselves just as in any great hotel. If you want anything you go to the Enquiry Office.
>
> The meals seem to get better from day to day, or maybe the sea air has improved our appetites. The service is excellent. We happen at table to have a little old man waiting on us who is quick, alert and attentive.
>
> To sum it up, my dear Maude, if one desires to spend a week in the greatest comfort – in fact, luxury – with little or no ship's motion, he

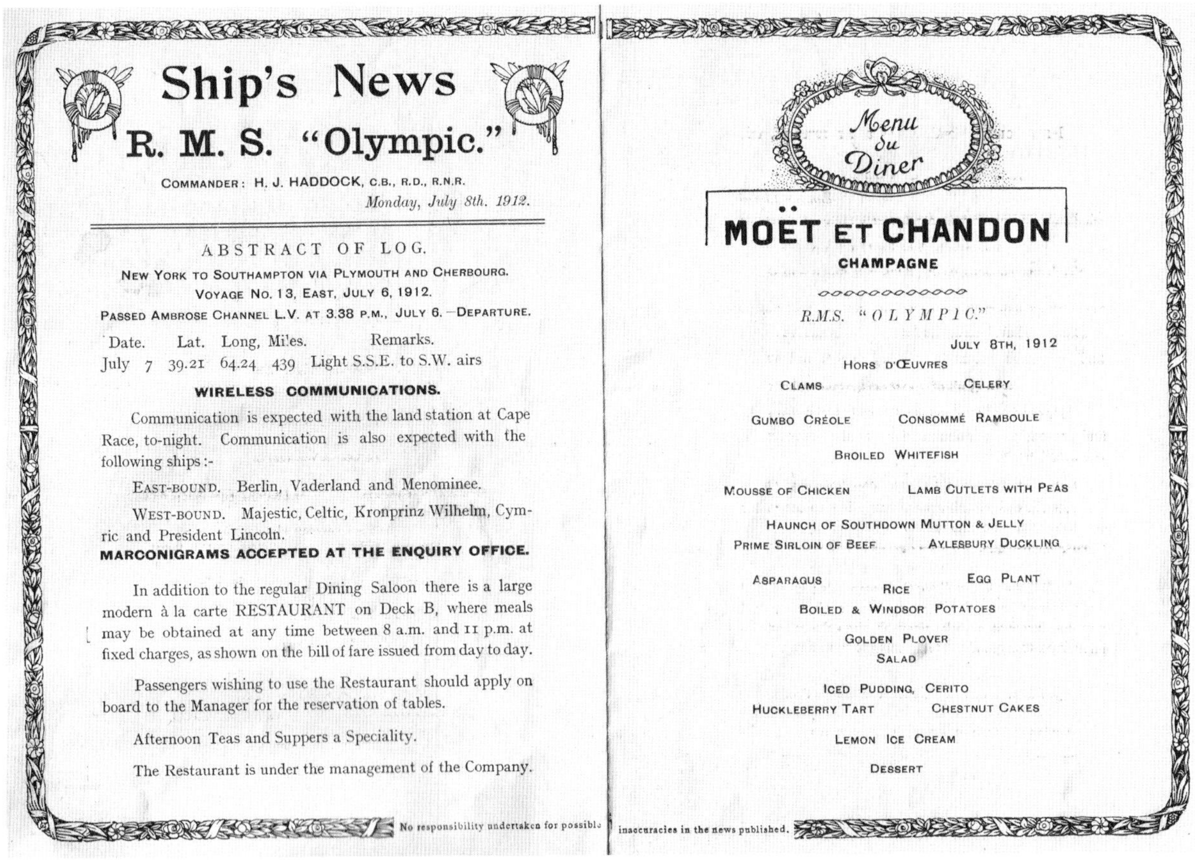

Ship's News
R. M. S. "Olympic."
COMMANDER: H. J. HADDOCK, C.B., R.D., R.N.R.
Monday, July 8th. 1912.

ABSTRACT OF LOG.
NEW YORK TO SOUTHAMPTON VIA PLYMOUTH AND CHERBOURG.
VOYAGE No. 13, EAST, JULY 6, 1912.
PASSED AMBROSE CHANNEL L.V. AT 3.38 P.M., JULY 6.—DEPARTURE.

Date.	Lat.	Long.	Miles.	Remarks.
July 7	39.21	64.24	439	Light S.S.E. to S.W. airs

WIRELESS COMMUNICATIONS.

Communication is expected with the land station at Cape Race, to-night. Communication is also expected with the following ships:-

EAST-BOUND. Berlin, Vaderland and Menominee.

WEST-BOUND. Majestic, Celtic, Kronprinz Wilhelm, Cymric and President Lincoln.

MARCONIGRAMS ACCEPTED AT THE ENQUIRY OFFICE.

In addition to the regular Dining Saloon there is a large modern à la carte RESTAURANT on Deck B, where meals may be obtained at any time between 8 a.m. and 11 p.m. at fixed charges, as shown on the bill of fare issued from day to day.

Passengers wishing to use the Restaurant should apply on board to the Manager for the reservation of tables.

Afternoon Teas and Suppers a Speciality.

The Restaurant is under the management of the Company.

No responsibility undertaken for possible inaccuracies in the news published.

Menu du Diner

MOËT ET CHANDON
CHAMPAGNE

R.M.S. "OLYMPIC."

JULY 8TH, 1912

HORS D'ŒUVRES
CLAMS CELERY
GUMBO CRÉOLE CONSOMMÉ RAMBOULE
BROILED WHITEFISH
MOUSSE OF CHICKEN LAMB CUTLETS WITH PEAS
HAUNCH OF SOUTHDOWN MUTTON & JELLY
PRIME SIRLOIN OF BEEF AYLESBURY DUCKLING
ASPARAGUS RICE EGG PLANT
BOILED & WINDSOR POTATOES
GOLDEN PLOVER
SALAD
ICED PUDDING, CERITO
HUCKLEBERRY TART CHESTNUT CAKES
LEMON ICE CREAM
DESSERT

'The Ship's News' on board *Olympic*, 8 July 1912. (Günter Bäbler collection)

should take such a steamer; but if he wants a real touch of the sea and a change from hotel life a smaller steamer should be chosen. For old people who travel like mummies and who want every comfort for their money, this is the ship. It is also good for nurses and children, as there is so much room and many ways and places to play.

Passengers settled into a routine:

The meals are quite good and we are very well served. At our table a passenger asked, 'Are your eggs fresh?' 'Oh, yes,' answered the steward, 'They just came aboard this morning.' He never cracked a smile – not bad for a Britisher.

A daily paper is published and we find it under our door every morning. The news and stock reports come by wireless. Sorry on Monday that 'steel' went down. Wall Street must be afraid of [Democratic Presidential candidate] Wilson.

There is a band, or rather orchestra, that plays morning, noon and night. After dinner we sit … drink coffee – gossip – smoke, and listen to the music. Some of the foreign ladies smoke cigarettes. This looks odd to Americans going over for the first time. They will become used to a good many things.

He saw the collapsible boats 'had been placed on the decks inside of the regular boats. I saw no equipment in these boats at all.' He contrasted the stewards on *Olympic*, who were 'not attentive or courteous', with the professionalism of their counterparts on the Atlantic Transport Line's *Minnewaska*.

Another criticism was directed at the leader of the band. On the Thursday night, a committee of passengers had arranged a concert, including music, games and prizes. The collection came to $269, but the leader of the band 'put it all in his pockets' and distributed it to his colleagues. Some of the money had been intended to go to seamen's charities but Captain Haddock advised the passengers that 'he had no authority over the band … they were not employees of the ship'. Three passengers 'demanded' the money and the bandmaster was 'persuaded' to give back $85 for the seamen's charities. Kinney's account indicated it had taken a physical altercation. He was critical of 'the tipping system' and likened some of the ship's staff to 'sharks'.

The passengers included an Italian aristocrat who was always well dressed:

> His costumes are dreams in harmony. When he wears a blue suit his shirt is a lighter shade of blue, his tie a little darker shade than the shirt, and his silk hose and jewelry ... all harmonise in blue with his tie. His eyes are also blue and I think all his shades of blue were carefully selected to harmonise with his eyes.

Then there were the Vanderbilts:

> Mrs. Reginald Vanderbilt and two maids, Miss Kathleen Vanderbilt and governess, are also with us. The V's are very quiet and do not wear any diamonds or pearls, that is, not in sight. Miss Kathleen is about nine years old – has a very good clear complexion, so I'm sure her diet is not all candy, ice cream and cake.
>
> [...]
>
> Friday afternoon we sighted the Scilly Islands, then Land's End. The sea was calm but it was hazy. We had not had a rough day the whole trip. One day it rained just a little.
>
> Following in the wake of the ship were hundreds of sea gulls. They always meet a steamer and escort her into port. Sea gulls are a sure sign of land.
>
> Dinner was served at six o'clock and soon after we slowly steamed into the harbour of Plymouth – after a pilot had scrambled up our rope ladder. All these pilots look more like bookkeepers than seamen. They wear Derby hats and cutaway coats. They should be more picturesque.
>
> After landing quite a lot of passengers we steamed away from Plymouth. To-morrow is Saturday and we will be landed at Cherbourg, France – just six days from New York. It does seem wonderful to leave New York Saturday noon and dine in Paris the following Saturday!

The *Western Daily Mercury* reported that *Olympic* had covered the 3,248 miles in five days, twenty-two hours and fifteen minutes. It quoted her average speed as 22.83 knots.

By the time Kinney disembarked at Cherbourg the weather had deteriorated, 'it drizzled rain'. He had to stand aboard *Nomadic* because there were not enough seats for the large number of passengers. Then there was a delay in the custom house shed and a 'long wait' for the train to Paris.

Olympic docked safely in Southampton, but she was hampered by 'a yacht with two anchors down in the radius of the swinging ground'. The White Star Line raised the issue and the Harbour Master advised that the yacht was, in fact, outside the swinging ground. It illustrated ongoing issues in navigating large ships in the port.

***

Olympic completed a successful westbound crossing to New York, where she arrived on 24 July. Her cargo included crates and cases of furniture and paintings from J.P. Morgan's collection, which was being shipped to the United States.[12] She carried the highest number of first-class passengers westbound since before the *Titanic* disaster, but second- and third-class bookings were relatively poor. Three days later, she left New York with another disappointing passenger list.

One of the passengers was George Leonard Barry, who wrote an account of his trip from a trade unionist perspective.[13] 'Everyone remained on deck until the setting of the sun, then gradually they adjourned to their cabins,' he wrote. The stewards were all ready to help and he 'wondered why they were so accommodating'. A seasoned passenger told him that the victualling staff relied heavily on tips from passengers.

Barry contrasted the 'trying circumstances' of the men in the boiler rooms with the 'comfort and pleasure of the more fortunate', writing that the *Titanic* disaster 'was one of the survival of the fittest'. The first-class passengers 'seemed to look down upon the deck of the second class [...] with a kind of patronising sympathy'. He was told J.P. Morgan travelled free of charge and one company official thought 'his presence attracted the better class of oceanic travellers'.

Barry compared *Olympic* and her doomed sister:

> Both ships were advertised to be non-sinkable and sister ships, with the greatest possible endurance. There was very little to choose between them. The same extravagance existed aboard the *Titanic* as was found aboard the *Olympic*, so far as space for the plutocratic traveller was concerned. Gymnasium, pool rooms, libraries, swimming tanks, saloons and parlours equal in size to any found in a modern hotel and furnished as elaborate as the most exclusive hostelry. The provisions for life-saving devices, for the prevention of accidents were given meager space aboard this floating palace, because there was presumably no income from such a source.

Olympic arrived at Plymouth shortly before noon on Friday, 2 August. Press reports stated cargo worth $220,000 was unloaded as well as hundreds of bags of mail, but only 140 passengers disembarked (the vast majority were destined for Cherbourg and Southampton, where she docked the next day).

***

The horror of the *Titanic* disaster was still fresh. Early in May 1912, J. Bruce Ismay had written 'a kind letter' to Anne Harrison, the widow of his private secretary, William Harrison, who had perished. Harrison's son, also named William, wrote back towards the end of the month and thanked Ismay for supporting the family and 'your great kindness to us'.[14]

Ismay wrote a heartfelt letter to Thomas Andrews' widow, Helen, on 31 May. 'It is impossible for me to express in words all I feel, or make you realise how truly sorry I am for you, or how my heart goes out to you,' he wrote. 'I knew your husband for many years, and had the highest regard for him, and looked upon him as a true friend … Words at such a time are useless.'

Senator Smith had completed his investigation and reported to the United States Senate. 'We shall leave to the honest judgment of England its painstaking chastisement of the British Board of Trade, to whose laxity of regulation and hasty inspection the world is largely indebted for this awful fatality,' he said on 28 May. 'Of contributing causes there were very many. In the face of warning signals, speed was increased, and messages of danger seemed to stimulate her to action rather than to persuade her to fear.' The Senator said that Captain Smith's 'indifference to danger was one of the direct and contributing causes of this unnecessary tragedy'. In the United Kingdom, the investigation was still ongoing.

J. Bruce Ismay testified before the British Wreck Commissioner's court on 4 and 5 June, followed by Harold Sanderson. They were both asked over 1,500 questions in total.

Several days later, Charles H. Fryer wrote a 'defence of Mr. Ismay' in a letter to *The Times*. Fryer thought Ismay was being unfairly criticised. He had recently been a passenger on *Olympic*. The senior steward in the first-class reception room had served Ismay on *Olympic*'s maiden voyage and 'on another crossing'. He told Fryer that the stewards had a common understanding that they should treat Ismay 'exactly as they would an ordinary passenger'. Ismay himself 'religiously observed all the rules that passengers are expected to adhere to'.

Fryer had crossed 'on nearly all the big liners' and felt that 'the attempts to discredit Mr. Ismay and the White Star Line are wholly unjust […] I have been on the Cunard, the German lines, and the American Line, as well as the White Star when ice was reported, and have never seen any signs of slowing down.' He concluded:

> I have no interest in the White Star Line, and am not even acquainted with Mr. Ismay. I am simply a business man who crosses several times a year on business. Notwithstanding the loss of the *Titanic*, I still believe that the White Star and Cunard lines in an equal degree are still the safest ships afloat.

On 30 July, Wreck Commissioner Lord Mersey issued his report, concluding 'that the loss of the [...] ship was due to collision with an iceberg, brought about by the excessive speed at which the ship was being navigated'. He criticised Captain Smith for failing to either divert course 'well to the southward' or 'reduce speed materially as night approached'. The common practice for liners to maintain their course and speed in clear weather 'and to trust a sharp lookout to enable them to avoid the danger', had proved to be a 'bad' practice.

Mersey thought Smith 'was only doing that which other skilled men would have done in the same position', but he had made 'a very grievous mistake'. He attributed the disaster to a 'mistake' rather than 'negligence', warning, 'What was a mistake in the case of the *Titanic* would without doubt be negligence in any similar case in the future.'

Mersey's recommendations included that the newly appointed Bulkhead Committee needed to investigate the best practice of watertight subdivision; lifeboat capacity should be related to 'the number of persons [...] carried in the ship', rather than the ship's size; lifeboat-manning and boat drills should be improved; lookouts should have eyesight tests at 'reasonable intervals'; and an international conference should consider common standards of safety at sea.

Mersey reluctantly addressed criticism of J. Bruce Ismay, 'It is no part of the business of the Court to enquire into such matters, and I should pass them by in silence if I did not fear that my silence might been misunderstood.' Some had suggested Ismay had a 'moral duty' as the chairman of the White Star Line to 'wait onboard until the vessel foundered':

> I do not agree. Mr. Ismay, after rendering assistance to many passengers, found C collapsible, the last boat on the starboard side, actually being lowered. No other people were there at the time. There was room for him and he jumped in. Had he not jumped in he would merely have added one more life, namely, his own, to the number of those lost.

Ismay had been anticipating criticism and there was a suggestion of writing a letter to the press, pointing out that he was not responsible for the ship's speed and citing the previous summer's discussions about *Olympic*. On 1 August, his solicitor, F.M. Radcliffe wrote to him arguing that Mersey's report gave a 'complete exoneration from the charge or suggestion that you were responsible for the speed of the *Titanic* at the time of the disaster'. He argued strongly against writing to the papers. The discussion the previous year about *Olympic*'s westbound arrival times in New York had a 'double edge':

> If you seek to use it as showing that you personally do not press for high speed, you must not be surprised if other people use it as showing that you are prepared to acquiesce in a speed higher than you yourself think necessary [...] and, since the determining voice is yours, as chairman, the responsibility is yours.[15]

There was 'no evidence that [Captain Smith] knew of the difference in the office on this question or of your individual views, but if he did, he knew equally that you waived those views ... If you start a correspondence, you would be giving any ill-disposed person pellets to fire at you.' Perhaps reluctantly, Ismay accepted the advice.

He was still busy with business matters. The idea of cooperation between the major shipping lines to spread the insurance risk of their largest and most valuable ships had not gone away. Ismay met with HAPAG, Norddeutscher Lloyd and Cunard in July 1912. They contacted Mr E.J. Spencer, from the insurance brokers Willis, Faber & Co., for suggestions. Ismay himself wrote about the 'difficulties there are in the way of placing insurance on steamers of very high value, and no doubt this will be increased owing to the unfortunate loss of the *Titanic*'.

Spencer thought it would be difficult to come to an arrangement which would suit all four shipping lines. All manner of details were being discussed into the autumn and winter of 1912. (There was also a shortfall for smaller ships. In 1912, Cunard's *Carpathia* was valued at £165,791 but insured for £140,000.)[16]

'SUGGESTIONS ... ON THE SUBJECT OF LIFESAVING APPLIANCES'

Many people wrote to the British government in the aftermath of the *Titanic* disaster to share their suggestions about making life at sea safer. Some of the suggestions were more practical than others!

- W.F.T. Salt suggested towing 'a small steamer behind, to be used in an emergency'.
- J. Dargue and W.H. Roper suggested ships should travel in pairs.
- 'A traveller' suggested ships should have two captains.
- D. Marschall suggested that the funnels should be used as lifeboats.
- W.J.C. Norris suggested that 'vessels should fire shells ahead'.
- 'Altruist' suggested that 'masters should be cautioned about icebergs'.
- W.J. Dickinson suggested that 'water which enters a leak may be forced out by air pressure'.
- E. Phillips suggested *Titanic*'s name should have been 'Titania'.
- E. Powell suggested that 'vessels on maiden voyages should be escorted' and 'there should be compulsory pilotage in iceberg zone'.
- J. Burn suggested a detachable promenade deck.

One of many ideas put forward following the *Titanic* disaster was for detachable decks that would form pontoons, floating off in the event that the ship sank. *Scientific American*, 1912. (Author's collection)

SAFETY AT SEA

'Subdivision by transverse and longitudinal bulkheads will do more for the safety of passengers and crew than the number of boats that may be carried,' wrote Leonard Peskett in a memo dated 15 May 1912. He thought, 'The placing of additional boats on many of the existing ships is a serious matter, as stability calculations were made on existing rules, and the addition of top weight might in some cases become a greater source of danger than an insufficient number of boats.' Shipping lines were certainly increasing the number of lifeboats, but the safest way to preserve life was to keep a ship afloat, if at all possible.

Three days later, the Admiralty wrote to Cunard saying that 'recent events' made it desirable to closely consider the watertight subdivision of their new *Aquitania*. 'Anything that may appear desirable [to increase] the efficiency of that subdivision' could be done at an early stage in construction. Independently, Cunard's chairman Alfred Booth wrote the same day to Sir Walter Howell as Chief of the Marine Department within the Board of Trade. 'The terrible disaster to the *Titanic* has naturally made us all go right back again to first principles,' Booth wrote. First, 'careful navigation'; second, 'bulkheads and watertight doors'; third, 'organisation and discipline on board'; and fourth, 'life saving appliances'. He had made it a 'practice to impress not only upon the captains but also upon the officers individually, how necessary it is to be constantly alive to the risks of the sea, and never to allow a false feeling of security to grow up'.*

Edward Wilding and Leonard Peskett produced a joint memo for the British Wreck Commissioner's Court to try and answer the hypothetical question about whether *Mauretania* would have remained afloat if she had sustained similar damage to *Titanic*:

> We have considered by approximate methods the flooding of the *Mauretania* in the event of an accident similar to that met with by the *Titanic*. We have assumed the watertight doors and hatches to be closed and similar deductions to those made in the calculations for the *Titanic*. From the calculations made, taking the vessel as damaged from the stem to the afterend of the forward boiler room (corresponding nearly – but not quite – to the length from the stem to the afterend of the No. 5 boiler room in the *Titanic*) the vessel would remain afloat with a considerable list, say 15 deg. to 20 deg., which, no doubt, could be slowly reduced by carefully flooding some after spaces on the opposite side. With the data available we do not think we can satisfactorily discuss flooding corresponding to the damage extending into No. 4 boiler room in the *Titanic*.

Their memo considered a specific scenario and it is not clear that *Mauretania* could have remained afloat. Wilding explained:

> In order that the water should not rise above the top of the bulkhead, *we had to assume the bunkers flooded on the other side* [author's emphasis]. It would be quite a practicable operation by raising the watertight doors, but they would have to be opened so that the water could get through to the port bunkers.

The longitudinal watertight bulkhead in the foremost boiler room would contain flooding, but at the cost of causing a considerable list to the starboard side. Peskett testified that he had no knowledge of any successful counter-flooding being carried out. Wilding thought it 'a dangerous thing to do, to try and remedy such a defect by letting water into a much larger compartment than you already have it in'. He also thought that with such a serious list, 'water might find some other way in', apart from the initial damage (such as through open portholes).

* Booth, Alfred, to Howell, Sir Walter J., 18 May 1912: Cunard Archives, C1/159.

Olympic's Refit

Olympic and *Titanic* were, fundamentally, very safe ships. The extent of the damage inflicted on *Titanic* by the iceberg resulted in the forepeak, three forward holds and Boiler Room 6 (representing 29 per cent of the ship's waterline length) flooding uncontrollably. It was the total loss of Boiler Room 6, representing that critical fifth compartment, which doomed the ship, let alone additional flooding in Boiler Room 5 (which the pumps were initially able to keep under control) and potential damage in Boiler Room 4 (where water was only observed later in the sinking).† All in all, she took on an estimated 13,500 tons of water in the first forty-five minutes after the collision.

Nonetheless the White Star Line's priority was to restore public confidence. 'After the accident to the *Titanic*, certain alterations were decided on, with a view to increasing the margin of safety,' Edward Wilding explained. 'We had arranged prior to that [...] for any two compartments flooded. The owners desired, in view of the character of the accident, to see that she would float with a large number of compartments flooded.'

They formulated a scheme to extend a number of the watertight bulkheads so high that *Olympic* would be divided into several groups of six watertight compartments and would 'float with any one of those groups flooded'. In those circumstances, the goal was for her to remain afloat at least long enough for assistance to reach her, if not permanently. It was a knee-jerk reaction, but it would provide an extraordinary degree of safety. Almost 40 per cent of the ship's length could be flooded and she would remain afloat.

The White Star Line also wanted an inner skin running along the length of the boiler and engine rooms, about 30in from the ship's side. Wilding said that the inner skin was intended partly for publicity purposes and partly to provide protection against the remote possibility of a puncturing blow similar to that received by *Titanic* (it was not intended to safeguard against a direct blow which penetrated deeply into the ship and Wilding did not recommend an inner skin on new ships).

On 9 December 1912, Leonard Peskett wrote a memo to Cunard's general manager. He 'understood that the following work is being carried out on board [*Olympic*], which is now at Belfast', reporting that the watertight bulkheads 'at the forward end of the ship are being carried up from the E deck to the under side of the D deck' and 'the inner shell is being fitted in the way of the boiler rooms by plating along the inside of the web frames. This will give about 2 feet 8 inches between the outer and inner shell.' Peskett did not know that the work was going to be much more extensive.

'Very Large Vessels'

In the aftermath of the *Titanic* disaster, shipbuilders and shipping lines opted to increase watertight subdivision and floatation capacity. Cunard raised several of *Aquitania*'s watertight bulkheads, confirming something that they had been considering before the disaster. (Estimates prepared by 23 September 1912 quoted an additional cost of £6,730 for alterations to watertight bulkheads and decks, as well as an additional £1,125 for 'extra Stone-Lloyd [watertight] doors'.)

Harland & Wolff adopted an approach that any 'very large vessels' they were building (over 20,000 gross tons) would be designed to remain afloat with any three compartments flooded, rather than two. Later, they turned to their existing ships: *Adriatic* was modified by carrying up her watertight bulkheads one deck higher, so that she could meet a three-compartment standard.

† Halpern, Samuel, et al., *Report into the Loss of the SS Titanic: A Centennial Reappraisal*, p. 112.

When *Olympic* returned to Harland & Wolff for her 1912–13 refit, her existing high standard of watertight subdivision was improved far beyond previous practice. Edward Wilding noted that the White Star Line specifically insisted on a new inner skin for publicity purposes. It featured prominently in this 1913 advertisement in *Harper's Magazine*. (Author's collection)

***

Four months after the *Titanic* disaster, *Olympic* seemed to be settling into a regular schedule, but she encountered her own share of bad luck. She left Southampton on 7 August 1912 as intended, but one of the starboard propeller blades came off after she left Queenstown. She arrived late in New York with 995 passengers and had to leave only two days later, on the 17th. This time, she had only 450 eastbound passengers. Bad weather was reported ('strong winds' and 'a good deal of rain').

She reached Plymouth on the afternoon of Saturday, 24 August. All the passengers disembarked and many of the crew were paid off so that she could proceed directly to Belfast for the propeller blade to be replaced. It took almost three hours to get everyone off.

The sight of *Olympic* anchored inside the breakwater attracted hundreds of people. The press reported that special trains were arranged to accommodate them. Passengers for Cherbourg had the option of travelling by the American Line's *New York* or a special cross-Channel steamer.

Olympic's log recorded her leaving Plymouth at 5.21 p.m. and she arrived in Belfast the following afternoon. She had to anchor off Grey Point until high tide on Monday, 26 August, when she could be moved into the dry dock. Captain Haddock and Captain Bartlett, White Star's Marine Superintendent, were on board as she was towed by tugs including *Hercules*, *Herculaneum*, *Alexandra* and *Wallasey*. There was a 'fairly stiff breeze from the north east' that morning, but she was manoeuvred with 'skill and finesse' into the dry dock.

That Monday evening, the *Liverpool Journal of Commerce* reported the White Star Line's announcement that *Olympic* would sail from Southampton a day late, on Thursday, 29 August. To make up time, the call at Queenstown was cancelled. They arranged for the first and second-class passengers booked from Queenstown to board at Southampton, but the third-class passengers would sail on *Celtic* instead on Friday, 30 August.

Olympic left Belfast on 27 August and arrived at Southampton the next day, giving very little time to prepare her for the westbound crossing to New York.

One of the first-class passengers was an American lady in her mid-forties, Bertha M. Colburn. She wrote in her travel diary that she and her travel companions went to the White Star Line's offices in Southampton only to learn that they would leave a day late. It provided them with an opportunity to walk up the High Street and do some shopping. They took a train to the village of Lyndhurst, where they had lunch and did some exploring in the New Forest, 'The purple heather stretched away on both sides and the distant hills were coloured with it.' On Thursday morning, they bought some candy and saw Southampton's Bar Gate, 'We ascended and stood over the gate where we had an excellent view of the harbour.' After calling at the White Star offices, they boarded and settled into 'our big stateroom'. They watched the 'great cranes swing the trunks onboard'.

Olympic left Southampton slightly late and a 'smooth crossing of the channel' allowed her to reach Cherbourg around 7 p.m., 'One tug [tender] on one side unloaded trunks and one on the other [side] passengers and luggage, till our empty decks had more life.' They left at 8.30 p.m. for New York.

It was approaching high season for westbound passenger traffic. There were 1,664 passengers on board, including 713 in first class.

Bertha herself was impressed. A few days later, she wrote:

> This boat is so immense that there is more than room to spare. Deck B is enclosed with glass windows. Deck A [is] open and has space for five or six persons to walk abreast, beyond the rows of deck chairs – which however are not yet even half-filled, perhaps because it is so comfortable withindoors for no one seems to be seasick. The boat moves so steadily that often one would not know we were in motion … Yesterday they said there was a heavy swell, but we never knew it. The reading and writing rooms are big and light

> and airy: the lounge where men may smoke is big and comfortable; just outside the dining room is another big lounge where men may smoke, and take after dinner coffee, and where there is music each evening by a good orchestra; around each stairway is also a large room with easy chairs, the top one being a music room: above is also a gymnasium, and below Turkish Baths, swimming pool, etc., etc. Elevators, three in number, run all the time. The dining room is immense, but divided by alcoves, and the windows are of stained glass. In fact portholes are covered, and everywhere one seems in a big luxurious hotel. The sun shines brightly and I can hardly realise I am returning to a land of sunshine – I have seen so little this summer, and I long for our blue blue sky. I expected to find England rainy, but all agree that this has been exceptionally rainy.

By the afternoon of Wednesday, 4 September, the crossing was almost over. 'We have had pleasant weather ever since we started,' she wrote, 'and fairly smooth seas, although there has been some swell at times.' The sea was:

> ... like glass [...] I have had a fine appetite for the past three days, and have all the time done ample justice to the long menu. The cooking is, however, too English – too unspiced – and not as good as in the *Zeeland*. Tomorrow morning we may dine at seven [o'clock] and I hope for an early landing, and a short customs ordeal.

Olympic reached the Ambrose Channel Lightship at 6.13 a.m. and docked some hours later, but the White Star Line had barely two days to turn her around for the eastbound departure on the 7th.

Once again, she had a disappointing passenger list, but she left New York as planned. She was scheduled to steam slightly slower than usual to economise on coal, because she had 'hundreds of tons less in her bunkers than usual' due to the rapid turnaround at Southampton. Nonetheless, she was expected to reach Plymouth around 9 p.m. on Friday, 13 September.

Many people thought Friday was unlucky if it fell on the thirteenth day of the month. Anyone who was superstitious had their fears confirmed when *Olympic* lost yet another propeller blade (this time from the port propeller) in the early hours of the morning. The *Western Daily Mercury* reported, 'It caused considerable vibration in the after portion of the ship until the revolutions of the engines were adjusted [...] The lost blade weighed about 7½ tons and before the accident the propeller had been making 75 revolutions per minute.' Some passengers were woken up 'and made their way on deck, but they were quickly assured that nothing more serious had happened'. First-class passengers were further away from the stern and 'most of them slept through it'.

Olympic arrived at Plymouth in the early hours of Saturday, 14 September, to drop off passengers and cargo. Some press reports called her the 'unfortunate *Olympic*' and referred to 'another' mishap. It was certainly an extraordinary run of bad luck. The White Star Line had already intended to withdraw *Olympic* from service for a major refit over the winter of 1912–13, but it was decided she should only make one more round trip in the meantime. Her round-trip sailing from Southampton near the end of October was cancelled.

Olympic's next westbound passenger list came to 1,710 passengers, representing her best since before the *Titanic* disaster. However, not everything was positive. The *Southampton Times* reported that a rumour had circulated in Southampton that '*Olympic* had sunk in mid ocean'.

On Tuesday, 24 September, people had contacted the paper to ask 'whether there was any truth in the story'. A number of crewmen's wives 'were greatly disturbed' and stayed waiting at White Star's office in Canute Road 'until a late hour'.

Olympic arrived safely in New York two days later. Joseph Chapman, a crewman who had survived the *Titanic* disaster, wrote a letter from New York to fellow survivor, Kate Buss:

> [*Olympic*] is always having accidents. Every trip something happens. We are now running with

> one broken proppeller [*sic*] and a damaged one. Only yesterday in a storm a large wave broke 3 ports in the saloon just as we were going to lunch and washed the saloon clean out … I am afraid she will have a serious accident some day.[17]

Plenty of people seemed to be on edge.

Another quick turnaround followed and she left New York on 28 September. Although the weather was 'fine' throughout the crossing, she averaged 18.1 knots. It was the natural consequence of running with a propeller blade missing. She reached Plymouth on the evening of Saturday, 5 October, called at Cherbourg and then arrived at Southampton the next day. It marked the end of her seventeenth round trip to New York.

Despite the many misfortunes since she had entered service, *Olympic* had already carried a total of 39,409 passengers. She left Southampton on Wednesday, 9 October, bound for Belfast. Harland & Wolff took her in hand for a major refit, allowing her to return to service in spring 1913.

THE 'OLD RELIABLE'

After her return to service in April 1913, *Olympic* proved popular with passengers. She rescued the crew of the doomed HMS *Audacious* in October 1914. She served as a troopship from September 1915, first in the Mediterranean and then on the Atlantic. Her eventful war service, including sinking *U-103* in May 1918 and surviving torpedo attacks, earned her the nickname 'Old Reliable'.

Fifteen years of successful service on the Atlantic ferry came to an end when, alongside so many other ships, she fell victim to the Great Depression. *Olympic* was mechanically sound and her engines were running very well, but she was no longer generating profits.

Her history illustrates what *Titanic* might have been but, as Philip Franklin said, the ship could be replaced. Human lives could not. It is impossible to comprehend the full horror of the disaster for the people whose potential was cut short or the impact that the survivors suffered. Their lives would never be the same again.

Olympic left her berth at Southampton just after 2 p.m. on 16 August 1933, then dropped her pilots, Mr Rowe and Mr Castell, off at the Nab Light Vessel about two hours later. It was the start of her 233rd round trip to New York. She logged a day's run of 566 miles on the westbound crossing and recorded a speed of 23.43 knots later in 1933. (Mark Chirnside/ Daniel Klistorner collection)

APPENDIX I

OLYMPIC AND *TITANIC*: CHRONOLOGY

	Olympic	*Titanic*
Yard Number	400	401
Order Date	30 April 1907	30 April 1907
'Design "D"' concept approved	31 July 1908	31 July 1908
Shipyard and engine works ordered to proceed, 'except with [propelling] machinery'	17 September 1908	17 September 1908
Shipyard and engine works ordered to proceed 'with boilers'	26 February 1909	26 February 1909
Shipyard and engine works ordered to proceed 'with remainder of machinery'	20 April 1909	20 April 1909
Keel laid	16 December 1908	22 or 31 March 1909
Framed to height of the double bottom	10 March 1909	15 May 1909
Fully framed	20 November 1909	6 April 1910
Fully plated	15 April 1910	19 October 1910
Launched	20 October 1910	31 May 1911
Delivered	31 May 1911	2 April 1912

APPENDIX II

OLYMPIC AND *TITANIC*: KEY CAPACITIES AND TECHNICAL SPECIFICATIONS

Harland & Wolff kept records containing *Olympic* and *Titanic*'s key dimensions, tonnage and loading conditions; passenger, crew and coal capacities; and propelling machinery data. This information is consolidated here.

DIMENSIONS, TONNAGE AND LOADING CONDITIONS

	Olympic	*Titanic*
Dimensions		
Length overall	882ft 8in	882ft 8in
Length between perpendiculars	850ft	850ft
Breadth moulded	92ft	92ft
Breadth extreme	92ft 6in	92ft 6in
Depth moulded	64ft 3in	64ft 3in
Tonnage		
Under-deck tonnage	17,871 tons	17,871 tons
Gross tonnage	45,324 tons	46,329 tons
Net tonnage	20,894 tons	21,831 tons
Load Conditions		
Draught	34ft 7in	34ft 7in
Displacement	52,310 tons	52,310 tons
Tons per inch	143.8	143.8
Deadweight	14,030 tons	13,550 tons
Block coefficient	0.684	0.684
Prism coefficient	0.705	0.705
Light Conditions		
Draught	26ft 4in	26ft 7½in
Tons per inch	138.8	138.8

The length of ships such as *Olympic* and *Titanic* can be given as 'between perpendiculars' or 'overall'. On official documentation, such as the ship's registration certificate, the ship's length was given as 'between perpendiculars'; however, that was measured on that document in two different ways:

1. 'Length from fore part of stem, under the bowsprit, to the aft side of the head of the stern post': 852ft 5 tenths [6in].
2. 'Length at quarter of depth from top of weather deck at side amidships to bottom of keel': 849ft 2 tenths [2.4in].

To complicate matters, shipbuilders such as Harland & Wolff tended to use the length 'between perpendiculars', but on their preferred method of measuring this then the ships were 850ft long. The result of all of this is that someone might see multiple figures quoted for the ship's length 'between perpendiculars'. It does not follow that any of these figures are wrong, but it is a question of being sure which measurement is being used. The length overall can also differ to one degree or another, being given as 882ft 8in or 882ft 9in.

PASSENGER CAPACITIES, CREW AND COAL

	Olympic	*Titanic*
First Class		
Rooms	381	416
Berths	844	928
Sofas	109	106
Second Class		
Rooms	162	162
Berths and Sofas	510	510
Third Class		
Rooms	786	982
Open Berths	164	40
Crew		
Deck (Sailing Department)	75	74
Engine Room (Engine Department)	325	326
Stewards (Victualling Department)	524	544
Coal		
Permanent bunkers	6,611 tons	6,611 tons
Reserve bunkers	1,201 tons	1,201 tons

(Sources: 'Particulars of Ships Built by Harland & Wolff', entries for *Olympic* and *Titanic*, and D2805/SHIP/4A, Public Record Office of Northern Ireland.)

Passenger capacity figures invariably differ according to source. This does not mean one particular number is wrong, but it is necessary to understand how it was calculated. It is important to recognise that *Olympic* and *Titanic*'s passenger capacities were variable, both by season (when some single-berth first-class staterooms became twin rooms) or whether interchangeable rooms were used as first, second or third class.

Olympic's first-class capacity varied in her early months of service (for example, four new first-class staterooms were added on the port side of the grand staircase on C-Deck by November 1911). *Titanic*'s first-class passenger accommodation was also expanded compared to *Olympic*.

Some sources appear to add up the maximum number of first-, second- and third-class passengers carried to obtain the maximum total number of passengers; however, it was not possible to maximise each simultaneously (if interchangeable stateroom accommodation was being used for first class then it could not be simultaneously used as second class).

Third-class accommodation plans for *Olympic* dated April 1911 show a total of 1,138 berths (including 112 which were interchangeable second or third class). Those plans included the portable Section 'S', but perhaps Harland & Wolff's count did not.

PROPELLING MACHINERY

	Olympic	*Titanic*
Boilers		
Double-ended	24	24
Single-ended	5	5
No. furnaces	159	159
Heating surface	144,142sq ft	144,142sq ft
Grate surface	3,430sq ft	3,430sq ft
Steam pressure	215lb/sq in	215lb/sq in
Reciprocating Engines		
Cylinders (diameter)	54in, 84in, 97in, 97in	54in, 84in, 97in, 97in
Stroke (length)	75in	75in
Propellers, Reciprocating Engines		
Diameter	23ft 6in	23ft 6in
Pitch of blades*	(33ft) 34ft 6in	(34ft 6in) 35ft
Area of blades	160sq ft	160sq ft
Number of blades	3	3
Propeller, Turbine		
Diameter	16ft 6in	17ft
Pitch of blades	14ft 6in	14ft 6in
Area of blades	120sq ft	120sq ft
Number of blades	4	3
Power		
Nominal Horsepower	6,906	6,906
Indicated Horsepower	50,000	50,000
Speed	21 knots	21 knots

*Note: It was Harland & Wolff's general practice to record the details of each ship as completed. *Olympic*'s propeller specifications do, however, reflect the increased pitch from 33ft to 34 ft 6in on the port and starboard propellers (which was a change made during the repairs after the *Hawke* collision). In *Titanic*'s case, the pitch of her port and starboard propellers was increased further to 35ft.

(Sources: 'Particulars of Ships Built by Harland & Wolff' [boilers] and D2805/SHIP/8A [reciprocating engines and propellers], Public Record Office of Northern Ireland.)

APPENDIX III

WHITE STAR LINE PROFITS 1898–1911

In 1908, the White Star Line had their profits certified independently by Messrs Price, Waterhouse & Co. Their figures for 1898–1907 were then made available to the investors who were lending money to the company to help finance its shipbuilding project. The data from 1908 to 1911 inclusive was later prepared on the same basis and is provided for added context. Shipping companies suffered a poor year in 1908, but White Star recovered and delivered record profits in 1910 and 1911.

30 September 1908

To the Directors of the Oceanic Steam Navigation Company, Limited,

We beg to report that we have examined the books of The Oceanic Steam Navigation Company, Limited, for the ten years ending December 31st 1907. We have regularly audited the books for the last five of these ten years and we have made certain adjustments as required in the profits of the earlier years, in order that the accounts of the whole ten years may be prepared upon the present basis.

We certify that the profits, after charging all outgoings, including repairs and renewals and all expenses of management, but before charging interest on borrowed money and income tax, and before providing for depreciation on steamships and other property, have been as under:–

1898	£523,838
1899	£444,590
1900	£780,900
1901	£454,024
1902 (11 months)	£451,670
1903 (13 months)	£651,622
1904	£502,730
1905	£712,679
1906	£805,309
1907	£848,486
Total of Ten Years	**£6,175,852**
Average of Ten Years	**£617,585**

1908	£298,941
1909	£613,054
1910	£1,057,519
1911	£1,073,752
Average of Four Years	**£760,817**

APPENDIX IV

IMM PROFITS 1903–13

IMM	1903	1904	1905	1906	1907	1908	1909	1910	1911	1912	1913
Gross Voyage Earnings	$29,677,756	$27,926,908	$32,435,512	$35,931,412	$38,253,588	$29,981,044	$33,461,485	$36,848,284	$38,193,546	$42,560,693	$47,519,404
Miscellaneous Earnings	$1,359,664	$920,085	$927,407	$1,227,800	$1,013,000	$576,023	$491,723	$1,225,312	$959,797	$1,164,816	$1,521,934
Total	$31,037,420	$28,846,993	$33,362,919	$37,159,213	$39,266,589	$30,557,068	$33,953,208	$38,073,596	$39,153,343	$43,725,508	$49,041,338
Gross Expenses	$27,036,898	$27,040,586	$27,456,174	$29,155,178	$32,242,543	$29,653,891	$29,257,755	$29,775,390	$31,070,784	$36,128,184	$39,474,291
Net Earnings	**$4,000,522**	**$1,806,407**	**$5,906,744**	**$8,004,035**	**$7,024,046**	**$903,176**	**$4,695,454**	**$8,298,206**	**$8,082,560**	**$7,597,325**	**$9,567,048**
Excess Premiums & Interest Credited to Insurance Fund Over Losses & Expenses			$864,159	$791,959	$441,661						
Profit on Purchase of Bonds for Sinking Fund				$28,597	$57,021						
Net Profit from Operation			**$6,770,904**	**$8,824,590**	**$7,522,728**	**$903,176**	**$4,695,454**	**$8,298,206**	**$8,082,560**	**$7,597,325**	**$9,567,048**
Interest on Bonds	$3,083,346	$3,312,151	$3,349,495	$3,420,099	$3,398,713	$3,452,484	$3,692,121	$3,895,943	$3,619,881	$3,599,883	$3,578,578
Interest on Loans	$441,217	$469,283	$482,167	$315,596	$36,339	$109,015	$91,884	$74,120	$180,602	$97,274	$115,165
Income Tax	$120,663	$64,123	$48,393	$60,141	$53,945	$89,631	$86,575	$51,570	$63,807	$112,257	$156,345
'Fixed Charges'											
Earnings	$355,295	($2,039,150)	$2,890,848	$5,028,754	$4,033,731	($2,747,954)	$824,874	$4,276,573	$4,218,270	$3,787,911	$5,716,959
Surplus Insurance Account	$1,442,502	$897,052				$1,062,344					
Payment on Account of Trade Agreement for 1904			$236,643								
Interest on Advances for Construction of Steamships						$44,373					
Transfer from Insurance Account							$357,461	$573,007	$291,000		
Total for Year (Deficit)	**$1,797,797**	**($1,142,098)**	**$2,654,205**	**$5,028,754**	**$4,033,731**	**($1,729,983)**	**$1,182,335**	**$4,849,581**	**$4,509,270**	**$3,787,911**	**$5,716,959**

APPENDIX V

LUSITANIA AND *MAURETANIA* PROFITS 1907–14

The operating profits generated by *Lusitania* and *Mauretania* provide some insight into the economics of the large North Atlantic passenger liners. These figures are presented as the average revenue (receipts) and average expenses (disbursements) for each round trip to New York. They do not include the subsidy received from the British government or the net profit, which would be significantly lower once depreciation charges and repairs were taken into account.

	Lusitania			***Mauretania***		
	Receipts	**Disbursements**	**Profit**	**Receipts**	**Disbursements**	**Profit**
1907	£40,524	£22,906	£17,618	£27,419	£23,399	£4,020
1908	£28,133	£21,432	£6,701	£27,036	£24,179	£2,857
1909	£31,382	£21,137	£10,245	£33,127	£20,559	£12,568
1910	£39,124	£21,637	£17,487	£41,515	£21,686	£19,829
1911	£38,863	£21,282	£17,581	£41,288	£21,853	£19,435
1912	£41,233	£24,179	£17,054	£42,865	£22,754	£20,111
1913	£39,817	£24,965	£14,852	£43,994	£22,955	£21,039
1914	£34,818	£22,998	£11,820	£35,744	£23,504	£12,240

(Source: 'The Cunard Steamship Company Limited: Abstract of Voyages No. 4': AC 14/1, Cunard Archives.)

Over the course of her entire career from 1907 to 1934, which included transatlantic crossings, war service (partially paid for by the government) and cruises, *Mauretania* earned a net profit after depreciation of £1,151,378. This helps to illustrate that exceptionally fast ships such as *Mauretania* were only made viable by state support, both in financing construction and with an ongoing additional subsidy.

APPENDIX VI

WHITE STAR LINE SOUTHAMPTON–NEW YORK SERVICE PASSENGERS 1907–14

The number of passengers carried by the White Star Line's Southampton to New York service helps set the context *Olympic* was operating in. They comprised 25 per cent first class, 22 per cent second class and 53 per cent third class. The breakdown between westbound and eastbound traffic was 51 to 49 per cent in both first and third class, whereas second class dominated westbound by 64 to 36 per cent.

The ships include *Adriatic* (1907–11), *Majestic* (1907–14), *Oceanic* (1907–14), *Olympic* (1911–14) and *Teutonic* (1907–11). The figures for 1907 are for part of the year, reflecting when the Southampton to New York service was established; 1908 to 1913 inclusive are full years; the figure for 1914 excludes crossings made after the outbreak of war in August 1914.

TOTAL PASSENGERS CARRIED 1907–14

Total Passengers	First Class	Second Class	Third Class
Westbound	53,810	59,919	116,491
Eastbound	50,757	34,014	110,211
Total	104,567	93,933	226,702

TOTAL PASSENGERS CARRIED ANNUALLY 1907–14

1907	First Class	Second Class	Third Class	Total	No. Crossings	Average
Westbound	6,378	6,604	11,715	24,697	29	852
Eastbound	5,161	3,524	10,325	19,010	32	594
				43,707	**61**	**717**
1908	**First Class**	**Second Class**	**Third Class**	**Total**	**No. Crossings**	**Average**
Westbound	7,448	6,820	10,121	24,389	49	498
Eastbound	6,644	4,403	24,282	35,329	50	707
				59,718	**99**	**603**
1909	**First Class**	**Second Class**	**Third Class**	**Total**	**No. Crossings**	**Average**
Westbound	6,735	8,246	20,115	35,096	50	702
Eastbound	5,781	4,052	14,514	24,347	50	487
				59,443	**100**	**594**

1910	First Class	Second Class	Third Class	Total	No. Crossings	Average
Westbound	7,580	9,188	17,364	34,132	44	776
Eastbound	6,345	4,334	11,335	22,014	43	512
				56,146	**87**	**645**
1911	**First Class**	**Second Class**	**Third Class**	**Total**	**No. Crossings**	**Average**
Westbound	7,891	8,132	12,335	28,358	36	788
Eastbound	7,780	5,091	15,106	27,977	36	777
				56,335	**72**	**782**
1912	**First Class**	**Second Class**	**Third Class**	**Total**	**No. Crossings**	**Average**
Westbound	6,031	7,233	12,465	25,729	34	757
Eastbound	6,142	3,921	9,902	19,965	34	587
				45,694	**68**	**672**
1913	**First Class**	**Second Class**	**Third Class**	**Total**	**No. Crossings**	**Average**
Westbound	7,770	10,257	23,035	41,062	42	978
Eastbound	7,615	5,558	13,902	27,075	42	645
				68,137	**84**	**811**
1914	**First Class**	**Second Class**	**Third Class**	**Total**	**No. Crossings**	**Average**
Westbound	3,977	3,439	9,341	16,757	19	882
Eastbound	5,289	3,131	10,845	19,265	18	1,070
				36,022	**37**	**974**

(Source: 'Report of the Trans-Atlantic Passenger Movement, Calendar Year' 1907–14 inclusive.)

APPENDIX VII

WHITE STAR LINE, CUNARD, HAPAG AND NORDDEUTSCHER LLOYD NEW YORK SERVICE PASSENGERS 1907–13

The White Star Line figures comprise their passenger carryings on both the Liverpool and Southampton to New York services; the Cunard figures comprise their passenger carryings on their Liverpool to New York service; the Norddeutscher Lloyd (NDL) and HAPAG figures comprise their passenger carryings on their 'express' and 'regular' services to New York. The average figures in each subcategory will not always be in alignment with the totals: for example, the White Star Line's total passengers carried were higher than Cunard's but they also had a larger number of sailings which reduced the average per crossing.

	Westbound			Eastbound			Westbound	Eastbound	
	First Class Total	Second Class Total	Third Class Total	First Class Total	Second Class Total	Third Class Total	Total Westbound	Total Eastbound	Total
White Star	100,852	145,386	277,347	90,358	83,298	209,897	523,585	383,553	907,138
Cunard	95,504	121,351	314,427	84,290	82,363	186,834	531,282	353,487	884,769
NDL	107,363	169,889	608,122	96,807	80,982	294,234	885,374	472,023	1,357,397
HAPAG	86,044	148,826	630,338	78,872	78,101	220,174	865,208	377,147	1,242,355

	Westbound			Eastbound			Westbound	Eastbound	
	First Class Average	Second Class Average	Third Class Average	First Class Average	Second Class Average	Third Class Average	Average	Average	Average
White Star	165	237	452	149	137	345	854	631	743
Cunard	209	265	687	188	184	417	1,160	789	977
NDL	144	229	818	154	129	469	1,192	753	991
HAPAG	145	251	1,061	151	149	420	1,457	720	1,111

(Source: 'Report of the Trans-Atlantic Passenger Movement, Calendar Year' 1907–13 inclusive.)

ACKNOWLEDGEMENTS

Although my name is on the cover, this book is the result of a collaborative effort between many people, public and private organisations. My family and friends are a source of appreciated, enduring support and encouragement. In alphabetical order, I would like to thank the following individuals:

Roy Anderson; Steve Anderson; Scott Andrews; Philip Armstrong; Mark Baber; Günter Bäbler; Raymond Bealing; Mike Beatty; George Behe; Bruce Beveridge; Rene Bergeron; Robin Beuting; Randy Bigham; Johannes Bo Larsen; Stephen Booth; Art Braunschweiger; Frank Braynard; Dawn Brooker; Christine M. Cacace; Joseph Carvalho; Giovanni Castro; Myles Chantler; Malcolm Cheape; Dan Cherry; Pavel Chlupac; Christian Cody; Cyril Codus; Lionel Codus; Ed Coghlan; Gary Cooper; John Creamer; Ralph Currell; Matt DeWinkeleer; Richard de Kerbrech; Philippe Delaunoy; Roy Derouin; Mark Doherty; Charles Dragonette; Shelley Dziedzic; John P. Eaton; Richard Edwards; Neil Egginton; David Hume Elkington; Mark Evans; Tad Fitch; Franck Gavard-Parret; Alan Geddes; Ioannis Georgiou; Jack Gibson; Sandra Gilpin; Dave Gittins; Joao Goncalves; Charles Haas; Cal Haines; Steve Hall; Sam Halpern; Nigel Hampson; Sean Hankins; Brian Hawley; Remco Hillen; Ken Hillier; Philip Hind; Pine Hodges; Brent Holt; David Hume Elkington; David Hutchings; Jennifer Irwin; Sheila Jemima; Ed and Karen Kamuda; Stuart Kelly; Arnold Kludas; Daniel Klistorner; Peter Kohler; J. Kent Layton; Mandy le Boutillier; Dr Paul Lee; Ray Lepien; Oliver Loerscher; Eric K. Longo; Michael Lowrey; Richard Lundgren; Thomas Lynskey; Stuart Lythgoe; Campbell McCutcheon; Janette McCutcheon; Ken Marschall; John Maxtone-Graham; Jenny McBriar; Ken McLeod; Aidan McMichael; Ryan McMillen; Jean-Philippe Marre; Olivier Mendez; Craig Mestach; Roy Mengot; Michail Michailakis; Susie Millar; Peter Mitchell; Norman Morse; Jill Neale; David Olivera; Jens Ostrowski; Dan Parkes; Brad Payne; James Penca; Henning Pfeifer; Mike Poirier; Trevor Powell; Bob Read; Una Reilly; Antoine Resche; Levi Rourke; Tom Rudderham; Brigitte Saar; Bill Sauder; Eric Sauder; Jean Scuissiato; Severin Schwendener; Inger Sheil; Andy Skinner; Joanne Smith; Jonathan Smith; Michael H. Standart; Tarn Stephanos; Parks Stephenson; Suzanne Stormer; Clive Sweetingham; Diana Teggart; Nick Thearle; Hilary Thomas; Brian Ticehurst; Thorsten Totzke; Timothy Trower; Steven Tuckwood; Marco Tulner; Rich Turnwald; Jerry N.J. Vondeling; Mark Warren; John D. ('Jack') Wetton; John White; Brandon Whited; Geoff Whitfield; Andrew Williams; Stuart Williamson; Russ Willoughby; Pat Winsip; Bill Wormstedt; Rich Woytowich. In the event I have missed anyone, please accept my apologies and let me know so that the omission can be corrected in any future editions.

Societies including the Belfast *Titanic* Society, British *Titanic* Society, French *Titanic* Society, German *Titanic* Society, *Titanic* Historical Society, *Titanic* International Society and Swiss *Titanic* Society have all been very helpful and their assistance is appreciated. The efforts of these societies and others do so much to preserve history.

The helpful and hardworking staff of the British Library; British Postal Museum & Archive; Cunard Archive, Sydney Jones Library, University of Liverpool; Glasgow City Archives; Glasgow University Archives Services; Library of Congress; Merseyside Maritime Museum; the Morgan Library and Museum; United Kingdom National Archives; Ulster Folk & Transport Museum, National Museums Northern Ireland; Public Records Office of Northern Ireland (PRONI), with thanks given to the Deputy Keeper of the Records at the Public Records Office of Northern Ireland and Harland & Wolff; Stadsarchief Den Haag; United Kingdom Parliamentary Archives; the Postal Museum, Royal Mail Group; United States Citizenship & Immigration Services, History Office & Library; United States National Archives and Records Administration (NARA); Oxford University Library Services. Organisations supplying images are also acknowledged in each caption.

My editor Amy Rigg and everyone at The History Press – project editor Jezz Palmer, designer Jemma Cox, cover designer Martin Latham, campaigns executive Graham Robson, and marketer Mollie Nattrass, as well as freelance copyeditor Sarah Wright and proofreader Paul Middleton – have worked tirelessly to make this book a reality.

NOTES

1: The North Atlantic Run

1 'Giant *Oceanic* Comes In', *New York Times*, 4 September 1899, p. 1.
2 Beaumont, Dr J.C.H., *Ships – and People* (Geoffrey Bles, 1926) pp. 93–94.
3 Anderson, Roy, *White Star* (T. Stephenson & Sons Ltd, 1964) p. 42.
4 'A Brief History of the White Star Line 1869–1934' (unpublished manuscript, 1947: Cunard Archives, PR3/23/6).
5 'Monarch of the Sea: The New Steamship *Oceanic*', *New York Herald*, 30 March 1871, p. 4.
6 'The Steamship *Oceanic*', *New York Times*, 30 March 1871, p. 8.
7 'Splendid Voyage', *Fort Wayne Daily Gazette*, 6 July 1871.
8 'Papers Relating to the Loss of the Steamship *Atlantic*', London: House of Commons, 31 July 1873, p. 164: The National Archives (UK), MT 9/123. The Cunard, Inman, National, Guion and White Star Line voyage durations were all calculated between Roche's Point (on departure from or arrival in Ireland) and Sandy Hook (on arrival in or departure from New York). The performance data presented in the table appears to have been transposed, with the 'inwards' (eastbound) passage quoted in all cases as being longer than the 'outwards' (westbound) passage. Other data (see p. 163) shows the westbound passages taking longer than the eastbound, which is logical when the currents are taken into account. Therefore, the average length of voyage quoted in the text has been amended accordingly.
9 Anderson, Roy, *White Star*, p. 58.
10 'Papers Relating to the Loss of the Steamship *Atlantic*', p. 82: The National Archives (UK), MT 9/123.
11 Ibid., pp. 65–66.
12 Ibid., pp. 42–43.
13 Ibid., p. 61.
14 Ibid., pp. 76–77.
15 Ibid., p. 86.
16 Hicks, C. Stansfeld, 'An Atlantic Chapter', White Star Magazine, Vol. 1 (1924) p. 403.
17 Anderson, Roy, *White Star*, pp. 50–51, 66–67, 70–72.
18 Oldham, Wilton J., *The Ismay Line* (The Journal of Commerce, 1961) pp. 90–91.
19 'An Atlantic Liner. A Chat with T.H. Ismay of the White Star Line', *The Pall Mall Gazette*, 26 September 1885. Reprinted verbatim in *Atlantic Daily Bulletin*, December 2014, pp. 23–24.
20 White Star Line, 'Presentation to the Managers', 1885.
21 'The *Teutonic* Launched', *New York Times*, 20 January 1889.
22 Warren, Mark D. (ed.), *The Cunard Royal Mail Twin-Screw Steamers Campania and Lucania* (Patrick Stephens Ltd, 1993) p. 4. According to a reference in *Engineering*: 'The *Teutonic*, for instance, in seven voyages in 1890 had an average of 1,400 passengers per trip, [...] while the *Majestic*, in nine trips, had an average of 1,357'.
23 'Beat the Maiden Record: First Trip of the New Steamship *Majestic*', *New York Times*, 11 April 1890, p. 1.
24 Kludas, Arnold, *Record Breakers of the North Atlantic: Blue Riband Liners 1838–1952* (Chatham Publishing, 2000) p. 147. Kludas explains in detail why so many listings of average speeds and Blue Riband records are inaccurate, as well as outlining a number of important points regarding the contest: it was never an official title, 'never endowed by any institution or authority'; while the rules were unwritten – it was understood that a liner 'had to cross the Atlantic at a higher average speed than the current holder of the Blue Riband. Second, the record crossing had to take place in the westbound direction, i.e. from Europe to America.' Speed was measured in terms of the average speed and not the time taken to cross the ocean, since it would be possible to cross the ocean in a shorter time over a smaller number of miles, yet with an average speed lower than that of a competitor. (In particular, see pp. 9–18 for a very worthwhile and comprehensive discussion of the development of the Blue Riband.) Kludas writes: 'Because the contest for the Blue Riband was never an official event, there is an almost complete lack of official archive material [...] As a result, virtually the only sources of information are shipping company documents and press reports. The inevitable errors and contradictions found in the sources has been described by Noel R.P. Bonsor in his standard work in which he provided plenty of examples.' As an example of one simple inaccuracy that has caused confusion, one archival document gave *Teutonic*'s average speed as 20.43 knots, even though the figures for mileage and time agreed with Kludas. The confusion is explained by a rounding error compounded by a typographical mistake: *Teutonic*'s average speed of exactly 20.349 knots was rounded to 20.34 knots, rather than 20.35 knots, and was then typed as 20.43 knots. See 'The Blue Riband of the Atlantic: List of Fastest Passages of Various Atlantic Liners, whether Westbound or Eastbound ...': Cunard Archives, PR3/11/1.
25 Oldham, Wilton J., *The Ismay Line*, p. 117.
26 'Mr. Bruce Ismay', *The Times*, 18 October 1937. As well as Ismay's obituaries, see also: 'Mr. J. Bruce Ismay Resigns L.M.S. Railway Directorship', *Liverpool Daily Post*, 10 July 1937.

27 Oldham, Wilton J., *The Ismay Line*, p. 64.
28 Ibid.
29 Ibid.
30 '*Titanic*: Extract from the Unpublished Autobiography of Lt. Colonel Frank Bustard, OBE'. Reproduced in *Titanic Commutator*, Vol. 39, No. 206 (2014) p. 86.
31 'Harold Sanderson, Shipping Man, Dies', *New York Times*, 27 February 1932.
32 'The Cunard Steamship Company Limited: Annual Reports' for 1892–98: Cunard Archives, D42/A4/1.
33 Ismay, J. Bruce, to J. Wierdsma, 6 February 1895. See: 'Koninklijke Nederlandsche Stoomboot Maatschappij en Nederlandsch- Amerikaansche Stoomvaart-Maatschappij, correspondentie betreffende mogelijke concurrentie in Amsterdam betreffende de handel op Noord-Amerika met exploitatieoverzichten en bijlagen; Overeenkomst Shaw, Savill & Albion Co. Ltd met Ismay, Imrie & Co. (Oceaan Steam Navigation Co. Ltd), 26 Juli 1887, 1887–1902': Stadsarchief Rotterdam, DB 726.
34 The profit figures for White Star are based on the 'net profit on working account after payment of insurance and all other outgoings' as detailed in a *Practical Engineer*, Vol. 30 (1904). The article also included information on dividends paid. I am grateful to Brad Payne for sharing this with me. The dividend information for Cunard was obtained from the company's annual reports. (At the time, it was common for dividend payments to be expressed as a percentage of the paid-up share capital that the company had raised. For example, White Star's shareholders had paid in full £1,000 for each of the 750 shares in the company by 1880, for a share capital of £750,000. A dividend of 10 per cent that year would have come to £75,000 or £100 per share. Another way to express dividend payments is to look at the amount paid out in dividends as a proportion of the year's net profits From 1890 to 1899, this averaged 23 per cent, and from 1896, was never more than 19 per cent, leaving plenty of money to be reinvested in the fleet without having to ask shareholders to contribute more capital.)
35 'In the Matter of an Investigation before Herbert Riversdale Mansel Jones, Esquire, Barrister-at-Law, J.H. Hallett, Esquire, and Captain E.M. Hughes, Inspectors for the Board of Trade, into the Circumstances Attending the Supposed Loss of the British Steamship *Naronic* of Liverpool, which has not been heard of since leaving Liverpool for New York on or about the 11th day of February 1893', 25 July 1893: The National Archives (UK), MT 9/507.
36 Ibid.
37 Ibid.
38 'The Cunard Steamship Company Limited: Annual Reports' for 1886–90: Cunard Archives, D42/A4/1.
39 McAuley, Rob, *The Liners: A Voyage of Discovery* (Boxtree, 1997) p. 41.
40 'To the Levant and the Land of the Pharoahs By the Steamers of the Norddeutscher Lloyd Steamship Co', pp. 11–13.
41 Brinnin, John Malcolm, *The Sway of the Grand Saloon: A Social History of the North Atlantic* (Delacorte Press, 1971) p. 316.
42 McAuley, Rob, *The Liners*, p. 42.
43 Griffiths, Denis, *Power of the Great Liners* (Patrick Stephens Ltd, 1990) p. 85.
44 'New Leviathan of the Sea', *New York Times*, 10 September 1899, p. 10.
45 'Thomas Henry Ismay Dead', *New York Times*, 24 November 1899, p. 7. An extensive obituary was published in the newspaper:

'Liverpool, Nov. 23 – Thomas Henry Ismay, the founder and Chairman of the Board of Directors of the White Star Line Steamship Company, died suddenly this evening.

'Thomas Henry Ismay was born in Maryport, Cumberland, England, Jan 7 1837, and was educated in the Croft House School, Carlisle. In 1832, he went to Liverpool, where he became an apprentice in the shipping firm of Imrie, Tomlinson & Co. After serving his term there he went in a sailing vessel to the west coast of South Africa, and on his return he directed his energies to developing the Australian wool trade. He became associated with the firm of Philip Nelson & Co., as junior partner, and in 1804 he became one of the directors of the National line.

'Two years later Mr. Ismay acquired the business of the old White Star Line of clipper ships, sailing to Australia, which dated from the old days of the gold fever in that continent. He substituted iron for wooden vessels, and in 1869 he formed the White Star Line of steamers. The following year he was joined by Mr. Imrie, the son of the senior partner of the firm with which he had served his articles, and who had also been his fellow apprentice. In 1881 W.S. Graves became a partner in the firm, and later Mr. Ismay's two sons, J. Bruce Ismay, who was the New York agent of the firm, and James H. Ismay, were also admitted to partnership. T.H. Ismay retired from active membership, but still retained a lively interest in the affairs of the concern.

'Mr. Ismay was Chairman of the Liverpool and London Steamship Protective Association, a Director of the London and Northwestern Railway Company, Deputy Chairman of the Royal Insurance Company, and served in several royal and departmental commissions. He was High Sheriff of Cheshire in 1892.

'After the Russian war scare in 1885 Mr. Ismay built two vessels of the *Teutonic* type, to be used by the government as fast auxiliary cruisers in case of war. He gave £20,000 as the nucleus of a fund for aged and indigent Liverpool sailors.

'John Lee, the agent of the White Star Line in this city, said last evening that Mr. Ismay died at 4 o'clock in the afternoon, and that word was received here two hours later. Mr. Ismay's health had been ailing since September.'
46 'The Liner of the Future', *Munsey's Magazine*, *c.* 1903, p. 204. Although the article's anonymous author was talking about *Cedric*, their comments are also a very apt description of *Celtic*.
47 'Says Ocean Racers Are Played Out', *Waterloo Daily Reporter*, 6 February 1901, p. 7.
48 'Largest Ship Afloat Arrives In Port', *New York Times*, 5 August 1901, p. 1.
49 'Liner *Cedric* In Port', *New York Times*, 21 February 1903, p. 1.
50 'No Fear for the *Cedric*', *New York* Times, 26 November 1903. The newspaper's editors apologised when it became clear that *Cedric* was safe: 'The report of the sinking of the White Star steamer *Cedric* in mid ocean caused great anxiety and distress of mind to many people, and that it should have found its way into print, even though accompanied with cautionary disclaimers, is to be regretted.' They described the original source as 'irresponsible', stating that 'the "faking" of news which derives its momentary value from its ability to strike consternation to the hearts of thousands of people on both sides of the ocean is beyond pardon'.
51 Beaumont, Dr J.C.H., *Ships – and People*, p. 92.

52 'Suggestions which I consider to be advantageous for the Company, taken from Personal Observations and Remarks of Passengers', p. 1. Cunard Archives, S7/1/23.
53 Ibid.
54 Ibid.
55 Strouse, Jean, *Morgan: American Financier* (Perennial, 2000) pp. vii–x.
56 Ibid., pp. 457–59.
57 'Preliminary Negotiations', p. 7. Cunard Archives, S10/2.
58 'Mr. Morgan's Shipping Deal', *Liverpool Daily Post*, 30 April 1901. Cunard Archives, Ca/9.
59 Ibid.
60 Navin, Thomas R., & Marian V. Sears, 'A Study in Merger: Formation of the International Mercantile Marine Company', *Business History Review*, Vol. 28, Issue 4 (December 1954) pp. 302–03.
61 'J. Bruce Ismay Returns to England', *New York Times*, 19 September 1901.
62 Strouse, Jean, *Morgan*, p. 464.
63 Ibid., pp. 459–60.
64 'Shareholders Agree to the Ship Combine. Consent of All but the White Star Stockholders Gained', *New York Times*, 1 May 1902.
65 'Annual Report of the Commissioner of Navigation for the Fiscal Year ended … June 30, 1902' (Washington: Government Printing Office 1902) p. 402.
66 'Preliminary Negotiations', p. 18. Cunard Archives, S10/2.
67 'Correspondence as to Sale of Shares in Oceanic Steam Navigation Company Ltd, Vol. 2', letter from J. Bruce Ismay dated 24 September 1902, p. 384. Cunard Archives, D42/A9/10.
68 The preliminary agreement of 4 February 1902 included the sale of 'the business, goodwill, assets and property of Ismay, Imrie & Co., including … the position of manager of the Oceanic Steam Navigation Co. Ltd., but excluding the properties and rights mentioned in the special contracts between the White Star vendors and the bankers'.
69 Ismay, J. Bruce, Limitation of Liability. Op. Cit., p. 36.
70 Oldham, Wilton J., *The Ismay Line*, p. 146.
71 Ashbrook, Claud (also named as Ashburn in the same report), 'International Mercantile Marine', *Washington Post*, 7 December 1904, p. 11.
72 'White Star Line. Royal and United States Mail Steamers: Liverpool New York. Liverpool Boston. New York Mediterranean. Boston Mediterranean', May 1905.
73 '*Amerika*, Hugest Liner, Arrives on First Trip', *New York Times*, 21 October 1905.
74 '*Hamburg-Amerika Linie. Amerika*', deck plan.
75 Flayhart, William H., & John H. Shaum, *Majesty at Sea: The Four Stackers* (Patrick Stephens Ltd, 1981) p. 58.
76 A comparison of the dates for construction of *Baltic*, *Amerika* and *Adriatic* is instructive. *Baltic*'s keel (Yard Number 352) was laid on No. 2 Slip on 7 June 1902 and the completed hull was launched on 21 November 1903. A notation '357 to follow' referred to Yard Number 357, which was assigned to *Amerika*, and her keel was laid less than a month later on the same slipway. *Adriatic* (Yard Number 358) had been laid down on No. 3 Slip almost a year before, and her double bottom was finished by 1 July 1903. In fact, *Adriatic* was fully framed a month after *Amerika*'s keel had been laid on the neighbouring slipway:

	Baltic	*Amerika*	*Adriatic*
Laid down	7 June 1902	15 December 1903	18 November 1902
Double bottom	1 October 1902	12 March 1904	1 July 1903
Fully framed	15 January 1903	16 June 1904	14 January 1904
Plated	20 August 1903	31 October 1904	29 May 1905
Launched	21 November 1903	20 April 1905	20 September 1906

Baltic and *Adriatic* began life on adjacent slipways; following *Baltic*'s launch, the new *Amerika* took her place alongside *Adriatic*.

The two ships were very similar in size and machinery. *Amerika* was 38ft shorter than *Adriatic* (between perpendiculars), yet the White Star ship was a foot wider and 4.5ft deeper. *Adriatic*'s engines were slightly more powerful than *Amerika*'s; both had eight double-ended boilers to provide steam, the difference being that *Adriatic* also had four single-ended boilers as well. Harland & Wolff gave *Amerika*'s speed as seventeen knots, with *Adriatic* half a knot better.

77 'Disaster at Last Befalls Captain Smith', *New York Times*, 16 April 1912, p. 7.
78 'Forty Years at Sea', *Poverty Bay Herald*, 27 July 1907, p. 4. I am indebted to Smith's biographer Gary Cooper and researcher Mark Baber for the reference.
79 Paul Lee has drawn attention to this quote from *The World's Work* in April 1909. It is a valuable source in that it was published prior to the *Titanic* disaster and quoted Smith using a qualifier that he would 'not assert that she is unsinkable'. Similarly, the article in the *Poverty Bay Herald* almost two years earlier paraphrased Smith, and the reporter wrote, 'He would not go so far as to say that the *Adriatic* was unsinkable.' By contrast, the *New York Times*' version, after the disaster, is the one that found its way into the history books. The paper did paraphrase Smith: 'Whatever happened, he contended, there would be time before the vessel sank to save the lives of every person on board,' therefore implicitly acknowledging that she could sink if she was damaged badly enough. Nonetheless, the qualifier is often forgotten. It also quoted him as saying, 'I will go a bit further. I will say that I cannot imagine any conditions which could cause a ship to founder. I cannot conceive of any vital disaster happening to this vessel. Modern shipbuilding has gone beyond that.' It was a more emphatic statement than the quotation three years earlier when he concluded that 'modern shipbuilding has reduced that danger to a minimum'.
80 'Forty Years at Sea', *Poverty Bay Herald*, 27 July 1907.
81 'Description of the Vessel', *Belfast News-Letter*, 21 October 1910, p. 8.
82 Pirrie, Margaret, to Lord Cadogan, July 1898: Public Records Office of Northern Ireland, T3449/8.
83 Hedgley, Alan, & George Lavery, 'The Story of Harland & Wolff, Ltd.: A Tribute to the Greatest Shipbuilding Firm in the World', *Titanic Commutator*, Vol. 3, No. 271980, p. 5.
84 Ibid., pp. 5–6.
85 Douglas, James, 'The Atlantic Hustlers', *London Opinion*, 27 April 1912, p.112.
86 Moss, Michael, & John R. Hume, *Shipbuilders to the World: 125 Years of Harland & Wolff, Belfast 1861–1986* (The Blackstaff Press, 1986) p. 19.
87 Oldham, Wilton J., *The Ismay Line*, pp. 84–85.

88 Louden-Brown, Paul, *The White Star Line: An Illustrated History 1869–1934* (*Titanic* Historical Society, 2001) p. 13.
89 Ibid., p. 33.
90 'War Compensation Court: Claim of The International Mercantile Marine Company and of The Oceanic Steam Navigation Company, Limited: Judgement, 1 April 1924', p. 8: The National Archives (UK), MT 40/184. Under the 4 February 1902 agreement 'between Messrs. Harland & Wolff and J.P. Morgan & Co. of New York', in light of the formation of IMM, the shipbuilder built several large ships for the White Star Line and the agreement ensured other IMM lines followed the same practice. According to Clause 4: 'the cost to the purchasers [is] to be (1) the cost as shown by the builders' books including wages, materials, direct expenses and a due proportion of fixed expenses and establishment charges (2) a commission on the whole as the builders' profit said commission in the case of new ships and their machinery to be reckoned at five percent [...] Payment is to be made by the purchasers as required by the builders on the basis of cash against expenditure during the progress of the work, and the ascertained balance on completion, and interest at five percent per annum to be allowed on expenditure in advance of payments or payments in advance of expenditure as the case may be but no large balance to stand over except by mutual arrangement.' The agreement ran for ten years, but could be terminated after five years' notice by either party.
91 'Memorandum of Commission Terms, Harland & Wolff Ltd': Stadsarchief Rotterdam, EB 547. Lord Pirrie sent the terms to the Holland America Line on 13 January 1911, because the Dutch company were discussing whether they would build ships on the same basis. He wrote that they were 'the terms on which we build, and have built for over half a century for our friends the White Star Line'.
92 'My dear Watson ...', 6 October 1902: Cunard Archives, Ca/20.
93 'Obituary: Viscount Pirrie', *The Engineer*, 13 June 1924, p. 655.
94 'Articles of Agreement, Entered Into This Seventh Day of August 1874': Public Records Office of Northern Ireland, D2805/HIST/3.
95 Moss, Michael, & John R. Hume, *Shipbuilders to the World*, pp. 36 and 47.
96 'Harland & Wolff Ltd. Issue at Par of 4,000,000 First Cumulative Preference Shares of £1 Each: Prospectus', July 1924; 'Harland & Wolff Ltd. Memorandum of Articles of Association', 22 December 1885: Public Records Office of Northern Ireland, D2805/MISC/3.
97 'Harland and Wolff's Works at Belfast', *The Engineer*, supplement, 18 June 1909, p. i.
98 Ibid.
99 Ibid.
100 Ibid.
101 Ibid.
102 'The Biggest Ship and the Biggest Dock', *Scientific American*, 6 May 1911, pp. 445–46.
103 'Wage Statistics from 1858 to 1911', unnumbered tables on notepaper covering 'Total Wages & Salaries Paid' ('private ledger not included'): Public Records Office of Northern Ireland, D2805/MISC/3.
104 'The *Titanic* Commutator', *Titanic Commutator*, Vol. 3, No. 27 (1980) p. 3.
105 'Preliminary Negotiations', p. 44: Cunard Archives, S10/2.
106 'The Cunard Steamship Company Limited: Issue of 40,000 Shares', London, 4 March 1880.
107 See Chirnside, Mark, *Oceanic: White Star's 'Ship of the Century'* (The History Press, 2018) pp. 9–11; letters to Vernon Brown, Cunard's agent in New York, dated 9 November 1894, 11 May 1905 and 22 April 1896: Cunard Archives, S5/1. (This refers to all the 1894–96 quotes.)
108 Watson, William, 'Statement by the Chairman of the Cunard Steamship Company, Limited', 1 June 1908: Cunard Archives, AC6/3/1.
109 'Correspondence as to Sale of Shares', letter from J. Bruce Ismay dated 4 July 1902, p. 384: Cunard Archives, D42/A9/10.
110 'Preliminary Negotiations', p. 58: Cunard Archives, S10/2.
111 Ibid., p. 57.
112 'Confidential: The Morgan Shipping Combination', The National Archives (UK), PRO 30/60/40.
113 Ibid., Appendix (B): Memorandum on the Situation Created by the Building of Four German Steamships for the Atlantic Trade of 23 knots and Upwards'.
114 Inverclyde, Lord, letter to shareholders, 30 September 1902. Cunard Archives, B1/5.
115 'Cunard Steamship Company Limited. Assistance for Acquisition of White Star Line & Building of New Liners', 1930. The National Archives (UK), T160/1313.
116 'Appendix (B): Memorandum of Interview Between Mr. Gerald Balfour and Mr. Pierpont Morgan and Sir Clifford Dawkins', 8 August 1902, pp. 3–4. The National Archives (UK), CAB 37/62.
117 'Memorandum: International Mercantile Marine Company of New Jersey', 8 August 1918, pp. 1–2; 'Ministry of Shipping: Position of the White Star and Associated Fleets', p. 1. The National Archives (UK), T1/12252. See also: 'Copy of An Agreement (dated 1 August 1903) between The Admiralty and the Board of Trade and the International Mercantile Marine Company and the Oceanic Steam Navigation Co., Limited', The National Archives (UK), MT23/163.
118 'This Agreement made in England the 18th day of May 1905 between John Brown & Company Limited [...] and the Cunard Steamship Company Limited'. Cunard Archives, S7/1/14.
119 'Specification of a Steel Quadruple-Screw Steamship with Turbine Engines'. Cunard Archives, S7/1/14.
120 'The Greatest Steamship Ever Built', *New York Times*, 12 May 1907, p. 4.
121 Letter from White, W.H., 'Dear Mr. Bain ...', dated 25 April 1908. Cunard Archives, B4/28.
122 Letters to Vernon Brown, dated 22 January and 15 February 1908. Cunard Archives, S5/3.
123 Letter from William Watson, Cunard's chairman, to Vernon Brown, Cunard's New York agent, dated 28 March 1908. Cunard Archives, S5/3.
124 Ballard, Robert, & Rick Archbold, *Lost Liners* (Madison Press, 1997) p.52: 'Before the war, *Lusitania* was the more popular ship, a fact easily forgotten because of her sister's longer life. Presumably this has something to do with her more airy interior décor.' The number of passengers carried clearly indicate otherwise. See my article, '*Lusitania* and *Mauretania*: Perceptions of Popularity', *Titanic Commutator*, Vol. 32, No. 184 (2008). I remain particularly grateful to Eric Longo for his considerable assistance with the article and John Maxtone-Graham for taking the time to assist me and putting me in touch with Norman Morse,

who was so generous in sharing material with me. If it was not for Norman's generosity, the article would never have been written.

125 Hyde, Francis E., *Cunard and the North Atlantic 1840–1973* (Macmillan Press, 1975) pp. 157–58.

126 'White Star Line Royal and United States Mail Steamers: Southampton–Cherbourg–New York. First Class Rates and Plans', April 1907. This brochure included the notice published in February 1907.

127 Oldham, Wilton J., *The Ismay Line*, p. 167.

128 For a detailed examination of the date issue, see: Bäbler, Günter, 'The Dinner at Lord Pirrie's in Summer 1907: Just a Legend?', *Titanic Post*, Vol. 32 (June 2000) p. 16. The earliest source for the 'legendary dinner' in the summer of 1907 appears to be Wilton J. Oldham's work (1961), who had extensive assistance from Florence Ismay – by then a widow of more than twenty years – during the course of his research. As Günter Bäbler speculates, perhaps she mistakenly recalled the year as 1907 when the occasion was actually earlier. Or it might have been the first occasion that she learned of the new ships, even if her husband and Lord Pirrie had been discussing them previously.

2: Yard Numbers 400 and 401

1 'Designing a White Star Liner: A Creative Miracle', *White Star Magazine*, Vol. 1, No. 1 (1923) p. 7.

2 Mersey, Lord J.B., 'Report of a Formal Investigation Ordered by the Board of Trade into the Loss of the SS *Titanic*' (London: His Majesty's Stationery Office, 1912), Question 21,313. See also: 'In Re.: The Oceanic Steamship Company [sic], Evidence of the Rt. Hon. Alexander Montgomery Carlisle', 3 April 1914, p. 2. National Archives and Records Administration – Northeast Region (New York).

3 'Carlisle, Designer of *Titanic*, is Dead', *New York Times*, 6 March 1926.

4 'Right Hon A.M. Carlisle Retires. A Notable Career', *Weekly Telegraph*, 13 August 1910.

5 Mersey, Lord J.B., 'Report of a Formal Investigation', Questions 21,267–21,269.

6 'Harland & Wolff Ltd. Minutes [of] Managing Directors', pp. 189 and 202. Public Record Office Northern Ireland, D/2805/MIN/A1.

7 Jefferson, Herbert, *Viscount Pirrie of Belfast* (William Mullan, 1948) p. 81.

8 Bullock, Shan F., *Thomas Andrews: Shipbuilder* (Maunsel & Company Ltd, 1912) pp. 26–27.

9 'Thomas Andrews, Junior', unidentified newspaper cutting from 27 April 1912. Public Record Office Northern Ireland, D/3655/A/6/2/1.

10 Pirrie, 'Dear Mr. Andrews', 2 November 1911. I am very grateful to Mark Doherty for sharing this and, in turn, Diana Teggart for sharing it with him.

11 'Deposition of Edward Wilding, In the Matter of the Petition of the Oceanic Steam Navigation Company, for Limitation of its Liability, as Owner of Steamship *Titanic*', 13–14 May 1915, p. 7. The National Archives and Records Administration – Northeast Region (New York). Although Wilding was describing his own duties, he had succeeded Thomas Andrews, who had previously occupied the same post.

12 Ibid.

13 Ibid., pp. 5–6.

14 'New Liverpool Firm: Mr. Edward Wilding, Consulting Naval Architect', *Journal of Commerce*, 12 March 1925.

15 Ibid.

16 'H&W Ltd. Minute Book B' (including minutes of general meetings and annual reports presented), pp. 29–30 and 65–66 (private collection). I am grateful to have had the opportunity to see this interesting documentation.

17 Peskett, Leonard, 'Memo to the General Manager: Proposed New steamer for New York Service', Liverpool, 19 January 1910. Cunard Archives, S7/3/36.

18 'Bigger than the *Lusitania*', *New York Times*, 12 September 1907.

19 'Biggest Liner Cuts Record', *New York Times*, 14 September 1907. The 'record' it referred to was *Lucania*'s best time from Daunt's Rock to Sandy Hook.

20 'The Thousand Foot Ship and What It Means', *New York Times*, 15 March 1908, p. 6.

21 '*Olympic*, 1,000-foot Liner: White Star Line Selects Name for its Leviathan – May Build Sister Ship', *New York Times*, 17 April 1908.

22 '*Olympic* and *Titanic*: Names of White Star Leviathans, to be More Than 840 feet Long', *New York Times*, 23 April 1908, p. 1. (See also: 'White Star's Two New Liners Named', *New York Times*, 16 September 1908, p. 1. Although it was old news, the newspaper reported the names of the two ships again, as if they were being reported for the first time.)

23 'Proposed Names for New Fast Steamers', Cunard Archives, S10/5. The names for any new ships were often a subject of considerable discussion. The names *Lusitania* and *Aquitania* were among those suggested by Lord Inverclyde. Although *Lusitania* was used for one of the 1907 steamers, the name *Aquitania* was not used until an even larger vessel was laid down in 1911. Some of the suggestions were cringeworthy, including: *Zenobia*, *Eboracia*, *Kashgaria* and *Draconia* (Lord Inverclyde), not to mention *Aetolia*, *Trezenia*, *Parnassia* and *Naupactia* (Sir W.B. Forwood).

24 'Naming of Ocean Liners: "Family" Appellations Distinguishing the Various Fleets', *New York Times*, 8 November 1903.

25 'No. 2 Orders to Proceed', pp. 26, 53–54, 82. Public Record Office of Northern Ireland, D2805/OTP/A/2. I have relied particularly on the White Star liners No. 807 (*Britannic*) and No. 896 (*Georgic*). Ismay testified he directed Harland & Wolff to build the ships on behalf of White Star. The orders date from the late 1920s. However, the format appears to have been fairly standard over the years. The information recorded was the most basic summary of each ship's dimensions and propelling machinery. When two near-identical sister ships were ordered, the two yard numbers were recorded on the same order. In the case of the motor ships *Britannic* and *Georgic*, ordered in April 1927 and November 1929 respectively, *Britannic* was ordered originally as a single vessel but her sister's order followed afterwards. The order for *Georgic* noted she was 'to be generally a duplicate of MV *Britannic* with such modifications as may be approved by the owners'.

26 See Ismay's testimony before the US Senate, April 1912. 'Report of the Committee on Commerce United States Senate Pursuant to S. Res. 283 Directing the Committee on Commerce to Investigate the Causes Leading to the Wreck of the White Star Liner *Titanic*' (Washington: Government Printing Office, 1912).

27 'In Re.: The Oceanic Steamship Company [sic], Evidence of the Rt. Hon. Alexander Montgomery Carlisle', 3 April 1914, pp. 6–8. The National Archives and Records Administration – Northeast Region (New York).

28 'Deposition of Edward Wilding, In the Matter of the Petition of the Oceanic Steam Navigation Company, for Limitation of its Liability, as Owner of Steamship *Titanic*', 13–14 May 1915, pp. 99–100. The National Archives and Records Administration – Northeast Region (New York).
29 Ibid.
30 Ibid., pp. 101–02.
31 Mersey, Lord J.B., 'Report of a Formal Investigation', Question 21,298.
32 Bäbler, Günter, 'The Dinner at Lord Pirrie's in Summer 1907: Just a Legend?', *Titanic Post*, Vol. 32 (June 2000).
33 'Queenstown Harbour Committee: Report from the Committee Appointed by the Postmaster-General to consider the Omission of the Call at Queenstown on the American Mail Service; Together with Minutes of Evidence, Appendices, and a Chart of Queenstown Harbour' (London: His Majesty's Stationery Office, 1914) p. 36.
34 'Memo to General Manager', 7 September 1910. Cunard Archives, S7/3/37.
35 The similarity in design is demonstrated by comparing the White Star liners with some of the Atlantic liners which preceded and succeeded them. *The Shipbuilder* published 'Fig. 6 – Equivalent Girders of Large Atlantic Steamers' as part of their *Mauretania* special number. They calculated the stress on the sheer strake plating, at the uppermost part of the hull at the side, assuming each ship was 'placed upon the crest of a wave having a length equal to her own length and a height equal to 1/20th of the length. The ship is supposed to be nearing the port of arrival with bunkers nearly empty, having been loaded at departure to the draught given in the table. In the case of actual bending moment calculations made at Wallsend, the actual value of the co-efficient "C" was found to be 29.25 and this value has been adopted for estimating the bending moments in the cases where their actual values were not available.'

Vessel	Length Between Perpendiculars	Displacement (tons)	Draught	Bending Moment (Tons)	Stress on Sheer Strake (Tons Per Square Inch)
Deutschland (1900)	662ft 9in	22,850	28ft 6½in	517,000	10.58
Campania (1893)	600ft	21,000	29ft	430,000	9.94
Oceanic (1899)	685ft	30,100	32ft 6in	704,000	8.83
Kaiser Wilhelm II (1903)	678ft	26,080	29ft 6½in	604,000	8.66

In January 1920, the Board of Trade were discussing *Imperator* (1913), the former German liner which became Cunard's *Berengaria* after the war. Cunard's Naval Architect, Leonard Peskett, wanted the ship's approved draught to be increased so that they could arrange additional ballast to improve her stability. The surveyor wrote, 'It appears very desirable that the vessel [*Imperator*] should be stiffened, and the scantlings [elements of the ship's hull structure, such as plating, beams, etc.] have been looked into as well as the short time has permitted [...] on the basis of similar passenger steamers of large size, such as the *Lusitania*, *Olympic* and *Aquitania*.' By approving a greater draught, *Imperator*'s displacement would be greater, increasing the bending moment and the stress on the sheer strake plating, so it was an important calculation. The surveyor was using the three other ships to compare strength to see whether *Imperator* would be strong enough with a deeper draught. He used a co-efficient C of 30 and assumed a vessel 'on a wave of her own length and height of 1/20 of her length', writing 'the assumption for the B.M. [bending moment] factor are very high for all these vessels':

Vessel	Length between perpendiculars	Displacement (tons)	Draught	Bending moment	Stress on sheer strake (tons per square inch)
Imperator (1913)	882ft 6in	60,610	37ft 9in	1,782,944	10.2
Lusitania (1907)	760ft	41,270	36ft	1,045,507	10.1
Olympic (1911) and *Titanic* (1912)	850ft	52,310	34ft 7in	1,473,333	9.9
Aquitania (1914)	866ft	51,700	35ft 4in	1,492,407	8.5

The Shipbuilder published significantly different figures for *Mauretania* in their special number, showing a stress of 7.95 tons per square inch on the sheer strake, but on a significantly lower draught and displacement (which resulted in a lower bending moment and thus a lower stress). The *Lusitania* figure seems more in line with normal service conditions (and was reported elsewhere as 10.6 tons). Both ships used high-tensile steel in the upper part of their hulls amidships, including the sheer strake, which was intended to allow them to bear a higher stress so that the scantlings could be reduced, to make them lighter and more stable. All these figures serve as an approximate comparison. The stresses varied depending on a particular ship's state of loading and displacement, but they show that all of these large North Atlantic liners were designed to a similar standard.
36 Attwood, Edward Lewis, *Textbook of Theoretical Naval Architecture* (Longman's, Green, & Co., 1899) pp. 206–8. I am grateful to Roy Mengot for sharing.
37 'SS *Titanic*. Correspondence Respecting The Assignment of Freeboard to the SS *Titanic*'. The National Archives (UK), MT 9/920/E.
38 Carruthers, F., 'General Remarks'. The National Archives (UK), MT 9/920/F. He wrote: 'The [rivet] roles in the keel and keel plate are drilled. The butt straps in the keel plate had the holes punched small and rimered out to full size. The butt straps in way of the bridge and the plates between them were dealt with in the same way as those in the keel plate. The keel rivets are 1¼ inch diameter riveted over into the countersink on the outside, with a full point ¼ inch above the plate. All the shell riveting, from keel to margin plate of the double bottom is hydraulic. The holes have the ordinary countersink to Lloyds' Rules and have full points ¼ inch above the plates [...] The shelter deck sheerstrake, bridge sheerstrake, and strake between these two in way of bridge, and the sheerstrake and strake below

at the break of the bridge, each end are hydraulic riveted. The holes are not countersunk beyond what is caused by punching the plates and the rivets have snap heads and points.'

39 Cunard's *Carmania* adopted the same approach. Her specifications called for 'all rivets to be of steel where hydraulic riveted, and of best iron for hand riveting'. Her riveting in general was 'to be according to Lloyd's Rules'. *Lusitania* and *Mauretania* were different in that they used steel rivets. Their specification, dated 6 January 1906, called for 'all rivets of be of ingot steel, except where H.T. [high-tensile] rivets are used'. However, E.W. De Rusett's article, 'Hydraulic Riveting in the *Mauretania*: How Four Million Rivets were Driven in the Big Steamship', which was published in *Scientific American*'s Supplement No. 1668 on 21 December 1907, noted that 'mild ingot steel was employed for all the rivets at the strong recommendation of the authorities of Lloyd's Register from their experience'. The result of this was that mild-steel rivets were used to connect the high-tensile steel plating in the sheer strake. There was some comment at the time that the material in the rivets was weaker than the hull plates they connected together (a particularly significant feature because this plating was in one of the areas of the hull exposed to the greatest stresses), however, the shipbuilders believed that the use of hydraulic riveting and the use of rivets 'at least equal to the requirements of Lloyd's rules for mild steel plates of equivalent strength to the high tensile steel used' mitigated this concern. See: 'Specification of a Steel Quadruple-Screw Steamship with Turbine Engines [*Lusitania* and *Mauretania* generic specification]', John C. Morrison, Naval Architect's Department, 6 January 1906, Merseyside Maritime Museum, B/CUN/3/4/3; and 'Specification. SS *Carmania*. No. 366', 10 September 1904, Merseyside Maritime Museum, B/CUN/3/4/1.

40 '*Olympic* and *Titanic*: Bulkheads'. The National Archives (UK), MT 9/920/F.

41 Sims, P., 'Comparative Naval Architecture of Passenger Ships', Transactions of the Society of Naval Architects, Vol. 111 (2003), pp. 233–58. My thanks to Sam Halpern for his assistance in obtaining a copy of this fascinating paper, which examines a number of aspects of naval architecture and *Titanic*'s design, in context with the progression of passenger ship design before and afterward.

42 Cooke, Anthony, '*Olympic* and *Titanic*: How Many Masts?', *Sea Lines*, Issue 42 (2006) p. 32 – an extremely interesting article. (See also Harris, James, letter to editor, *Atlantic Daily Bulletin*, 2006.) I appreciate David Hill's assistance.

43 Letter to Alfred Booth from Lord Inverclyde, 14 October 1910. Cunard Archives, S7/3/4.

44 Mersey, Lord J.B., 'Report of a Formal Investigation', Questions 23,851–54 and 23,963.

45 Ibid., Questions 20,649–53.

46 'Harland & Wolff Ltd. Minutes [of] Managing Directors', pp. 39–40. Public Record Office Northern Ireland, D/2805/MIN/A1.

47 'Contracts No. 7 Iron & Steel Department', Public Record Office of Northern Ireland, D2805/MISC/1. Pages are un-numbered, however, the orders are recorded chronologically.

48 Ibid.

49 Ibid.

50 See Louden-Brown, Paul, '*Titanic*'s "Brittle" Steel?', online article, *Titanic* Historical Society (2001); and 'Contracts No. 7. Iron & Steel Dept'. Public Record Office Northern Ireland, D2805/MISC/1.

51 See '*Titanic* Harland & Wolff's 401 Boiler Plates by D. Colville & Sons', March 1910: The National Archives (UK), MT 15/212/3; 'The Rivet, Bolt & Nut Company (Ltd.), 12 May 1910, p. 155: The National Archives (UK), MT 15/212/4; '*Titanic* Harland & Wolff's 401. Boiler – Designs Submitted', 10 March 1909, p. 41: The National Archives (UK), MT 15/212.

52 'Chain Cables. Steam Ships. *Olympic* and *Titanic*', 24 June 1909. The National Archives (UK), MT 15/229.

53 I am grateful to Steven Tuckwood and Ralph Currell for drawing my attention to the red annotations on the keel plate plan, which stated, 'All bars marked in red refer to No. 400', and then showed the relevant keel bars in red with the reference to 'ex. 391'.

Richard de Kerbrech and Dave Williams researched *Europa*'s history in some detail and found no evidence that her keel was laid. Harland & Wolff's Order Book has no entry confirming that a contract was ever formally agreed. According to an article in the *New York Times* on 18 July 1908, the order was cancelled in January 1908. Harland & Wolff's records for the cancelled ship showed that she was intended to have been 715ft long between perpendiculars and 80ft in width, driven by quadruple-expansion reciprocating engines. They envisaged eight double-ended and four single-ended boilers working at 215lb per square inch. HAPAG's annual report for 1907 stated, 'Den Bau des der Werft von Harland & Wolff in Auftrag gegebenen Schwesterschiffes der "*Amerika*" haben wir im Hinblick auf die gegenwärtige Geschäftslage vorläufig sistiert' ['In view of the current business situation, we have temporarily suspended the construction of the sister ship of the *Amerika* ordered by the Harland & Wolff shipyard']. I am grateful to Günter Bäbler for the translation.

German maritime historian Arnold Kludas and Dr Herbert Bischoff found evidence that plans for *Europa* came to envisage a significantly larger ship of about 44,000 gross tons and with combination reciprocating and turbine machinery. In other words, she grew from being an enlarged sister ship to *Amerika* to a much larger vessel altogether – a slightly smaller version of *Olympic*. See Williams, Dave, and de Kerbrech, Richard, *Great Passenger Ships That Never Were: Damned by Destiny Revisited* (The History Press; 2021).

54 'Belfast Shipbuilding: Progress of the White Star Liners', *Belfast Evening Telegraph*, 5 November 1909, p. 4.

55 'White Star Mammoths. Chairman Inspects New Liners', *Liverpool Daily Post*, 25 March 1910.

56 Currie, J., & A. Galbraith, 'Technical Subjects Bearing on Atlantic Shipping', Memorandum from the Superintendent Engineer, the Cunard Steamship Company Limited, to the General Manager, Liverpool, 24 August 1910, in 'The Cunard Steamship Company Limited: Minutes of Executive Committee Meetings', p. 334. Cunard Archives, B4/32.

57 Ibid.

58 'General Post Office: Transatlantic Sea Post Offices. Proposed Accommodation On Board SS *Titanic* and SS *Olympic*. Consideration of Plans. Superseded by Revised Plans … 1909–10'. British Postal Museum & Archive, POST 29/1117.

59 Ibid.

60 'City Inquests: The Queen's Island Fatality and How It Happened', *The Irish News and Belfast Morning News*, 16 October 1909, p. 8; 'Serious Queen's Island Accident', *The Irish News and Belfast Morning News*, 10 January 1910, p. 4.

61 'Shipyard Fatality: Inquest on the Victims', *Belfast Weekly Telegraph*, 19 February 1910, p. 8.
62 'Harland & Wolff Ltd. Minutes [of] Managing Directors', p. 165. Public Record Office Northern Ireland, D/2805/MIN/A1.
63 'Belfast Police Courts: Stolen Paint Brushes', *Belfast Evening Telegraph*, 5 November 1909, p. 4.

3: 'Unsinkable'

1 'Safety on the Ocean', *New York Times*, 30 May 1891.
2 Ibid.
3 'Description of the Vessel', *New York Times*, 13 November 1894.
4 'Laurels for St. Louis', *New York Times*, 30 May 1895.
5 'No News of the Liner St. Louis', *New York Times*, 16 January 1903.
6 'Ship Can Save Herself', *New York Times*, 11 November 1903.
7 Warren, Mark D. (ed.), *Distinguished Liners from The Shipbuilder 1906–14, Volume 1* (Blue Riband Publications, 1997) p. 39.
8 Ibid., p. 76.
9 Ibid., p. 16.
10 The number of copies printed is not known and there are some errors in the text, including a reference to '*Titantic*', which indicate it was not proofread very well. The brochure was apparently printed for the first time after its initial production in the *Titanic* Historical Society's *Titanic Commutator*. It was contributed by Geoff Robinson and appeared in an article by Don Lynch. See: Lynch, Don, with Geoff Robinson, 'The "Unsinkable" *Titanic*, Ad Advertised: Another Great Myth is Dispelled', *Titanic Commutator*, Vol. 16, No. 4 (1993) pp. 4–6.
11 'Departmental Committee on Bulkheads and Watertight Compartments: Digest of Evidence', pp. 9 and 17. The National Archives (UK), MT 15/266.
12 'Report of the Committee Appointed by the President of the Board of Trade to Consider and Report upon The Spacing and Construction of Watertight Bulkheads in Ships of the Mercantile Marine with Appendices and Diagrams' (London: Her Majesty's Stationery Office, 1891) The National Archives (UK), MT 15/145.
13 Ibid., p. 6.
14 Ibid., p. 7.
15 See: 'Specification of a Steel Quadruple-Screw Steamship with Turbine Engines … [*Lusitania* and *Mauretania* generic specification]', John C. Morrison, Naval Architect's Department, 6 January 1906, Merseyside Maritime Museum, B/CUN/3/4/3; and 'Specification. SS *Carmania*. No. 366', 10 September 1904: Merseyside Maritime Museum, B/CUN/3/4/1. Leonard Peskett also confirmed in his testimony before Lord Mersey that *Lusitania* and *Mauretania* were built to that standard and 'they will float with more [compartments flooded]'. See also my article, 'Thomas Andrews in Court, in Rotterdam, in Belfast and Standing-in: November 1911 to March 1912', *Titanic Commutator*, Winter No. 247 (October 2024) pp. 18–28.
16 'Departmental Committee on Bulkheads and Watertight Compartments: Digest of Evidence', p. 63. The National Archives (UK), MT 15/266.
17 Mersey, Lord J.B., 'Report of a Formal Investigation', Questions 20,290–96.
18 Ibid., Question 20,063.
19 Ibid., Questions 20,877–80.
20 Ibid., Questions 21,000–11.
21 Ibid., Questions 21,187–94.
22 Comparison data obtained from Walker, J. Bernard, 'The "Unsinkable" Ship: Longitudinal Coal Bunkers and Higher Bulkheads Might Have Saved the *Titanic*', *Scientific American*, 11 May 1912, pp. 417–18. *Adriatic*, *Oceanic* and *Teutonic* were not included in Walker's article, but they have been added as a useful comparison using published data. The comparison was between the number of compartments and the length of a ship between perpendiculars, not the overall length.
23 'SS *Titanic*. Correspondence Respecting The Collision Bulkhead of SS *Olympic*'. The National Archives (UK), MT 9/920/E.
24 Ibid.
25 Ibid.
26 'The Assignment of Freeboard to the SS *Titanic*. The National Archives (UK), MT 9/920/E.
27 Ibid.
28 'Departmental Committee on Bulkheads and Watertight Compartments: Digest of Evidence', pp. 9 and 17. The National Archives (UK), MT 15/266.
29 Carruthers, F., 'Dimensions of Vessel', pp. 400–01. The National Archives (UK), MT 9/920/C. Carruthers' statements in this particular report were prepared in reference to *Titanic*, but applied equally to *Olympic*.
30 'Departmental Committee on Bulkheads and Watertight Compartments: Digest of Evidence', pp. 8, 11, 18, 24 and 75. The National Archives (UK), MT 15/266.
31 'Safety of Life at Sea: Draft Lifesaving Appliances Rules, With memorandum on Safety of Life at Sea by the President of the Board of Trade' (London: His Majesty's Stationery Office, 1912), p. 135: The National Archives (UK), RAIL 1053/295.
32 Ibid., p. 59.
33 Ibid., pp. 34–37.
34 'Wireless Telegraphy on Ships' and 'Statement as to the number of British ships fitted with Wireless Telegraphy'. The National Archives (UK), MT 15/149.
35 Mersey, Lord J.B., 'Report of a Formal Investigation', Question 22,095.
36 'Report to the Board of Trade on the Question of Boats, Rafts, and Lifesaving Apparatus carried by Seagoing Merchant Ships' (London: Her Majesty's Stationery Office, 1887) p. 4: The National Archives (UK), MT 9/280.
37 Ibid.
38 Ibid.
39 'Report from the Select Committee on Saving Life at Sea; together with the Proceedings of the Committee, Minutes of Evidence, and Appendix' (London: Her Majesty's Stationery Office, 1887) p. iii. The National Archives (UK), MT9/304A.
40 Ibid., p. 18.
41 Ibid., p. 19; Kennedy, Captain Charles William, 'Aboard White Star's *Germanic*: A Commander's Recollections', *Titanic Commutator*, Vol. 40, No. 211 (2015) p. 133.
42 Ibid., p. 33.
43 Peskett, Leonard, 'Memo to the General Manager', Liverpool, 15 May 1912: Cunard Archives, S7/3/20.
44 In 'Re.: The Oceanic Steamship Company, Evidence of the Rt. Hon. Alexander Montgomery Carlisle', 3 April 1914, p. 12: The National Archives and Records Administration – Northeast Region (New York).
45 Haas, Charles, 'The Hero Who Was Not On Board', *Voyage* 104 (2018) pp. 153–54.

46 Ibid., p. 156.
47 Mersey, Lord J.B., 'Report of a Formal Investigation', Question 21,275.
48 In Warren, Mark D. (ed.), *Distinguished Liners from The Shipbuilder 1907–14, Volume 2* (Blue Riband Publications, 1997) p. 110. Welin's paper, 'The Arrangement of Boat Installations On Modern Ships', was presented to the Institution of Naval Architects in March 1912.
49 Mersey, Lord J.B., 'Report of a Formal Investigation', Question 21,297.
50 'In Re.: The Oceanic Steamship Company [sic], Evidence of the Rt. Hon. Alexander Montgomery Carlisle', 3 April 1914, p. 8: The National Archives and Records Administration – Northeast Region (New York). Carlisle recalled that the meeting 'in the beginning of 1910' – the January 1910 meeting, according to his earlier 1912 account – took place 'at Harland & Wolff's office in Cockspur Street'.
51 Mersey, Lord J.B., 'Report of a Formal Investigation', Question 21,288.
52 Ibid., Question 21,291.
53 Ibid., Question 21,294.
54 Ibid., Question 21,401.
55 'In Re.: The Oceanic Steamship Company [sic], Evidence of the Rt. Hon. Alexander Montgomery Carlisle'. 3 April 1914, p. 14: The National Archives and Records Administration – Northeast Region (New York).
56 'A. Welin. Double-acting boat gear "DA2" davits. Letter and designs (6) submitted for approval as a Standard', 9 March 1910; 'Engineer Surveyor in Chief. Life Saving Appliances Mr. Welin's Double Acting boat gear DA2 davits for *Titanic* and *Olympic*. Observations, letter and designs', 16 March 1910: United Kingdom National Archives, MT 15/114.
57 'In Re.: The Oceanic Steamship Company [sic], Deposition of Edward Wilding', 13 and 14 May 1915, p. 131: The National Archives and Records Administration – Northeast Region (New York).
58 Peskett, Leonard, 'Memo to the General Manager: SS *Aquitania*', Liverpool, 9 August 1911: Cunard Archives, S7/3/36.
59 Mersey, Lord J.B., 'Report of a Formal Investigation', Question 20,826.
60 The blueprint is in the Merseyside Maritime Museum's collections (reference B/CUN/8/1911.1/1/11) and is drawn to a scale of 1/8in to 1ft. There is a note on the plan towards the top left, about a third of the way in, which states, 'BOARD OF TRADE requires 16 boats under davits. Capacity 9625 cubic ft'. To the right and below, it noted the specifications of fourteen lifeboats and the two emergency cutters, which provided a combined total of 9,702 cubic feet of capacity. Beneath, a further note read, 'PROPOSED ADDITIONAL BOATS 16 lifeboats 28'0" X 8'6" X 3'6" – 8000 cubic feet'. The boats required by the Board of Trade regulations were calculated as having a cubic capacity equivalent to 970 persons, then the 'proposed additional boats' were calculated as having a cubic capacity equivalent to 800 persons, for a total of 1,770 people.

The blueprint was shown in the provocatively titled television programme *A Question of Murder*, which was first broadcast in 1983. The museum's catalogue record dates the blueprint to about March 1910. This is, in all likelihood, correct. The first-class gymnasium was not yet shown on the Boat Deck, although it was included on plans from the end of June 1910, which Harland & Wolff submitted to the Board of Trade. If the plan appears to date from about March 1910, who was it produced by? It does not have Harland & Wolff's stamp on it and the headings indicate it was produced by Welin. Oddly, the first line reads 'S.Ss. *Titanic* & *Olympic*', which is unusual as *Olympic*'s name was always given first. However, a letter exists from Axel Welin himself, which is dated 9 March 1910 and addressed to the Engineer Surveyor-in-Chief of the Board of Trade at 54 Victoria Street in London. Welin, too, transposed the ship names and referred to 'the *Titanic* and *Olympic* now building by Messrs. Harland & Wolff'.

It appears the blueprint must have come to the museum from White Star Line records which were passed on to the merged Cunard White Star company in 1934. However, neither Ismay or Sanderson said that they recalled seeing it prior to the *Titanic* disaster. The implication is it was provided to White Star by Harland & Wolff. Wilding did testify in London that, 'I think there was a plan submitted to us by Welin showing three boats. I do not think we made any plan ourselves.' He corrected himself later, testifying before the Limitation of Liability hearings in 1915, 'I subsequently referred to the plan again, and found that it was not as I had testified; it was two drawn and one dotted, giving the impression of three boats.' If he was referring to the same blueprint, then in looking at the plan after 1912 it would seem to imply it remained with Harland & Wolff at that time. Asked whether 'such plans' were submitted to White Star, he said, 'We, of course, have been making a search in all our records, and as far as I can ascertain they were not.' (See Mersey, Lord J.B., 'Report of a Formal Investigation', *Questions* 20,757–65.)

Compared to the initial 'Design "D"' concept from July 1908, the 'proposed additional boats' provided for an 82 per cent increase in lifeboat accommodation. However, the increase was 50 per cent when compared to the final lifeboat configuration *Olympic* and her sister were completed with. A lifeboat capacity of 1,770 people would have been around 50 per cent of the maximum number of passengers and crew carried, compared to about 35 per cent for the ship as she was actually completed. See: Chirnside, Mark, '*Titanic*'s Lifeboats: An Increased Capacity', *Atlantic Daily Bulletin*, September 2018, pp. 28–29.
61 Ibid., Questions 20,757–72.
62 Ibid., Questions 19,375–76. Two answers have been collapsed into one.
63 Ibid., Questions 18,663 and 18,668.
64 Ibid., Question 18,758.
65 Ibid., Question 19,131.
66 Ibid., Question 19,103. It is worthwhile noting some of the points that stood out in Harold Sanderson's mind. He mentioned the powerful Marconi apparatus; the number of strongly constructed watertight bulkheads; the features of the double bottom; and the pumping arrangements. He was asked in 1912:

19103. Can you give me in a couple of sentences what you say the special features are to which you attach importance, and leave Mr. Wilding to explain them in detail?
Very well. I should mention that she had a specially powerful wireless installation, long distance. She was built with an unusual number of watertight bulkheads, 15 in all; those bulkheads were of special construction, carried up as much as possible in one fair line, and they were built in excess of the requirements of Lloyd's. The plating and connections were also of special strength, and in excess,

I am told, of what Lloyd's requirements would have been. She had a double bottom, which was carried nine-tenths of the way; that double bottom was divided for the midship section of the ship into four tanks athwartships, and the rest of the distance into two tanks athwartships; and the double bottom, instead of terminating at the bottom of the bilge, which is ordinarily the case, was carried up to the top of the bilge. In regard to the wireless installation, she had got the power for that from three different sources; in the first place, from the ordinary electric light plant; in the second place, from an emergency plant which could be put in use if the ordinary electric light plant in the engine room were flooded; and, in the third place, she had storage batteries in the Marconi Room. The pumping arrangements were exceptional, each boiler compartment having its own equipment, which is quite an unusual thing of the kind.

67 Peskett, Leonard, 'Memo to the General Manager: SS *Aquitania*', Liverpool, 14 September 1911: Cunard Archives, S7/3/36.

68 'The New Vessel's Davits. Some Interesting Innovations', *Belfast Telegraph*, 21 October 1910.

69 Mersey, Lord J.B., 'Report of a Formal Investigation', Question 20,491.

70 Eaton, John P., 'The Saga of *Titanic*'s Lifeboats', *Voyage* 63 (2008) p. 137.

71 The McAlister order book contains the entries for *Olympic* and *Titanic*'s collapsible boats, pp. 99–100 and 105–06: Glasgow City Archives, TD1347/10/14.

72 Eaton, John P., 'The Saga of *Titanic*'s Lifeboats', p. 136.

73 As an added curiosity, following *Titanic*'s loss, the United States Senate Report gave another set of variations: fourteen boats carrying sixty-five persons each; two 'emergency boats' carrying twenty-five persons each; and the four Engelhardt collapsibles carrying forty-nine persons each: 'Total lifeboat capacity, 1,176'. The most common figures are those using Stirling's Rule, which was preferred by the Board of Trade. See: Beveridge, et al. (eds), *Titanic: The Ship Magnificent: Volume One: Design & Construction* (The History Press, 2008).

74 For the full details and methodology behind these calculations, see Chirnside, Mark, '*Titanic*'s Lifeboats: An Increased Capacity', *Atlantic Daily Bulletin*, September 2018, pp. 28–29. There is a further article: Chirnside, Mark, '*Titanic*'s Lifeboats: Fact & Fiction', *Atlantic Daily Bulletin*, March 2019, pp. 29–31. This explores the inaccurate claims made in the television series *The Liners* and an accompanying hardback book in the late 1990s. The programme claimed incorrectly that amendments in a Harland & Wolff drawing-office notebook were evidence that Harland & Wolff had planned originally for enough lifeboats to accommodate all the passengers and crew on board, but that these plans had changed during the design and construction period. In fact, the lifeboat configuration which was crossed out was the specification of the lifeboats fitted to *Olympic* in 1911 and it was crossed out because it was no longer up to date following the changes to her lifeboat provision made during the 1912–13 refit. The amendments simply updated the shipbuilder's record for *Olympic* to reflect the changes made in 1912–13, but were completely mischaracterised.

75 'Steamship *Olympic* (Provision of Lifeboats)', *Hansard* 22 November 1910, p. 262.

76 The series of articles in *John Bull* are: 'The White Star's *Olympic*. A serious Insufficiency of Lifeboats', 19 November 1910, p. 789; 'The White Star's *Olympic*. Antiquated Board of Trade Regulations Hazard Thousands of Lives', 3 December 1910, p. 1,010; 'The White Star's *Olympic*', 10 December 1910, p. 923; 'The White Star's *Olympic*. Our Rejoinder to An "Official Reply"', 3 December 1910, p. 884.

77 Robinson, Geoff, 'Lifeboats and the *Titanic* (Part One)', *Atlantic Daily Bulletin*, No. 4 (1999) p. 6.

78 Mersey, Lord J.B., 'Report of a Formal Investigation', pp. 49–50.

79 Robinson, Geoff, 'Lifeboats and the *Titanic* (Part One)', p. 6.

80 Mersey, Lord J.B., 'Report of a Formal Investigation', Question 22,875.

81 Ibid., Question 23,357.

82 Ibid., Questions 23,364–71.

83 Ibid., Question 23,356a.

84 Ibid., Questions 23,390–91.

85 Ibid., Questions 24,608–14.

86 Ibid., Questions 24,672–73.

87 'Loss of SS *Titanic*. Lord Mersey's Report', *Hansard*, 7 October 1912, p. 106. The quotation is from Liberal MP William Pringle. Thanks are due to Richard Ward and the staff of the UK Parliamentary Archives for their assistance.

88 Mersey, Lord J.B., 'Report of a Formal Investigation', Question 21,517.

89 'Estimate for Q.S. [Quadruple Screw] Atlantic Liner for Cunard Steamship Co., 850' x 95.2' by 64': 26 November 1910'; and Peskett, Leonard, 'Memorandum to the General Manager' 13 August 1912. The latter noted, 'it is now proposed to fit: 81 lifeboats (including collapsible), 4 motor boats, 8 rafts, the total cost of which will be approximately £9,700'. A pro rata estimate for twenty-eight sets of Welin davits 'would amount to £9,240, giving a total of £18,940 for boats and davits' or '£15,400 approximate extra for boats and davits' compared to the original estimates. The specification dated March 1911 included sixteen lifeboats weighing a total of 35 tons, lifeboat davits weighing a total of 60 tons and disengaging gear weighing a total of 2 tons: Cunard Archives, D42/S7/3/41.

90 On 9 August 1911, Cunard's Naval Architect Leonard Peskett visited London, along with a representative from Clydebank, and spoke with Archer at the Board of Trade. Archer told them that the board was 'formulating a scheme to increase the total capacity of boats on all passenger steamers'. In discussing the new *Aquitania*, Archer indicated that it was 'advisable that a plan showing [the] position and capacity of boats, should be sent in to the Board of Trade; plan to be by way of information only. By doing this, some statement may be elicited from the Committee.' See Peskett, Leonard, 'Memorandum to the General Manager', 9 August 1911: Cunard Archives, D42/S7/3/37. Leonard Peskett also visited Sir Walter Howell and Captain Young at the Board of Trade's offices in Whitehall, London, on 14 May 1912: 'Owing to recent events, it was thought prudent to leave with the Board of Trade an 1/8" scale plan of the boat deck plan of the *Aquitania* in order that it might be shown that it was the intention of the Cunard company, previous to the recent disaster, to carry sufficient boats on the *Aquitania* to accommodate all the passengers and crew in case of emergency.' They then discussed the question of lifeboats generally. Peskett noted the reasons behind *Lusitania* and *Mauretania* being fitted with sixteen lifeboats, and continued: 'Regarding the *Olympic* and *Titanic*, the builders made provision for carrying double the number of boats required by the Rules. These ships were considered to be well subdivided by transverse bulkheads and thought by

the builders to be unsinkable, and the ships were allowed to sail with the number of boats required by the Rules. The question of carrying boats sufficient for all on board has been under consideration not only by the Board but by the steamship owners themselves. As above stated, the White Star [Line] had provided the means of dealing with double the number of boats, and in the early part of this year, the Cunard Company laid before the Board of Trade a plan showing a proposal for their new steamer *Aquitania* to carry sufficient boats which in case of emergency could accommodate all the passengers and crew, which was approved. Subdivision by transverse and longitudinal bullheads will do more for the safety of passengers and crew than the number of boats that may be carried. The placing of additional boats on many of the existing ships is a serious matter, as stability calculations were made on existing rules, and the addition of top-weight might in some cases become a greater source of danger than an insufficient number of boats.' See Peskett, Leonard, 'Memo to the General Manager', 15 May 1912: Cunard Archives, D42/S7/3/20.

91 Alfred Booth to Sir Walter J. Howell, 18 May 1912: Cunard Archives, C1/159.

4: Forging Ahead

1 'Correspondence as to Sale of Shares', p. 386: Cunard Archives, D42/A9/10.

2 'Safety of Life at Sea', p. 58: The National Archives (UK), RAIL 1053/295.

3 Pounder, C.C., 'Some Notable Belfast-Built Engines', Belfast Association of Engineers, Sixth Meeting, 57th Session, 3 March 1948: Public Record Office of Northern Ireland, D4063/9/63.

4 On 24 September 1902, Ismay wrote, 'If these steamers are to be continued in the Atlantic trade as mail and passenger steamers, it will be necessary to re-boiler them, and there would be many advantages in re-engining them, but doing the latter is not absolutely necessary, as the engines are in excellent condition. To carry out the entire scheme would, it is estimated, cost £300,000 per steamer, and entail their withdrawal from the berth for a considerable period.' See 'Correspondence as to Sale of Shares', p. 384: Cunard Archives, D42/A9/10.

5 See, for example: 'Particulars of Ships', Harland & Wolff Archives, Queen's Road, Belfast. These indicated 51,000 horsepower and 'estimated consumption 825 [tons]' with 'estimated speed' 21 knots. A discussion of this and other related sources can be found in Beveridge, Bruce, et al. (eds), *Titanic: The Ship Magnificent*.

6 Parsons, the Hon. Charles A., & R.J. Walker, 'The Institution of Naval Architects. Combination of Reciprocating Engines and Turbines', *Engineering*, 17 April 1908, pp. 502–03 and 511–14. I am grateful to Ralph Currell for drawing my attention to this article.

7 Bain, J., 'Untitled Memorandum from the General Superintendent, the Cunard Steamship Company Limited, to the Secretary, Liverpool,' 5 August 1909. In: 'The Cunard Steamship Company Limited: Minutes of Executive Committee Meetings' p. 308: Cunard Archives, B4/30.

8 Moss, Michael, & John R. Hume, *Shipbuilders to the World*, pp. 133–35.

9 'Ship's Log: HM Ship *Laurentic*. Commencing 25 November 1914 Ending to 5 September 1915': The National Archives (UK), ADM 53/46282.

10 '9 Liners Building for Ship Combine', *New York Times*, 10 August 1907.

11 '*Mauretania* Goes 30.5 Miles an Hour', *New York Times*, 22 September 1907.

12 'The Thirty Knot An Hour Liner and What it Means', *New York Times*, 15 September 1907.

13 'The *Laurentic*'s Trials. Successful Experiments with Turbine Machinery', *Belfast Evening Telegraph*, 1 February 1909, p. 3. See also: Ed Coghlan, 'It Said In The Papers', White Star Journal, Vol. 17, No. 1 (2009) pp. 10–11.

14 Bain, J., 'Untitled Memorandum from the General Superintendent'.

15 'Combination of Reciprocating and Turbine Machinery', *Engineering*, 30 April 1909, pp. 598–99. I am grateful to Ralph Currell for sharing with me a copy of this article.

16 Currie, J., & A. Galbraith, 'Technical Subjects Bearing on Atlantic Shipping', 24 August 1910. In: 'The Cunard Steamship Company Limited: Minutes of Executive Committee Meetings', p. 334: Cunard Archives, B4/32.

17 Willett Bruce, W.J., 'Marine Propulsion: Evolution in North Atlantic Liners' (abstract of a lecture given before the Engineering Section of Liverpool University), Supplement to the *Journal of Commerce and Shipping Telegraph*, 14 February 1929.

18 Pounder, C.C., 'Some Notable Belfast-Built Engines', p. 23. Comments from Superintendent Engineer W.J. Willett Bruce were reported in *Transactions* (Institution of Naval Architects) in 1928. He was a strong advocate of the combination system, which Harland & Wolff used on the second *Laurentic*, completed in 1927: 'As far back as 1911–12 we had comparative data of two steamers of similar construction and displacement running in the same service, but one having quadruple expansion engines and the other with four crank triple engines exhausting into a centre low-pressure turbine, resulting in a 21 per cent gain in economy and an accelerated speed of 1 knot.' They were cited by Kohler, Peter C., in 'White Star & Maple Leaf: RMS *Laurentic*', an excellent online monograph published on his blog 'Wanted On Voyage: Ocean Liner Monographs', 14 March 2024. An article by W.O. Horsnaill, 'Machinery for Passenger Liners: Recent Ships Built and Building', published in the Liverpool *Journal of Commerce* on 18 August 1927, argued that, 'It is only the *Laurentic* [the second ship of the name, completed that year], which burns coal, that has any chance of competing with existing motorships in fuel economy. A consumption of 1.2lbs of coal per indicated horsepower per hour for main engines and auxiliaries has been achieved by plain quadruple expansion engines with ordinary boiler pressures. With a turbine for the low pressure stage better results should be obtained. A figure that can reasonably be expected for the machinery in the *Laurentic* is 1.1lb. per indicated horsepower per hour.'

19 I am grateful to Scott Andrews for a fascinating discussion on this subject, 29 June 2014.

20 Letter from William Watson, Cunard's chairman, to Vernon Brown, Cunard's New York agent, dated March 1908: Cunard Archives, S5/3.

21 'Notebook containing propeller data for ship numbers 387–390 (*Beagle*, *Bulldog*, *Foxhound*, *Bristol*), *Indefatigable* (391), *Brisk* (395), *Petersburg* (397), *Yarmouth* (398), *Argyllshire* (399) also the centre propeller (turbine driven) for *Olympic* and *Titanic* built by Harland & Wolff', 1909–11: University of Glasgow Archive Services, GB 248 UGC 100/3/7. I am grateful to Joao Goncalves for bringing this source material

to my attention in January 2020. For further reading, see: Chirnside, Mark, '*Titanic*'s Centre Propeller: The Stephen Pigott Evidence', *Atlantic Daily Bulletin*, June 2020, pp. 14–19.

22 Details of the *Titanic* turbine inspection, taken from the Board of Trade Survey, 19 July 1910: 'J. Brown & Co.'s Low Pressure Turbine H.W. 4 for Messrs. Harland & Wolff's SS 401 *Titanic*': The National Archives (UK), MT 15/212.

23 See Beveridge, et al., *Titanic: The Ship Magnificent: Volume One: Design & Construction*, p. 283: table headed 'Comparative Fuel Consumption – Common Propulsion Systems *c.* 1912'. The sources cited in compiling the data include: Griffiths, Denis, *Power of the Great Liners* (Patrick Stephens Ltd, 1990) and Baker, G.S., *Ship Form, Resistance and Screw Propulsion* (D. Van Nostrand Co., 1915).

There are inevitable differences in coal-consumption data depending on the source material. Denis Griffiths, for example, also cites (p.105) a consumption of 1.43lb/SHP/hr for *Lusitania* (direct-drive compound turbines), which matches Thomas Bell's data presented in 1908 for her actual consumption on trials; 1.4lb/SHP/hr for *Olympic* (combination triple-expansion reciprocating and turbine); and 1.38lb/SHP/hr for *Aquitania* (1914) (direct-drive triple-expansion turbines). On this basis, the coal consumption for *Olympic* would be only slightly lower than *Lusitania* on a unit basis, even if her total consumption would be significantly lower because her propelling machinery was not as powerful. Griffiths cited as a source, T.W.F. Brown, 'Marine Engineering Review', *Transactions Royal Institute of Naval Architects* (22 March 1960), Tables 4 and 6. This does seem at odds with the actual performance data from *Olympic* in service (see the coverage of her maiden voyage elsewhere in this book). On the basis of Harland & Wolff's original expectation that 51,000 horsepower would be required for *Olympic* to achieve 21 knots (with a total coal consumption of 825–850 tons per twenty-four hours), she would have consumed 1.48lb/SHP/hr. She did substantially better, averaging a consumption of 629 tons and a speed of 21.43 knots on her maiden voyage.

Turbine engine power output was expressed in shaft horsepower (SHP), whereas reciprocating-engine power output was expressed in terms of indicated horsepower (IHP). This creates complications when dealing with a ship with combination propelling machinery. Many references simply refer to 'horsepower' – apparently on the basis of adding up the IHP of the reciprocating engines and the SHP of the turbine and without clarification. Cuthbert Coulson Pounder provided Harland & Wolff data for a round voyage *Olympic* made in July 1927, when the propelling machinery developed '58,000 IHP + SHP'; he assumed 'an equivalent total of 54,500 SHP' with an 'all-purposes fuel rate' of 0.97lb/per hour/per 1 SHP (this is in line with earlier *Olympic* coal-consumption data, allowing for the conversion to oil fuel in 1919–20, which was more efficient in terms of energy per fuel unit). Pounder's figures reflect that the IHP of a reciprocating engine will be somewhat less in SHP terms, so *Olympic*'s power output on a SHP equivalent basis would be lower.

24 'Memo to the General Manager: Proposed New Steamer', 23 August 1910. Cunard Archives, S7. These statistics should be taken as an approximate indication only. They were based on a model experiment conducted by John Brown & Co. on a ship assumed to be 850ft long between perpendiculars (the same as *Olympic*) and 93ft in breadth (6in more than *Olympic*). The ship was assumed to have a 35ft draft and a displacement of about 47,500 tons. However, the model's lines were 'an expanded form of those used for *Lusitania* and *Mauretania*', which were expressly designed for speed and were finer than *Olympic*: 'It was found that under these conditions it would require about 55,000 horsepower for a speed of 23 knots.'

25 The figures for *Celtic* are taken from 'The White Star Liner *Celtic*', *Scientific American*, 17 August 1901, p. 103. *Adriatic*'s are from a count of the crew engaged for her westbound crossing leaving Southampton on 23 February 1910. The *Olympic* figures are based on the number of Engine Department crew carried on her maiden voyage, but Lester Mitcham's research into *Titanic*'s crew confirmed a total of 325 in her case (see, for example: Halpern, Sam, et al., *Report into the Loss of SS Titanic: A Centennial Reappraisal* (History Press, 2011) p. 67.) There were always slight variations from voyage to voyage, however, the trend is very clear.

26 Mersey, Lord J.B., 'Report of a Formal Investigation', Question 19,081.

5: 'The Passengers Have Never Been More Comfortable'

1 'The *Oceanic* at Queenstown', *Irish Times*, 8 September 1899, p. 5.

2 Letter from 'Secretary' to H.A. Peto, Esq., Bradford on Avon, dated 13 September 1905, 'Dear Sir…': Cunard Archives, S7/1/16. Cunard were far from extravagant: 'Everything in connection with the designing and decorating of the first class accommodation is to be of the best character, workmanship and material, but the directors do not desire anything of an extravagant nature. The simpler the work, so long as it is good, the better, and the more the cost can be kept down the more satisfactory it will be to the Cunard company provided that the general effect is tasteful and pleasing.

'In preparing designs the structural construction of the ship must have the first consideration – as this cannot be interfered with except with the consent of Messrs. Swan, Hunter & Wigham Richardson and the Cunard Company. Great importance is attached to keeping down weight as much as possible.'

3 'Mr dear Ismay', from William Watson, 5 December 1908: Cunard Archives, S5/3.

4 Ibid., pp. 111–13.

5 Thanks to Daniel Klistorner and Jerry N.J. Vondeling. See: Lobell, Hakan, 'Foreign Exchange Rates 1804–1914' in Jacobson, Tor, Rodney Edvinsson & Daniel Waldenstrom (eds), *Historical Monetary and Financial Statistics for Sweden* (Ekerlids Forlag, 2010); Koninklijke Nederlandsche Meubelenfabriek, H.P. Mutters & Zoon, Den Haag, Behangerij en Stoffeerderij, 'Begrooting voor Holland Amerika Lyn, Rotterdam, betreffende: Betimmering & Meubileering [Panelling & Furniture] SS *Statendam*', 15 Februari 1913: Stadsarchief Rotterdam, EB 1133. The prices that the firm quoted in February 1913 to outfit thirty-two deluxe staterooms on the Holland America Line's new *Statendam*, then being built by Harland & Wolff, amounted to 136,000 Dutch guilders. At the exchange rate with the British pound then, it was over £11,300. The price per stateroom would be over £350.

6 'Remarks in Connection with the Outline Specification of a Steel Twin-Screw Steamer to be Built for the

Holland-America Line by Messrs. Harland & Wolff Ltd': Stadsarchief Rotterdam, EB 1125.

7 'Harland & Wolff Ltd. Minutes [of] Managing Directors', pp. 250 and 292: Public Record Office Northern Ireland, D/2805/MIN/A1.

8 Ibid., p. 246.

9 Munsey, Frank A., 'Giant Liners of the North Atlantic', *Munsey's Magazine*, Vol. LII, No. 1 (June 1914) p. 22.

10 This is based on the 'Design "D"' concept plans which were approved in July 1908, compared with later general arrangement plans. I am grateful to the helpful staff of National Museums Northern Ireland, including Michael McCaughan, George Wright and many others.

11 'SS Nos 400–1: General Arrangement: 850 feet x 92 feet x 64 feet 6 inches to Shelter Deck. Scale 1/16 = 1 foot', 31 May 1910. Harland & Wolff Ltd., Shipbuilders and Engineers. Ulster Folk & Transport Museum.

12 I am grateful to Roy Mengot for sharing his analysis of citations on the various plans that David Livingstone shared with him.

13 'Extracts from a Letter': Cunard Archives, S7/1/4.

14 'Specification of a Steel Quadruple-Screw Steamship with Turbine Engines [*Lusitania* and *Mauretania* generic specification]', John C. Morrison, Naval Architect's Department, 6 January 1906: Merseyside Maritime Museum, B/CUN/3/4/3.

15 'Cunard Line: New York Liverpool Service', March 1912.

16 Sauder, Eric, & Ken Marschall, *RMS Lusitania: Triumph of the Edwardian Age* (Transatlantic Designs, 1991) p. 22.

17 See, for example, 'The *George Washington*. The New Norddeutscher Lloyd Steamer', *Scientific American*, Supplement No. 1747 (26 June 1909) pp. 408–09.

18 Peskett, Leonard. 'Rough Notes for Consideration when discussing proposals for new steamers for Liverpool and New York service'. January 1909. Cunard archives. S7, pp. 8–9.

19 Jones, B. Sydney, 'RMS Ivernia. 24 May 1904': Cunard Archives, S7/1/23.

20 There were additional suggestions that were carried through, such as 'a lift for passengers from the lower to each deck above'. Cunard were not necessarily keen on other amenities. In May 1912, the company decided to remove the Turkish baths from *Aquitania*'s accommodation, but she would have a swimming pool.

21 All of these recommendations and observations were submitted to Cunard's Board of Directors after they requested feedback from their staff. They were fed back so that they could be considered and/or incorporated into the designs of *Lusitania* and *Mauretania*. See correspondence in S7/1/23, Cunard Archives.

22 'Reports of the Immigration Commission: Steerage Conditions' (Vol. 37) (Washington: Government Printing Office, 1911). Under the leadership of William Paul Dillingham, the Republican Senator from Vermont, the United States Immigration Commission is often known as the Dillingham Commission. Their findings helped provide the foundation for immigration restrictions in the 1920s. Although Anna Herkner and *Cedric* are not named in the report itself, the available evidence seems clear that she was on board the White Star liner. The narrative is based primarily on pages 6 to 29.

23 'Reports of the Immigration Commission: Immigration Legislation' (Vol. 39) (Washington: Government Printing Office, 1911) pp. 413–29. The quotes from pp. 125–30 are all sourced from here.

24 Gerahty, G.M., 'Crossing the Atlantic – How the Days Are Spent', *The Sphere*, 19 August 1905, p. 178.

25 See Stormer, Susanne, 'Marketing-Kommunikation Zur Markteinfuhrung der Schiffe der *Olympic*-Klasse 1908–1911' (Bachelor Thesis, BA Bachelor of Arts, an der Europaischen Fernhochschule Hamburg, 24 May 2007) p. 25.

26 One crude method to illustrate the difference is to compare the number of passengers per ton of commercial space on each vessel (the space available for passenger accommodation and cargo):

	Net Tonnage	Total No. Passengers	Passengers Per Net Ton	Ratio of First & Second Class to Third Class
Majestic (1890)	4,443	1,490	0.33	0.49
Oceanic (1899)	6,917	1,710	0.25	0.71
Mauretania (1907)	12,797	2,335	0.18	0.80
Baltic (1904)	15,295	2,875	0.18	0.44
Aquitania (1914)	21,993	3,230	0.15	0.62
Olympic (1908 'Design "D"' concept)	20,894	3,104	0.15	0.73
Olympic (1911)	20,894	2,435	0.12	1.37
Britannic (1914)	24,592	2,579	0.10	1.71

27 Watson, William, Cunard Archives, S5/3, p. 99.

28 'Report from Supt. Catering Department': Cunard Archives, GM29/1.

29 'My dear Sir', undated letter by Adolphe Smith, written after November 1908: Cunard Archives, GM29/1.

30 'No More "Ship Diet," Says *Amerika* Caterer', *New York Times*, 22 October 1905.

31 J. Bruce Ismay was writing to G.C. Cigolini, Esq. at Claridge's Hotel Restaurant, London. His letter was in response to one he had received from them three days earlier. They had apparently been offering the White Star Line advice: 'I note your view that the only way in which such a restaurant can be kept up-to-date is by having it run by a large company, such as the Savoy, and when we come to consider its organisation, I will bear your suggestion in mind. I further note that the managing director of the Savoy company would place himself at my disposal, and should we wish to consult him, we will not hesitate to ask his advice.'

32 Quimby, Harriet, 'How the Ocean Traveller is Fed', *Leslie's Illustrated Weekly Newspaper*, 13 June 1912. Thanks to Ioannis Georgiou for providing a copy of the original article from his collection.

6: 'We Have Reached the Limit'

1 'The Next Thing in Liners', *Belfast News-Letter*, 18 May 1909. The article was cited by Alan Geddes.
2 Franklin, P.A.S., 'New York and the Atlantic Shipping', *New York Times*, 10 January 1909, p. 16.
3 'Sixth Annual Report of the International Mercantile Marine Company, for the Fiscal Year Ended December 31st 1908', Office of the International Mercantile Marine Co., 51 Newark Street, Hoboken, N.J. [New Jersey], 30 June 1909.
4 'Installments Due': Public Records Office of Northern Ireland, D2805/SHIP/2.
5 For example, on 12 May 1908 the bonds were selling for 72.25 cents on the dollar (*Sheffield Daily Telegraph*); on 7 July 1908, about 70 cents on the dollar (*Yorkshire Post*).
6 'Seventh Annual Report of the International Mercantile Marine Company, for the Fiscal Year Ended December 31st 1908', Office of the International Mercantile Marine Co., 51 Newark Street, Hoboken, N.J. [New Jersey], 30 June 1909.
7 Ibid., pp. 5–6.
8 'The Oceanic Steam Navigation Co. Ltd.: [...] £1,250,000 4½ I First Mortgage Debentures', 5 October 1908: The National Archives (UK), BT 31/30783/4540.
9 Ibid.
10 'Seventh Annual Report of the International Mercantile Marine Company', 30 June 1910.
11 Pirrie, William, to Lord Dufferin, June 1898: Public Records Office of Northern Ireland, D1071/H/B/P/268/1-24.
12 These entries are found in a 'Book of Monthly Office Accounts of J. Bruce Ismay in account with the Oceanic Steam Navigation Co.', 'Monthly Office Accounts No. 3': Merseyside Maritime Museum, D/B/174/1.
13 Financial comparison data has been obtained from IMM's annual reports for 1903 through to 1910 inclusive. The comparison for 1910 comes from White Star's reported net earnings for that year, in British pounds, converted into US dollars at an exchange rate of £1 to $4.86, divided by IMM's reported net earnings.
14 See Navin, Thomas R., & Marian V. Sears, 'A Study in Merger', pp. 291–328. Their detailed study of IMM is full of valuable information and data from a financial history perspective. They argued that White Star's shareholders 'would have been much better off if they had not sold'. The healthy dividends they received each year, including when companies such as Cunard were suspending their dividends, came to an end with the company's absorption into the IMM combine. Moreover, the proportion of the sale price which was paid in IMM securities did not turn out well: 'White Star's owners would have been much better off had they not sold. They had a property easily worth $32,000 a share and believed they were being paid for that property about $50,000 a share in cash and securities. But the securities turned out to be worth considerably less than expected and so the value received was nearer $25,000 (at average 1907–13 market quotations). Had they *not* [original emphasis] sold they might easily have continued to receive their regular dividends and might even, in view of their company's remarkable earnings record, have been able eventually to liquidate their holdings for as much as or more than the nominal $50,000 paid for their shares by IMM. As it was they received half of what they had bargained for and no dividends at all; White Star paid dividends to IMM but IMM paid nothing.'
15 Clarke, John, & Margaret, 'The International Mercantile Marine Company: A Financial Analysis', *The American Neptune*, Vol. 57, No. 2 (Spring 1997) p. 147.
16 Ibid. In 1912, IMM's net earnings per gross ton stood at $7.10; Norddeutscher Lloyd's at $14.27; HAPAG's at $16.88; and Cunard's at $18.63. My own estimate for White Star is $12.27. In 1912, IMM's bonded debt stood at $72.19 per ton, followed by Cunard at $65.44; Norddeutscher Lloyd at $22.69 and HAPAG at $17.97. My own estimate for White Star as a standalone entity is $17.33.
17 Saliers, Earl A., 'Some Financial Aspects of the International Mercantile Marine Company', *Journal of Political Economy* (1915) pp. 910–25; 'Atlantic Rate War. White Star "Cut" Cunard's Response, *Lloyd's List & Shipping Gazette*, 10 January 1908; 'The Atlantic Rate War', *Journal of Commerce*, 11 February 1908; Watson, William, to Brown, Vernon H., 23 October 1907 (copy of J. Bruce Ismay's letter to Dr Wiegand attached): Cunard Archives, S5/2.
18 'New Queens of the Atlantic To Be Marvels of Luxury', *New York Times*, 3 March 1907, p. 9.
19 'American Shipping Notes. Groundings at New York', *Lloyd's List & Shipping Gazette*, 30 May 1907, p. 13.
20 Extract from a letter, dated 31 May 1910, from Charles P. Sumner, Esq., manager of the New York office, to A.A. Booth, Esq., chairman, through to correspondence dated 17 December 1910: Cunard Archives, S7/3/4.
21 '*Mauretania* Stuck in Ambrose Channel', *New York Times*, 4 November 1910, pp. 1–2.
22 Franklin gave a number of press interviews and reports about his efforts to get the piers extended were covered in a number of newspapers such as the *New York Times*. For further details, see: 'Asks National Aid for Montauk Port', *New York Times*, 9 December 1910; 'May Yet Get Piers for New Big Ships', *New York Times*, 4 January 1911; 'Pier Extensions Up to Harbour Board', *New York Times*, 24 January 1911; 'The Future of New York as a World Harbour', *New York Times*, 29 January 1911; 'May Grant Chelsea Pier Extensions', *New York Times*, 12 February 1911; 'Permits Long Piers for the New Liners', *New York Times*, 7 March 1911.
23 Southampton Harbour Board, Minute Books and Reports: 1908–09, pp. 236–37: Southampton City Archives, SHB 1/2/13.
24 Ibid., pp. 270–71.
25 Southampton Harbour Board, Minute Books and Reports: 1909–10, p. 69: Southampton City Archives, SHB 1/2/14.
26 Ibid., p. 238.
27 Ibid., p. 240.
28 Ibid., pp. 240–42.
29 Ibid., pp. 242–43.
30 Ibid., p. 244.
31 Ibid., pp. 245–49.
32 Southampton Harbour Board, Minute Books and Reports: 1910–11, p. 219: Southampton City Archives, SHB 1/2/15.
33 Ibid., pp. 282–83.
34 Wentworth-Shields, Francis Ernest, 'The Construction of the "White Star" Dock and Adjoining Quays at Southampton' (London: Institution of Civil Engineers, 1914) pp. 4, 14, 45, 47, 76 and 77.
35 'London & South Western Railway: Inspection of New Deep Water Dock Southampton, Thursday, 8 June 1911': The National Archives (UK), RAIL 411/661.

36 Ibid.
37 'Southampton's New Dock: London and South-Western Enterprise', *The Mariner*, Vol. XXXV, No. 9 (15 June 1911) p. 221.
38 Dodd, George H., 'Supplementary Report: Report of Cherbourg', 12 August 1908, in: 'The Cunard Steamship Company Limited: Minutes of Executive Committee Meetings', p. 18: Cunard Archives, B4/29.
39 'American Mail Service: Omission of Call at Queenstown by Cunard Packetts' (London: His Majesty's Stationery Office, 1914) pp. 6–7.
40 Ibid. See 'Appendix B: Particulars of the Entrances to 12 of the Principal Deep Water Harbours in the World'.

7: 'Something Awe-Inspiring'

1 'Building of Giant Liners', *New York Times*, 10 July 1910, p. 4.
2 'New Marine Giants: *Olympic* and *Titanic* to be Launched at Belfast', *Manitoba Free Press*, 30 September 1910, p. 14.
3 'Harland & Wolff Ltd. Minutes [of] Managing Directors', p. 121: Public Record Office Northern Ireland, D/2805/MIN/A1.
4 Ibid., pp. 124–25.
5 Ibid., pp. 134, 136–39.
6 Currie, W.A. ,'Belfast Harbour Notice to Mariners', 11 October 1910: Public Record Office Northern Ireland, HAR 1A/2/220.
7 'Memo for Launches', c. 1910: Public Record Office Northern Ireland, D/2805/C1/1.
8 The launch was covered in detail by many newspapers and quotations are often attributed directly in the text. However, I have relied particularly on: 'Largest Ship in the World', *Daily Mail*, 21 October 1910, p. 5; 'Launch of the World's Biggest Ship', *Irish Times*, 29 October 1910, p. 6; 'The World's Biggest Ship', *Belfast News-Letter*, 21 October 1910, p. 6; 'The World's Biggest Ship', *Belfast Weekly News*, 27 October 1910, p.8. See also: Warren, Mark D. (ed.), *Distinguished Liners from The Shipbuilder 1906–1914, Volume 1*, p. 128.
9 See *Olympic* Launching Curves Diagram and Displacement Scale: National Museums Northern Ireland, HOYFM.2010.73.84.
10 'Lowering Boilers into the *Olympic*', *The Irish Times*, 30 December 1910, p. 7.
11 'Harland & Wolff Ltd. Minutes [of] Managing Directors', p. 164: Public Record Office Northern Ireland, D/2805/MIN/A1.
12 See 'Obituary: Joseph Bell, Chief Engineer', *The Marine Engineer & Naval Architect*, Vol. XXIV (May 1912) p. 381. Another source is 'The Engine Room As A Training Ground', *Shipbuilding & Shipping Record*, June 1914, p. 737. J. Bruce Ismay testified about Bell being sent over to Belfast. See also Cunard Archives, EN1/1.
13 Engine Works Management to Shipyard, 'No. 400 ss', 2 March 1911. This document was made available by White Star Memories in July 2021.
14 On 15 February 1911, Harland & Wolff wrote to W.A. Currie, Secretary of the Belfast Harbour Commissioners. They advised that their plan at that stage was to dry dock *Olympic* on Thursday, 30 March 1911 and desired 'to have everything in readiness in good time. We shall require by the 27 March, for placing in position, 80 cast iron blocks for use as bilge blocks, i.e. 40 each side, in addition to the 322 keel blocks already provided for the centre line'. Following correspondence in May and June 1910, the shipbuilders thought that the requirements were fully understood, but they were worried that there was 'still some question as to the number of metal blocks that will be available for bilge blocks'. Unfortunately, it transpired that no orders had been placed for cast-iron bilge blocks. Seven days later, Harland & Wolff had discussed the problem with the engineer and harbour master, and at his request, the firm said they were willing 'to supply 66 complete sets of keel block castings (wood not included)'. The engineer confirmed that fourteen sets could be supplied, so together with the sixty-six that Harland & Wolff were making, there were enough for the task: Public Record Office Northern Ireland, HAR/1E/1/2/4.
15 'Dry-docking the *Olympic*. Gigantic Undertaking', *The Irish Times*, 3 April 1911.
16 'Inquests in Belfast: Fatalies at the Queen's Island', *Belfast News Letter*, 7 June 1911, p. 9.
17 '*Olympic* in Dry Dock', *The Irish Times*, 15 April 1911, p. 9.
18 Ibid.
19 *The Shipbuilder: The White Star Liners Olympic & Titanic Souvenir Number*, Vol VI (Midsummer 1911) p. 129. E.C Parker & Company (Services) Ltd, 2009.
20 Mersey, Lord J.B., 'Report of a Formal Investigation', Question 20,491.
21 'Harland & Wolff Ltd. Minutes [of] Managing Directors', p. 207: Public Record Office Northern Ireland, D/2805/MIN/A1.
22 'Big Shipowning', *Syren & Shipping*, 24 May 1911, p. 166.
23 'The *Olympic* and the *Titanic*: Two Giant Ocean Steamships', *Scientific American*, Supplement No. 1850 (17 June 1911) p. 380.
24 'Vessels Leaving', *c.* 1910: Public Record Office Northern Ireland, D/2805/C1/1.
25 Fifth Officer Lowe described *Titanic*'s trials in April 1912, much of which were apparently the same as her older sister but over a shorter period. It appears Harland & Wolff tended to use the same practices on trials for many different ships. See: 'Report of the Committee on Commerce United States Senate Pursuant to S. Res. 283 Directing the Committee on Commerce to Investigate the Causes Leading to the Wreck of the White Star Liner *Titanic*' (Washington: Government Printing Office, 1912).
26 This description is based upon Edward Wilding's testimony at the Limitation of Liability Hearings into *Titanic*'s trials in April 1912. During his testimony, he described *Titanic*'s manoeuvres in some detail, including the comparative speeds of both ships when they were tested to see how long it took for them to stop; where *Titanic*'s test was performed, which he said was the same place as *Olympic*; how the engines were handled on both occasions; and how long each ship took to stop. He stated, 'The engines were rather more smartly handled on the occasion of the *Olympic* than the *Titanic*, and she stopped in just about the same distance; rather less, although the speed was higher.'
27 *The Shipbuilder: The White Star Liners Olympic & Titanic Souvenir Number*, Vol. VI (Midsummer 1911).
28 'Departure of SS *Olympic* and Launch of SS *Titanic*', 23 May 1911: Public Record Office Northern Ireland, T3893/1.
29 'Belfast Shipbuilding Triumph', *The Northern Whig*, 1 June 1911, p. 7
30 *The Northern Whig*, 1 June 1911, p. 8.
31 'Launching Curves SS. No. 400': National Museums Northern Ireland, HOYFM.2010.73.17; and 'Launching

Curves' from 418–34, National Museums Northern Ireland, HOYFM.2010.73.75.

32 'Inquests in Belfast: Fatalies at the Queen's Island', *Belfast News-Letter*, 7 June 1911, p. 9.

33 'Half-Yearly Agreement and Account of Voyages and Crew of A Ship Engaged In the Home Trade Only', *Olympic*, Official Number 131,346, 30 June 1911: The National Archives (UK), BT 100/245.

34 'White Star RMS *Olympic*. Elkington & Co., Ltd', *The Journal of Commerce*, 1 June 1911, p. 6.

35 'Hocking & Co., Ltd. Liverpool', *The Journal of Commerce*, 1 June 1911, p. 7.

36 'Making Ships' Telegraphs', *The Journal of Commerce*, 31 May 1911, p. 19. William Joseph Willett Bruce joined the White Star Line in 1881 at the age of 20, where he enjoyed a successful engineering career. After fourteen years' seagoing experience, he became Assistant Superintendent Engineer, selected from a field of 400 applicants. By 30 September 1904, he had earned promotion to Superintendent Engineer, devising a new system for training young engineers, 'which was highly thought of', while his Automatic Whistle Control was 'generally adopted by many leading steamship lines'. The device, which was controlled electrically from the bridge, had a timing mechanism so the whistles could be blown automatically at regular intervals during fog. Willett Bruce was also an advocate of the combination of reciprocating engines and a low-pressure turbine. In June 1915, a profile looked back on his career and concluded that he 'carries his many responsibilities as he has done his many strenuous tasks, with a virile and seemingly boundless energy, one feeling always the effect of his urbane and courteous personality when in contact with him. Always willing to impart his wise counsel or advice to those in need of it, he has also made it a principle to listen to any suggestion from any of his men; this faculty, so developed in his career, has done much to crown it with such marked success and usefulness to the mercantile marine, to the British flag afloat and to marine engineering throughout the world.' From 1914 to 1927, he was a member of the Institution of Naval Architects (INA). Awarded the Order of the British Empire (OBE) in 1918, he was promoted to Engineer Captain in the Royal Naval Reserve in 1923, before retiring in 1927. He died in West Sussex in 1953. (See 'Superintendent Engineer: White Star Line', *The Marine Engineer and Naval Architect*, June 1915, p. 323, and Sear, Ruth, 'William Joseph Willett Bruce', *Atlantic Daily Bulletin*, September 2011, pp. 14–17.)

37 'Boltless Improved Boiler Furnace Front'; 'The Utley Sidelights and Patent Windows'; 'Immense Evaporators of the Quiggin Type'; 'Silent Ash Hoists and other work by Railton, Campbell and Crawford', *The Journal of Commerce*, 31 May 1911, pp. 18 and 19.

38 'The Utley Sidelights and Patent Windows', ibid.

39 Peskett, Leonard, 'Memorandum to the General Manager. SS *Olympic*', Executive Committee, 28 June 1911: Cunard Archives, S7/3/75.

40 Peskett also noted, 'When passing through the [à la carte] restaurant it was observed that temporary shores had been fitted between the decks, which appeared to indicate that some trouble had been experienced in that particular part of the ship.' Although *Olympic* proved generally free from vibration, even at high speed, it seems probable that some local stiffening was being applied to the existing deck beams overhead (the underside of A-Deck). The restaurant was, after all, one of the most exclusive apartments on board ship and it was imperative that first-class passengers who had paid an extra tariff enjoyed their meals in comfortable and relaxing surroundings.

41 Currie, J., & A. Galbraith, 'Memorandum to the General Manager', Executive Committee, 21 June 1911: Cunard Archives, S7/3/75.

42 'Harbour Dues Log Book'. Cunard archives. Gm 13/6.

43 Bowyer, George, *Lively Ahoy: Reminiscences of 58 Years in the Trinity House Pilotage Service* (H.B. Broadbere, 1930) p. 35.

8: 'The Greatest Ship in the World'

1 'Movements of Liners', *The Globe*, 3 June 1911, p. 2. The full entry read, 'White Star – *Olympic*, from Liverpool, arrived Southampton 11.35 p.m. yesterday'. The definition of her 'arrival' was not clear and it may be when she came into the Solent, as most reports had her arriving in the early hours of 3 June 1911.

2 *Southampton Times* coverage, June 1911.

3 'The Largest Ship in the World: Motherwell's Share in its Construction', *Motherwell Times*, 16 June 1911, p. 7.

4 Beaumont, Dr J.C.H., *Ships – and People* (Geoffrey Bles, 1926) pp. 96–7.

5 'The Troubles With the *Olympic*', *Justice*, 17 June 1911, p. 5.

6 'Statement by Mr. Russell Rea, MP', *Daily Express*, 12 June 1911.

7 Curry's comments to the *Southern Daily Echo* were also reported elsewhere. See 'The Case of the *Olympic*', *Northern Daily Mail*, 15 June 1911.

8 'The Victualling of an Ocean Liner', *Railway & Travel Monthly*, Vol. 3, No. 20 (December 1911), p. 477.

9 'No. 8 – White Star and American Lines' Laundry', *Southampton and District Pictorial*, 29 May 1912, p. 5.

10 Hyslop, Donald, Alistair Forsyth & Sheila Jemima, *Titanic Voices: Memories from the Fateful Voyage* (Chancellor Press, 2000) pp. 48–50.

11 'The Cunard Steamship Company Limited: Minutes of Board Meetings', 1910, p. 154: Cunard Archives, B1/8. Although Captain Smith seems to have earned considerably more than his Cunard counterpart, there is other evidence to suggest that his reported annual earnings of £1,250 may be correct. In November 1914, his successor as *Olympic*'s commander, Captain Haddock, had been selected 'to be Commodore Commanding the 10th Squadron for special service' by the Admiralty. A draft letter to him confirming his appointment stated, 'Your emoluments while employed on this service will be your pay as captain of the *Olympic* + 15%', but an official had amended it in green to suggest using a 'per annum' figure, noting '£1,500 is what Commodore Haddock should get'. It is not clear if the £1,500 figure is precisely the same as Haddock's previous salary plus 15 per cent. However, if it had been, then his income as *Olympic*'s commander in 1914 would have been about £1,300. (See: 'RNR: Captain H.J. Haddock RNR. Pay and Conditions: Question of Emoluments', November 1914: The National Archives (UK), ADM 1/8402/410.) The assistant commander role, which was introduced early in 1914, was paid £500 (about £42 monthly), but it was specified as per annum rather than monthly. (In December 1913, Captain Charles Bartlett had recommended that the assistant commander role have a £400 annual salary plus a £100 performance-related bonus.)

12 Bäbler, Günter, *Guide to the Crew of Titanic: The Structure of Working Aboard the Legendary Liner* (The History Press, 2017) pp. 122–27.
13 'Master's Certificate No. 14102 Edward John Smith': Guildhall Library, MS 18569/35.
14 Mersey, Lord J.B., 'Report of a Formal Investigation', Question 19,105.
15 Chirnside, Mark, *The 'Big Four' of the White Star Fleet: Celtic, Cedric, Baltic & Adriatic* (The History Press, 2016) pp. 76–78.
16 'Official Log Book. SS *Adriatic*', pp. 31–40: The National Archives (UK) BT 165/469.
17 'Long Sea Service: Death of Captain J. Evans, Caernarvon – Well Known Master Mariner', *Caernarvon and Denbigh Herald*, 26 March 1937, p. 10. Thanks to Inger Sheil for kindly sharing this article.
18 'Master's Certificate No. 028211 Robert Hume': Guildhall Library, MS 18569/19.
19 The majority of the White Star Line officer appointments for the junior officers are taken from the White Star Line's volumes containing their officer records. However, there are often slight discrepancies in dates of appointment or resignation, compared to their papers at the Guildhall Library. The main objective here has been to demonstrate career progression. See Cunard Archives, ASC/12/3/1 and ASC 12/3/2.
20 Bäbler, Günter, *Guide to the Crew of Titanic*, pp. 36, 63–64 and 106–07.
21 'An "Orphaned" *Olympic*: Captain & Chief Engineer Retire Together', unidentified newspaper cutting, c. December 1929, kindly provided with other information from John H. Thearle's grandson, Nick Thearle. See also: Chirnside, Mark, *RMS Olympic: Titanic's Sister* (The History Press, 2015).
22 Charles William McKimm (1876–1945) was the third child of Charles McKimm (1848–1907), who was the head gardener and curator for the Royal Belfast Botanic Gardens for twenty-nine years. His older brother, John Jephson McKimm, born on 15 December 1873, also went into engineering. I am grateful to Jenny McBriar for sharing information about her family's history.
23 See 'Obituary. J.H. Hesketh' and 'Arthur Ward', *The Marine Engineer & Naval Architect*, May 1912, p. 382; and 'Obituary. Peter Sloan', *The Marine Engineer & Naval Architect*, June 1912, p. 424.
24 Jessop, Violet, & John Maxtone-Graham (ed.), *Titanic Survivor: The Memoirs of Violet Jessop Stewardess* (Sutton Publishing, 1997) pp. 102–03.
25 Bäbler, Günter, *Guide to the Crew of Titanic*, pp. 28–30.
26 I am grateful to Günter Bäbler for drawing my attention to two articles covering this topic: 'Maritime Wireless Telegraphy', *The Marconigraph*, June 1911, p. 28, and 'The Equipment of SS *Olympic*', *The Marconigraph*, July 1911, pp. 27–28.
27 Bruce, Mr, 'Sea Post Offices: Report of Inspection by Mr. Bruce, Controller London Postal Service', p. 49, January 1909: British Postal Museum & Archive, POST 43/321.
28 'General Post Office: Transatlantic Sea Post Offices. White Star and American Lines. Quarters for Sorters: Suggested Provision of Second Instead of First Class Accommodation. No Action Taken. 13–17 June 1911': British Postal Museum & Archive, POST 29/1117.
29 'General Post Office: Transatlantic Sea Post Offices: White Star Line: SS *Olympic*. Inspection of Accommodation. Satisfactory. 8 June 1911': British Postal Museum & Archive, POST 29/1117.
30 'For Sale by Auction Steamship Collectibles, 2 p.m., 5 May 2012, p. 26, Stephen Booth, *Atlantic Crossing Auctions* (2012).

9: 'One Cannot Help Being Impressed'

1 Bowyer, George, *Lively Ahoy*, p. 35.
2 *Southampton Times*.
3 Jessop, Violet, & John Maxtone-Graham, *Titanic Survivor*, pp. 103–04.
4 Oldham, Wilton J., *The Ismay Line*, p. 174.
5 It appears that a complete set of the notes Thomas Andrews made on *Olympic*'s maiden voyage is not publicly available at this time. There seem to have been a number of copies made, perhaps for Andrews' personal use and for distribution to Harland & Wolff and the White Star Line. They are numbered individually, with the numbers running from 1 to 56. However, numbers 28 to 39 and 47 to 55 are missing at the time of writing. Pages from Andrews' notes are available from various private collections and have also been sold at public auctions over the years. I am grateful to researchers including Scott Andrews, Bruce Beveridge, Robin Beuting, Günter Bäbler, Joao Goncalves, Ray Lepien and Bill Sauder, for sharing information where available and supporting with my interpretation of either the handwriting or a particular detail. Any errors are mine alone. Efforts to complete the set and make them publicly available are ongoing. (See also: Lepien, Ray, '*Olympic* The Maiden Voyage', *Titanic Commutator*, Vol. 27, No. 162, 2003; Marre, Jean-Philippe, *Thomas Andrews: Architecte du Titanic*, Lulu, 2015.)
6 'The Giant *Olympic* A Luxurious Floating Hotel', *New York Times*, 25 June 1911, p. 5.
7 'Queenstown Harbour Committee: Report from the Committee Appointed by the Postmaster-General to consider the Omission of the Call at Queenstown on the American Mail Service; Together with Minutes of Evidence, Appendices, and a Chart of Queenstown Harbour', p. 68: Cunard Archives, C1/2/161.
8 Bell, Chief Engineer J., 'SS *Olympic* Voy. 1 to Messrs. Ismay, Imrie & Co., Liverpool, 24 June 1911': Merseyside Maritime Museum Archives, DX/504/1-3.
9 Chirnside, Mark, & Sam Halpern, '*Olympic* & *Titanic*: Maiden Voyage Mysteries', *Voyage*, 59 (2007).
10 '*Olympic*'s Fine Voyage', *Irish Times*, 20 June 1911, p. 7; 'World's Biggest Liner *Olympic* Nearing New York', *Irish Times*, 21 June 1911, p. 7.
11 Bell's quotes from his report, available in the Merseyside Maritime Museum Archives: 'SS *Olympic* Voy. 1 to Messrs. Ismay, Imrie & Co., Liverpool', 24 June 1911: DX/504/1-3.
12 Munsey, Frank, 'The Heritage of the Locust', *Munsey's Magazine*, Vol XLV, April to September 1911, pp. 815–16.
13 The first officer's duties are outlined in 'Ships' Rules and Uniform Regulations, International Mercantile Marine Company', 1 July 1907, pp. 35–36 (courtesy Günter Bäbler). The White Star Line specifically described Murdoch's actions, writing, 'He *ordered* [my emphasis] the engines to be stopped, which order was promptly obeyed'. See 'Peter Cahill V. The Steamship *Olympic* and Oceanic Steam Navigation Company, Ltd.', Case 55-24, US District Court for the Southern District of New York: National Archives & Records Administration, National Archives Identifier: 579537; and 'Oceanic Steam Navigation Company, Ltd. V. Steamtug *O.L. Hallenbeck* her engines, etc. Peter Cahill, Claimant', Case 55-227, US

District Court for the Southern District of New York: National Archives & Records Administration, National Archives Identifier: 579537.

14 Ibid. The events leading up to the collision, the collision itself and the aftermath are described in detail in the testimonies from named individuals.

15 Carton, Chester, 'Docking a Big Liner', *Technical World Magazine* (*c.* 1912) pp. 276–79.

16 The press made sure the maiden voyage was well covered. The New York newspapers, in particular, gave very good coverage to both the westbound arrival and eastbound departure, as well as details from the crossing itself. See: '*Olympic* Sails from Queenstown', *Irish Times*, 24 June 1911; 'Biggest Liner Nears US', *Chicago Daily Tribune*, 21 June 1911; 'The Mighty *Olympic* Ends Her First Voyage, Entering Port in Triumph, While Other Steamships Seem Dwarfs Beside Her', *New York Herald*, 22 June 1911; 'Biggest Ship Arrives', *New York Tribune*, 22 June 1911; 'Biggest of Liners Gets Noisy Welcome', *New York Times*, 22 June 1911; '4,000 Visit the *Olympic*', *New York Times*, 24 June 1911; 'Thousands Visit the *Olympic*', *New York Times*, 25 June 1911; '*Olympic* Preparing to Sail', *New York Times*, 27 June 1911; '*Olympic* Off Today with Record List', *New York Times*, 28 June 1911; '10,000 See *Olympic* Sail, *New York Times*, 29 June 1911; '$175,000 Profits in Single Trip', *Medford Mail Tribune*, 29 June 1911; and 'Giant New Liner Ends First Trip', *Trenton Evening News*, 22 June 1911.

17 See Chirnside, Mark, & Sam Halpern, '*Olympic* & *Titanic*: Maiden Voyage Mysteries'. I am grateful to Stuart Lythgoe for sharing with me a reference from Thomas Ward's records. After *Olympic* was scrapped at their Jarrow yard, their publicity material used her as an example of their capabilities, because she was the largest ship yet to be scrapped. The accompanying short description of the ship gave the correct time of five days, fifteen hours (albeit missing off the two minutes) and correct average speed. Brent Holt also deserves thanks for drawing my attention to a November 1937 article which also used the corrected figure (see: 'The *Olympic*'s Last Voyage: Passing of a Famous Ocean Greyhound', *The Meccano Magazine*, November 1937, p. 661).

18 'The Cunard Steamship Company Limited: Atlantic Homeward and Outward Passages', p. 10: Merseyside Maritime Museum, B/CUN/4/4/5.

19 See 'Particulars of Ships', extract released by Harland & Wolff in the mid-1990s, which gave an 'estimated consumption' of 825 tons and an 'estimated speed' of 21 knots; Carruthers, F., Board of Trade Surveyor's Office, Belfast, to Archer, W. David, Principle Ship Surveyor, London, 'SS *Olympic* – Deadweight at end of passage', 5 January 1911: The National Archives (UK), MT 9/920/E.

20 Oldham, Wilton J., *The Ismay Line*, pp. 172–75.

21 MacNewman, Laura (Associate Archivist, Cranbrook Archives), 'Eastward ho! On RMS *Olympic*', 30 August 2024: 'Cranbrook Kitchen Sink: The Weekly Blog of the Cranbrook Center for Collections and Research'. And see: 'Abstract of Engineer's Log of the Cunard Twin screw Steamship *Campania* from Liverpool to New York', *Voyage*, 218 (June 1911): Merseyside Maritime Museum, B/CUN/4/2/21.

22 Data from *Olympic*'s third round-trip voyage put the revenue earned by the restaurant at an estimated £1,250, with 160 individual passengers using it on the westbound crossing and 105 on the eastbound crossing. On that basis, the average total expenditure was £4.72 ($23.58) per passenger. It is not known how many of these individuals used the restaurant exclusively throughout their voyage or used it as a one-off.

23 'High-Priced Suites Pay On Steamships: The Past Season Saw an Exceptional Demand for Fine Accommodations', *New York Times*, 21 September 1913.

24 'White Star Line Royal and United States Mail Steamers: Triple-Screw RMS *Olympic* 45,000 tons, the Largest and Finest Steamer in the World, and Triple-Screw SS *Titanic*, 45,000 tons (Building), Regular Sailings between Southampton–Cherbourg–Queenstown–New York. Second Class Rates and Plans', March 1911. I am indebted to Daniel Klistorner for sharing the contents of this brochure from his own collection.

25 'The Cunard Steamship Company Limited: Abstract of Voyages No. 4', p. 92: Cunard Archives, AC 12/1.

26 'The First Ocean Smoker', by Clive Livingstone Du Val in *History of the class of 1903, Yale College* by Yale University. I am grateful to Mike Poirier for sharing with me.

10: The Height of the Season

1 See letter from Blake to Messrs Ismay, Imrie & Co., dated 6 July 1911: Merseyside Maritime Museum, DX/504/12/3; and Pounder, Cuthbert Coulson, 'Some Reflective Comments on Engines and Engineers', Belfast Association of Engineers, 7 February 1968, p.6: Public Record Office Northern Ireland, D4063/9/133.

2 This letter is one of a series of letters from July to September 1911 where extracts have survived. See: Franklin, Philip, to Ismay, J. Bruce, 'Extract of Letter', 19 July 1911, through to Ismay, J. Bruce, to Franklin, Philip, 'Extract from Letter', 7 September: Merseyside Maritime Museum, DX/504/19/1-16. The extracts from correspondence in this chapter are all taken from this series of letters unless stated otherwise.

3 See, for example, 'White Star to Fishguard?', *New York Times*, 29 June 1911.

4 'Round the World Racer Here on Time: Will Sail Today on the *Olympic* and Complete the Girdle of the Earth in Forty Days', *New York Times*, 19 August 1911.

5 See file of memos in Cunard Archives, S7/3/36.

11: 'Spinning Right Round': *Hawke* Strikes

1 Duncan Nash, Zetie, 'The Shortest Voyage on the Longest Ship', *Titanic Commutator*, Vol. 42, No. 219 (2017) pp. 133–35. Zetie wrote to Father Roberto Pirrone in 1977, after reading a story in the *Los Angeles Times* about Pirrone's cutaway model of *Titanic*. He then shared the story with the *Titanic* Historical Society. It is interesting to note that her stepfather and mother sailed on *Olympic* again in September 1912: neither the *Hawke* collision nor *Titanic*'s loss was enough to put them off. Zetie married William F. Nash, passing away in California in 1983.

2 In the High Court of Justice. Probate, Divorce, and Admiralty Division. Stating the case between the Oceanic Steam Navigation Company, Limited, Owners of the Steamship *Olympic* verses Commander William F. Blunt, of H.M. Cruiser Hawke and the Commissioners for Executing the Office of Lord High Admiral of the United Kingdom verses Owners of Steamship *Olympic*. London: commencing 16 November 1911. National Archives of the United Kingdom. TS 36/175-76.

3 Mersey, Lord J.B., 'Report of a Formal Investigation', Question 20,447.
4 '*Olympic* Hit By A Cruiser; Badly Damaged', *New York Times*, 21 September 1911.
5 A little over one month after the collision, an investigation as to why *Hawke*'s helm had jammed just before the collision did not reach any firm conclusion. The steering gear had worked satisfactorily both before and after the collision. It was suggested that 'the cause [...[was primarily due to the [ship's] wheel being worked rapidly on the order "hard a port" being given, and so overriding the steering engine'. Another problem might have been the rapid swing to port, as 'this violent swing against the helm may have been a contributory cause of the helm jamming'. They felt that the collision would have happened anyway.
6 '*Olympic*'s Repairs to Take Three Months', *New York Times*, 22 September 1911, pp. 1–2.
7 Ibid.
8 Duncan Nash, Zetie, 'The Shortest Voyage on the Longest Ship', pp. 133–35.
9 I am grateful to Charles Haas for sharing the original article.
10 'SS *Olympic*. Damage to Machinery caused by collision with HMS Hawke. Inspection of ...', Director of Dockyards, 19 October 1911: The National Archives (UK), ADM116/ 1163A.
11 The majority of this account is drawn from the depositions given by each man, contained in '*The King vs. Edward John Wilson*: Depositions', Belfast Spring Assizes, 1912. Public Record Office Northern Ireland, BELF/1/1/2/37/5. I have also used newspaper reports of proceedings, noting the newspaper name in the main text.
12 The court actions related to the crewmen's wages dispute were covered in some detail by the press at the time. In particular, see: '*Olympic* Collision: Men's Claim for Wages', *Daily Telegraph*, 30 September 1911, p. 3; 'The *Olympic* Action', *Western Evening Herald*, 29 September 1911; '*Olympic–Hawke* Collision. Seamen's Wages Claim. White Star Line Test Case', *Liverpool Daily Echo*, 5 March 1912; '*Olympic* Wages Case', *Hampshire Independent*, 9 March 1912; '*Olympic* A Wreck', *Tewkesbury Register*, 6 April 1912.
13 High Court of Justice, Probate, Divorce and Admiralty Division, 'Writ in person issued 21 September 1911 between the Oceanic Steam Navigation Company Limited, the owners of the Steamship *Olympic*, plaintiffs, and Commander W.F. Blunt of HM Cruiser Hawke, Defendant: HMS Hawke: Statement of Claim: The National Archives (UK), ADM 116/1163a, and High Court of Justice, Probate, Divorce and Admiralty Division, 'Writ issued 28 September 1911. The Commissioners for Executing the Office of Lord High Admiral of the United Kingdom, verses the owners of SS *Olympic*: Statement of Claim': The National Archives (UK), ADM 116/1163a.
14 'The SS *Olympic*: Advice on Evidence', 17 October 1911: The National Archives (UK), ADM 116/1163a.
15 Sir Samuel Evans (1859–1918) became a Queen's Counsel in 1901, then the Recorder of Swansea from 1906–08. He died in Brighton, aged 59. 'Death of Sir Samuel Evans', *Liverpool Journal of Commerce*, 14 September 1918.
16 '*The Oceanic Steam Navigation Company, Limited, Owners of the Steamship Olympic v. Commander William F. Blunt, of HM Cruiser Hawke*', Probate, Divorce & Admiralty Division, Folio 400 (1911) O. No. 1166: The National Archives (UK), ADM 116/1163a.
17 High Court of Justice, Probate, Divorce, and Admiralty Division, 'Stating the case between the Oceanic Steam Navigation Company, Limited, Owners of the Steamship *Olympic* verses Commander William F. Blunt, of H.M. Cruiser Hawke and the Commissioners for Executing the Office of Lord High Admiral of the United Kingdom verses Owners of Steamship *Olympic*', p. 768: UK National Arcives, TS 36/175-76. (The block quotation from Evans' judgement preceded his earlier comment. They have been inverted for editorial reasons, which does not alter the meaning.)
18 Royal Courts of Justice, 'In His Majesty's Court of Appeal. On Appeal from the High Court of Justice (Admiralty Division). Between the Oceanic Steam Navigation Company Limited, owners of the Steamship *Olympic*, Appellants, and Commander William F. Blunt of HM Cruiser Hawke, Respondent: Report of Proceedings as to Calling Further Evidence': The National Archives (UK), TS 36/183.

12: 'The Stormiest Trip'

1 I am grateful to Mark Baber and Inger Sheil for confirming details of Holme's career. Holme was serving as *Suevic*'s chief officer in March 1907, when she grounded and became stranded. He attained his first command, *Cufic*, in 1912.
2 Her draft was 34ft 2in forward and 34ft 34in aft. Cuthbert Coulson Pounder noted, '18,700 IHP port, 18,900 IHP starboard, 18,500 SHP centre; 56,100 IHP + SHP total; 78 r.p.m. port, 79.3 r.p.m. starboard, 169 r.p.m. centre; boiler pressure 205 lbs sq. in. by gauge; turbine initial steam pressure 7 in. vacuum, terminal pressure 27.5 in. vacuum, condenser vacuum 28.26 in. (barometer 30.3 in); HP cylinder cut-off 69.6%.' See: 'Some Notable Belfast-Built Engines', p. 24.
3 'Hustle on the *Olympic*', *New York Times*, 9 December 1911, p. 4.
4 Piek was born in Amsterdam on 16 March 1874 and married Johanna van Rees on 10 July 1910. In 1911, he was working as an accountant for the Holland America Line; in 1912, he became their head agent in New York; and in 1916, he became a director of the company, serving until 1935. For biographical details of Willem Frederik Piek, I have relied on Brugman, H., & N. Japikse, *Persoonlijkheden in het koninkrijk der Nederlanden in woord en beeld (Nederlanders en hun werk* Amsterdam, 1938) p. 1,734.
5 Schoonderbeek, Captain Albert, 'The Interiors of Holland America Line's Six *Statendams*: Part One – the first three *Statendams*', *Sea Lines*, Issue 111 (Summer 2023) p. 28.
6 'Piek quotes are from Piek, W.F. aan board van ss *Olympic*, rapport met bijlagen, 1911': Stadsarchief Rotterdam, EB 1278.
7 See Chirnside, Mark, 'A Voyage on *Olympic*: Willem Frederik Piek Jr's Notes for the Holland America Line, December 1911', *Voyage*, 128 (July 2024) pp. 158–62. I'm grateful to Daniel Klistorner for his expertise, clarifications and comments on a number of points; Marco Tulner for his help in deciphering otherwise indecipherable words; translator William Richmond and colleagues at Kings of Translation Ltd for working with me to understand specific nautical or *Olympic*-related terms and providing a fully certified translation, ensuring I had understood everything correctly; and Marie-Christine Engels, Coördinator Studiezaal/ Informatiedienstverlening Stadsarchief Rotterdam, and her colleagues, for their assistance.
8 Lome, Herbert M., 'The Sanitation of a Modern Ocean Liner', *Modern Sanitation*, Vol. 8 (1911) pp. 445–51.
9 I am grateful to Mike Poirier for sharing Fred L. Boalt, 'Do you Know How It Feels to Face Death on One of the

World's Mightiest Steamships?', *The Seattle Star*, 16 April 1912. It is possible Boalt's dramatic account was influenced by the aftermath of the *Titanic* disaster.

10 '*Lusitania* Battered by 80-Foot Wave', *New York Times*, 16 January 1910.

11 The *New York Times* cited, '21,600 first class passengers from European ports to America, and 21,314 from America to Europe'. It is assumed that 'America' refers to the United States only but, even including Canada, the numbers do not match with the figures reported by the North Atlantic Passenger Conference. It is the conference figures cited here for the different lines and services in 1911. See 'Record for White Star Line', *New York Times*, 26 January 1912.

12 Within IMM, the American Line and White Star competed on the Southampton to New York service: the American Line's average passenger list fell by 8 per cent in 1911; White Star's rose by 21 per cent.

13 This is based on an analysis of passenger numbers reported by the North Atlantic Passenger Conference and includes the nearest comparable Cunard, Holland America, HAPAG, Norddeutscher Lloyd sailings ranked by arrival date and departure date at New York. It should be taken as a crude comparison only. All ships were on the North Atlantic service to New York and back, but some of their ports of call differed and the nearest sailing date of another line could be five or six days apart.

14 Oldham, Wilton J., *The Ismay Line*, pp. 177–83.

15 See Ismay, Booth and Ballin's correspondence from August 1911 to March 1912: Cunard Archives, C1/2/146.

16 'How the market works: Insuring the *Titanic*', Lloyd's summary booklet, 2019. See also Eaton, John P., & Charles A. Haas, *Titanic: Triumph & Tragedy* (Patrick Stephens Ltd, 1994) p. 43. One of the companies providing insurance coverage was Atlantic Mutual.

17 'First Annual Report of the International Mercantile Marine Company, for the Fiscal Year Ended 31 December 1903', Office of the International Mercantile Marine Co., 51 Newar Street, Hoboken, New Jersey, June 1904, p. 6. See also Philip Franklin's testimony before the United States Senate.

18 See Ismay, Booth and Ballin's correspondence from February 1912 to March 1912: Cunard Archives, C1/1/31.

19 This figure was 'including payments on account of new tonnage, after deducting depreciation' and did not include other company assets, such as 'movable plant and stores'; 'trade investments'; 'general investments'; 'sundry debtors'; 'pending voyages' and 'cash at banks, on hand and in transit', which took the company's reported assets on the balance sheet dated 31 December 1911 to £9,394,254 14*s* 4*d*.

20 There are five letters between Captain Smith and Frank Hancock in the private collection of the Blunt White Library, Mystic Seaport, Mystic, Connecticut. I am grateful to Shelley Dziedzic for sharing and drawing these letters to my attention. See: Smith, Captain Edward John, to Hancock, Frank, 'My Dear Frank', 14 February 1912.

21 'Mishap to the *Olympic*', *Belfast Weekly News*, 29 February 1912, p. 7; 'Arrival of the *Olympic*', *Westminster Gazette*, 28 February 1912, p. 10; 'Mishap to the *Olympic*', *Western Daily Mercury*, 29 February 1912, p. 8; and 'The *Olympic* at Belfast', *Belfast Weekly News*, 7 March 1912. See also: Hutchings, David, 'The Lost Blade', *Atlantic Daily Bulletin*, spring 2021, pp. 38–41.

22 See Chirnside, Mark, *Oceanic: White Star's 'Ship of the Century'* (History Press, 2018) p. 86.

23 I am grateful to Dan Cherry for sharing a report from an American newspaper, published in Michigan, 'Much Distress Among Workmen of England'.

24 See 'Coal Situation' and cable dated 28 March 1912: Merseyside Maritime Museum, DX 504/12/4. On that date, *Olympic* was five days into her eastbound crossing and *Oceanic* lay idle at Southampton.

25 I am grateful to Inger Sheil for sharing extracts of Moody's correspondence with me. His comment about 'any unforeseen diversion' has a horribly ironic connotation in light of his death on *Titanic* less than a month later. See also Chirnside, Mark, *Oceanic: White Star's 'Ship of the Century'*, p. 86.

26 'Trial Trip of the *Patriotic*'. Unidentified newspaper cutting, 28 March 1912, in the diary of Robert H.H. Baird, 1912: Public Records Office of Northern Ireland, D/3330/B/1/12.

13: 'A Contest of Sea Giants'

1 See correspondence files from 23 May 1911 to 16 December 1911: Public Record Office of Northern Ireland, T3893/1. (These papers have been cited by a number of other researchers, including Tim Silcocks.)

2 See Bell's record in Cunard Archives, EN 1/1.

3 Klistorner, Daniel, & Steve Hall, with Bruce Beveridge, Art Braunschweiger & Scott Andrews, *Titanic in Photographs* (The History Press, 2011) pp. 54–55.

4 H.P. Mutters & Zoon to Messrs. Harland & Wolff Ltd, 'No. 401. SS *Titanic*', 30 September 1911. I am grateful to Daniel Klistorner for drawing my attention to this letter, which was auctioned by Henry Aldridge & Son Ltd.

5 Bäbler, Günter, 'Indienststellung der *Titanic*: 1911', *Titanic Post*, No. 52 (2005) pp. 8–12.

6 I am grateful to Charles Haas for sharing this information. See also Klistorner, Daniel, et al., *Titanic in Photographs*, pp. 55–56.

7 'White Star Line United States and Royal Mail Steamers: First Class Passage Rates', January 1912. The comparison of first-class ticket prices is based on the figures from January 1912. I appreciate Günter Bäbler sharing this information with me.

8 See 'Deposition of Edward Wilding, In the Matter of the Petition of the Oceanic Steam Navigation Company, for Limitation of its Liability, as Owner of Steamship *Titanic*', 13–14 May 1915: The National Archives and Records Administration – Northeast Region (New York). The early *Aquitania* structural design seems to have been particularly conservative compared to other similar ships, with very heavy structural members which were simply not needed to give her the necessary strength. On 5 January 1911, Cunard's Leonard Peskett and representatives of shipbuilders John Brown, met with representatives of Lloyd's Register and discussed a number of changes. They agreed with the shipbuilder's suggestion to fit the heavy web frames every three spaces apart in the engine room, rather than two as Cunard had proposed; they slightly reduced the thickness of the middle flat keel plate; and discussed a proposal from the shipbuilder to reduce the length of the doubled shell plating on B-Deck (which was rejected). On 4 September 1911, Leonard Peskett reported his observations of *Olympic*. He compared a midship section showing *Olympic*'s structural elements compared to the midship section for *Aquitania*, noting her structure (scantlings) was 'somewhat lighter than that approved for the *Aquitania*, the greatest difference being in the width and depth of the shell and deck doublings'.

Shipbuilders sought to achieve an optimum balance of the necessary structural strength, without making the ship unduly heavy (or using unduly thick plating, which could be harder to rivet). Accordingly, they drew up a revised midship section for *Aquitania* in January 1912, highlighting proposed modifications to reduce unduly heavy plating. The *Aquitania* midship section published in 1914 shows the key changes were implemented and her design brought closer into line with *Olympic*. See 'Midship Section No. 409', 20 March 1911: University of Glasgow Archive Services, UCS 1/110/409/20; and 'No. 409 [*Aquitania*] Midship Section' ('showing proposed reduction in scantlings of topside double plating and plating of B, C and D decks'), Lloyds Register Foundation. (See also De Rusett, E.W., 'Hydraulic Riveting in the *Mauretania*: How Four Million Rivets Were Driven in the Big Steamship', *Scientific American*, supplement, 21 December 1907, p. 388. He noted that hydraulic riveting was used to a large extent because of the great thickness of the hull plates and the large size of the rivets used.)

In January 1925, Rosoe & Little assessed the necessary repairs and modifications for strengthening the former German liner *Bismarck* (1914), now White Star's *Majestic* (1922), following a serious structural failure in December 1924. Again, *Olympic* was cited as a benchmark example of a strong ship. One of the more minimalist schemes of repair being proposed was expected to leave *Majestic* 20 per cent weaker than *Olympic* (see Merseyside Maritime Museum, B/CUN/3/16/1 and Chirnside, Mark, *RMS Majestic: The 'Magic Stick'*, History Press, 2024).

9 Chirnside, Mark, 'The Enclosure of *Titanic*'s Forward A-deck Promenade: Popular Myth?', *Atlantic Daily Bulletin*, March 2016, pp. 6–11; and Robinson, Andrew, 'The Partial Enclosure of the *Titanic*'s A-deck Promenade', *Atlantic Daily Bulletin*, No. 1 (2000) p. 12. The same conclusions are drawn in Klistorner, Daniel, et al., *Titanic in Photographs*.

10 'No. 436 – SS *Statendam* Modifications Involving Extras and Credits for Which Prices Have Not Been Submitted to Owners', 16 August 1917: Stadsarchief Rotterdam, EB 1133.

11 The great length of both ships meant that their hulls were subject to additional shearing stresses about a quarter of the way aft from the bow and a quarter of the forward of the stern. The rivets which needed repairing on *Olympic* included 'about 160 rivets' which were 'slack and were drilled out and renewed' between frames 63 to 74 forward; below the landing, in the tank bar, from frames 71 to 75, 50 slack rivets were drilled out and replaced on the starboard side, and 'about 90 rivets were showing a little slack and were caulked' on the port side from frames 71 to 78. 'On both sides aft, in way of the turbine room, in the shell landing of K & L strakes from about frames 52 to 69 [aft], 100 rivets found slack were drilled out and replaced'. Similar, more serious, repairs than those needed by *Olympic* were undertaken on other large liners of the period. Shortly after *Bismarck*/*Majestic* (1922) entered service, thousands of rivets in the bottom of the hull needed caulking, and similar issues were reported on *Empress of Britain* (1905) and *Aquitania* (1914).

The new 'straps' on *Titanic* extended from forward frames 81 to 63 and aft frames 50 to 73 at the riveted landing between the J and K strakes of plating, more than covering the affected area. (Harland & Wolff started amidships and then numbered frames fore and aft, so there could be two frames with the same number.) It was calculated that the shearing stress on rivets in quadruple-riveted landings was reduced to 4.9 tons per square inch, from 6.5 tons for treble-riveted landings and 9.8 tons for double-riveted landings. Similar or identical changes were made to *Olympic* and there were no further reports of any issues at all. Edward Wilding also noted a change to the specifications of their younger sister on 5 February 1912, specifying the need for treble riveting 'between frames 39 & 85 forward & aft (about)'.

The key documentation concerning the rivets repaired on *Olympic* and the modifications made to *Titanic* comes from letters and reports written by Francis Carruthers, dated from 13 February 1912 to 6 March 1912. See The National Archives (UK) MT 15/504. I first published information about these changes to *Titanic* in an online article for the *Titanic* Research & Modelling Association (TRMA) in summer 2005. See Chirnside, Mark, '*Olympic* & *Titanic*: Refining a Design', *Atlantic Daily Bulletin*, December 2019, pp. 18–22, which was an updated and expanded version of the TRMA article. For an analysis of these changes and their potential impact on *Titanic* as related to the collision with the iceberg, see Parks Stephenson's article, 'What Caused *Titanic* to Sink?' in the *Titanic* Historical Society's *Titanic Commutator*, Vol. 39, No. 206 (2014) pp. 92–100. See, also Rudi Newman's 'A "Riveting" Article – an Historical Rejoinder to Metallurgical Studies of the *Titanic* Disaster' in the British *Titanic* Society's *Atlantic Daily Bulletin* (2012) pp. 18–30.

When I published my original article in 2005, my argument was that these minor changes demonstrated the fundamental strength of *Olympic*. She came through one of the worst North Atlantic storms in her lengthy career and did not show any significant weakness. Unfortunately, a number of conspiracy theorists took this information out of context and portrayed these modifications as evidence that Harland & Wolff had not built *Titanic* to be strong enough. For further detail, see Chirnside, Mark, '*Titanic*: Allegations and Evidence', *Voyage*, 94 (December 2015) pp. 55–60.

It is likely that minor fractures seen in *Olympic*'s deckhouse side plating at the corners of rectangular windows near to the expansion joints, following the same severe storm in January 1912, were why Harland & Wolff opted to increase the number of expansion joints on *Britannic*. It was not a significant issue as far as the ship's strength was concerned, because these deckhouses were part of the superstructure and not the hull. Nor was it anything uncommon. It happened on all large ships including *Queen Mary* (1936) in her first two years of service. Moreover, *Olympic*'s expansion joints served her well throughout her career. However, improving the number of expansion joints was beneficial. Continual improvement and refinement of their designs was a hallmark of Harland & Wolff. See Chirnside, Mark, '*Olympic*'s Expansion Joints', *Titanic Commutator*, Vol. 31, No. 178, pp. 84–86.

12 Harland & Wolff kept a number of volumes of engineering notebooks which documented key statistics for each ship they completed, including propeller and engine details. The entries on the relevant pages include *Lapland*, *Laurentic*, *Median*, *Memphian*, *Minnewaska*, *Mercian*, *Megantic*, *Olympic*, *Titanic* and *Leopoldville*. The purpose was to document each ship's as-built configuration and, generally speaking, the entries for each ship were not kept up to date throughout her service. For example, the *Olympic* record only reflects her as-built configuration in 1911 and the modification

to her port and starboard propellers later that year. The various other propeller configurations from 1913 onwards were not included. See Public Record Office Northern Ireland. D/2805/SHIP/8A-E. These records were deposited between 1972 and 1994. This material was first made public in Chirnside, Mark, 'The Mystery of *Titanic*'s Centre Propeller', *Voyage*, 63 (2008) pp. 123–28.

There are no known photos of *Titanic* in dry dock with her propellers fitted and visible. All too often, photos of *Olympic* are used to portray her sister and either miscaptioned as *Titanic*, or used to represent her without the caption making this clear. For decades, many secondary-source publications also used the *Olympic* propeller specification details as if they were *Titanic*'s, on the assumption that they were the same. However, the contemporary primary-source documentation from Harland & Wolff's own records makes it clear that their propellers were different.

13 Mersey, Lord J.B., 'Report on a Formal Investigation', Questions 19,052–53.
14 'Loadline Certificate of Approval', February 1912, p. 28: The National Archives (UK), MT 15/212/1.
15 This narrative is based on the records of the Belfast Harbour Commissioners. I am grateful to Alan Geddes for bringing these to my attention. See, in particular: Public Record Office Northern Ireland, HAR 1/A/2/234 to HAR/1/A/2/237.
16 'Harland & Wolff Ltd. Minutes [of] Managing Directors', pp. 287–88 and 299–300: Public Record Office Northern Ireland, D/2805/MIN/A1.
17 Currie, W.A. Esq., Secretary, Belfast Harbour Commissioners, 'No. 401 SS *Titanic*', 20 March 1912: Public Record Office Northern Ireland, HAR 1/A/2/234 to HAR/1/A/2/237.
18 'Half Yearly Agreement and Account of Voyages and Crew for a Ship Engaged In The Home Trade Only': Public Record Office Northern Ireland, TRANS. 2A/45/381A.
19 Public Record Office Northern Ireland, D2805/SHIP/3.

14: 'The Dangers of the Seas'

1 Mersey, Lord J.B., 'Report on a Formal Investigation', Question 23,996.
2 'Chain Cables. Steam Ships. *Olympic* and *Titanic*', 24 June 1909: The National Archives (UK), MT 15/229.
3 Bullock, Shan, *Thomas Andrews: Shipbuilder*, pp. 56–57.
4 Whited, Brandon, 'A "Quay" Question: Rethinking *Titanic*'s Alleged Belfast Passenger', *Voyage*, 127 (2024) pp. 95–100, and Whited, Brandon, 'A "Quay" Addendum: Putting the 'Van Der Hoef Myth' To Rest', *Voyage*, 131, (2025) pp. 96–97.
5 'Deposition of Edward Wilding', p. 113: The National Archives and Records Administration – Northeast Region (New York).
6 I am grateful to Ioannis Georgiou, Steve Hall and Paul Lee for sharing their research.
7 I am grateful to Günter Bäbler for drawing my attention to the Southampton Harbour Master's Log, retained by Associated British Ports.
8 Lehrer, Stanley, '*Titanic* at Southampton: A Time of Great Expectations', *Titanic Commutator*, Vol. 27, No. 161 (2003) p. 22.
9 'Half Yearly Agreement and Account of Voyages and Crew for a Ship Engaged In The Home Trade Only': Public Record Office Northern Ireland, TRANS. 2A/45/381A.
10 Kamuda, Ed, '*Titanic*'s Only Completed Voyage', *Titanic Commutator*, No. 244, Winter 2024, p. 46.
11 'Memorandum: *Titanic*', 11 April 1912: British Postal Museum & Archive, POST 43/321.
12 The original ledger, reference D/Z 627, covers 1909–16. Details taken from a display facsimile at Southampton's SeaCity Museum, April 2017. The supplies to the restaurant were accounted for as a separate order. Thanks to Learning & Engagement Officer Andy Skinner, SeaCity Museum, Southampton, and Joanne Smith, Archivist, Southampton City Libraries for confirming further details in August 2024.
13 Kamuda, Ed, '*Titanic*'s Only Completed Voyage', pp. 54–55.
14 Boland, Cathalijne, *Reuchlin Reis: De Holland – Amerika Lijne en de landverhuizers* (Alfabet Uitgevers, 2023) p. 276.
15 Mersey, Lord J.B., 'Report of a Formal Investigation', Question 1,911.
16 See Chirnside, Mark, *The 'Olympic' Class Ships: Olympic, Titanic & Britannic* (The History Press, 2011, revised and expanded edition). The appendices included a comprehensive analysis of *Titanic*'s coal consumption based on available coal consumption data from *Olympic*, which was expanded from an analysis completed in 2003 and published in the 2004 edition. Sam Halpern developed a detailed mathematical model of coal consumption, taking into account the ship's engine revolutions and state of loading. This model, tested against the known coal consumption data for *Olympic*'s maiden voyage, was accurate to within 1 per cent of the known outcome. The conclusions from the updated analysis, based on coal consumption data and this modelling, are based on conservative assumptions. *Titanic* was set to have over 1,300 tons of coal remaining, equivalent to about two days' spare consumption at a speed of over 21 knots.
17 This analysis is based on data sourced from 'The Cunard Steam Ship Company Limited: Atlantic Outward and Homeward Passages': Merseyside Maritime Museum, B/CUN/4/4/5. The data used in the analysis includes *Mauretania*'s westbound Voyage 50 (May 1911) through to westbound Voyage 61 (April 1912) inclusive. Over the course of these twelve westbound crossings, she had a total of 7,960 tons of coal reported to be left over on arrival in New York, which works out as an average of 663 tons on each crossing. The amount of coal remaining ranged from 500 tons to 950 tons, but even the higher quantity equated to about a day's steaming at her normal average speed.
18 Bullock, Shan, *Thomas Andrews: Shipbuilder*.
19 I am grateful to Tom Rudderham for providing his transcript of the original letter.
20 There was 'little doubt that the lines would, without the loss of the *Titanic*, have [...] altered the Southern Route to the southward, but unhappily the first ice of the season was not reported unusually far south until the 11th April, the day upon which *Titanic* left Queenstown'. See 'Safety of Life at Sea: Draft Lifesaving Appliances Rules, With memorandum on Safety of Life at Sea by the President of the Board of Trade', p. 36: The National Archives (UK), RAIL 1053/295.
21 Hill, Brian T., & Alan Ruffman, 'E.J. Smith of the RMS *Titanic*: His Previous Ice Experience', *Atlantic Daily Bulletin* (2014) pp. 28–41; and Newman, Dr Rudi, 'Among the Icebergs: *Titanic*'s Place in the Evolution of Ice Encounters', *Voyage*, 126 (2024).
22 Buss Willis, Kate, 'A Story of the *Titanic* Disaster' (February 1932), *Titanic Commutator*, Vol. 30, No. 176, p. 189.

23 I am grateful to Paul Lee for sharing his research, including his transcription of a 19 June 1955 letter from George Kemish to Walter Lord. This is available as part of the Lord MacQuitty collection, see National Maritime Museum, LMQ/2/4/147 LMQ/7/1/49.
24 I am grateful to George Behe for sharing his extensive research with me.
25 Behe, George, *The Triumvirate: Captain Edward J. Smith, Bruce Ismay, Thomas Andrews and the Sinking of Titanic* (The History Press, 2024) pp. 84 and 91.
26 Ryerson, Emily. Affidavit 16 May 1912. 'Report of the Committee on Commerce United States Senate Pursuant to S. Res. 283 Directing the Committee on Commerce to Investigate the Causes Leading to the Wreck of the White Star Liner *Titanic*' (Washington: Government Printing Office, 1912).
27 Ibid.
28 See Buley, Edward John. He testified before the US Senate on 25 April 1912. 'Report of the Committee on Commerce United States Senate Pursuant to S. Res. 283 Directing the Committee on Commerce to Investigate the Causes Leading to the Wreck of the White Star Liner *Titanic*' (Washington: Government Printing Office, 1912).
29 Bisset, Sir James, *Tramps & Ladies: My Early Years in Steamers* (Patrick Stephens Ltd, 1988) pp. 286–87.
30 Behe, George, 'How the *Titanic* Became Unsinkable'.

15: 'A Mournful Voyage'

1 'Proces-Verbal SS *Olympic*. Voyage 10. The Marconi International Marine Communication Co. Ltd', April 1912: The Bodleian Library, Oxford University Library Services, MS. Marconi (T6).
2 His death was reported in various newspapers and Beaumont's entry in the log. I am also grateful to George Behe for sharing his extensive research into professional gamblers on board *Titanic*. See Behe, George, *Fate Deals a Hand: The Slippery Fortunes of Titanic's Professional Gamblers* (The History Press, 2023).
3 Bisset, Sir James, *Tramps & Ladies: My Early Years in Steamers*, pp. 265–66. There are some odd details in Bisset's account. For example, he mistakenly recalled *Titanic*'s trials as taking place on 1 April 1912. He also critiqued the watertight bulkhead design, but did not provide any comparison with contemporary liners.
4 The material in this narrative concerning *Olympic*'s wireless communications are taken from various sources, including messages submitted to the United States Senate investigation; the ship's wireless log (the summaries of some messages differ slightly from the full transcripts); testimony by Captain Haddock and Ernest James Moore; and a later manuscript written by Alec Bagot. See, for example: 'Proces-Verbal SS *Olympic*. Voyage 10. The Marconi International Marine Communication Co. Ltd', April 1912: The Bodleian Library, Oxford University Library Services, MS. Marconi (T6); Bagot, Alec, 'Roaming Around: RMS *Olympic* Marconi Operator'. Bagot's manuscript and papers are available at the National Library of Australia. His material has been cited by a number of researchers, including John Godl of the *Titanic* Historical Society: https://titanichistoricalsociety.org/i-heard-titanics-call/ (accessed August 2024) and Shelley Dziedzic of the *Titanic* International Society (December 2005): www.encyclopedia-titanica.org/roaming-around-memoirs-marconi-operator.html (accessed August 2024).
5 Samuel Halpern calculated the distance as 502 miles in comparing *Olympic*'s position as given in the wireless log for 11.24 p.m., New York Time, with the report that she was 310 miles away at 7.45 a.m. That would mean *Olympic* covered 192 miles in eight hours, twenty-one minutes at an average speed of about 23 knots, reflecting that she was going slower than normal when the distress call was received and that it took time to increase the ship's speed.
6 Shannon, Cecile, '*Titanic* interest: Rare White Star Line notice to *Olympic* passengers and a *Carpathia* bronze medal steam ahead at auction', *Antiques Trade Gazette*, 18 July 2024. Thank you to Dan Parkes for sharing.
7 I am grateful to Sam Halpern for sharing his navigational analysis and calculations.
8 Beaumont, Dr J.C.H., 'To the Board of Trade, Southampton', 15 May 1912: The National Archives (UK), BT 100/248.

16: The 'Scallywags' of Portsmouth

1 Clarke, Maurice Harvey, 'Report re.: Emigration Clearance and survey of SS *Olympic* O.N. [official number] 131,346', 28 April 1912: The National Archives (UK), MT 15/150. The majority of the narrative is based directly on Clarke's own comments. He included detailed remarks about the chronology of events and what happened at each stage.
2 Beaumont, Dr J.C.H., *Ships – and People*, pp. 100–01.
3 Turner, J.H., 'A Three Months' Tour in the United States and Canada', pp. 5–8. I am grateful to Rene Bergeron for sharing a very rare copy with me and allowing me to quote from it.
4 Goodenough, Captain W.E., 'Report from HMS *Cochrane* at Portsmouth, to Rear Admiral Commanding, 2nd Cruiser Squadron', 26 April 1912: The National Archives (UK), MT 15/150.
5 See 'Peter Cahill V. The Steamship *Olympic* and *Oceanic* Steam Navigation Company, Ltd', Case 55-24, US District Court for the Southern District of New York: National Archives & Records Administration: National Archives Identifier: 579537, and 'Oceanic Steam Navigation Company, Ltd. V. Steamtug *O.L. Hallenbeck* her engines, etc. Peter Cahill, Claimant', Case 55-227, US District Court for the Southern District of New York: National Archives & Records Administration: National Archives Identifier: 579537. I am grateful to Charles Haas for sharing 'The *Olympic*. The *O.L. Hallenbeck*', District Court, SD New York, 4 November 1914: *221 The Federal Reporter*, pp. 296–98, and 'The *Olympic*', Circuit Court of Appeals, Second Circuit, 12 May 1915: *140 US Circuit Courts of Appeals Reports*, pp. 130–31.

17: 'The Shadow of a Great Calamity'

1 'White Star Line. Special Meeting, 22 April 1912', *Titanic Commutator*, Vol. 22, No. 4 (1999), pp. 27–29.
2 'General Post Office: Transatlantic Sea Post Offices. White Star Line: SS *Olympic*. Unsuitable Sleeping Accommodation. Memorial for Staff at Southampton. 1912': British Postal Museum & Archive, POST 29/1117.
3 'General Post Office: Transatlantic Sea Post Offices. White Star Line: SS *Titanic* and SS *Olympic*. SS *Titanic* Lost – April 1912. Alterations On Board SS *Olympic*. Sleeping Accommodation on Deck "E" Provided. 1912': British Postal Museum & Archive, POST 29/1117.

4 The longstanding track agreements from 1899 were altered in April 1912, July 1912 and September 1912. A comprehensive track agreement was signed by the major shipping lines following a meeting in Brussels on 27 March 1913. See 'Track Chart Agreements': Cunard Archives, PR3/6/18.
5 See correspondence, including letters and telegrams, between J. Bruce Ismay and the Earl of Derby, May 1912: Merseyside Maritime Museum, DX 504/8/1.
6 Finance Committee, p. 69: Southampton City Archives, SHB1-2-17.
7 Ibid., p. 91.
8 I am grateful to Mike Poirier for sharing this with me: 'The "*Olympic*" An Unusual Voyage', *The American Marine Engineer*, June 1912, pp. 20–21. An invaluable account.
9 See Weaver, Maurice, & Edwin Steel, with thanks from Brian Ticehurst, 'As *Titanic*'s Sister Ship Headed for the Rocks …', *Titanic Commutator*, Vol. 12, No. 1, pp. 40–41. Edwin Steel was the son of Captain Benjamin Steel, who was the White Star Line's Marine Superintendent at Southampton in 1912. Edwin himself was only 7 years old, but his father told him about the incident in confidence some years later. He stated that he only shared it when he felt he was released from his promise, seventy-five years after the *Titanic* disaster and many years after Benjamin Steel's death in 1944. It was reportedly subject to an internal White Star Line investigation with the result that Captain Haddock was 'always monitored'. Trying to verify such details is extremely difficult. The incident was reportedly seven weeks after the disaster and on a June evening, which would place it at the end of the eastbound crossing departing New York on 25 May 1912. However, Edwin recalled it being 10 p.m. on a Sunday evening, whereas *Olympic* had already arrived at Plymouth earlier than that and on the Friday evening, 31 May 1912. Another discrepancy is that he said *Olympic* docked at Southampton on the following day, a Monday, but she actually docked on a Saturday, 1 June 1912. It is possible he got the days of the week mixed up. Human memory is not infallible, particularly after many decades. (On the other hand, the reported distance covered on the eastbound crossing is in line with what would be expected. If correct, it seems hard to reconcile with an error taking her significantly off course. Harold Sanderson later described Captain Haddock as the best man they had and he continued to command *Olympic*.)
10 'Queenstown Harbour Committee: Report from the Committee Appointed by the Postmaster-General to consider the Omission of the Call at Queenstown on the American Mail Service; Together with Minutes of Evidence, Appendices, and a Chart of Queenstown Harbour', pp. 24–25: Cunard Archives, C1/2/161.
11 Kinney, Mike, 'On Board SS *Olympic*', *The Gimlet*, Vol. V, No. 6 (July 1912) pp. 333–69. His account was published as a series of letters recording his voyage. I am grateful to George Behe for sharing with me his transcription. Daniel Klistorner and Mike Poirier also deserve credit for bringing it to my attention and ensuring I did not forget about it. Kinney's account is not widely known but it was previously published in *Voyage* in 1996.
12 There were various shipments throughout 1912, including 'jewels and crystals' shipped on 15 May 1912 by *Olympic* and paintings shipped by *Olympic* and *Majestic* from 26 June to 3 July 1912. *Celtic*, *Cedric*, *Baltic*, *Adriatic* and *Oceanic* were among the other White Star liners carrying cargoes for J.P. Morgan. I am grateful to Günter Bäbler for this information. See 'Pierpont Morgan Papers (ARC 1196)', p. 51: The Morgan Library & Museum, The Pierpoint Morgan Library.
13 Barry, George Leonard, 'The Environments of an Ocean Liner' in *Labor Conditions Abroad: A Review of the Social, Economic, and Political Conditions of the Workers of France, Germany, England, Switzerland, Italy and Spain* (Technical Trade School Printing Pressmen and Assistants' Union, 1913). I am grateful to Mike Poirier for sharing.
14 Behe, George, *The Triumvirate*, pp. 221–22.
15 Radcliffe, F.M., to Ismay, J. Bruce, letter dated 1 August 1912.
16 See file of correspondence between J. Bruce Ismay, Alfred Booth, E.J. Spencer and others: Cunard Archives, C1/1/31.
17 Lynch, Don (ed.), '*Titanic* Survivor Accounts: Kate Buss Willis – Her Story', *Titanic Commutator*, Vol. 30, No. 176, p. 184.

BIBLIOGRAPHY

NOTE ON SOURCES

One of the most significant barriers to us understanding *Olympic* and *Titanic*'s history accurately is the proliferation of inaccurate information in secondary source material (such as articles, books, television programmes or websites). Many aspects of their popular history are misunderstood. These misunderstandings tend to be repeated from one secondary source to another. It is such a broad subject to study that it is inevitable that some secondary sources are required, but many of these misunderstandings could be cleared up easily merely by using the available primary source material.

Primary sources (documents produced by those who were there) are generally superior. An example is the popular belief that the '*Olympic*'-class ships were financed by American capital from either IMM or J.P. Morgan. It is certainly true that IMM was, ultimately, the controlling shareholder in the White Star Line. However, the capital raised by IMM was for the purpose of acquiring that ownership stake through purchasing shares in the White Star Line. Having purchased that controlling interest, it was IMM that was dependent on the dividends paid by the White Star Line to its shareholders. IMM needed those dividends to support itself and pay its debts.

When it came to financing construction, White Star opted to borrow the money by issuing bonds ('mortgage debentures') with its own fleet as security. As we have seen in Chapter 6, the prospectus they issued to prospective investors still exists and a statement explaining what they were doing was even included in IMM's Annual Report for 1908, signed off by J. Bruce Ismay as president. Once issued, the bonds could be traded on the stock market. The official data from the London Stock Exchange contains quotations of the prices. There are therefore multiple sources confirming how the finance was raised. Researchers such as Earl A. Saliers (1915), J.H. Isherwood (1956), Roy Anderson (1964) and Paul Louden-Brown (2001) have all documented that White Star raised the finance. Nonetheless, all too many secondary sources simply repeat unsubstantiated claims to the contrary.

In 1973, Lieutenant Colonel Frank Bustard wrote to the maritime historian John Maxtone-Graham. He reflected, 'It is extremely doubtful whether they would ever have been built but for the finances made available by the so-called Morgan Combine [IMM].' Bustard had started an apprenticeship with the White Star Line in 1902 and rose through the ranks to passenger traffic manager, leaving when the company merged with Cunard in 1934. It might seem reasonable to suppose that he was in a position to know how the ships' construction was financed; however, he was not part of the decision-making process in 1908. While his reflections decades later were undoubtedly well intended, they are at odds with the facts. It is a good example of the need for critical analysis and cross-referencing of material. See Bustard, Lieutenant Colonel Frank, 'An Unpublished Autobiography': Extract (Chapter 10) – 'The Sinking of the *Titanic* in 1912 and the Reputation of Mr. J. Bruce Ismay', *Titanic Commutator*, No. 245 (April–June 2024) p. 28.

Technical details are frequently given inaccurately in secondary sources. A classic example is the figure of '66,000 tons displacement' that is commonly quoted for *Olympic* and *Titanic*. This figure dates back to at least 1910, because it appeared on a postcard commemorating *Olympic*'s launch. However, the accurate figure is 52,310 tons (assuming both ships were loaded to their designed draught of 34ft 7in). This is confirmed in numerous primary source documents: Harland & Wolff's records; *Olympic*'s displacement scale; Board of Trade files; figures that Thomas Andrews provided in November 1911; and by benchmarking with similar liners such as *Aquitania*. It is not something that is debatable as far as the evidence is concerned: the figure of 66,000 tons is simply wrong.

We often see specifications quoted for *Titanic*'s propellers which are simply the *Olympic* propeller configuration as published in articles and journals in 1911, which modern authors had assumed would also be accurate for *Titanic*. All too many people have been willing to take this secondary source information as if it were a fact, yet Harland & Wolff's own records state that *Titanic*'s propellers were different to her older sister.

Even primary sources require cross-referencing where possible. Captain Smith gave his age as 59 when he signed on to *Titanic*, but he was actually 62. He gave his age correctly as 61 when he signed on to *Olympic* the previous year!

I have used the figures provided by the North Atlantic Passenger Conference (NAPC) Annual Reports for the number of passengers a ship carried on a particular voyage or during the year, unless noted otherwise. However, figures do not always tally with other sources An example is *Olympic*'s third eastbound crossing, where the figures Leonard Peskett gave differed from the NAPC data.

Many individual details might seem trivial on their own, but the study of history requires a great effort to get the facts right. Even with that effort, not everything will be correct. Our interpretation of history always needs to be open to revision to reflect the best available evidence.

PUBLISHED SOURCES

Books

Anderson, Roy, *White Star* (T. Stephenson & Sons Ltd, 1964).
Attwood, Edward Lewis, *Textbook of Theoretical Naval Architecture* (Longman's, Green & Co., 1899).
Bäbler, Günter, *Guide to the Crew of Titanic: The Structure of Working Aboard the Legendary Liner* (The History Press, 2017).
Beaumont, Dr J.C.H., *Ships – and People* (Geoffrey Bles, 1926).
Behe, George, *Fate Deals a Hand: The Slippery Fortunes of Titanic's Professional Gamblers* (The History Press, 2023).
—*The Triumvirate: Captain Edward J. Smith, Bruce Ismay, Thomas Andrews and the Sinking of Titanic* (The History Press, 2024).
Beveridge, Bruce, Scott Andrews, Steve Hall, Daniel Klistorner & Art Braunschweiger (eds), *Titanic: The Ship Magnificent: Volume One: Design & Construction* (The History Press, 2008).
Bisset, Sir James, Tramps and Ladies (Patrick Stephens Ltd, 1988).
Boland, Cathalijne, *Reuchlin Reis: De Holland – Amerika Lijne en de landverhuizers* (Alfabet Uitgevers, 2023).
Bowyer, George, *Lively Ahoy: Reminiscences of 58 Years in the Trinity House Pilotage Service* (H.B. Broadbere, 1930).
Brinnin, John Malcolm, *The Sway of the Grand Saloon: A Social History of the North Atlantic* (Delacorte Press, 1971).
Bullock, Shan F., *Thomas Andrews: Shipbuilder* (Maunsel & Company Ltd, 1912).
Chirnside, Mark, *Olympic, Titanic, Britannic: An Illustrated History of the 'Olympic' Class Ships* (The History Press, 2012).
—*RMS Olympic: Titanic's Sister* (The History Press, 2015).
—*The 'Big Four' of the White Star Fleet: Celtic, Cedric, Baltic & Adriatic* (The History Press, 2016).
—*Oceanic: White Star's 'Ship of the Century'* (The History Press, 2018).
—*RMS Majestic: The 'Magic Stick'* (The History Press, 2024).
Chirnside, Mark, & Sam Halpern, *The Sting of the Hawke: Collision in the Solent* (Createspace, 2015).
Eaton, John P., & Charles A. Haas, *Titanic: Triumph & Tragedy* (Patrick Stephens Limited; 1994).
Fitch, Tad, J. Kent Layton & Bill Wormstedt, *On A Sea of Glass: The Life & Loss of the RMS Titanic* (Amberley Publishing, 2012).
Flayhart, William H., & John H. Shaum, *Majesty at Sea: The Four Stackers* (Patrick Stephens Ltd, 1981).
Griffiths, Denis, *Power of the Great Liners* (Patrick Stephens Ltd, 1990).
Halpern, Sam, with Cathy Akers-Jordan, George Behe, Bruce Beveridge, Mark Chirnside, Tad Fitch, Dave Gittins, Steve Hall, Lester J. Mitcham, Captain Charles Weeks and Bill Wormstedt (with foreword by J. Kent Layton), *Report into the Loss of SS Titanic: A Centennial Reappraisal* (History Press, 2011).
Hyde, Francis E., *Cunard and the North Atlantic 1840–1973* (Macmillan Press, 1975).
Hyslop, Donald, Alistair Forsyth & Sheila Jemima, *Titanic Voices: Memories from the Fateful Voyage* (Chancellor Press, 2000).
Jefferson, Herbert, *Viscount Pirrie of Belfast* (William Mullan, 1948).
Jessop, Violet, & John Maxtone-Graham (eds), *Titanic Survivor: The Memoirs of Violet Jessop Stewardess* (Sutton Publishing, 1997).
Jordan, Humfrey, *Mauretania: Landfalls and Departures of Twenty-Five Years* (Patrick Stephens Ltd, 1988).
Klistorner, Daniel, & Steve Hall, with Bruce Beveridge, Art Braunschweiger & Scott Andrews, *Titanic in Photographs* (The History Press, 2011).
Kludas, Arnold, Record Breakers of the North Atlantic: Blue Riband Liners 1838–1952 (Chatham Publishing, 2000).
Layton, J. Kent, Tad Fitch, Thomas Lynskey, Mike Poirier & Levi Rourke, with Stuart Williamson, *Lusitania: An Illustrated Biography: Volume One Life of a Greyhound* (The History Press, 2024).
Louden-Brown, Paul, The White Star Line: An Illustrated History 1869–1934 (*Titanic* Historical Society, 2001).
Lynch, Don, & Ken Marschall, *Titanic: An Illustrated History* (Hodder & Stoughton, 1992).
McAuley, Rob, *The Liners: A Voyage of Discovery* (Boxtree, 1997).
McCulloch, Hugh, *Men and Measures of Half a Century* (C. Scribner's Sons, 1889).
Moss, Michael, & John R. Hume, *Shipbuilders to the World: 125 Years of Harland & Wolff, Belfast 1861–1986* (The Blackstaff Press, 1986).
Oldham, Wilton J., *The Ismay Line* (The Journal of Commerce, 1961).
Sauder, Eric, & Ken Marschall, *RMS Lusitania: Triumph of the Edwardian Age* (Transatlantic Designs, 1991).
Sisson, Wade, *Racing Through the Night: Olympic's Attempt to Reach Titanic* (Amberley Publishing, 2012).
Strouse, Jean, *Morgan: American Financier* (Perennial, 2000).
The Shipbuilder: The White Star Liners Olympic & Titanic Souvenir Number, Vol. VI (Midsummer 1911) (E.C Parker & Company (Services) Ltd, 2009).
Tute, Warren, *Atlantic Conquest: The Ships and the Men of the North Atlantic Passenger Services 1816–1961* (Cassell, 1962).
Warren, Mark D. (ed.), *The Cunard Royal Mail Twin-Screw Steamers Campania and Lucania* (Patrick Stephens Ltd, 1993).
—*Distinguished Liners from The Shipbuilder 1906–1914, Volume 1* (Blue Riband Publications, 1995).
—*Distinguished Liners from The Shipbuilder 1907–1914, Volume 2* (Blue Riband Publications, 1997).
Williams, Dave, & Richard de Kerbrech, *Great Passenger Ships That Never Were: Damned by Destiny Revisited* (The History Press, 2021).

Articles and Periodicals

'4,000 Visit the *Olympic*', *New York Times*, 24 June 1911.
'9 Liners Building for Ship Combine', *New York Times*, 10 August 1907.
'10,000 See *Olympic* Sail', *New York Times*, 29 June 1911.
'$175,000 Profits in Single Trip', *Medford Mail Tribune*, 29 June 1911.
'American Shipping Notes. Groundings at New York', *Lloyd's List & Shipping Gazette*, 30 May 1907, p. 13.
'*Amerika*, Hugest Liner, Arrives on First Trip', *New York Times*, 21 October 1905.
'An Atlantic Liner. A Chat with T.H. Ismay of the White Star Line', *The Pall Mall Gazette*, 26 September 1885. (Reprinted verbatim in *Atlantic Daily Bulletin*, December 2014.)
'Arrival of the *Olympic*', *Westminster Gazette*, 28 February 1912.
Ashbrook, Claud (also named as Ashburn in the same report), 'International Mercantile Marine', *Washington Post*, 7 December 1904.
'Asks National Aid for Montauk Port', *New York Times*, 9 December 1910.
'Atlantic Rate War. White Star "Cut" Cunard's Response', *Lloyd's List & Shipping Gazette*, 10 January 1908.

Bäbler, Günter, 'The Dinner at Lord Pirrie's in Summer 1907: Just a Legend?', *Titanic Post*, Vol. 32 (June 2000).
'Beat the Maiden Record: First Trip of the New Steamship *Majestic*', *New York Times*, 11 April 1890.
'Belfast Police Courts: Stolen Paint Brushes', *Belfast Evening Telegraph*, 5 November 1909.
'Belfast Shipbuilding: Progress of the White Star Liners', *Belfast Evening Telegraph*, 5 November 1909.
'Big *Mauretania* a Sprinter Again', *New York Times*, 30 January 1909.
'Big *Oceanic* Made $90,360 on Last Trip', *New York World*, 14 October 1901.
'Big Shipowning', *Syren & Shipping*, 24 May 1911.
'Bigger than the *Lusitania*', *New York Times*, 12 September 1907.
'Biggest Liner Cuts Record', *New York Times*, 14 September 1907.
'Biggest Liner Nears US', *Chicago Daily Tribune*, 21 June 1911.
'Biggest of Liners Gets Noisy Welcome', *New York Times*, 22 June 1911.
'Biggest Ship Arrives', *New York Tribune*, 22 June 1911.
'Building of Giant Liners', *New York Times*, 10 July 1910.
'Carlisle, Designer of *Titanic*, is Dead', *New York Times*, 6 March 1926.
Chirnside, Mark, '*Olympic*'s Expansion Joints', *Titanic Commutator*, Vol. 31, No. 178 (2007) pp. 84–86.
—'*Titanic*: Allegations and Evidence', *Voyage* 94 (2015) pp. 55–60.
—'The Enclosure of *Titanic*'s Forward A-deck Promenade: Popular Myth?', *Atlantic Daily Bulletin* (March 2016).
—'*Titanic*'s Lifeboats: An Increased Capacity', *Atlantic Daily Bulletin* (September 2018).
—'*Titanic*'s Lifeboats: Fact & Fiction', *Atlantic Daily Bulletin* (March 2019).
—'*Olympic* & *Titanic*: Refining a Design', *Atlantic Daily Bulletin* (December 2019) pp. 18–22.
Chirnside, Mark, & Sam Halpern, '*Olympic & Titanic*: Maiden Voyage Mysteries', *Voyage 59* (2007).
'City Inquests: The Queen's Island Fatality and How It Happened', *The Irish News and Belfast Morning News*, 16 October 1909.
Clarke, John, & Margaret, 'The International Mercantile Marine Company: A Financial Analysis', *The American Neptune* (Spring 1997), Vol. 57, No. 2.
Coghlan, Ed, 'It Said In The Papers', *White Star Journal*, Vol. 17, No. 1 (2009) pp. 10–11.
'Combination of Reciprocating and Turbine Machinery', *Engineering*, 30 April 1909.
Cooke, Anthony, '*Olympic* and *Titanic*: How Many Masts?', *Sea Lines*, Issue 42 (2006).
'Description of the Vessel', *Belfast News-Letter*, 21 October 1910.
'Description of the Vessel', *New York Times*, 13 November 1894.
'Designing a White Star Liner: A Creative Miracle', *White Star Magazine*, Vol. 1, No. 1 (1923).
'Disaster at Last Befalls Captain Smith', *New York Times*, 16 April 1912.
Douglas, James, 'The Atlantic Hustlers', *London Opinion*, 27 April 1912.
'Dry-docking the *Olympic*. Gigantic Undertaking', *The Irish Times*, 3 April 1911.
Eaton, John P., 'The Saga of *Titanic*'s Lifeboats', *Voyage* 63 (2008).
'First Turbine Steamship to Cross the Atlantic to New York', *New York Times*, 10 December 1905.
Franklin, P.A.S., 'New York and the Atlantic Shipping', *New York Times*, 10 January 1909.
Gerahty, G.M., 'Crossing the Atlantic – How the Days Are Spent', *The Sphere*, 19 August 1905.
'Giant New Liner Ends First Trip', *Trenton Evening News*, 22 June 1911.
'Giant *Oceanic* Comes In', *New York Times*, 4 September 1899.
'Greatest Ship Ever Built Comes to Port', *New York Times*, 9 July 1904.
Haas, Charles, 'The Hero Who Was Not On Board', *Voyage* 104 (2018) pp. 153–54.
'Harland and Wolff's Works at Belfast', *The Engineer*, Supplement, 18 June 1909.
'Harold Sanderson, Shipping Man, Dies', *New York Times*, 27 February 1932.
Hedgley, Alan, & George Lavery, 'The Story of Harland & Wolff, Ltd: A Tribute to the Greatest Shipbuilding Firm in the World', *Titanic Commutator*, Vol. 3, No. 27 (1980).
Hicks, C. Stansfeld, 'An Atlantic Chapter', *White Star Magazine*, Vol. 1 (1924).
'High-Priced Suites Pay On Steamships: The Past Season Saw an Exceptional Demand for Fine Accommodations', *New York Times*, 21 September 1913.
Hill, Brian T., & Alan Ruffman, 'E.J. Smith of the RMS *Titanic: His Previous Ice Experience*', *Atlantic Daily Bulletin* (2014).
'Hocking & Co., Ltd. Liverpool', *The Journal of Commerce*, 1 June 1911.
Hutchings, David, 'The Lost Blade', *Atlantic Daily Bulletin* (Spring 2021) pp. 38–41.
'Inquests in Belfast: Fatalities at the Queen's Island', *Belfast News-Letter*, 7 June 1911.
'J. Bruce Ismay Returns to England', *New York Times*, 19 September 1901.
Kamuda, Ed., '*Titanic*'s Only Completed Voyage', *Titanic* Commutator, No. 244 (Winter 2024).
Kennedy, Captain Charles William, 'Aboard White Star's *Germanic*: A Commander's Recollections', *Titanic Commutator*, Vol. 40, No. 211 (2015).
Kinney, Mike, 'On Board SS *Olympic*', *The Gimlet* Vol. V, No. 6 (July 1912) pp. 333–69.
'Largest Ship Afloat Arrives In Port', *New York Times*, 5 August 1901.
'Largest Ship in the World', *Daily Mail*, 21 October 1910.
'Launch of the World's Biggest Ship', *Irish Times*, 29 October 1910.
'Laurels for *St. Louis*', *New York Times*, 30 May 1895.
Lehrer, Stanley, '*Titanic* at Southampton: A Time of Great Expectations', *Titanic Commutator*, Vol. 27, No. 161 (2003).
Lepien, Ray, '*Olympic*: The Maiden Voyage', *Titanic Commutator*, Vol. 27, No. 162 (2003).
'Liner *Cedric* In Port', *New York Times*, 21 February 1903.
Lome, Herbert M., 'The Sanitation of a Modern Ocean Liner', *Modern Sanitation*, Vol. 8 (1911) pp. 445–51.
'Long Sea Service: Death of Captain J. Evans, Caernarvon – Well Known Master Mariner', *Caernarvon and Denbigh Herald*, 26 March 1937.
Louden-Brown, Paul, 'Oceanic House: White Star's London West End Offices', *Titanic Commutator*, Vol. 36, No. 196 (2011).
'Lowering Boilers into the *Olympic*', *The Irish Times*, 30 December 1910.
Lynch, Don, with Geoff Robinson, 'The "Unsinkable" *Titanic*, Ad Advertised: Another Great Myth is Dispelled', *Titanic Commutator*, Vol. 16, No. 4 (1993).
Lynch, Don (ed.), '*Titanic* Survivor Accounts: Kate Buss Willis – Her Story', *Titanic Commutator*, Vol. 30, No. 176 (2006).
'Making Ships' Telegraphs', *The Journal of Commerce*, 31 May 1911.
'Maritime Wireless Telegraphy', *The Marconigraph*, June 1911.
'*Mauretania* Goes 30.5 Miles an Hour', *New York Times*, 22 September 1907.

'*Mauretania* Stuck in Ambrose Channel', *New York Times*, 4 November 1910.
'May Grant Chelsea Pier Extensions', *New York Times*, 12 February 1911.
'May Yet Get Piers for New Big Ships', *New York Times*, 4 January 1911.
'Mishap to the *Olympic*', *Belfast Weekly News*, 29 February 1912.
'Mishap to the *Olympic*', *Western Daily Mercury*, 29 February 1912.
'Monarch of the Sea: The New Steamship *Oceanic*', *New York Herald*, 30 March 1871.
'Movements of Liners', *The Globe*, 3 June 1911.
'Mr. Bruce Ismay', *The Times*, 18 October 1937.
'Mr. J. Bruce Ismay Resigns L.M.S. Railway Directorship', *Liverpool Daily Post*, 10 July 1937.
'Mr. Morgan's Shipping Deal', *Liverpool Daily Post*, 30 April 1901.
Munsey, Frank, 'The Heritage of the Locust', *Munsey's Magazine*, Vol. XLV (April to September 1911) pp. 815–16.
—'Giant Liners of the North Atlantic', *Munsey's Magazine*, Vol. LII, No. 1 (June 1914) p. 22.
'Naming of Ocean Liners: "Family" Appellations Distinguishing the Various Fleets', *New York Times*, 8 November 1903.
Navin, Thomas R., & Marian V. Sears, 'A Study in Merger: Formation of the International Mercantile Marine Company', *Business History Review*, Vol. 28, Issue 4 (December 1954).
Nash, Zetie Duncan, 'The Shortest Voyage on the Longest Ship', *Titanic Commutator*, Vol. 42, No. 219 (2017) pp. 133–35.
'New Giant Liner Sails', *New York Times*, 12 October 1905.
'New Leviathan of the Sea', *New York Times*, 10 September 1899.
'New Marine Giants: *Olympic* and *Titanic* to be Launched at Belfast', *Manitoba Free Press*, 30 September 1910.
'New Queens of the Atlantic To Be Marvels of Luxury', *New York Times*, 3 March 1907.
'No. 8 – White Star and American Lines' Laundry', *Southampton and District Pictorial*, 29 May 1912, p.5.
'No Fear for the *Cedric*', *New York Times*, 26 November 1903.
'No More "Ship Diet," Says *Amerika* Caterer', *New York Times*, 22 October 1905.
'Obituary: Joseph Bell, Chief Engineer', *The Marine Engineer & Naval Architect*, Vol. XXIV (May 1912) p. 381.
'Obituary: Viscount Pirrie', *The Engineer*, 13 June 1924.
'*Olympic*, 1,000-foot Liner: White Star Line Selects Name for its Leviathan – May Build Sister Ship', *New York Times*, 17 April 1908.
'*Olympic* and *Titanic*: Names of White Star Leviathans, to be More Than 840 feet Long', *New York Times*, 23 April 1908.
'*Olympic* A Wreck', *Tewkesbury Register*, 6 April 1912.
'*Olympic* Collision: Men's Claim for Wages', *Daily Telegraph*, 30 September 1911, p. 3.
'*Olympic–Hawke* Collision. Seamen's Wages Claim. White Star Line Test Case', *Liverpool Daily Echo*, 5 March 1912.
'*Olympic* Hit By A Cruiser; Badly Damaged', *New York Times*, 21 September 1911.
'*Olympic* in Dry Dock', *Irish Times*, 15 April 1911.
'*Olympic* Off Today with Record List', *New York Times*, 28 June 1911.
'*Olympic* Preparing to Sail', *New York Times*, 27 June 1911.
'*Olympic* Sails from Queenstown', *Irish Times*, 24 June 1911.
'*Olympic*'s Fine Voyage', *Irish Times*, 20 June 1911, p. 7.
'*Olympic*'s Repairs to Take Three Months', *New York Times*, 22 September 1911.
'*Olympic* Wages Case', *Hampshire Independent*, 9 March 1912.
'On the *Olympic*. A Secret Inspection at Southampton', *Daily News*, 7 May 1912.
Parsons, the Hon. Charles A., & R.J. Walker, 'The Institution of Naval Architects. Combination of Reciprocating Engines and Turbines', *Engineering*, 17 April 1908.
'Permits Long Piers for the New Liners', *New York Times*, 7 March 1911.
'Pier Extensions Up to Harbour Board', *New York Times*, 24 January 1911.
Quimby, Harriet, 'How the Ocean Traveller is Fed', *Lesliel's Illustrated Weekly Newspaper*, 13 June 1912.
'Right Hon A.M. Carlisle Retires. A Notable Career', *Weekly Telegraph*, 13 August 1910.
Robinson, Andrew, 'The Partial Enclosure of the *Titanic*'s A-deck Promenade', *Atlantic Daily Bulletin*, No. 1 (2000).
Robinson, Geoff, 'Lifeboats and the *Titanic*' (Part One), *Atlantic Daily Bulletin*, No. 4 (1999).
'Safety on the Ocean', *New York Times*, 30 May 1891.
Saliers, Earl A., 'Some Financial Aspects of the International Mercantile Marine Company', *Journal of Political Economy* (1915) pp. 910–25.
'Says Ocean Racers Are Played Out', *Waterloo Daily Reporter*, 6 February 1901.
Schoonderbeek, Captain Albert, 'The Interiors of Holland America Line's Six *Statendams*: Part One – the first three *Statendams*', *Sea Lines*, Issue 111 (Summer 2003).
Sear, Ruth, 'William Joseph Willett Bruce', *Atlantic Daily Bulletin*, September 2011.
'Serious Queen's Island Accident', *The Irish News and Belfast Morning News*, 10 January 1910.
'Shareholders Agree to the Ship Combine. Consent of All but the White Star Stockholders Gained', *New York Times*, 1 May 1902.
'Ship Can Save Herself', *New York Times*, 11 November 1903.
'Shipyard Fatality: Inquest on the Victims', *Belfast Weekly Telegraph*, 19 February 1910.
'Silent Ash Hoists and other work by Railton, Campbell and Crawford', *The Journal of Commerce*, 31 May 1911.
Sims, P., 'Comparative Naval Architecture of Passenger Ships', *Transactions of the Society of Naval Architects*, Vol. 111 (2003).
'Southampton's New Dock: London and South-Western Enterprise', *The Mariner*, Vol. XXXV, No. 9 (15 June 1911).
'Splendid Voyage', *Fort Wayne Daily Gazette*, 6 July 1871.
'Statement by Mr. Russell Rea, MP', *Daily Express* 12 June 1911.
'Superintendent Engineer: White Star Line', *The Marine Engineer and Naval Architect* (June 1915).
'The Biggest Ship and the Biggest Dock', *Scientific American*, 6 May 1911.
'The Case of the *Olympic*', *Northern Daily Mail*, 15 June 1911.
'The Engine Room As A Training Ground', *Shipbuilding & Shipping Record* (June 1914) p. 737.
'The Equipment of SS *Olympic*', *The Marconigraph* (July 1911).
'The Future of New York as a World Harbour', *New York Times*, 29 January 1911.
'*The George Washington*. The New Norddeutscher Lloyd Steamer', *Scientific American* Supplement, No. 1747, 26 June 1909.
'The Giant *Olympic* A Luxurious Floating Hotel', *New York Times*, 25 June 1911, p. 5.
'The Greatest Steamship Ever Built', *New York Times*, 12 May 1907.
'The Largest Ship in the World: Motherwell's Share in its Construction', *Motherwell Times*, 16 June 1911.
'The *Laurentic*'s Trials. Successful Experiments with Turbine Machinery', *Belfast Evening Telegraph*, 1 February 1909, p. 3.
'The Liner of the Future', *Munsey's Magazine* (c. 1903).
'The Mighty *Olympic* Ends Her First Voyage, Entering Port in Triumph, While Other Steamships Seem Dwarfs Beside Her', *New York Herald*, 22 June 1911.

'The Next Thing in Liners', *Belfast News-Letter*, 18 May 1909.
'The New Unsinkable Vessel', *New York Times*, 27 July 1902.
'The New Vessel's Davits. Some Interesting Innovations', *Belfast Telegraph*, 21 October 1910.
'The *Olympic* Action', *Western Evening Herald*, 29 September 1911.
'The *Olympic* and the *Titanic*: Two Giant Ocean Steamships', *Scientific American* Supplement, No. 1850 (17 June 1911).
'The *Olympic* at Belfast', *Belfast Weekly News*, 7 March 1912.
'The *Oceanic* at Queenstown', *Irish Times*, 8 September 1899.
'The *Olympic*'s Boilers', *Irish Times*, 14 November 1910.
'The Sister Ship, "*Mauretania* the Magnificent"', *New York Times*, 10 November 1907.
'The Steamship *Oceanic*', *New York Times*, 30 March 1871.
'The *Teutonic* Launched', *New York Times*, 20 January 1889.
'The Thirty Knot An Hour Liner and What it Means', *New York Times*, 15 September 1907.
'The Thousand Foot Ship and What It Means', *New York Times*, 15 March 1908.
'The Troubles With the *Olympic*', *Justice*, 17 June 1911
'The Utley Sidelights and Patent Windows', *The Journal of Commerce*, 31 May 1911.
'The Victualling of an Ocean Liner', *Railway & Travel Monthly*, Vol. 3, No. 20 (December 1911).
'The White Star's *Olympic*', *John Bull*, 10 December 1910, p. 923.
'The White Star's *Olympic*. Antiquated Board of Trade Regulations Hazard Thousands of Lives', *John Bull*, 3 December 1910, p. 1010.
'The White Star's *Olympic*. A Serious Insufficiency of Lifeboats', *John Bull*, 19 November 1910, p. 789.
'The White Star's *Olympic*. Our Rejoinder to An "Official Reply"', *John Bull*, 3 December 1910, p. 884.
'The World's Biggest Ship', *Belfast News-Letter*, 21 October 1910.
'The World's Biggest Ship', *Belfast Weekly News*, 27 October 1910.
'Thomas Henry Ismay Dead', *New York Times*, 24 November 1899.
'Thousands Visit the *Olympic*', *New York Times*, 25 June 1911.
'Thursday's Launchings', *New York Times*, 24 September 1906.
'*Titanic*: Extract from the Unpublished Autobiography of Lt. Colonel Frank Bustard, OBE'. Reproduced in *Titanic Commutator*, Vol. 39, No. 206 (2014;).
Trower, Timothy, 'Book Reviews by Tim Trower. *The "Olympic" Class Ships* by Mark Chirnside', *Titanic Commutator*, Vol. 30, No. 176 (2006).
'Unsinkable Passenger Craft', *New York Times*, 3 June 1908.
Walker, J. Bernard, 'The "Unsinkable" Ship: Longitudinal Coal Bunkers and Higher Bulkheads Might Have Saved the *Titanic*', *Scientific American*, 11 May 1912.
Wentworth-Shields, Francis Ernest, 'The Construction of the "White Star" Dock and Adjoining Quays at Southampton', London Institution of Civil Engineers (1914).
'White Star Mammoths. Chairman Inspects New Liners', *Liverpool Daily Post*, 25 March 1910.
'White Star's Newest Liner: On Board *Republic*', *Titanic Commutator*, Vol. 31, No. 180 (2008).
'White Star RMS *Olympic*. Elkington & Co., Ltd', *The Journal of Commerce*, 1 June 1911.
'World's Biggest Liner *Olympic* Nearing New York', *Irish Times*, 21 June 1911, p. 7.
'World's Biggest Ship, *Amerika*, Coming Soon', *New York Times*, 17 September 1905.

Select Official Publications

Mersey, Lord J.B., 'Report of a Formal Investigation Ordered by the Board of Trade into the Loss of the SS *Titanic*' (London: His Majesty's Stationery Office, 1912).
'Papers Relating to the Loss of the Steamship *Atlantic*' (London: House of Commons, 31 July 1873).
'Report of the Committee on Commerce United States Senate Pursuant to S. Res. 283 Directing the Committee on Commerce to Investigate the Causes Leading to the Wreck of the White Star Liner *Titanic*' (Washington: Government Printing Office, 1912).
'Reports of the Immigration Commission: Immigration Legislation' (Vol. 39) (Washington: Government Printing Office, 1911).
'Reports of the Immigration Commission: Steerage Conditions' (Vol. 37) (Washington: Government Printing Office, 1911).

ARCHIVAL SOURCES

Glasgow City Archives

Order book of R. McAlister & Sons: TD1345/10/14.

Guildhall Library

'Master's Certificate No. 028211, Robert Hume': MS 18569/19.
'Master's Certificate No. 14102, Edward John Smith': MS 18569/35.

National Archives & Records Administration (NARA, New York)

'Deposition of Edward Wilding, In the Matter of the Petition of the Oceanic Steam Navigation Company, for Limitation of its Liability, as Owner of Steamship *Titanic*', May 13–14th 1915: The National Archives and Records Administration – Northeast Region (New York).
'In Re.: The Oceanic Steamship Company [*sic*], Evidence of the Rt. Hon. Alexander Montgomery Carlisle', 3 April 1914: The National Archives and Records Administration – Northeast Region (New York).
'Peter Cahill V. The Steamship *Olympic* and Oceanic Steam Navigation Company, Ltd.', Case 55-24, U.S. District Court for the Southern District of New York: National Archives & Records Administration.
'Oceanic Steam Navigation Company, Ltd. V. Steamtug *O.L. Hallenbeck* her engines, etc. Peter Cahill, Claimant', Case 55-227, U.S. District Court for the Southern District of New York: National Archives & Records Administration.

The National Archives of the UK (formerly the Public Records Office)

'American Mail Service: Omission of Call at Queenstown by Cunard Packetts' (London: His Majesty's Stationery Office, 1914).
'Appendix (B): Memorandum of Interview Between Mr. Gerald Balfour and Mr. Pierpont Morgan and Sir Clifford Dawkins, 8 August 1902', pp. 3–4: CAB 37/62.
Beaumont, Dr J.C.H., 'To the Board of Trade, Southampton …', 15 May 1912: BT 100/248.

Carruthers, F., 'Dimensions of Vessel': MT 9/920/C.
Carruthers, F., 'General Remarks': MT 9/920/F.
'Certificate of British Registry', Official Number 124,061, *Adriatic*: BT 110/976.
'Certificate of British Registry', Official Number 135,583, *Aquitania*: BT 110/1312
'Certificate of British Registry', Official Number 137,490, *Britannic*: BT 110/260.
'Certificate of British Registry', Official Number 113,476, *Celtic*: BT 110/1005.
'Certificate of British Registry', Official Number 124,093, *Mauretania*: BT 110.
'Certificate of British Registry', Official Number 146,555, *Majestic*: BT 110/1084.
'Certificate of British Registry', Official Number 110,596, *Oceanic*: BT 110/377.
'Certificate of British Registry', Official Number 131,346, *Olympic*: BT 110/1265.
'Certificate of British Registry', Official Number 97,763, *Teutonic*: BT 110/356.
Clarke, Maurice Harvey, 'Report re.: Emigration Clearance and survey of SS *Olympic* O.N. [official number] 131,346', 28 April 1912: MT 15/150.
'Confidential: The Morgan Shipping Combination': PRO 30/60/40.
'Copy of An Agreement (Dated 1 August 1903) between The Admiralty and the Board of Trade and the International Mercantile Marine Company and the Oceanic Steam Navigation Co., Limited' (His Majesty's Stationery Office, 1903): MT23/163.
Court of Appeal, 'On Appeal from the High Court of Justice, Probate, Divorce, and Admiralty Division. Stating the case between Folio 400 (The Oceanic Steam Navigation Company, Limited, Owners of the SS *Olympic* verses Commander William F. Blunt, of H.M. Cruiser *Hawke*) and Folio 382 (The Commissioners for Executing the Office of Lord High Admiral of the United Kingdom verses Owners of SS *Olympic*), London, commencing 21 January 1913: TS 36/185.
'Cunard Steamship Company Limited. Assistance for Acquisition of White Star Line & Building of New Liners', 1930: T160/1313.
Goodenough, Captain W.E., 'Report from HMS *Cochrane* at Portsmouth, to Rear Admiral Commanding, 2nd Cruiser Squadron', 26 April 1912: MT 15/150.
High Court of Justice: Probate, Divorce, and Admiralty Division, 'Writ in person issued 21 September 1911 between the Oceanic Steam Navigation Company Limited, the owners of the Steamship *Olympic*, plaintiffs, and Commander W.F. Blunt of HM Cruiser *Hawke*, Defendant: HMS *Hawke*, Statement of Claim': ADM 116/1163a.
High Court of Justice: Probate, Divorce, and Admiralty Division, 'Writ issued 28 September 1911. The Commissioners for Executing the Office of Lord High Admiral of the United Kingdom, verses the owners of SS *Olympic*: Statement of Claim': ADM 116/1163a.
High Court of Justice: Probate, Divorce, and Admiralty Division, 'Stating the case between the Oceanic Steam Navigation Company, Limited, Owners of the Steamship *Olympic* verses Commander William F. Blunt, of H.M. Cruiser *Hawke* and the Commissioners for Executing the Office of Lord High Admiral of the United Kingdom verses Owners of Steamship *Olympic*, London', commencing 16 November 1911: TS 36/175-76.
'London & South Western Railway: Inspection of New Deep Water Dock Southampton, Thursday, 8 June 1911': RAIL 411/661.
'Memorandum: International Mercantile Marine Company of New Jersey. 8 August 1918', pp. 1–2: Ministry of Shipping: Position of the White Star and Associated Fleets': T1/12252.
'Official Log Book. SS *Adriatic*': BT 165/469.
'Report of the Committee Appointed by the President of the Board of Trade to Consider and Report upon The Spacing and Construction of Watertight Bulkheads in Ships of the Mercantile Marine with Appendices and Diagrams' (London: Her Majesty's Stationery Office, 1891): MT 15/145.
'RNR: Captain H.J. Haddock RNR. Pay and Conditions: Question of Emoluments', November 1914: ADM 1/8402/410.
Royal Courts of Justice. In His Majesty's Court of Appeal: 'On Appeal from the High Court of Justice (Admiralty Division). Between the Oceanic Steam Navigation Company Limited, owners of the Steamship *Olympic*, Appellants, and Commander William F. Blunt of HM Cruiser Hawke, Respondent: Report of Proceedings as to Calling Further Evidence': TS 36/183.
'Safety of Life at Sea: Draft Lifesaving Appliances Rules, With memorandum on Safety of Life at Sea by the President of the Board of Trade' (London: His Majesty's Stationery Office, 1912): RAIL 1053/295.
'SS *Olympic*. Damage to Machinery caused by collision with HMS Hawke. Inspection of …', Director of Dockyards, 19 October 1911: ADM116/ 1163A.
'SS *Titanic*. Correspondence Respecting The Collision Bulkhead of SS *Olympic*': MT 9/920/E.
'The Assignment of Freeboard to the SS *Titanic*': MT 9/920/E.
'The Oceanic Steam Navigation Co. Ltd.: … £1,250,000 4½ I First Mortgage Debentures', 5 October 1908: BT 31/30783/4540.
'The SS *Olympic*: Advice on Evidence', 17 October 1911: ADM 116/1163a.
'*Titanic* Harland & Wolff's 401 Boiler Plates by D. Colville & Sons', March 1910: MT 15/212/3.
'War Compensation Court: Claim of The International Mercantile Marine Company and of The Oceanic Steam Navigation Company, Limited: Judgement, 1 April 1924, p. 8: MT 40/184.

National Museums Liverpool (Maritime Archives & Library, Merseyside Maritime Museum)

'Abstract of Engineer's Log SS *Campania*': B/CUN/4/2/13.
Bell, Chief Engineer J., 'SS *Olympic* Voy. 1' to Messrs. Ismay, Imrie & Co., Liverpool, 24 June 1911: DX/504/1-3.
'Coal Situation' and cable dated 28 March 1912: DX 504/12/4.
Franklin, Philip, to Ismay, J. Bruce, 'Extract of Letter …', 19 July 1911, through to Ismay, J. Bruce, to Franklin, Philip, 'Extract from Letter …', 7 September: DX/504/19/1-16.
'Monthly Office Accounts No. 3': D/B/174/1.
'Specification. Twin Screw Steamer, No. 857 [*Franconia*]', 20 September 1909: B/CUN/3/4/6.
'Specification SS *Aquitania* No. 409', Mr. C. Garrard: B/CUN/3/4/4.
'Specification of a Steel Quadruple-Screw Steamship with Turbine Engines … [*Lusitania* and *Mauretania* generic specification]', John C. Morrison, Naval Architect's Department, 6 January 1906: B/CUN/3/4/3.

'Specification. SS *Carmania*. No. 366', 10 September 1904: B/CUN/3/4/1.
'The Cunard Steam Ship Company Limited: Atlantic Outward and Homeward Passages': B/CUN/4/4/5.

Public Record Office of Northern Ireland (PRONI)

'Articles of Agreement, Entered Into This Seventh Day of August 1874': D2805/HIST/3.
'Contracts No. 7 Iron & Steel Department': D2805/MISC/1.
Currie, W.A., 'Belfast Harbour Notice to Mariners', 11 October 1910: HAR 1A/2/220.
'Departure of SS *Olympic* and Launch of SS *Titanic*', 23 May 1911: T3893/1.
'Half Yearly Agreement and Account of Voyages and Crew for a Ship Engaged In The Home Trade Only ...': TRANS. 2A/45/381A.
'Harland & Wolff Ltd. Issue at Par of 4,000,000 First Cumulative Preference Shares of £1 Each: Prospectus', July 1924: D2805/MISC/3.
'Harland & Wolff Ltd. Memorandum of Articles of Association', 22 December 1885: D2805/MISC/3.
'Harland & Wolff Ltd. Minutes [of] Managing Directors': D/2805/MIN/A1.
'Installments Due': D2805/SHIP/2.
'Memo for Launches', c. 1910: D/2805/C1/1.
Pirrie, Margaret, to Lord Cadogan, July 1898: T3449/8.
Pirrie, William, to Lord Dufferin, June 1898: D1071/H/B/P/268/1-24.
Pounder, C.C., 'Some Notable Belfast-Built Engines', Belfast Association of Engineers, Sixth Meeting, 57th Session, 3 March 1948: D4063/9/63.
'The King vs. Edward John Wilson: Depositions', Belfast Spring Assizes 1912: BELF/1/1/2/37/5.
'Trial Trip of the *Patriotic*', unidentified newspaper cutting, 28 March 1912. Diary of Robert H.H. Baird, 1912: D/3330/B/1/12.
'Wage Statistics from 1858 to 1911': D2805/MISC/3.

Stadsarchief Rotterdam

'Memorandum of Commission Terms, Harland & Wolff Ltd.': EB 547.
'Koninklijke Nederlandsche Meubelenfabriek. H.P. Mutters & Zoon, Den Haag. Behangerij en Stoffeerderij. 'Begrooting voor Holland Amerika Lyn, Rotterdam, betreffende: Betimmering & Meubileering SS *Statendam*', 15 Februari 1913: EB 1133.
'Koninklijke Nederlandsche Stoomboot Maatschappij en Nederlandsch-Amerikaansche Stoomvaart-Maatschappij, correspondentie betreffende mogelijke concurrentie in Amsterdam betreffende de handel op Noord-Amerika met exploitatieoverzichten en bijlagen; Overeenkomst Shaw, Savill & Albion Co. Ltd. met Ismay, Imrie & Co. (Oceaan Steam Navigation Co. Ltd.), 26 Juli 1887, 1887–1902': DB 726.
'No. 436 – SS *Statendam* Modifications Involving Extras and Credits for Which Prices Have Not Been Submitted to Owners', 16 August 1917: EB 1133.
'Piek, W.F., aan board van ss *Olympic*, rapport met bijlagen, 1911': EB 1278.
'Remarks in Connection with the Outline Specification of a Steel Twin-Screw Steamer to be Built for the Holland-America Line by Messrs. Harland & Wolff Ltd': EB 1125.

Southampton Archives Services, Civic Centre, Southampton

Southampton Harbour Board, Minute Books and Reports, 1908–09: SHB 1/2/13; 1909–10: SHB 1/2/14; 1910–11: SHB 1/2/15; 1912–13: SHB 1/2/17; 1913–14: SHB 1/2/18.

The British Postal Museum and Archive

Bruce, Mr, 'Sea Post Offices: Report of Inspection by Mr. Bruce, Controller London Postal Service', p. 49, January 1909: POST 43/321.
'General Post Office: Transatlantic Sea Post Offices. Proposed Accommodation On Board SS *Titanic* and SS *Olympic*. Construction of Revised Plans. 1910–11': POST 29/1117.
'General Post Office: Transatlantic Sea Post Offices. Proposed Accommodation On Board SS *Titanic* and SS *Olympic*. Consideration of Plans. Superseded by Revised Plans ... 1909–10': POST 29/1117.
'General Post Office: Transatlantic Sea Post Offices. White Star Line: SS *Olympic*. Unsuitable Sleeping Accommodation. Memorial for Staff at Southampton. 1912': POST 29/1117.
'General Post Office: Transatlantic Sea Post Offices. White Star Line: SS *Titanic* and SS *Olympic*. SS *Titanic* Lost – April 1912. Alterations On Board SS *Olympic*. Sleeping Accommodation on Deck "E" Provided. 1912': POST 29/1117.
'General Post Office: Transatlantic Sea Post Offices. White Star and American Lines. Quarters for Sorters: Suggested Provision of Second Instead of First Class Accommodation. No Action Taken. 13–17 June 1911': POST 29/1117.
'General Post Office: Transatlantic Sea Post Offices: White Star Line: SS *Olympic*. Inspection of Accommodation. Satisfactory. 8 June 1911': POST 29/1117.
'Memorandum: *Titanic*', 11 April 1912: POST 43/321.

The Cunard Archive, Sydney Jones Library, Liverpool University

'A Brief History of the White Star Line 1869–1934' (unpublished manuscript, 1947): PR 3/23/6.
Bain, J., 'Untitled Memorandum from the General Superintendent, the Cunard Steamship Company Limited, to the Secretary, Liverpool, 5 August 1909'. In: 'The Cunard Steamship Company Limited: Minutes of Executive Committee Meetings', p. 308: B4/30.
Currie, J., & A. Galbraith, 'Memorandum to the General Manager', Executive Committee, 21 June 1911: S7/3/75.
Currie, J., & A. Galbraith, 'Technical Subjects Bearing on Atlantic Shipping', Memorandum from the Superintendent Engineer, the Cunard Steamship Company Limited, to the General Manager, Liverpool, 24 August 1910. In: 'The Cunard Steamship Company Limited: Minutes of Executive Committee Meetings', p. 334: B4/32.
'Correspondence as to Sale of Shares in Oceanic Steam Navigation Co. Ltd, Vol. 2': D42/A9/10.
Dodd, George H., 'Supplementary Report: Report of Cherbourg', 12 August 1908. In: 'The Cunard Steamship Company Limited: Minutes of Executive Committee Meetings', p. 18: B4/29
'Extract from a letter, dated 31 May 1910, from Charles P. Sumner, Esq., manager of the New York office, to A.A. Booth, Esq., chairman', through to correspondence dated 17 December 1910: S7/3/4.

Inverclyde, Lord, letter to shareholders, 30 September 1902: B1/5.
Letter from 'Secretary' to H.A. Peto, Esq., Bradford on Avon, dated 13 September 1905, 'Dear Sir ...', untitled: S7/1/16.
Letters to Vernon Brown, dated 22 January and 15 February 1908: S5/3.
Memorandum: 'SS *Mauretania*', 7 November 1907.
'Official Trials of First *Mauretania*': PR3/7/13.
'Preliminary Negotiations': S10/2
'Proposed Names for New Fast Steamers': S10/5.
Peskett, Leonard, 'Memorandum to the General Manager', 15 May 1912: S7/3/20.
Peskett, Leonard, 'Memorandum to the General Manager', 9 August 1911: S7/3/37.
Peskett, Leonard, 'Memorandum to the General Manager. SS *Aquitania*', Naval Architect's Department, 1 October 1912: S7/3/20.
Peskett, Leonard, 'Memorandum to the General Manager. SS *Olympic*', Executive Committee, 28 June 1911: S7/3/75.
'Queenstown Harbour Committee: Report from the Committee Appointed by the Postmaster-General to consider the Omission of the Call at Queenstown on the American Mail Service; Together with Minutes of Evidence, Appendices, and a Chart of Queenstown Harbour' (London: His Majesty's Stationery Office, 1914).
Smith, Adolphe, 'My dear Sir', November 1908: GM29/1.
'Specification of a Steel Quadruple-Screw Steamship with Turbine Engines': S7/1/1.
'The Blue Riband of the Atlantic: List of Fastest Passages of Various Atlantic Liners, whether Westbound or Eastbound ...': PR3/11/1.
'The Cunard Steamship Company Limited: Annual Reports' for 1892–98: D42/A4/1.
'The Cunard Steamship Company Limited: Issue of 40,000 Shares ...', London, 4 March 1880.
'The Cunard Steamship Company Limited: Minutes of Board Meetings', 1910: B1/8.
'This Agreement made in England the 18th day of May 1905 between John Brown & Company Limited [...] and the Cunard Steamship Company Limited ...': S7/1/14.
Watson, William, 'Statement by the Chairman of the Cunard Steamship Company, Limited', 1 June 1908: AC6/3/1.
'Watson, William, to Brown, Vernon H.', 23 October 1907 (copy of J. Bruce Ismay's letter to Dr Wiegand attached): S5/2.

The Morgan Library and Museum

'Pierpont Morgan Papers (ARC 1196)', p. 51, The Pierpoint Morgan Library.

University of Glasgow Archive Services

'Notebook containing propeller data for ship numbers 387–390 (*Beagle*, *Bulldog*, *Foxhound*, *Bristol*), *Indefatigable* (391), *Brisk* (395), *Petersburg* (397), *Yarmouth* (398), *Argyllshire* (399), also the centre propeller (turbine driven) for *Olympic* and *Titanic* built by Harland & Wolff', 1909–11: GB 248 UGC 100/3/7.

INDEX

BY THE SAME AUTHOR

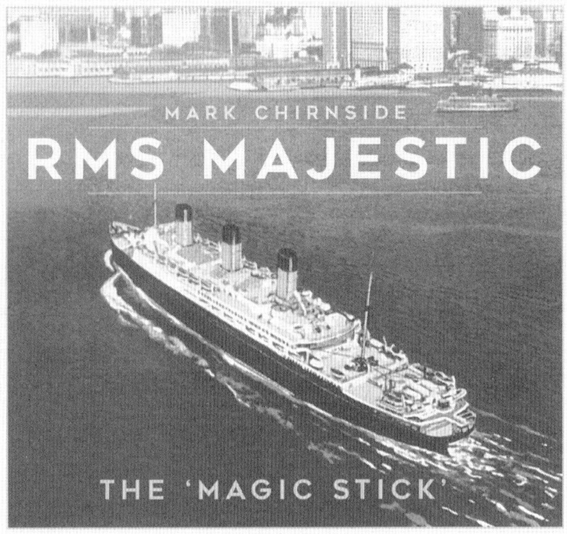

978 1 80399 337 9

978 0 7509 5623 9

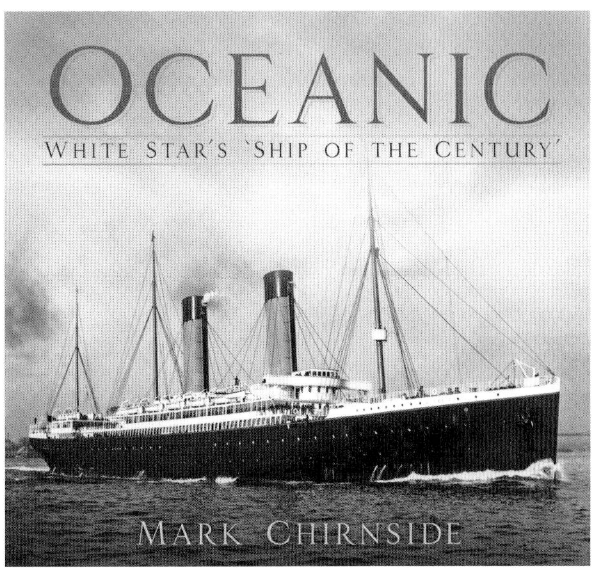

978 0 7509 8578 9